Handbook of Environmental Engineering

Volume 20

Series Editors

Lawrence K. Wang
PhD., Rutgers University, New Brunswick, NJ, USA
MS, University of Rhode Island, Kingston, RI, USA
MSCE, University of Missouri, Rolla, MO, USA
BSCE, National Cheng Kung University, Tainan, Taiwan, ROC

Mu-Hao Sung Wang
PhD., Rutgers University, New Brunswick, NJ, USA
MS, University of Rhode Island, Kingston, RI, USA
BSCE, National Cheng Kung University, Tainan, Taiwan, ROC

The past 30 years have seen the emergence of a growing desire worldwide to take positive actions to restore and protect the environment from the degrading effects of all forms of pollution: air, noise, solid waste, and water. The principle intention of the Handbook of Environmental Engineering (HEE) series is to help readers formulate answers to the fundamental questions facing pollution in the modern era, mainly, how serious is pollution and is the technology needed to abate it not only available, but feasible. Cutting-edge and highly practical, HEE offers educators, students, and engineers a strong grounding in the principles of Environmental Engineering, as well as providing effective methods for developing optimal abatement technologies at costs that are fully justified by the degree of abatement achieved. With an emphasis on using the Best Available Technologies, the authors of these volumes present the necessary engineering protocols derived from the fundamental principles of chemistry, physics, and mathematics, making these volumes a must have for environmental pollution researchers.

More information about this series at http://www.springer.com/series/7645

Lawrence K. Wang • Mu-Hao Sung Wang
Yung-Tse Hung • Nazih K. Shammas
Editors

Integrated Natural Resources Management

Springer

Editors
Lawrence K. Wang
Lenox Institute of Water Technology
Lenox, MA and Newtonville, NY, USA

Mu-Hao Sung Wang
Lenox Institute of Water Technology
Lenox, MA and Newtonville, NY, USA

Yung-Tse Hung
Department of Civil and Environmental Engineering, Cleveland State University
Cleveland, OH, USA

Nazih K. Shammas
Lenox Institute of Water Technology
and Krofta Engineering Corporation
Lenox, MA and Newtonville, NY, USA

ISSN 2512-1359 ISSN 2512-1472 (electronic)
Handbook of Environmental Engineering
ISBN 978-3-030-55171-1 ISBN 978-3-030-55172-8 (eBook)
https://doi.org/10.1007/978-3-030-55172-8

This Springer imprint is published by the registered company Springer Nature Switzerland AG
The registered company address is: Gewerbestrasse 11, 6330 Cham, Switzerland

The Handbook of Environmental Engineering *(HEE), Volume 20, Integrated Natural Resources Management, is dedicated to the memory of Dr. Nazih Shammas – loving husband, father, and grandfather – who passed away on Friday, May 29, 2020, at age 81. Dr. Nazih K. Shammas contributed significantly as an author/coeditor in the HEE series. Dr. Shammas was an Editorial Board*

member of HEE, a Professor, and an Ex-Dean of the Lenox Institute of Water Technology (LIWT) for over two decades. His former colleagues of HEE and LIWT, and his professional friends, have timely completed this HEE (Volume 20) to salute and forever remember Dr. Shammas.

Dr. Shammas had a lifelong love for scientific discovery and cutting-edge innovation, which fueled his many accomplishments in his life. He earned his Civil Engineering degree with distinction from the American University of Beirut, MS in Sanitary Engineering from the University of North Carolina, and PhD in Civil Engineering from the University of Michigan. His academic career began as an Assistant Professor of Civil Engineering at the American University of Beirut and continued as an Associate Professor at the King Saud University in Riyadh, and most recently a Professor of Environmental Engineering and Dean of Education at the Lenox Institute of Water Technology in Lenox, Massachusetts, USA. His achievements were recognized in multiple editions of the Marquis Who's Who in America. He authored, coauthored, edited, and coedited numerous academic publications and textbooks used at universities worldwide, including several in this series, to pass on his knowledge and love of science to future generations.

Dr. Shammas thoroughly enjoyed life, had a passion for spending time with family and friends, and frequently orchestrated adventurous trips to travel around the world with his beautiful wife. He loved spending time with his two devoted sons and their wives, as well as his grandson, who shares his love of science and discovery. Dr. Shammas was known for his kind heart, infectious smile, distinctive laugh, love of family, and compassionate spirit. He was a loyal friend to his many acquaintances around the world. Most of all, he will be remembered for being a wonderful, loving, and devoted husband of over 50 years. His example was an inspiration to all those around him.

Dr. Shammas is survived by his wife Norma, two sons, grandson, brother, two sisters, and their families, as well as numerous cousins, nieces, and nephews.

His life was well lived because he is loved by so many. He will forever stay with us in our hearts.

Preface

The past seventy-five years have seen the emergence of a growing desire worldwide that positive actions be taken to restore and protect the environment from the degrading effects of all forms of pollution – air, water, soil, thermal, radioactive, and noise. Since pollution is a direct or indirect consequence of waste, the seemingly idealistic demand for "zero discharge" can be construed as an unrealistic demand for zero waste. However, as long as waste continues to exist, we can only attempt to abate the subsequent pollution by converting it to a less noxious form. Three major questions usually arise when a particular type of pollution has been identified: (1) How serious are the environmental pollution and water resources crisis? (2) Is the technology to abate them available? and (3) Do the costs of abatement justify the degree of abatement achieved for environmental protection and natural resources conservation? This book is one of the volumes of the *Handbook of Environmental Engineering* series. The principal intention of this series is to help readers formulate answers to the above three questions.

The traditional approach of applying tried-and-true solutions to specific environmental and water resources problems has been a major contributing factor to the success of environmental engineering, and has accounted in large measure for the establishment of a "methodology of pollution control." However, the realization of the ever-increasing complexity and interrelated nature of current environmental problems renders pollution control imperative that intelligent planning of pollution abatement systems be undertaken. Prerequisite to such planning is an understanding of the performance, potential, and limitations of the various methods of environmental protection available for environmental scientists and engineers. In this series of handbooks, we will review at a tutorial level a broad spectrum of engineering systems (natural environment, processes, operations, and methods) currently being utilized, or of potential utility, for pollution abatement, environmental protection, and natural resources conservation. We believe that the unified interdisciplinary approach presented in these handbooks is a logical step in the evolution of environmental engineering.

Treatment of the various engineering systems presented will show how an engineering formulation of the subject flows naturally from the fundamental principles and theories of chemistry, microbiology, physics, and mathematics. This emphasis on fundamental science recognizes that engineering practice has, in recent years, become more firmly based on scientific principles rather than on its earlier dependency on empirical accumulation of facts. It is not intended, though, to neglect empiricism where such data lead quickly to the most economic design; certain engineering systems are not readily amenable to fundamental scientific analysis, and in these instances, we have resorted to less science in favor of more art and empiricism.

Since an environmental natural resources engineer must understand science within the context of applications, we first present the development of the scientific basis of a particular subject, followed by exposition of the pertinent design concepts and operations, and detailed explanations of their applications to natural resources conservation or environmental protection. Throughout the series, methods of mathematical modeling, system analysis, practical design, and calculation are illustrated by numerical examples. These examples clearly demonstrate how organized, analytical reasoning leads to the most direct and clear solutions. Wherever possible, pertinent cost data or models have been provided.

Our treatment of environmental natural resources engineering is offered in the belief that the trained engineer should more firmly understand fundamental principles, should be more aware of the similarities and/or differences among many of the engineering systems, and should exhibit greater flexibility and originality in the definition and innovative solution of environmental system problems. In short, the environmental and natural resources engineers should by conviction and practice be more readily adaptable to change and progress.

Coverage of the unusually broad field of environmental natural resources engineering has demanded an expertise that could only be provided through multiple authorships. Each author (or group of authors) was permitted to employ, within reasonable limits, the customary personal style in organizing and presenting a particular subject area; consequently, it has been difficult to treat all subject materials in a homogeneous manner. Moreover, owing to limitations of space, some of the authors' favored topics could not be treated in great detail, and many less important topics had to be merely mentioned or commented on briefly. All authors have provided an excellent list of references at the end of each chapter for the benefit of the interested readers. As each chapter is meant to be self-contained, some mild repetition among the various texts was unavoidable. In each case, all omissions or repetitions are the responsibility of the editors and not the individual authors. With the current trend toward metrication, the question of using a consistent system of units has been a problem. Wherever possible, the authors have used the British system (fps) along with the metric equivalent (mks, cgs, or SIU) or vice versa. The editors sincerely hope that this redundancy of units' usage will prove to be useful rather than being disruptive to the readers.

The goals of the *Handbook of Environmental Engineering* series are (1) to cover entire environmental fields, including air and noise pollution control, solid waste processing and resource recovery, physicochemical treatment processes, biological treatment processes, biotechnology, biosolids management, flotation technology, membrane technology, desalination technology, water resources, natural control processes, radioactive waste disposal, hazardous waste management, and thermal pollution control; and (2) to employ a multimedia approach to environmental conservation and protection since air, water, soil, and energy are all interrelated.

This book (Volume 20, Integrated Natural Resources Management) and its two current sister books (Volume 17, Natural Resources and Control Processes and Volume 19, Environmental and Natural Resources Engineering) of the *Handbook of Environmental Engineering* series have been designed to serve as natural resources engineering reference books as well as supplemental textbooks. We hope and expect they will prove of equal high value to advanced undergraduate and graduate students, to designers of natural resources systems, and to scientists and researchers. The editors welcome comments from readers in all of these categories. It is our hope that the three natural resources engineering books will not only provide information on natural resources engineering, but also serve as a basis for advanced study or specialized investigation of the theory and analysis of various natural resources systems. The third sister book, *Integrated Natural Resources Research*, is now in the planning stage.

This book *Integrated Natural Resources Management* (Volume 20) covers the topics on the effect of global warming and climate change on glaciers and salmons; village-driven latrines with "engineers without borders – USA"; analysis for surface water quality management; treatment of electrical and electronic components manufacturing wastes; water quality control of tidal rivers and estuaries; geographic information systems and remote sensing applications in environmental and resources engineering; investigation and management of water losses from wet infrastructure; lake restoration and acidic water control; biohydrogen production through mixed culture dark anaerobic fermentation of industrial waste; agricultural wastes–derived adsorbents for decontamination of heavy metals; removal of heavy metal ions using magnetic materials; and biohydrogen production from lignocellulosic biomass by extremely halotolerant bacterial communities from a salt pan and salt-damaged soil.

This book's first sister book, *Natural Resources and Control Processes* (Volume 17), covers the topics on the management of agricultural livestock wastes for water resources protection; application of natural processes for environmental protection; proper deep-well waste disposal, treating and managing industrial dye wastes; health effects and control of toxic lead in the environment; municipal and industrial wastewater treatment using plastic trickling filters for BOD and nutrient removal; chloride removal for recycling fly ash from municipal solid waste incinerator; recent evaluation of early radioactive disposal and management practice; recent trends in the evaluation of cementitious material in radioactive waste disposal; extensive monitoring system of sediment transport for reservoir sediment management; and land and energy resources engineering glossary.

This book's second sister book, *Environmental and Natural Resources Engineering* (Volume 19), covers the topics on understanding, conservation, and protection of precious natural resources – bees; waste reclamation for reuse, biological processes for water resources protection and water reuse, and removal of endocrine disruptors for environmental protection; cooling and reuse of thermal discharges; basic hydrology, water resources, and DAF boat plant for lake restoration; cadmium detoxification by sintering with ceramic matrices; treatment of vegetable oil–refining wastes; environmental engineering education; environmental control of pests and vectors; new book reviews; and glossary of environmental and natural resources engineering.

The editors are pleased to acknowledge the encouragement and support received from Mr. Aaron Schiller, Executive Editor of the Springer Nature, Switzerland, and his colleagues, during the conceptual stages of this endeavor. We wish to thank the contributing authors for their time and effort, and for having patiently borne our reviews and numerous queries and comments. We are very grateful to our respective families for their patience and understanding during some rather trying times.

Newtonville, NY, USA — Lawrence K. Wang
Newtonville, NY, USA — Mu-Hao Sung Wang
Cleveland, OH, USA — Yung-Tse Hung
Newtonville, NY, USA — Nazih K. Shammas

Contents

1 Effect of Global Warming and Climate Change on Glaciers and Salmons 1
Lawrence K. Wang, Mu-Hao Sung Wang, Nai-Yi Wang, and Josephine O. Wong

2 Village-Driven Latrines with "Engineers Without Borders USA" 37
Joshua Knight, Melissa Montgomery, Debbie Heuckeroth, Eugene Lendzemo, and David Sacco

3 Surface Water Quality and Analysis 63
Hamidi Abdul Aziz, Salem S. Abu Amr, and Yung-Tse Hung

4 Treatment of Electrical and Electronic Component Manufacturing Wastes 115
Omotayo Sarafadeen Amuda, Yung-Tse Hung, and Lawrence K. Wang

5 Geographic Information Systems and Remote Sensing Applications in Environmental and Water Resources 197
Samuel O. Darkwah, Michael D. Scoville, and Lawrence K. Wang

6 Investigation and Management of Water Losses from Wet Infrastructure 237
Nazih K. Shammas, Lawrence K. Wang, Mohamed A. Khadam, and Yousef Al-Feraiheedi

7 Lake Restoration and Acidic Water Control 257
Lawrence K. Wang, Mu-Hao Sung Wang, Nazih K. Shammas, R. Derrick I. Kittler, and Donald B. Aulenbach

8 Biohydrogen Production Through Mixed Culture Dark Anaerobic Fermentation of Industrial Waste 323
Abdollah Hajizadeh, Noori M. Cata Saady, Sohrab Zendehboudi, Rajinikanth Rajagopal, and Yung-Tse Hung

9 Agricultural Waste-Derived Adsorbents for Decontamination of Heavy Metals 371
Soh-Fong Lim, Siti Kartina Abdul Karim, S. N. David Chua, and Bee-Huah Lim

10 Removal of Heavy Metal Ions Using Magnetic Materials 393
Soh-Fong Lim, Agnes Yung-Weng Lee, S. N. David Chua, and Bee-Huah Lim

11 Biohydrogen Production from Lignocellulosic Biomass by Extremely Halotolerant Bacterial Communities from a Salt Pan and Salt-Damaged Soil 411
Dyah Asri Handayani Taroepratjeka, Tsuyoshi Imai, Prapaipid Chairattanamanokorn, Alissara Reungsang, and Yung-Tse Hung

Index 429

About the Editors

Lawrence K. Wang has more than 28 years of experience in facility design, environmental sustainability, natural resources, resources recovery, global pollution control, construction, plant operation, and management. He has expertise in water supply, air pollution control, solid waste disposal, water resources, waste treatment, and hazardous waste management. He is a retired dean/director/VP of the Lenox Institute of Water Technology, Krofta Engineering Corporation, and Zorex Corporation, respectively, and was an Assistant Professor/Associate Professor/Professor of Rensselaer Polytechnic Institute, Stevens Institute of Technology, and the University of Illinois, respectively, in the USA. He has represented the US Government to serve the United Nations Industrial Development Organization (UNIDO) as a Senior Advisor in Vienna, Austria. Dr. Wang is the author of over 700 papers and 45 books, and is credited with 24 US patents and 5 foreign patents. He received his BSCE from National Cheng-Kung University, Taiwan, ROC; his two MS degrees from the University of Missouri (Missouri University of Science and Technology) and the University of Rhode Island, USA; and his PhD degree from Rutgers University, USA. Currently, he is the chief series editor of the *Advances in Industrial and Hazardous Wastes Treatment* series (CRC Press of Taylor & Francis Group), the *Handbook of Environmental Engineering* series (Springer), the *Evolutionary Progress in Science, Technology, Engineering, Arts, and Mathematics* series (Lenox Institute Press), and the *Environmental Science, Technology, Engineering and Mathematics* series (Lenox Institute Press). He is also the coeditor of the *Handbook of Environment and Waste Management* series (World Scientific) and the coauthor of the *Water and Wastewater Engineering* series (John Wiley).

Mu-Hao Sung Wang has been an engineer, an editor, and a professor serving private firms, governments, and universities in the USA and Taiwan for over 25 years. She is a licensed professional engineer and a Diplomate of American Academy of Environmental Engineers. Her publications have been in the areas of water quality, modeling, environmental sustainability, waste management, NPDES, flotation, and analytical methods. Dr. Wang is the author of over 50 publications and an inventor of

14 US and foreign patents. She received her BSCE degree from National Cheng Kung University, Taiwan, ROC; her MSCE degree from the University of Rhode Island, USA; and her PhD degree from Rutgers University, USA. She was an Adjunct Associate Professor at Stevens Institute of Technology, NJ, USA, an Adjunct Professor at National Cheng Kung University, Taiwan, ROC, and an Adjunct Professor at the Lenox Institute of Water Technology, MA, USA. She is the co-series editor of the *Handbook of Environmental Engineering* series (Springer), and the coeditor of the *Environmental Science, Technology, Engineering and Mathematics* (STEM) series (Lenox Institute Press). She is a member of AWWA, NEWWA, WEF, and OCEESA.

Yung-Tse Hung has been Professor of Civil Engineering at Cleveland State University, Cleveland, Ohio, USA, since 1981. He is a Fellow of the American Society of Civil Engineers, professional engineer in Ohio and North Dakota, and Diplomate of American Academy of environmental Engineers. He has taught at 16 universities in 8 countries. His research interests and publications have been involved with biological treatment processes and industrial waste treatment. Dr. Hung is credited with over 470 publications and presentations, 28 books, and 159 book chapters in water and wastewater treatment. He received his BSCE and MSCE from National Cheng-Kung University, Taiwan, and his PhD from the University of Texas at Austin, USA. He is the editor-in-chief of *International Journal of Environment and Waste Management*, *International Journal of Environmental Engineering*, and *International Journal of Environmental Engineering Science*, and coeditor of the *Advances in Industrial and Hazardous Wastes Treatment* series (CRC Press of Taylor & Francis Group) and the *Handbook of Environmental Engineering* series (Springer). Dr. Hung is also the chief editor of the *Handbook of Environment and Waste Management* series (World Scientific) and the permanent executive director and ex-president of OCEESA (Overseas Chinese Environmental Engineers and Scientists Association).

Nazih K. Shammas was an environmental consultant and professor for over 45 years. He is an ex-dean/director of the Lenox Institute of Water Technology, and an advisor to Krofta Engineering Corporation, USA. Dr. Shammas was the author of over 250 publications and 15 books in the field of environmental engineering. He had experience in environmental planning, curriculum development, teaching, and scholarly research; and has expertise in water quality control, wastewater reclamation and reuse, physicochemical and biological processes, and water and wastewater systems. He received his BE degree from the American University of Beirut, Lebanon; his MS degree from the University of North Carolina at Chapel Hill; and his PhD degree from the University of Michigan, USA. He was the coeditor of the *Advances in Industrial and Hazardous Wastes Treatment* series (CRC Press of Taylor & Francis Group), the *Handbook of Environmental Engineering* series (Springer), the *Environmental Science, Technology, Engineering and Mathematics* series (Lenox Institute Press), and the *Handbook of Environment and Waste Management* series (World Scientific). In addition, Dr. Shammas was the coauthor of the *Water and Wastewater Engineering* series (John Wiley).

Contributors

Salem S. Abu Amr School of Civil Engineering, Engineering Campus, Universiti Sains Malaysia, Penang, Malaysia

Yousef Al-Feraiheedi Ar-Riyadh Development Authority (ADA), Ar-Riyadh, Saudi Arabia

Omotayo Sarafadeen Amuda, Ph.D. Analytical/Environmental Chemistry Unit, Department of Pure and Applied Chemistry, Ladoke Akintola University of Technology, Ogbomoso, Nigeria

Donald B. Aulenbach, Ph.D., P.E., D.EE. Rensselaer Polytechnic Institute, Troy, NY, USA
Lenox Institute of Water Technology, Lenox, MA and Newtonville, NY, USA

Hamidi Abdul Aziz, Ph.D. School of Civil Engineering, Solid Waste Management Cluster, Engineering Campus, Universiti Sains Malaysia, Penang, Malaysia

Noori M. Cata Saady, M.Sc., Ph.D., P.Eng. Department of Civil Engineering, The Memorial University of Newfoundland, St. John's, NLA, Canada

Prapaipid Chairattanamanokorn Department of Environmental Technology and Management, Faculty of Environment, Kasetsart University, Bangkok, Thailand

S. N. David Chua Faculty of Engineering, Universiti Malaysia Sarawak, Kota Samarahan, Sarawak, Malaysia

Samuel O. Darkwah, PhD Lenox Institute of Water Technology, Lenox, MA and Newtonville, NY, USA

Abdollah Hajizadeh, M.Sc. Department of Process Engineering, The Memorial University of Newfoundland, St. John's, NLA, Canada

Debbie Heuckeroth, P.E., ENV SP Engineers Without Borders-USA Philadelphia Professional Chapter, Philadelphia, PA, USA

Yung-Tse Hung, Ph.D., P.E., D.EE, F-ASCE. Department of Civil and Environmental Engineering, Cleveland State University, Cleveland, OH, USA

Tsuyoshi Imai, DR. ENG Graduate School of Sciences and Technology for Innovation, Yamaguchi University, Tokiwadai, Ube, Yamaguchi, Japan

Siti Kartina Abdul Karim Faculty of Applied Sciences, Universiti Teknologi MARA, Kota Samarahan, Sarawak, Malaysia

Mohamed A. Khadam, Ph.D. Civil Engineering Department, University of Khartoum, Khartoum, Sudan

R. Derrick I. Kittler, ME Lenox Institute of Water Technology, Lenox, MA and Newtonville, NY, USA

Joshua Knight, Ph.D., P.E. Global Hope Network International, Orlando, FL, USA

Agnes Yung-Weng Lee Faculty of Engineering, Universiti Malaysia Sarawak, Kota Samarahan, Sarawak, Malaysia

Eugene Lendzemo, PG Greenery Association, Kumbo, Cameroon

Bee-Huah Lim Fuel Cell Institute, Universiti Kebangsaan Malaysia, 43600 UKM, Bangl, Selangor Darul Ehsan, Malaysia

Soh-Fong Lim, PhD Faculty of Engineering, Universiti Malaysia Sarawak, Kota Samarahan, Sarawak, Malaysia

Melissa Montgomery, P.E. Engineers Without Borders-USA, Denver, CO, USA

Rajinikanth Rajagopal, M.Eng., Ph.D. Agriculture and Agri-Food Canada, Science and Technology Branch, Sherbrooke Research and Development Centre, Sherbrooke, QC, Canada

Alissara Reungsang Department of Biotechnology, Faculty of Technology, Khon Kaen University, Khon Kaen, Thailand

David Sacco, PE, EWB-USA Yale University Student Chapter, New Haven, CT, USA

Michael D. Scoville Blasland, Bouck & Lee, Inc., Syracuse, NY, USA

Nazih K. Shammas, Ph.D. Lenox Institute of Water Technology, Lenox, MA and Newtonville, NY, USA
Krofta Engineering Corporation, Lenox, MA, USA

Dyah Asri Handayani Taroepratjeka Department of Environmental Engineering, Faculty of Civil Engineering and Planning, Institut Teknologi Nasional, Bandung, Bandung, Jawa Barat, Indonesia

Lawrence K. Wang, Ph.D., P.E., D.EE. Lenox Institute of Water Technology, Lenox, MA and Newtonville, NY, USA
Krofta Engineering Corporation, Lenox, MA, USA
Rensselaer Polytechnic Institute, Troy, NY, USA

Mu-Hao Sung Wang, Ph.D., P.E., D.EE. Lenox Institute of Water Technology, Lenox, MA and Newtonville, NY, USA

Nai-Yi Wang, Ph.D. Department of Chemistry, State University of New York, Buffalo, NY, USA

Josephine O. Wong, BS, MSCE City of San Francisco, San Francisco, CA, USA

Sohrab Zendehboudi, M.Sc., Ph.D., P.Eng. Department of Process Engineering, The Memorial University of Newfoundland, St. John's, NLA, Canada

Chapter 1
Effect of Global Warming and Climate Change on Glaciers and Salmons

Check for updates

Lawrence K. Wang, Mu-Hao Sung Wang, Nai-Yi Wang, and Josephine O. Wong

Contents

1 Environment and Ecosystem Protection 2
2 Weather, Climate, Climate Change, and Air Pollution Control 3
3 Greenhouse Gases, Greenhouse Effect, Global Warming, Global Warming Potential 4
3.1 Main Contributors to Greenhouse Gases 4
3.2 Effect of Greenhouse Gases on Global Warming and Climate Change 8
4 Effect of Global Warming and Climate Change on Glaciers 11
4.1 Glacier Protection 11
4.2 Mendenhall Glacier 12
4.3 Glacier Bay National Park and Reserve 14
5 Effect of Global Warming and Climate Change on Salmon and Ecosystem 17
5.1 Salmon Protection 17
5.2 Effect on Ecosystem 18
5.3 Effect on Salmon 19
5.4 Carbon Dioxide Changes in the Environment 21
6 Summary 22
Glossary 24
References 34

Abstract Scientific studies of global warming, climate change, glacier melting, and salmon protection conducted by international researchers are reviewed, presented, and discussed. The topics covered in this book chapter include technical terminologies, climate change, global warming, greenhouse gases, global warming potential, rising temperature trend in the environment, glacier melting, glacier protection, tidewater glaciers, Glacier Bay National Park and Reserve, Mendenhall Glacier,

L. K. Wang (✉) · M.-H. S. Wang
Lenox Institute of Water Technology, Newtonville, NY, USA

N.-Y. Wang
University of Buffalo, Buffalo, NY, USA
e-mail: NaiYiWang@comcast.net

J. O. Wong
City of San Francisco, San Francisco, CA, USA

L. K. Wang, M. -H. S. Wang, Y. -T. Hung, N. K. Shammas (eds.), *Integrated Natural Resources Management*, Handbook of Environmental Engineering 20,
https://doi.org/10.1007/978-3-030-55172-8_1

Mendenhall Lake, salmon protection, salmon life cycle, fire frequency, carbon dioxide stabilization, actions for environmental protection, and Macaulay Salmon Hatchery, Alaska, USA.

Keywords Global warming · Climate change · Glacier melting · Salmon protection · Glacier Bay National Park and Reserve · Mendenhall Glacier · Salmon life cycle · Macaulay Salmon Hatchery · Greenhouse gases

Acronym and Nomenclature

BC	Black carbon
CCl_4	Carbon tetrachloride
CCS	Carbon capture and sequestration
CFC	Chlorofluorocarbon
CH_4	Methane
CO_2	Carbon dioxide
GHG	Greenhouse gas
GWP	Global warming
HBFC	Hydrobromofluorocarbon
HC	Hydrocarbon
HCFCs	Halogenated fluorocarbons
HFCs	Hydrofluorocarbons
IPCC	Intergovernmental Panel on Climate Change
N_2O	Nitrous oxide
NMVOCs	Non-methane volatile organic compounds
O_3	Ozone
PFCs	Perfluorinated carbons
PM	Particulate matter
UNCCD	United Nations Convention to Combat Desertification
USEPA	US Environmental Protection Agency
USGS	US Geological Survey
WMO	World Meteorological Organization

1 Environment and Ecosystem Protection

Environment is an envelope or complex physical, chemical, and biotic factors (as climate, air, soil, light, temperature, living things, etc.) that act upon an ecological community (a collection of living things) and ultimately determine its form and survival.

Ecosystem is a natural unit or entity including living and nonliving parts that interact to produce a stable system through cyclic exchange of materials.

Protection of our environment and ecosystem is important to our human survival. This is the theme of this book chapter [1–42].

2 Weather, Climate, Climate Change, and Air Pollution Control

Weather is the specific condition of the atmosphere at a particular place and time. It is measured in terms of such things as wind, temperature, humidity, atmospheric pressure, cloudiness, and precipitation. In most places, weather can change from hour to hour, day to day, and season to season. Climate in a narrow sense is usually defined as the "average weather" or, more rigorously, as the statistical description in terms of the mean and variability of relevant quantities over a period of time ranging from months to thousands or millions of years. The classical period is 30 years, as defined by the World Meteorological Organization (WMO). These quantities are most often surface variables such as temperature, precipitation, and wind. Climate in a wider sense is the state, including a statistical description, of the climate system. A simple way of remembering the difference is that climate is what you expect (e.g., cold winters) and "weather" is what you get (e.g., a blizzard).

Climate in a narrow sense is usually defined as the "average weather" or, more rigorously, as the statistical description in terms of the mean and variability of relevant quantities over a period of time ranging from months to thousands of years. The classical period is three decades, as defined by the World Meteorological Organization (WMO). These quantities are most often surface variables such as temperature, precipitation, and wind. Climate in a wider sense is the state, including a statistical description, of the climate system. It is the average weather (usually taken over a 30-year time period) for a particular region and time period. Climate is not the same as weather, but rather, it is the average pattern of weather for a particular region. Weather describes the short-term state of the atmosphere. Climatic elements include precipitation; temperature; humidity; sunshine; wind velocity; phenomena such as fog, frost, and hail storms; and other measures of the weather.

Climate change means the changes in average weather conditions that persist over multiple decades or longer. Climate change encompasses both increases and decreases in temperature, as well as shifts in precipitation, changing risk of certain types of severe weather events, and changes to other features of the climate system. Climate change refers to any significant change in the measures of climate lasting for an extended period of time. In other words, climate change includes major changes in temperature, precipitation, or wind patterns, among others, that occur over several decades or longer. It is believed that air pollutants, known as greenhouse gases, cause the climate change. One of the objectives of this study is to determine (a) the effect of climate change and (b) whether or not proper air pollution control may prevent the climate change.

The addition into natural air of any harmful or objectionable substances, or energy (heat, chill, radioactivity, wavelength), or noise, in large enough quantities to adversely affect the natural air's usefulness is air pollution. We must control these objectionable substances, or energy, or noise, in order to protect our precious environment and its natural resources. We neither need excessive global warming nor climate change.

3 Greenhouse Gases, Greenhouse Effect, Global Warming, Global Warming Potential

A greenhouse gas (GHG) is any gas that absorbs infrared radiation in the atmosphere. Greenhouse gases include water vapor, carbon dioxide (CO_2), methane (CH_4), nitrous oxide (N_2O), halogenated fluorocarbons (HCFCs), ozone (O_3), perfluorinated carbons (PFCs), and hydrofluorocarbons (HFCs), water vapor and sulfur hexafluoride. Gases absorb heat in the atmosphere near the earth's surface, preventing it from escaping into space. If the atmospheric concentrations of these gases rise, the average temperature of the lower atmosphere will gradually increase, a phenomenon known as the greenhouse effect.

Specifically greenhouse effect is produced as greenhouse gases allow incoming solar radiation to pass through the earth's atmosphere, but prevent most of the outgoing infrared radiation from the surface and lower atmosphere from escaping into outer space. This process occurs naturally and has kept the earth's temperature about 60 degrees Fahrenheit warmer than it would otherwise be. Current life on earth could not be sustained without the natural greenhouse effect. The *greenhouse effect* trap and buildup heat in the atmosphere (troposphere) near the earth's surface. Some of the heat flowing back toward space from the earth's surface is absorbed by water vapor, carbon dioxide, ozone, and several other gases in the atmosphere and then reradiated back toward the earth's surface. If the atmospheric concentrations of these greenhouse gases rise, the average temperature of the lower atmosphere will gradually increase.

Global warming is known due to the recent and ongoing global average increase in temperature near the earth's surface. It is the observed increase in average temperature near the earth's surface and in the lowest layer of the atmosphere. In common usage, "global warming" often refers to the warming that has occurred as a result of increased emissions of greenhouse gases from human activities. Global warming is a type of climate change; it can also lead to other changes in climate conditions, such as changes in precipitation patterns.

Global warming potential (GWP) is a measure of the total energy that a gas absorbs over a particular period of time (usually 100 years), compared to carbon dioxide. GWP is a number that refers to the amount of global warming caused by a substance. The GWP is also the ratio of the warming caused by a substance to the warming caused by a similar mass of carbon dioxide (CO_2). Thus, the GWP of CO_2 is 1.0. Chlorofluorocarbon (CFC)-12 has a GWP of 8500; CFC-11 has a GWP of 5000; hydrochlorofluorocarbons and hydrofluorocarbons have GWPs ranging from 93 to 12,100; and water has a GWP of 0.

3.1 *Main Contributors to Greenhouse Gases*

It has been known that any gas that absorbs infrared radiation in the atmosphere is GHG, which may include water vapor, carbon dioxide (CO_2), methane (CH_4), nitrous oxide (N_2O), halogenated fluorocarbons (HCFCs), ozone (O_3), perfluorinated carbons (PFCs), and hydrofluorocarbons (HFCs), water vapor, and sulfur hexafluoride. Each of them and its related process or technologies are discussed below.

3.1.1 Carbon Dioxide

Carbon dioxide is a naturally occurring gas and also a by-product of burning fossil fuels and biomass, as well as land-use changes and other industrial processes. It is the principal human-caused greenhouse gas that affects the earth's radiative balance. It is the reference gas against which other greenhouse gases are measured and therefore has a global warming potential of 1.

Carbon dioxide's ability to absorb heat is characterized by the molecular structure, the wavelength, and radiative properties. Visible light from the sun is able to pass the carbon dioxide molecules without its energy being absorbed since the frequency of visible light does induce a dipole moment on the atmospheric CO_2 molecules. Carbon dioxide does however absorb infrared radiation (heat frcm the earth's surface) and also reemits that energy at the same wavelength as what was absorbed (also as heat) [6]. As for its molecular structure, "Carbon dioxide doesn't have a molecular dipole in its ground state. However, some CO_2 vibrations produce a structure with a molecular dipole. Because of this, CO_2 strongly absorbs infrared radiation." [7]

On the electromagnetic spectrum, infrared lies in the range of 700 nm to 1 mm (1000,000 nm). Carbon dioxide has absorption wave numbers of 667 cm^{-1} and 2349 cm^{-1} [11]. When converted to wavelengths, it is equal to ~15,000 nm and ~ 4257 nm, respectively, and is well within the infrared range. Coincidently 15,000 nm also corresponds to the maximum intensity of the Planck function [6].

The energy of a molecule can change due to a change in the energy state of the electrons of which it is composed. Thus, the molecule also has electronic energy. The energy levels are quantized and take discrete values only. Absorption and emission of radiation takes place when the atoms or molecules undergo transitions from one energy state to another. In general, these transitions are governed by selection rules. Atoms exhibit line spectra associated with electronic energy levels.

The dipole moment is determined by the magnitude of the charge difference and the distance between the two centers of charge. If there is a match in frequency of the radiation and the natural vibration of the molecule, absorption occurs and this alters the amplitude of the molecular vibration. This also occurs when the rotation of asymmetric molecules around their centers results in a dipole moment change, which permits interaction with the radiation field. Dipole moment is a vector quantity and depends on the orientation of the molecule and the photon electric vector [12].

Carbon dioxide equivalent is a metric measure used to compare the emissions from various greenhouse gases based upon their global warming potential (GWP). Carbon dioxide equivalents are commonly expressed as "million metric tons of carbon dioxide equivalents ($MMTCO_2Eq$)." The carbon dioxide equivalent for a gas is derived by multiplying the tons of the gas by the associated global warming potential, i.e., $MMTCO_2Eq$ = (million metric tons of a gas) * (GWP of the gas).

Carbon dioxide fertilization is the enhancement of the growth of plants as a result of increased atmospheric CO_2 concentration. Depending on their mechanism of photosynthesis, certain types of plants are more sensitive to changes in atmospheric CO_2 concentration.

Carbon footprint is the total amount of greenhouse gases that are emitted into the atmosphere each year by a person, family, building, organization, or company. A person's carbon footprint includes greenhouse gas emissions from fuel that an individual burns directly, such as by heating a home or riding in a car. It also includes greenhouse gases that come from producing the goods or services that the individual uses, including emissions from power plants that make electricity, factories that make products, and landfills where trash gets sent.

Carbon sequestration is a process involving the storage of carbon through natural or technological processes in biomass or in deep geological formations. Terrestrial or biological carbon sequestration is the process by which trees and plants absorb carbon dioxide, release the oxygen, and store the carbon. Geologic sequestration is one step in the process of carbon capture and sequestration (CCS) and involves injecting carbon dioxide deep underground where it stays almost permanently.

3.1.2 Methane

Methane (CH_4) is a colorless, odorless, flammable hydrocarbon (CH_4) gas. It is a product of decomposition of organic matter and of the carbonization of coal. Methane is one of the greenhouse gas chemical compounds. It is one of greenhouse gases with a global warming potential most recently estimated at 25 times that of carbon dioxide (CO_2). Methane is produced through anaerobic (without oxygen) decomposition of waste in landfills, animal digestion, decomposition of animal wastes, production and distribution of natural gas and petroleum, coal production, and incomplete fossil fuel combustion.

Natural gas is the underground deposits of gases consisting of 50–90 percent methane (CH_4) and small amounts of heavier gaseous hydrocarbon compounds such as propane (C_3H_8) and butane (C_4H_{10}).

3.1.3 Nitrous Oxide

Nitrous oxide (N_2O) is a powerful greenhouse gas with a global warming potential of 298 times that of carbon dioxide (CO_2). Major sources of nitrous oxide include soil cultivation practices, especially the use of commercial and organic fertilizers, fossil fuel combustion, nitric acid production, and biomass burning. The GWP is from the IPCC's Fourth Assessment Report (AR4). Natural emissions of N_2O are mainly from bacteria breaking down nitrogen in soils and the oceans. Nitrous oxide is mainly removed from the atmosphere through destruction in the stratosphere by ultraviolet radiation and associated chemical reactions, but it can also be consumed by certain types of bacteria in soils.

3.1.4 Ozone

Ozone is the triatomic form of oxygen (O_3). It is a gaseous atmospheric constituent. In the troposphere, it is created by photochemical reactions involving gases resulting both from natural sources and from human activities (photochemical smog). In high concentrations, tropospheric ozone can be harmful to a wide range of living organisms. Ozone is a bluish gas that is harmful to breathe. Tropospheric ozone acts as a greenhouse gas. Nearly 90% of the earth's ozone is in the stratosphere and is referred to as the ozone layer. In the stratosphere, ozone is created by the interaction between solar ultraviolet radiation and molecular oxygen (O_2). Stratospheric ozone plays a decisive role in the stratospheric radiative balance. Since ozone absorbs a band of ultraviolet radiation called UVB that is particularly harmful to living organisms, the ozone layer prevents most UVB from reaching the ground. Depletion of stratospheric ozone, due to chemical reactions that may be enhanced by climate change, results in an increased ground-level flux of ultraviolet (UV-) B radiation.

Ozone "Hole" is a large area of the stratosphere with extremely low amounts of ozone.

Ozone-depleting substance (ODS) is a family of man-made compounds that include, but are not limited to, chlorofluorocarbons (CFCs), bromofluorocarbons (halons), methyl chloroform, carbon tetrachloride, methyl bromide, and hydrochlorofluorocarbons (HCFCs). These compounds have been shown to deplete stratospheric ozone and therefore are typically referred to as ODSs.

Ozone layer depletion is a chemical reaction involving destruction of ozone molecules in the ozone layer. Depletion of this ozone layer by ozone-depleting substances will lead to higher UVB levels (a band of ultraviolet radiation), which in turn will cause increased skin cancers and cataracts and potential damage to some marine organisms, plants, and plastics.

3.1.5 Fluorinated Gases

Fluorinated gases are very powerful synthetic greenhouse gases such as hydrofluorocarbons, perfluorocarbons, and sulfur hexafluoride that are emitted from a variety of industrial processes. Fluorinated gases are sometimes used as substitutes for stratospheric ozone-depleting substances (e.g., chlorofluorocarbons, hydrochlorofluorocarbons, and halons) and are often used in coolants, foaming agents, fire extinguishers, solvents, pesticides, and aerosol propellants. These gases are emitted in small quantities compared to carbon dioxide (CO2), methane (CH_4), or nitrous oxide (N_2O), but because they are potent greenhouse gases, they are sometimes referred to as high global warming potential gases (high GWP gases).

Fluorocarbons are carbon-fluorine compounds that often contain other elements such as hydrogen, chlorine, or bromine. Common fluorocarbons include chlorofluorocarbons (CFCs), hydrochlorofluorocarbons (HCFCs), hydrofluorocarbons (HFCs), and perfluorocarbons (PFCs).

3.2 Effect of Greenhouse Gases on Global Warming and Climate Change

The authors concur with Michael Robert [2] that there are some effects of greenhouse gases on global warming before human activities and after human activities.

Before the human activity, it is assumed that (a) sunlight penetrates the atmosphere and warms the earth's surface; (b) the earth's surface radiates heat to the atmosphere, and some escapes into space; and (c) the natural activities such as volcanic activity and natural forest fires would emit greenhouse gases in the atmosphere, and so greenhouse gases in the atmosphere absorb some of the heat and trap it near the earth's surface keeping the earth warm enough to sustain life. So the main reason for the earth's warming is due to the greenhouse effect. A microcosm of this balance can be seen in the relationship between humans, animals, and trees and plants. Humans and animals take in oxygen and release carbon dioxide (CO_2), while trees and other plants take in carbon dioxide and release oxygen, thereby creating a balance in nature.

After the human activity, it is assumed that (a) sunlight still penetrates the atmosphere and warms the earth's surface; (b) the earth's surface still radiates heat to the atmosphere, and some escapes into space as usual; and (c) the natural activities such as volcanic activity and natural forest fires would still emit greenhouse gases in the atmosphere, but, due to increased human activity (from electricity, transportation, industry, and population increase), large amounts of CO_2 and other greenhouse gases have been released into the atmosphere, and also as trees get cut down and deforested, there are less resources using up the CO_2, causing levels to rise beyond natural levels. So, the excess CO_2 and other greenhouse gases trap the extra radiation near the earth's surface, causing global temperatures to rise or global warming due to in-balance of carbon intake and output. In turn, the climate on the earth is changed due to global warming. There is a chain reaction.

The definitions of greenhouse gases and other technical terms have been introduced previously. The main contributor to the greenhouse effect is carbon dioxide. Following carbon dioxide is methane gas (CH_4) and nitrous oxide (N_2O) and the halocarbons as the leading greenhouse gases. Figure 1.3 shows the major climate changing agents of greenhouse gases and their radiative forcing (W m^{-2}), showing their emission ability to retain heat and their great amounts on the earth's surface [4]. Carbon dioxide has the greatest amount where it absorbs heat at a radiative forcing above 1.5 W m^{-2} of increased and retained solar radiation at the earth's surface. Although the halocarbons, methane, and nitrous oxide have a greater warming potential than carbon dioxide, the larger quantity of carbon dioxide has a greater impact [28]. Ozone depends on the location. In the troposphere where people live is a greenhouse gas where it absorbs heat; however, ozone in the stratosphere actually absorbs UV radiation and holds back the radiation from hitting the earth's surface, thereby keeping the earth cooler. Water does not affect the warming of the earth too much since the concentration levels are fairly constant. Land use goes both ways where dark forested areas or black carbon on snow or from diesel engines in the troposphere would absorb heat, whereas planting lighter colored plants on arid regions where light reflects back to space would actually cool the earth.

Aerosol is a small droplet or particle suspended in the atmosphere, typically containing sulfur. Aerosols are emitted naturally (e.g., in volcanic eruptions) and as the result of human activities (e.g., by burning fossil fuels). There is no connection between particulate aerosols and pressurized products also called aerosols. Aerosol is also a product that relies on a pressurized gas to propel substances out of a container. Consumer aerosol products in the United States have not used ozone-depleting substances (ODS) since the late 1970s because of voluntary switching followed by federal regulation. The Clean Air Act and the US Environmental Protection Agency (USEPA) regulations further restricted the use of ODS for non-consumer products. All consumer products, and most other aerosol products, now use propellants that do not deplete the ozone layer, such as hydrocarbons and compressed gases. In summation, aerosols are fine solid or liquid particles, caused by people or occurring naturally, that are suspended in the atmosphere. Aerosols can cause cooling by scattering incoming radiation or by affecting cloud cover. Aerosols can also cause warming by absorbing radiation. Accordingly aerosols' effects are uncertain. The concentration of aerosols can be monitored based on the brightness of clouds, the higher the concentration of aerosols are, the brighter the clouds are. The reason is the aerosols feed the water droplets that contribute to the clouds, and the more water droplets there are, the more the droplets reflect light more. Aerosols in the stratosphere from volcanic activity block the radiation to help cool the earth. There is still much to investigate and discover about aerosols [2, 4].

There has been a total of 2.29 W m^{-2} increase in the amount of solar energy absorbed at the surface of the earth since 1975 [2, 4]. Because carbon dioxide shows the greatest quantity and has the greatest impact, carbon dioxide will be the main focus and should often be a reference point to compare to other greenhouse gases. Radiative forcing (RF) is defined as a change in the balance between incoming solar radiation and outgoing infrared radiation or a measure of the influence of a particular factor (e.g., greenhouse gas (GHG), aerosol, or land-use change) on the net change in the earth's energy balance. The strength of drivers is quantified as radiative forcing (RF) in units watts per square meter (W m^{-2}).

Intergovernmental Panel on Climate Change (IPCC) was established jointly by the United Nations Environment Programme and the World Meteorological Organization in 1988. The purpose of the IPCC is to assess information in the scientific and technical literature related to all significant components of the issue of climate change. The IPCC draws upon hundreds of the world's expert scientists as authors and thousands as expert reviewers. Leading experts on climate change and environmental, social, and economic sciences from some 60 nations have helped the IPCC to prepare periodic assessments of the scientific underpinnings for understanding global climate change and its consequences. With its capacity for reporting on climate change, its consequences, and the viability of adaptation and mitigation measures, the IPCC is also looked to as the official advisory body to the world's governments on the state of the science of the climate change issue. For example, the IPCC organized the development of internationally accepted methods for conducting national greenhouse gas emission inventories.

The Kyoto Protocol is an international treaty that sets obligations and industrialized countries to lower their emissions of greenhouse gases. In the earlier sections of this chapter, global warming potential (GWP) is a measurement of how well heat is absorbed by greenhouse gases. The IPCC defines global warming potential (GWP) as "the ratio of the time integrated radiative forcing from a pulse emission of 1 kg of a substance, relative to that of 1 kg of carbon dioxide, over a fixed horizon period. GWP is a relative index used to compare the climate impact of an emitted greenhouse gas, relative to an equal amount of Carbon Dioxide" [10]. Also, the IPCC examines the GWP for 1 g of carbon dioxide at a 20-, 100-, and 500-year time horizon in comparison to other greenhouse gases [9]. The six major greenhouse gases determined by the Kyoto Protocol are carbon dioxide (CO_2), methane (CH_4), nitrous oxide (N_2O), HFCs, PFCs, and sulfur hexafluoride (SF_6). The lifetime among the six gases ranges from 12 to 10,000 years. The value closest to the median is sulfur hexafluoride, with a lifetime of 3200 years. At a time horizon of 500 years, while carbon dioxide releases 1 g, sulfur hexafluoride releases 11,200 g of carbon dioxide (11,200 times more!) for the same time horizon. A stronger greenhouse gas can easily leak and create a major impact.

As a result of increased human activity, more greenhouse gases are warming the earth, resulting in an increased temperature trend around the world. The following graphs suggest the increase in emissions have led to the increase in CO_2 in the atmosphere, thereby increasing the temperature of the earth's surface on land. Table 1.1 shows the global per capita carbon emission estimates versus years. It appears that the global per capita carbon emission increases significantly after year 2000 [15, 41, 42].

As land temperatures rise, we find ocean temperatures rise as well. "The global average combined land and ocean surface temperature data as calculated by a linear trend, show a warming of 0.85 [0.65–1.06] °C, over the period 1880–2012, when multiple independently produced datasets exist. The total increase between the average of the 1850–1900 period and the 2003–2012 period is 0.78 [0.72–0.85] °C, based on the single longest dataset available. Table 1.2 shows an increase of carbon dioxide in the atmosphere from 1958 to 2012 [4], while Fig. 1.1 shows the annual temperature anomalies from land ocean during the period 1880–2012 [16]. Based on the presented figures, an increase of global carbon emissions shown in Table 1.1 leads to an increase of carbon dioxide in the atmosphere shown in Table 1.2 and finally results in a temperature increase on land and ocean shown in Fig. 1.1.

Table 1.1 Global per capita carbon emission estimates versus years [15]

Per capita carbon emission	Year metric ton/capita
0.65	1950
0.85	1960
1.10	1970
1.14	1975
1.19	1980
1.11	1985
1.16	1990
1.13	1995
1.11	2000
1.23	2005
1.30	2010

Table 1.2 Increase of carbon dioxide in the atmosphere [4]

Carbon dioxide	Year concentration, ppm
316	1960
318	1965
324	1970
330	1975
335	1980
345	1985
350	1990
355	1995
364	2000
375	2005
385	2010
400	2015

4 Effect of Global Warming and Climate Change on Glaciers

4.1 *Glacier Protection*

Briefly speaking, glacier is an extended mass of ice formed from snow falling and accumulating over the years and moving very slowly, either descending from high mountains, as in valley *glaciers*, or moving outward from centers of accumulation, as in continental *glaciers.* According to the USGS (US Geological Survey), a glacier is a large, perennial accumulation of crystalline ice, snow, rock, sediment, and often liquid water that originates on land and moves down slope under the influence of its own weight and gravity. Typically, glaciers exist and may even form in areas where:

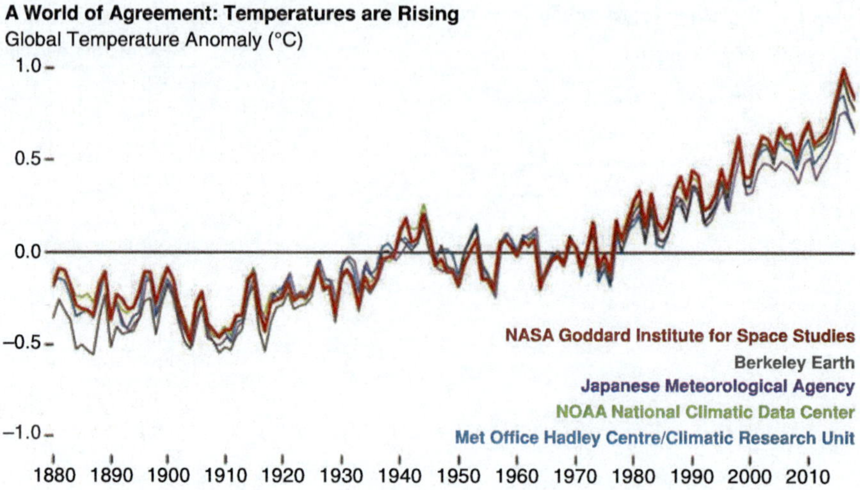

Fig. 1.1 Land and ocean temperature increase: Annual temperature anomalies from land and ocean, 1880–2012 [16]

1. Mean annual temperatures are close to the freezing point.
2. Winter precipitation produces significant accumulations of snow.
3. Temperatures throughout the rest of the year do not result in the complete loss of the previous winter's snow accumulation.

Over multiple decades this continuing accumulation of snow results in the presence of a large enough mass of snow for the metamorphism from snow to glacier ice process to begin. Glaciers are classified by their size (i.e., ice sheet, ice cap, valley glacier, cirque glacier), location, and thermal regime (i.e., polar vs. temperate). Glaciers are sensitive indicators of changing climate. Prevention of global warming may protect the existing glaciers.

Both snow cover and glaciers are melting. Many glaciers have already disappeared, while existing glaciers are retreating. Some are retreating faster than predicted rates; according to a climate-based computer model, glaciers will vanish by 2030 [22].

4.2 *Mendenhall Glacier*

A few figures are presented here for the purpose of illustration. Figure 1.2 shows a closer winter view of Mendenhall Glacier (also Sitaantaagu), which is about 13.6 miles (21.9 km) long located in *Mendenhall Valley*, about 12 miles (19 km) from downtown *Juneau*, Alaska, USA. The glacier has retreated 1.75 miles (2.82 km) since 1929, when Mendenhall Lake was created, and over 2.5 miles

Fig. 1.2 Closer view of Mendenhall Glacier and frozen Mendenhall Lake in winter (https://en.wikipedia.org/wiki/Mendenhall_Glacier#External_links)

Fig. 1.3 Closer view of Mendenhall Glacier and Mendenhall Lake in July 2019. (Credit: Lawrence K. Wang and Mu-Hao Sung Wang)

(4.0 km) since 1500. The end of the glacier currently has a negative *glacier mass balance* and will continue to retreat due to that fact that the annual temperatures are currently increasing, and the outlook is for this trend to continue due to global warming. It is noted that the white Mendenhall Glacier face has now almost totally pulled out of beautiful Mendenhall Lake in July 2019, as shown in Fig. 1.3.

Without global warming, a normal glacier is stable and its mass is sustainable. This is because in cold weather, the entire glacier is frozen, but in warm weather, moist air will be carried up to the head of the glacier's ice field, where colder ambient temperatures at the mountain top will cause it to precipitate as snow. The increased amount of snow will feed the ice field, enough to offset the melting at the *glacier's terminus* in warm weather. However, with global warming, this sustainable phenomenon will fade away if temperatures continue to climb, since the head of the glacier will no longer have cold enough ambient temperatures to cause snow to precipitate.

Although there are many negative effects of the recession of the Mendenhall Glacier and glaciers in general, there are also a few positive outcomes. With the retreat of a glacier, such as Mendenhall Glacier, a lake such as Mendenhall Lake (Fig. 1.3) has formed. The lake is a result of the runoff from the glacier and is increasing in size as the glacier continues to retreat. The lake began to form and has continued to grow becoming a unique ecosystem or a nursery for a variety of aquatic animals and plants, including several types of salmon, Dolly Varden char, and cutthroat trout.

4.3 Glacier Bay National Park and Reserve

Glacier Bay National Park and Preserve is located in Southeast Alaska west of Juneau, USA . The park and preserve cover a total of 3,223,384 acres (5037 sq. mi; 13,045 km^2), with 2,770,000 acres (4328 sq. mi; 11,210 km^2) being designated as a wilderness area. Ecosystems in the park are wet tundra, Alpine tundra, coastal forest, and glaciers and ice fields. Sport hunting and trapping are also allowed in the preserve. Wildlife in Glacier Bay includes both brown and black bear species, timber wolf, coyote, moose, black-tailed deer, red fox, porcupine, marmot, dall sheep, beaver, lynx, two species of otter, mink, wolverine, and mountain goat. Birds that nest in this park include the bald eagle, golden eagle, five species of woodpecker, two species of hummingbird, raven, four species of falcon, six species of hawk, osprey, and ten species of owl. Marine mammal species that swim offshore are the sea otter, harbor seal, Steller sea lion, Pacific white-sided dolphin, orca, minke whale, and humpback whale.

To hunt and trap, you must have all required licenses and permits and follow all other state regulations. The National Park Service and the State of Alaska cooperatively manage the wildlife resources of the preserve.

The Park is named for its abundant tidewater and terrestrial glaciers, numbering 1045 in total. There are seven tidewater glaciers in the park: Margerie Glacier, Grand Pacific Glacier, McBride Glacier, Lamplugh Glacier, Johns Hopkins Glacier, Gilman Glacier, and LaPerouse Glacier. In general, tidewater and terrestrial glaciers in the Park have been thinning and slowly receding over the last several decades, although some glaciers continue to advance, including Johns Hopkins Glacier and glaciers in Lituya Bay.

The rapid advance and subsequent retreat of the glaciers in the Glacier Bay National Park and Preserve area during several substages from 1680 to the present time are recorded and illustrated by Fig. 1.4a, b.

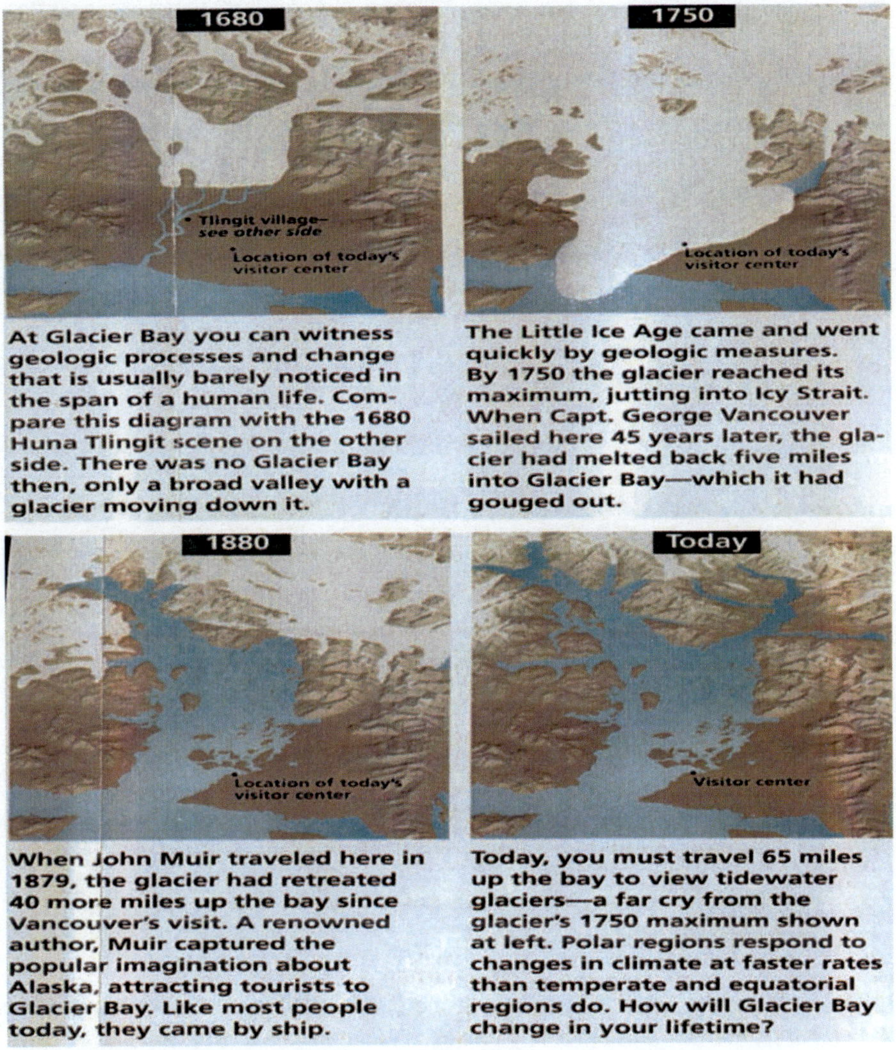

Fig. 1.4 (**a**) Rapid advance of glacier at Glacier Bay from 1680 to 1750. (Credit: State of Alaska, USA). (**b**) Subsequent retreat of glacier at Glacier Bay from 1880 to today. (Credit: State of Alaska, USA)

There are no roads leading to the park and it is most easily reached by air travel or sea travel. Despite the lack of roads, the Glacier Bay National Park and Preserve has received an average of about 470,000 recreational visitors annually from 2007 to 2016, with 520,171 visitors in 2016. Most of the visitors arrive via cruise ships from 1890 to the present time, shown in Fig. 1.5. The number of ships that may arrive each day is limited by regulation. Other travelers come on white-water rafting trips. Figure 1.6 presents a closer view of one of the great glaciers of Glacier Bay National

Fig. 1.5 Most of the visitors arrive at Glacier Bay via cruise ships from the 1890s to the present time. (July 2019; Golden Princess cruise ship; USA)

Fig. 1.6 Tidewater glacier of Glacier Bay National Park and Preserve, Alaska, USA. (Credit: LK Wang, MHS Wang)

Park and Preserve. Scientists work with the park rangers in the park and preserve hoping to learn how glacial activity relates to *climate change* (Fig. 1.7).

Scientists who observe earth's climate change have documented warming temperatures around the world (31–32). Of the more than 100,000 glaciers in Alaska, USA, 95% are currently thinning, stagnating, or retreating, and Glacier Bay's glaciers follow this trend. The combination of wind, tidal fluctuation, and moderate maritime temperatures keeps the Glacier Bay's seawater from freezing over. Much of the Glacier Bay water is over 1000 feet deep [32].

Tidewater glaciers shown in Fig. 1.6 are being eaten away on both ends as global warming worsens, suggesting faster sea level rise and ice melt that can alter ocean ecosystems [31]. Until now, scientists had limited understanding of what happens under the water at the point where land-based glaciers meet the seawater. Using a combination of radar, sonar and time-lapse photography, a team of scientists

Fig. 1.7 Scientists who investigated the Glacier Bay National Park and Preserve, Alaska, USA, in July 2019. (Note: Dr. Lawrence K. Wang 王抗曝, Dr. Mu-Hao Sung Wang 宋慕浩, Mrs. Sandra Ma Wang 馬曉蘭, and Dr. Nai-Yi Wang 王乃彝, from left to right)

including Nina Pullano has now provided the first detailed measurements of the underwater changes over time. Their scientific findings suggest that the theories and principles currently used to gauge tidewater glacier changes are significantly underestimating glaciers' ice loss because the overall trend of glacier retreat around the world is caused by both warming air and warming oceans [31].

5 Effect of Global Warming and Climate Change on Salmon and Ecosystem

5.1 Salmon Protection

Salmon is the common name for several species of ray-finned fish in the family Salmonidae. Salmon are native to tributaries of the North Atlantic and Pacific Ocean. Many species of salmon have been introduced into non-native environments such as the Great Lakes of North America and Patagonia in South America. Salmon are intensively farmed in many parts of the world.

Typically, salmon are anadromous: they hatch in freshwater, migrate to the ocean, then return to freshwater to reproduce. However, populations of several species are restricted to freshwater through their lives. Folklore has it that the fish return to the exact spot where they hatched to spawn. Tracking studies have shown this to be mostly true. A portion of a returning salmon run may stray and spawn in different freshwater systems; the percent of straying depends on the species of salmon. Homing behavior has been shown to depend on olfactory memory [38–40]. Salmon is a precious natural resource that needs to be protected.

5.2 *Effect on Ecosystem*

When temperatures increase and glaciers retreat, forests are more susceptible to fires and may be more frequent [33]. And as glaciers recede and ice melts in the Arctic, polar bears can lose their natural habitats, starve from malnutrition, and drown [30] if their food source, such as salmon, is depleted. Animals listed as endangered due to climate change include the Mediterranean monk seal, sea turtles, and more. Another example of the ecosystem becoming unstable is seen by a visiting manatee found in the Hudson River, New York, USA. The problem is manatees usually reside in Florida's warm waters. Evidently, the water temperatures were comfortable enough that the manatee continued up to New York without turning back to Florida [24]. Figure 1.8 shows that fire frequency may increase as glaciers retreat at Glacier Bay National Park and Reserve. Figure 1.9 shows how bears rely on salmon as their food source.

Fig. 1.8 Global warming and climate change increase forest fire frequency. (Credit: LK Wang and MHS Wang)

Fig. 1.9 Salmon is a major food source for bears. (Credit: LK Wang ad MHS Wang)

5.3 *Effect on Salmon*

A warming climate will alter both freshwater and marine communities, affecting resources for both fishers and endangered fish species. For Pacific salmon, climate has diverse affects. Changes in stream temperature and flow alter fish survival, swimming performance, and metabolic rates, which in turn determine energetic costs and growth [34–36].

Climate also affects salmon's habitat abundance, diversity, and access. The physical environment affects all species, most notably for salmon, both prey and predators. In response to higher temperatures, predators will likely consume more prey. Furthermore, warmwater invasive predators such as bass will grow more prevalent.

A review of the scientific literature of climate impacts on salmon reveals the many pathways these influences can take [34–36].

In the Ecosystems Analysis Program of Northwest Fisheries Science Center (NWFSC), the scientists' goal is to quantify climate influences on salmon in freshwater and marine environments over all life stages. The NWFSC supports the conservation and management of living marine resources and their habitats in the Northeast Pacific Ocean. Their research assists resource managers in making sound decisions that build sustainable fisheries, recover endangered and threatened species, sustain healthy ecosystems, and reduce risks to human health [34].

Frequently NWFSC research requires developing novel statistical tools. Their scientists then incorporate these relationships into models that can be used to assess extinction risk. Their comprehensive, multilateral approach has the following specific objectives: (a) Identify relationships between the environment and population responses such as migration timing and rates of survival. Using these relationships, develop life cycle models that can help predict the response of endangered salmon populations to climate change [35]; (b) monitor the state of the science in annual

Fig. 1.10 Salmon ladders of Macaulay Salmon Hatchery of Douglas Island Pink and Chum, Inc., Juneau, AK, USA. (Credit: LK Wang and MHS Wang)

syntheses of the global literature on climate effects on salmon (35); (c) provide scientific support for resource management actions to protect endangered species [36]; and (d) collaborate with other groups on the Pacific coast, the nation, and the world to enhance our resilience to climate variability and climate change.

To combat the loss of Pacific salmon, Macaulay Salmon Hatchery was established in Juneau, AK, USA. Figures 1.10 and 1.11 partially introduce the hatchery facility. In the wild, salmon have to beat all the odds to leave a home stream as juveniles to begin their life journey at seas or oceans. Only a very small fraction ranging from 5% to 10% of salmon eggs survive from fertilization to outmigration into the big ocean environment. At Macaulay Salmon Hatchery, 80–90% success rate from egg to juvenile stage has been consistently achieved due to their research. Once salmon juveniles leave the hatchery, they face very dangerous life at ocean as wild salmon. Figures 1.10 and 1.11 show the hatchery that rear and release salmon into the wild to enhance the opportunities for additional research, commercial use, personal use, sport, and subsistence fishing. Douglas Island Pink and Chum Inc. (DIPCI) owns the hatchery with a mission to sustain and enhance the valuable salmon resources, since salmon, as well as many other animals, have been directly or indirectly affected by the climate changes. Salmon does need protection [37, 41]. The hatchery's salmon management effort is presented in its process steps presented below.

In Step 1 of the hatchery's salmon management, millions of fertilized eggs develop in dark incubation rooms beginning in the late summer.

In Step 2, while in the incubation trays, eggs hatch into alevin and have a yolk sac on their bellies they feed off of all winter.

In Step 3, fish outmigrate in the spring as smolt and are transported to saltwater net pens where they spend 1 to 3 months growing and imprinting.

Fig. 1.11 Internal facility of Macaulay Salmon Hatchery, Douglas Island Pink and Chum, Inc., Juneau, AK, USA. (Credit: LK Wang and MHS Wang)

In Step 4, salmon leave the net pens in late spring to early summer and spend their lives at sea. After 1 to 6 years in the ocean, they return to the place they imprinted upon with the goal of reproducing in freshwater.

In Step 5, salmon return to spawn in late June through September. All Pacific salmon naturally die after spawning in the wild, and their dead bodies act as a natural fertilizer for the stream beds of their birth. This phenomenon gives the Tongass National Forest in the Southeast of Alaska its nickname of " America's Salmon Forest." The specific salmon released by Macaulay Salmon Hatchery are imprinted in their brains knowing where their home is, and they will swim to the hatchery's entrance ladders (Fig. 1.10), gradually jumping up to the hatchery internal facility (Fig. 1.11 Step 5).

In Step 6 inside the internal facility, mature salmon are separated by species for the separate egg collection and sperm collection. By mixing the eggs with the sperms, the salmon eggs are fertilized. Fertilized eggs are sent to the hatchery's freshwater incubation rooms, and carcasses of the adult salmon are delivered to processors. After Step 6 is over, Step 1 is repeated again.

5.4 Carbon Dioxide Changes in the Environment

As atmospheric temperatures increase and land temperatures increase, ocean temperatures increase, causing snow and glaciers to melt and sea levels to rise. Surely, there will be negative effects as a result of changing the earth's composition. In order to stay away from the most severe consequences, the concentration levels of CO_2 in the atmosphere must be stabilized.

Currently, the concentration of carbon dioxide is 430 parts per million (ppm) CO_2, whereas the concentration of CO_2 before the industrial revolution was only 280 ppm. Concentrations of CO_2 have caused the world to warm by more than half a degree Celsius and will warm by another half a degree over the next few decades, because of the inertia in the climate system. "Even if the annual flow of emissions did not increase beyond today's rate, the stock of greenhouse gases in the atmosphere would reach double pre-industrial levels by 2050 – that is 550 ppm CO_2 – and would continue growing thereafter.

But the annual flow of emissions is accelerating, as fast-growing economies invest in high-carbon infrastructure and as demand for energy and transport increases around the world. The level of 550 ppm CO_2 could be reached as early as 2035. At this level there is at least a 77% chance (and perhaps up to a 99% chance, depending on the climate model used)of a global average temperature rise exceeding 2 °C [25].

So as temperatures increase, there will be the consequences of hunger, droughts, and extinction. Crops will yield less during harvest especially in developing countries, water will become scarce as glaciers disappear, and as the ocean waters get warmer, animals will lose their habitats, leading to negative effects of the ecosystem and extinction.

Obviously, the greatest negative impacts are costly. It is clear we need to take action sooner rather than later. The longer we debate about what to do, the more costly it will be. A particular study tried to stabilize the concentration of greenhouse gases in the atmosphere targeting the year 2040; if we start in 2010, we can decrease emissions at a rate of 3.2% per year, but if we wait 10 years later in 2020 and allow emissions to increase during that time, it would require decreasing emissions at a rate of 8.2% per year to arrive to the same goal in the year 2040 [2, 41].

6 Summary

This chapter covers the wide fields of meteorology, chemistry, ecology, biology, geology, engineering, technology, mathematics, and arts and is a typical STEAM (science, technology, engineering, arts, and mathematics) project.

The importance and controversy of global warming, climate change, glacier, salmon, etc. are without saying. The authors present the scientific studies and their findings for the readers to judge.

It is now certain that greenhouse gases (GHGs), such as carbon dioxide, methane, and nitrous oxide, absorb heat and emit heat. Since GHGs get trapped in the atmosphere, the heat gets trapped in the atmosphere and warms the earth. Before the human activity, it is understood that (a) sunlight penetrates the atmosphere and warms the earth's surface, (b) the earth's surface radiates heat to the atmosphere and some escapes into space, and (c) the natural activities such as volcanic activity and natural forest fires would emit greenhouse gases in the atmosphere, and so greenhouse gases in the atmosphere absorb some of the heat and trap it near the earth's surface keeping the earth warm enough to sustain life. So the main reason for the earth's warming is due to the greenhouse effect. A microcosm of this balance can be seen in the relationship between humans, animals, and trees and plants. Humans and animals take in oxygen and release carbon dioxide (CO_2), while trees and other plants take in carbon dioxide and release oxygen, thereby creating a balance in nature.

After the human activity, it is now understood that (a) sunlight still penetrates the atmosphere and warms the earth's surface; (b) the earth's surface still radiates heat to the atmosphere, and some escapes into space as usual; and (c) the natural activities such as volcanic activity and natural forest fires would still emit greenhouse gases in the atmosphere, but, due to increased human activity (from electricity, transportation, industry, and population increase), large amounts of CO_2 and other greenhouse gases have been released in to the atmosphere, and also as trees get cut down and deforested, there are less resources using up the CO_2, causing levels to rise beyond natural levels. So, the excess CO_2 and other greenhouse gases trap the extra radiation near the earth's surface, causing global temperatures to rise or global warming due to in-balance of carbon intake and output. In turn, the climate on the earth is changed due to global warming.

There are chain reactions. Due to global warming and climate changes, glaciers are melting and salmons are adversely affected, needing our protection.

The uncertainty is the natural events. The natural activities such as volcanic activity and natural forest fires have been emitting greenhouse gases into the atmosphere for million years, when there have been many cycles of climate changes due to periodical global warming or global cooling (i.e., ice ages). Is the earth due for global warming naturally anyway regardless whether or not there would be any human's industrial activities? What is the current ratio of natural activities and human activities that causes the global warming and climate change? More scientific studies are needed for protection of our precious resources, such as glaciers, salmons, and others.

What is certain? It is certain that:

- The increasing amount of human activity is changing the composition of the atmosphere with overwhelming supporting data.
- Carbon dioxide, methane, and nitrous oxide are increasing dramatically as a result of human activities.
- Greenhouse gases absorb heat and emit heat; since they get trapped in the atmosphere, the heat gets trapped in the atmosphere and warms the earth.
- "The lifetime of major greenhouse gases from human activities can remain in the atmosphere for decades to centuries.
- Average global temperatures have risen over 1 degree Fahrenheit, and up to 4 degrees Fahrenheit in some regions, over the last century."

What is uncertain? It is uncertain that:

- Forecasting exact impacts to health, agriculture, water resources, forests, wildlife and coastal areas in regional basis is difficult.
- There is also uncertainty in quantifying the exact magnitude and extent of adverse effects, projecting magnitude of sea level rise.
- Quantifying the indirect effects of aerosol particles to the earth's energy balance (i.e., cloud formation and its radiative properties, precipitation efficiencies).

Adaptation/mitigation for the effects of climate change is necessary because evidence shows it is too late for complete prevention. The responsible thing to do is to start preparing now.

Climate change has become a contentious political issue, which is unfortunate if only because it distracts society and our policy makers from necessary discussions and decisions about how to respond to the impacts of climate change on our communities.

Over a full century has passed since the beginning of climate science; climate science is older than the atom bomb, the discovery of penicillin, trans-Atlantic jet flights, digital computers, and moon rockets. Since 1827, our understanding and certainty of the earth warming is a proven result of human activities.

But, as always, we must try and separate the science we know from the policy we must create. Perhaps the solution to climate change will something environmentalists hate. If that's our best bet, so be it. But whatever bet we make on our future in terms of climate action, it must truly be our best and our most informed. Pretending climate science hasn't been saying much the same thing for a long, long time now is nothing more than wishful thinking.

Glossary

Aerosol (a) A small droplet or particle suspended in the atmosphere, typically containing sulfur. Aerosols are emitted naturally (e.g., in volcanic eruptions) and as the result of human activities (e.g., by burning fossil fuels). There is no connection between particulate aerosols and pressurized products also called aerosols. (b) A product that relies on a pressurized gas to propel substances out of a container. Consumer aerosol products in the USA have not used ozone-depleting substances (ODS) since the late 1970s because of voluntary switching followed by federal regulation. The Clean Air Act and the US Environmental Protection Agency (USEPA) regulations further restricted the use of ODS for non-consumer products. All consumer products, and most other aerosol products, now use propellants that do not deplete the ozone layer, such as hydrocarbons and compressed gases. (c) Aerosols are fine solid or liquid particles, caused by people or occurring naturally, that are suspended in the atmosphere. Aerosols can cause cooling by scattering incoming radiation or by affecting cloud cover. Aerosols can also cause warming by absorbing radiation. (d) Small particles or liquid droplets in the atmosphere that can absorb or reflect sunlight depending on their composition.

Air pollution The addition into natural air of harmful or objectionable substances or extreme energy (heat, chill, noise, radioactivity) in large enough quantities to adversely affect the natural air's usefulness.

Air pollution control The limitations, prevention, reduction, or management of air pollution, so there will be no harmful or objectionable substances or extreme energy (heat, chill, noise, radioactivity) in large enough quantities to be added to natural air, in turn, not to adversely affect the natural air's usefulness.

Anthropogenic Made by people or resulting from human activities. Usually used in the context of emissions that are produced as a result of human activities.

Atmosphere The gaseous envelope surrounding the earth. The dry atmosphere consists almost entirely of nitrogen (78.1% volume mixing ratio) and oxygen (20.9% volume mixing ratio), together with a number of trace gases, such as argon (0.93% volume mixing ratio), helium, radiatively active greenhouse gases such as carbon dioxide (0.035% volume mixing ratio), and ozone. In addition the atmosphere contains water vapor, whose amount is highly variable but typically 1% volume mixing ratio. The atmosphere also contains clouds and aerosols. The atmosphere can be divided into a number of layers according to its mixing or chemical characteristics, generally determined by temperature. The layer nearest the earth is the troposphere, which reaches up to an altitude of about 8 km (about 5 miles) in the polar regions and up to 17 km (nearly 11 miles) above the equator. The stratosphere reaches to an altitude of about 50 km (31 miles) and lies above the troposphere. The mesosphere extends up to 80–90 km and is above the stratosphere, and finally, the thermosphere, or ionosphere, gradually diminishes and forms a fuzzy border with outer space. There is very little mixing of gases between layers.

Black carbon Soot produced from coal burning, diesel engines, cooking fires, wildfires, and other combustion sources. These particles absorb solar energy and have a warming influence on the climate.

Black carbon aerosol Black carbon (BC) is the most strongly light-absorbing component of particulate matter (PM) and is formed by the incomplete combustion of fossil fuels, biofuels, and biomass. It is emitted directly into the atmosphere in the form of fine particles (PM2.5).

Carbon capture and sequestration (CCS) Carbon capture and sequestration (CCS) is a set of technologies that can greatly reduce carbon dioxide emissions from new and existing coal- and gas-fired power plants, industrial processes, and other stationary sources of carbon dioxide. It is a three-step process that includes capture of carbon dioxide from power plants or industrial sources, transport of the captured and compressed carbon dioxide (usually in pipelines), and underground injection and geologic sequestration, or permanent storage, of that carbon dioxide in rock formations that contain tiny openings or pores that trap and hold the carbon dioxide.

Carbon capture and storage The process of capturing carbon dioxide and injecting it into geologic formations underground for long-term storage. "Carbon capture and storage" is same as "carbon capture and sequestration."

Carbon cycle (a) All parts (reservoirs) and fluxes of carbon. The cycle is usually thought of as four main reservoirs of carbon interconnected by pathways of exchange. The reservoirs are the atmosphere, terrestrial biosphere (usually includes freshwater systems), oceans, and sediments (includes fossil fuels). The annual movements of carbon, the carbon exchanges between reservoirs, occur because of various chemical, physical, geological, and biological processes. The ocean contains the largest pool of carbon near the surface of the earth, but most of that pool is not involved with rapid exchange with the atmosphere. (b) Circulation of carbon atoms through the earth systems as a result of photosynthetic conversion of carbon dioxide into complex organic compounds by plants, which are consumed by other organisms, and return of the carbon to the atmosphere as carbon dioxide as a result of respiration, decay of organisms, and combustion of fossil fuels.

Carbon dioxide A naturally occurring gas and also a by-product of burning fossil fuels and biomass, as well as land-use changes and other industrial processes. It is the principal human-caused greenhouse gas that affects the earth's radiative balance. It is the reference gas against which other greenhouse gases are measured and therefore has a global warming potential of 1.

Carbon dioxide equivalent A metric measure used to compare the emissions from various greenhouse gases based upon their global warming potential (GWP). Carbon dioxide equivalents are commonly expressed as "million metric tons of carbon dioxide equivalents ($MMTCO_2Eq$)." The carbon dioxide equivalent for a gas is derived by multiplying the tons of the gas by the associated global warming potential, i.e., $MMTCO_2Eq$ = (million metric tons of a gas) * (GWP of the gas).

Carbon dioxide fertilization The enhancement of the growth of plants as a result of increased atmospheric CO_2 concentration. Depending on their mechanism of photosynthesis, certain types of plants are more sensitive to changes in atmospheric CO_2 concentration.

Carbon footprint The total amount of greenhouse gases that are emitted into the atmosphere each year by a person, family, building, organization, or company. A persons carbon footprint includes greenhouse gas emissions from fuel that an individual burns directly, such as by heating a home or riding in a car. It also includes greenhouse gases that come from producing the goods or services that the individual uses, including emissions from power plants that make electricity, factories that make products, and landfills where trash gets sent.

Carbon sequestration (a) Storage of carbon through natural or technological processes in biomass or in deep geological formations. (b) Terrestrial, or biologic, carbon sequestration is the process by which trees and plants absorb carbon dioxide, release the oxygen, and store the carbon. Geologic sequestration is one step in the process of carbon capture and sequestration (CCS) and involves injecting carbon dioxide deep underground where it stays almost permanently.

Carbon tetrachloride (CCl_4) A compound consisting of one carbon atom and four chlorine atoms. Carbon tetrachloride was widely used as a raw material in many industrial uses, including the production of chlorofluorocarbons and as a solvent. Solvent use ended when it was discovered to be carcinogenic. It is also used as a catalyst to deliver chlorine ions to certain processes. Its ozone depletion potential is 1.2.

Chlorofluorocarbon (CFC) A compound consisting of chlorine, fluorine, and carbon. Chlorofluorocarbons (CFCs) are very stable in the troposphere. They move to the stratosphere and are broken down by strong ultraviolet light, where they release chlorine atoms that then deplete the ozone layer. CFCs were commonly used as refrigerants, solvents, and foam blowing agents. The most common CFCs were CFC-11, CFC-12, CFC-113, CFC-114, and CFC-115. CFCs have been phased out in the USA, with a few exceptions. CFCs are the gases covered under the 1987 Montreal Protocol and used for refrigeration, air conditioning, packaging, insulation, solvents, or aerosol propellants. Since they are not destroyed in the lower atmosphere, CFCs drift into the upper atmosphere where, given suitable conditions, they break down ozone. These gases are being replaced by other compounds: hydrochlorofluorocarbons, an interim replacement for CFCs that are also covered under the Montreal Protocol, and hydrofluorocarbons, which are covered under the Kyoto Protocol. All these substances are also greenhouse gases.

Class I ozone-depleting substance One of several groups of chemicals with an ozone depletion potential of 0.2 or higher. Class I ozone-depleting substances listed in the Clean Air Act include chlorofluorocarbons, halons, carbon tetrachloride, methyl chloroform hydrobromofluorocarbons, and methyl bromide.

Class II ozone-depleting substance A chemical with an ozone depletion potential of less than 0.2. Currently, all hydrochlorofluorocarbons are Class II ozone-depleting substances.

Climate (a) Climate in a narrow sense is usually defined as the "average weather" or, more rigorously, as the statistical description in terms of the mean and variability of relevant quantities over a period of time ranging from months to thousands of years. The classical period is three decades, as defined by the World Meteorological Organization (WMO). These quantities are most often surface variables such as temperature, precipitation, and wind. Climate in a wider sense is the state, including a statistical description, of the climate system. (b) It is the average weather (usually taken over a 30-year time period) for a particular region and time period. Climate is not the same as weather, but rather, it is the average pattern of weather for a particular region. Weather describes the short-term state of the atmosphere. Climatic elements include precipitation; temperature; humidity; sunshine; wind velocity; phenomena such as fog, frost, and hail storms; and other measures of the weather.

Climate change (a) Changes in average weather conditions that persist over multiple decades or longer. Climate change encompasses both increases and decreases in temperature, as well as shifts in precipitation, changing risk of certain types of severe weather events, and changes to other features of the climate system. (b) Climate change refers to any significant change in the measures of climate lasting for an extended period of time. In other words, climate change includes major changes in temperature, precipitation, or wind patterns, among others, that occur over several decades or longer.

Deforestation The change of forested lands to non-forest uses. This is often cited as one of the major causes of the enhanced greenhouse effect for two reasons: (a) trees that are burned or decompose release carbon dioxide; and (b) trees that are cut no longer remove carbon dioxide from the atmosphere.

Desertification Land degradation in arid, semiarid, and dry subhumid areas resulting from various factors, including climatic variations and human activities. Further, the UNCCD (the United Nations Convention to Combat Desertification) defines land degradation as a reduction or loss, in arid, semiarid, and dry subhumid areas, of the biological or economic productivity and complexity of rain-fed cropland, irrigated cropland, or range, pasture, forest, and woodlands resulting from land uses or from a process or combination of processes, including processes arising from human activities and habitation patterns, such as (a) soil erosion caused by wind and/or water; (b) deterioration of the physical, chemical, and biological or economic properties of soil; and (c) long-term loss of natural vegetation, such as conversion of forest to non-forest.

Ecosystem (a) All the living things in a particular area as well as components of the physical environment with which they interact, such as air, soil, water, and sunlight. (b) Any natural unit or entity including living and nonliving parts that interact to produce a stable system through cyclic exchange of materials. (c) The complex of a community of organisms and the community's environment functioning as an ecological unit.

Emissions The release of a substance (usually a gas when referring to the subject of climate change) into the atmosphere.

Enhanced greenhouse effect The concept that the natural greenhouse effect has been enhanced by increased atmospheric concentrations of greenhouse gases (such as CO_2 and methane) emitted as a result of human activities. These added greenhouse gases cause the earth to warm.

Environment	The complex of physical, chemical, and biotic factors (as climate, soil, and living things) that act upon an organism (a living thing) or an ecological community (a collection of living things) and ultimately determine its form and survival. The circumstances, objects, and conditions that surround each of us.
Fluorinated gases	Powerful synthetic greenhouse gases such as hydrofluorocarbons, perfluorocarbons, and sulfur hexafluoride that are emitted from a variety of industrial processes. Fluorinated gases are sometimes used as substitutes for stratospheric ozone-depleting substances (e.g., chlorofluorocarbons, hydrochlorofluorocarbons, and halons) and are often used in coolants, foaming agents, fire extinguishers, solvents, pesticides, and aerosol propellants. These gases are emitted in small quantities compared to carbon dioxide (CO2), methane (CH4), or nitrous oxide (N2O), but because they are potent greenhouse gases, they are sometimes referred to as high global warming potential gases (high GWP gases).
Fluorocarbons	Carbon-fluorine compounds that often contain other elements such as hydrogen, chlorine, or bromine. Common fluorocarbons include chlorofluorocarbons (CFCs), hydrochlorofluorocarbons (HCFCs), hydrofluorocarbons (HFCs), and perfluorocarbons (PFCs).
Forcing	Factors that affect the earth's climate. For example, natural factors such as volcanoes and human factors such as the emission of heat-trapping gases and particles through fossil fuel combustion.
Fossil fuel	(a) A general term for a fuel that is formed in the earth from plant or animal remains, including coal, oil, natural gas, oil shales, and tar sands. (b) A general term for organic materials formed from decayed plants and animals that have been converted to crude oil, coal, natural gas, or heavy oils by exposure to heat and pressure in the earth's crust over hundreds of millions of years.
Glacier	(a) A very large body of ice moving slowly down a slope or valley or spreading outward on a land surface. (b) A multiyear surplus accumulation of snowfall in excess of snowmelt on land and resulting in a mass of ice at least 0.1 km^2 in area that shows some evidence of movement in response to gravity. A glacier may terminate on land or in water. Glacier ice is the largest reservoir of fresh water on earth and second only to the oceans as the largest reservoir of total water. Glaciers are found on every continent except Australia.
Global average temperature	An estimate of earth's mean surface air temperature averaged over the entire planet.
Global change	Changes in the global environment that may alter the capacity of the earth to sustain life. Global change encompasses climate change, but it also includes other critical drivers of environmental change that may interact with climate change, such as land-use change, the alteration of the water cycle, changes in biogeochemical cycles, and biodiversity loss.
Global climate models (GCM)	Mathematical models that simulate the physics, chemistry, and biology that influence the climate system.
Global warming	(a) The recent and ongoing global average increase in temperature near the earth's surface. (b) The observed increase in average temperature near the earth's surface and in the lowest layer of the atmosphere. In common usage, "global warming" often refers to the warming that has occurred as a result of increased emissions of greenhouse gases from human activities. Global warming is a type of climate change; it can also lead to other changes in climate conditions, such as changes in precipitation patterns.
Global warming potential (GWP)	(a) A measure of the total energy that a gas absorbs over a particular period of time (usually 100 years), compared to carbon dioxide. (b) A number that

refers to the amount of global warming caused by a substance. The GWP is the ratio of the warming caused by a substance to the warming caused by a similar mass of carbon dioxide (CO_2). Thus, the GWP of CO_2 is 1.0. Chlorofluorocarbon (CFC)-12 has a GWP of 8500; CFC-11 has a GWP of 5000; hydrochlorofluorocarbons and hydrofluorocarbons have GWPs ranging from 93 to 12,100; and water has a GWP of 0.

Greenhouse effect (a) The effect produced as greenhouse gases allow incoming solar radiation to pass through the earth's atmosphere, but prevent most of the outgoing infrared radiation from the surface and lower atmosphere from escaping into outer space. This process occurs naturally and has kept the earth's temperature about 60 degrees Fahrenheit warmer than it would otherwise be. Current life on earth could not be sustained without the natural greenhouse effect. (b) Trapping and buildup of heat in the atmosphere (troposphere) near the earth's surface. Some of the heat flowing back toward space from the earth's surface is absorbed by water vapor, carbon dioxide, ozone, and several other gases in the atmosphere and then reradiated back toward the earth's surface. If the atmospheric concentrations of these greenhouse gases rise, the average temperature of the lower atmosphere will gradually increase.

Greenhouse gas (GHG) (a) Any gas that absorbs infrared radiation in the atmosphere. Greenhouse gases include water vapor, carbon dioxide (CO_2), methane (CH_4), nitrous oxide (N_2O), halogenated fluorocarbons (HCFCs), ozone (O_3), perfluorinated carbons (PFCs), and hydrofluorocarbons (HFCs), water vapor and sulfur hexafluoride. (b) Gases that absorb heat in the atmosphere near the earth's surface, preventing it from escaping into space. If the atmospheric concentrations of these gases rise, the average temperature of the lower atmosphere will gradually increase, a phenomenon known as the greenhouse effect.

Habitat The place or environment where a plant or animal naturally or normally lives and grows.

Hydrobromofluorocarbon (HBFC) An ozone-depleting substance consisting of hydrogen, bromine, fluorine, and carbon.

Hydrocarbon (HC) A compound consisting of carbon and hydrogen. Hydrocarbons include methane, ethane, propane, cyclopropane, butane, and cyclopentane. Although they are flammable, hydrocarbons (HCs) may offer advantages as substitutes to ozone-depleting substances because they have zero ozone depletion potential, low toxicity, and, with the exception of methane, low global warming potentials. Fossil fuels are made up of hydrocarbons.

Hydrochlorofluorocarbon (HCFC) A compound consisting of hydrogen, chlorine, fluorine, and carbon. Hydrochlorofluorocarbons (HCFCs) are one class of chemicals that was used to replace the chlorofluorocarbons (CFCs). They contain chlorine and thus deplete stratospheric ozone, but to a much lesser extent than CFCs. HCFCs have ozone depletion potentials (ODPs) ranging from 0.01 to 0.1. In conclusion, although ozone-depleting substances, they are less potent at destroying stratospheric ozone than chlorofluorocarbons (CFCs). They have been introduced as temporary replacements for CFCs and are also greenhouse gases. Production of HCFCs is being phased out.

Hydrofluorocarbon (HFC) A compound consisting of hydrogen, fluorine, and carbon. Hydrofluorocarbons (HFCs) are a class of replacements for chlorofluorocarbons. Because they do not contain chlorine or bromine, they do not deplete the ozone layer. All HFCs have an ozone depletion potential of 0. Some HFCs have high global warming potentials. In conclusion, they were introduced as alternatives to ozone-depleting substances in serving many industrial, commercial, and personal needs. HFCs are emitted as by-products of industrial processes and are also used in manufacturing. They do not significantly deplete the stratospheric ozone layer, but they are powerful greenhouse gases with global warming potentials ranging from 140 (HFC-152a) to 11,700 (HFC-23).

Hydrofluoroolefin (HFO) A compound consisting of hydrogen, fluorine, and carbon. Hydrofluoroolefins (HFOs) are alternatives to ozone-depleting substances that typically have very low global warming potentials.

Hydrologic cycle The process of evaporation, vertical and horizontal transport of vapor, condensation, precipitation, and the flow of water from continents to oceans. It is a major factor in determining climate through its influence on surface vegetation, the clouds, snow and ice, and soil moisture. The hydrologic cycle is responsible for 25 to 30 percent of the midlatitudes' heat transport from the equatorial to polar regions.

Hydrosphere The component of the climate system comprising liquid surface and subterranean water, such as oceans, seas, rivers, freshwater lakes, underground water, etc.

Ice core A cylindrical section of ice removed from a glacier or an ice sheet in order to study climate patterns of the past. By performing chemical analyses on the air trapped in the ice, scientists can estimate the percentage of carbon dioxide and other trace gases in the atmosphere at a given time. Analysis of the ice itself can give some indication of historic temperatures.

Indicator An observation or calculation that allows scientists, analysts, decision-makers, and others to track environmental trends, understand key factors that influence the environment, and identify effects on ecosystems and society.

Indirect emissions Indirect emissions from a building, home, or business are those emissions of greenhouse gases that occur as a result of the generation of electricity used in that building. These emissions are called "indirect" because the actual emissions occur at the power plant which generates the electricity, not at the building using the electricity.

Industrial revolution A period of rapid industrial growth with far-reaching social and economic consequences, beginning in England during the second half of the eighteenth century and spreading to Europe and later to other countries including the USA. The industrial revolution marks the beginning of a strong increase in combustion of fossil fuels and related emissions of carbon dioxide.

Infrared radiation Infrared radiation consists of light whose wavelength is longer than the red color in the visible part of the spectrum, but shorter than microwave radiation. Infrared radiation can be perceived as heat. The earth's surface, the atmosphere, and clouds all emit infrared radiation, which is also known as terrestrial or long-wave radiation. In contrast, solar radiation is mainly short-wave radiation because of the temperature of the Sun.

Intergovernmental Panel on Climate Change (IPCC) The IPCC was established jointly by the United Nations Environment Programme and the World Meteorological Organization in 1988. The purpose of the IPCC is to assess information in the scientific and technical literature related to all significant components of the issue of climate change. The IPCC draws upon hundreds of the world's expert scientists as authors and thousands as expert reviewers. Leading experts on climate change and environmental, social, and economic sciences from some 60 nations have helped the IPCC to prepare periodic assessments of the scientific underpinnings for understanding global climate change and its consequences. With its capacity for reporting on climate change, its consequences, and the viability of adaptation and mitigation measures, the IPCC is also looked to as the official advisory body to the world's governments on the state of the science of the climate change issue.

For example, the IPCC organized the development of internationally accepted methods for conducting national greenhouse gas emission inventories.

Kyoto Protocol An international treaty that sets obligations and industrialized countries to lower their emissions of greenhouse gases.

Mesosphere Earth's atmosphere is divided into five main layers: the exosphere, the thermosphere, the mesosphere, the stratosphere, and the troposphere. Stratosphere is the region of the atmosphere above the troposphere. Mesosphere is the region of the atmosphere above the stratosphere, but below thermosphere.

Methane (CH_4) (a) Colorless, odorless, flammable hydrocarbon (CH_4) that is a product of decomposition of organic matter and of the carbonization of coal. Methane is one of the greenhouse gas chemical compounds. (b) A hydrocarbon that is a greenhouse gas with a global warming potential most recently estimated at 25 times that of carbon dioxide (CO_2). Methane is produced through anaerobic (without oxygen) decomposition of waste in landfills, animal digestion, decomposition of animal wastes, production and distribution of natural gas and petroleum, coal production, and incomplete fossil fuel combustion.

Natural gas Underground deposits of gases consisting of 50 to 90 percent methane (CH_4) and small amounts of heavier gaseous hydrocarbon compounds such as propane (C_3H_8) and butane (C_4H_{10}).

Natural variability Variations in the mean state and other statistics (such as standard deviations or statistics of extremes) of the climate on all time and space scales beyond that of individual weather events. Natural variations in climate over time are caused by internal processes of the climate system, such as El **Niño**, as well as changes in external influences, such as volcanic activity and variations in the output of the sun.

Nitrogen cycle The natural circulation of nitrogen among the atmosphere, plants, animals, and microorganisms that live in soil and water. Nitrogen takes on a variety of chemical forms throughout the nitrogen cycle, including nitrous oxide (N_2O) and nitrogen oxides (NOx).

Nitrogen oxides (NOx) Gases consisting of one molecule of nitrogen and varying numbers of oxygen molecules. Nitrogen oxides are produced in the emissions of vehicle exhausts and from power stations. In the atmosphere, nitrogen oxides can contribute to formation of photochemical ozone (smog), can impair visibility, and have health consequences; they are thus considered pollutants.

Nitrous oxide (N_2O) A powerful greenhouse gas with a global warming potential of 298 times that of carbon dioxide (CO_2). Major sources of nitrous oxide include soil cultivation practices, especially the use of commercial and organic fertilizers, fossil fuel combustion, nitric acid production, and biomass burning. The GWP is from the IPCC's Fourth Assessment Report (AR4). Natural emissions of N_2O are mainly from bacteria breaking down nitrogen in soils and the oceans. Nitrous oxide is mainly removed from the atmosphere through destruction in the stratosphere by ultraviolet radiation and associated chemical reactions, but it can also be consumed by certain types of bacteria in soils.

Non-methane volatile organic compounds (NMVOCs) Organic compounds, other than methane, that participate in atmospheric photochemical reactions.

Nutrients Chemicals (such as nitrogen and phosphorus) that plants and animals need to live and grow. At high concentrations, particularly in water, nutrients can become pollutants.

Ocean acidification (a) Increased concentrations of carbon dioxide in seawater causing a measurable increase in acidity (i.e., a reduction in ocean pH). This may lead to reduced calcification rates of calcifying organisms such as corals, mollusks, algae, and crustaceans. (b) The process by which ocean waters have become more acidic due to the absorption of human-produced carbon dioxide, which interacts with ocean water to form carbonic acid and lower the ocean's pH. Acidity reduces the capacity of key plankton species and shelled animals to form and maintain shells.

Ozone Ozone is the triatomic form of oxygen (O_3). It is a gaseous atmospheric constituent. In the troposphere, it is created by photochemical reactions involving gases resulting both from natural sources and from human activities (photochemical smog). In high concentrations, tropospheric ozone can be harmful to a wide range of living organisms. Ozone is a bluish gas that is harmful to breathe. Tropospheric ozone acts as a greenhouse gas. Nearly 90% of the earth's ozone is in the stratosphere and is referred to as the ozone layer. In the stratosphere, ozone is created by the interaction between solar ultraviolet radiation and molecular oxygen (O_2). Stratospheric ozone plays a decisive role in the stratospheric radiative balance. Since ozone absorbs a band of ultraviolet radiation called UVB that is particularly harmful to living organisms, the ozone layer prevents most UVB from reaching the ground. Depletion of stratospheric ozone, due to chemical reactions that may be enhanced by climate change, results in an increased ground-level flux of ultraviolet (UV-) B radiation.

Ozone "Hole" A large area of the stratosphere with extremely low amounts of ozone.

Ozone-depleting substance (ODS) A family of man-made compounds that include, but are not limited to, chlorofluorocarbons (CFCs), bromofluorocarbons (halons), methyl chloroform, carbon tetrachloride, methyl bromide, and hydrochlorofluorocarbons (HCFCs). These compounds have been shown to deplete stratospheric ozone and therefore are typically referred to as ODSs.

Ozone layer depletion Chemical destruction of ozone molecules in the ozone layer. Depletion of this ozone layer by ozone-depleting substances will lead to higher UVB levels (a band of ultraviolet radiation), which in turn will cause increased skin cancers and cataracts and potential damage to some marine organisms, plants, and plastics.

Particulate matter (PM) Very small pieces of solid or liquid matter such as particles of soot, dust, fumes, mists, or aerosols. The physical characteristics of particles, and how they combine with other particles, are part of the feedback mechanisms of the atmosphere.

Photosynthesis The process by which plants take CO_2 from the air (or bicarbonate in water) to build carbohydrates, releasing O_2 in the process. There are several pathways of photosynthesis with different responses to atmospheric CO_2 concentrations.

Phytoplankton Microscopic plants that live in salt and freshwater environments.

Precession The wobble over thousands of years of the tilt of the earth's axis with respect to the plane of the solar system.

Precipitation Rain, hail, mist, sleet, snow, or any other moisture that falls to the earth.

Radiation Energy transfer in the form of electromagnetic waves or particles that release energy when absorbed by an object.

Radiative forcing (a) A change in the balance between incoming solar radiation and outgoing infrared radiation. (b) A measure of the influence of a particular factor (e.g., greenhouse gas (GHG), aerosol, or land-use change) on the net change in the earth's energy balance.

Salmon It is the common name for several species of ray-finned fish in the family Salmonidae. Salmon are native to tributaries of the North Atlantic and Pacific Ocean. Many species of salmon have been introduced into non-native environments such as the Great Lakes of North America and Patagonia in South

America. Salmon are intensively farmed in many parts of the world. Typically, salmon are anadromous: they hatch in freshwater, migrate to the ocean, then return to freshwater to reproduce. However, populations of several species are restricted to freshwater through their lives. Folklore has it that the fish return to the exact spot where they hatched to spawn. Tracking studies have shown this to be mostly true. A portion of a returning salmon run may stray and spawn in different freshwater systems; the percent of straying depends on the species of salmon. Homing behavior has been shown to depend on olfactory memory.

Soil Complex mixture of inorganic minerals (i.e., mostly clay, silt, and sand), decaying organic matter, water, air, and living organisms.

Soil carbon A major component of the terrestrial biosphere pool in the carbon cycle. The amount of carbon in the soil is a function of the historical vegetative cover and productivity, which in turn is dependent in part upon climatic variables. The amount of carbon in the soil is a function of the historical vegetative cover and productivity which in turn is dependent in part upon climatic variables.

Solar energy Also called solar radiation. Energy from the Sun. Also referred to as short-wave radiation. Of importance to the climate system, solar radiation includes ultraviolet radiation, visible radiation, and infrared radiation. It also includes indirect forms of energy such as wind falling or flowing water's hydropower, ocean thermal gradients, and biomass, which are produced when direct solar energy interact with the earth

Solar radiation Radiation emitted by the Sun. It is also referred to as short-wave radiation. Solar radiation has a distinctive range of wavelengths (spectrum) determined by the temperature of the Sun. Of importance to the climate system, solar radiation includes ultraviolet radiation, visible radiation, and infrared radiation.

Source Any process or activity that releases a greenhouse gas, an aerosol, or a precursor of greenhouse gas into the atmosphere

Stratosphere The region of the atmosphere above the troposphere and between the troposphere and mesosphere. The stratosphere extends from about 8–50 km (6–31 miles) in altitude. Specifically it has a lower boundary of approximately 8 km at the poles to 15 km at the equator and an upper boundary of approximately 50 km. Depending upon latitude and season, the temperature in the lower stratosphere can increase, be isothermal, or even decrease with altitude, but the temperature in the upper stratosphere generally increases with height due to absorption of solar radiation by ozone. So the stratosphere gets warmer at higher altitudes. In fact, this warming is caused by ozone absorbing ultraviolet radiation. Warm air remains in the upper stratosphere, and cool air remains lower, so there is much less vertical mixing in this region than in the troposphere. Commercial airlines fly in the lower stratosphere.

Terrestrial radiation The total infrared radiation emitted by the earth and its atmosphere in the temperature range of approximately 200 to 300 Kelvin. Terrestrial radiation provides a major part of the potential energy changes necessary to drive the atmospheric wind system and is responsible for maintaining the surface air temperature within limits of livability.

Troposphere (a) The region of the atmosphere closest to the earth. The troposphere extends from the surface up to about 10 km (6 miles) in altitude, although this height varies with latitude. Almost all weather takes place in the troposphere. Mt. Everest, the highest mountain on earth, is only 8.8 km (5.5 miles) high. Temperatures decrease with altitude in the troposphere. As warm air rises, it cools, falling back to earth. This process, known as convection, means there are huge air movements that mix the troposphere very efficiently. (b) The lowest part of the atmosphere from the surface to about 10 km in altitude in midlatitudes (ranging from 9 km in high latitudes to 16 km in the tropics on average) where clouds and "weather" phenomena occur. In the troposphere temperatures generally decrease with height. All weather processes take place in the troposphere. Ozone that is formed in the troposphere plays a significant role in both the greenhouse gas effect and urban smog. The troposphere contains about 95 percent of the mass of air in the earth's atmosphere.

Weather Weather is the specific condition of the atmosphere at a particular place and time. It is measured in terms of such things as wind, temperature, humidity, atmospheric pressure, cloudiness, and precipitation. In most places, weather can change from hour to hour, day to day, and season to season. Climate in a narrow sense is usually defined as the "average weather," or more rigorously, as the statistical description in terms of the mean and variability of relevant quantities over a period of time ranging from months to thousands or millions of years. The classical period is 30 years, as defined by the World Meteorological Organization (WMO). These quantities are most often surface variables such as temperature, precipitation, and wind. Climate in a wider sense is the state, including a statistical description, of the climate system. A simple way of remembering the difference is that climate is what you expect (e.g., cold winters) and "weather" is what you get (e.g., a blizzard).

References

1. Nodvin SC. Global warming. Encyclopedia of Earth. 9 May 2010: n. pag. Web. 13 Jan 2014. http://www.eoearth.org/view/article/153038
2. Robert M. Climate change overview. Engineering for Professionals. Johns Hopkins University. 2010. Lecture. 13 Jan 2014. http://media.ep.jhu.edu/EPmedia/play/?file=epmedia/courses/online/575_720/Module12/Module12Part1.mp4
3. "Carbon Dioxide Emissions." *EPA*. United States Environmental Protection Agency. 9 Sep 2013. Web. 13 Jan 2014. http://www.epa.gov/climatechange/ghgemissions/gases/co2.html
4. IPCC (2013) Summary for Policymakers. In: Stocker TF, Qin D, Plattner G-K, Tignor M, Allen SK, Boschung J, Nauels A, Xia Y, Bex V, Midgley PM (eds) Climate change 2013: the physical science basis. Contribution of Working Group I to the fifth assessment report of the intergovernmental panel on climate change. Cambridge University Press, Cambridge, UK; New York
5. NASA. Evidence – climate change: how do we know? Global climate change. National Aeronautics and Space Administration. N.d. Web. 13 Jan 2014. http://climate.nasa.gov/evidence
6. Menzel WP. Absorption, emission, reflection, and scattering. Remote Sensing Applications with Meteorological Satellites. Madison: University of Wisconsin, 2005. Web http://www.ssec.wisc.edu/library/coursefiles/03_abs_emiss_ref.pdf
7. Shapley P. Greenhouse gasses. Chemistry 102. University of Illinois. 12 Oct 2011. Lecture. 13 Jan 2014. http://butane.chem.uiuc.edu/pshapley/GenChem1/L15/web-L15.pdf
8. Simeonova K, et al. Greenhouse gas inventory-related requirements. Kyoto Protocol Reference Manual on accounting of emissions and assigned amount. Bonn: United Nations Framework Convention on Climate Change. 2008. p.50. Web. 13 Jan 2014. http://unfccc.int/resource/docs/publications/08_unfccc_kp_ref_manual.pdf
9. IPCC. Direct global warming potentials. IPCC Fourth Assessment Report: Climate Change 2007: Working Group I: The Physical Science Basis. United Nations Framework Convention on Climate Change, n.d. Web. 13 Jan 2014. http://www.ipcc.ch/publications_and_data/ar4/wg1/en/ch2s2-10-2.html

10. IPCC Chemical and radiative effects of halocarbons and their replacement compounds. In: Metz B, Kuijpers L, Solomon S, Andersen SO, Davidson O, Pons J, de Jager D, Kestin T, Manning M, Meyer L (eds) IPCC/TEAP special report: safeguarding the ozone layer and the global climate system. Cambridge University Press, Cambridge, UK and New York. IPCC/TEAP, 2005. p.157. Web. 13 Jan. 2014 http://www.ipcc.ch/pdf/special-reports/sroc/sroc_full.pdf
11. USSC. Carbon dioxide (12C16O2). NIST Standard Reference Data. U.S. Secretary of Commerce. 2011. Web. 13 Jan 2014. http://webbook.nist.gov/cgi/cbook.cgi?ID=B4000020&Mask=800
12. UC-Davis, Osibanjo, Richard, Rachael Curtis, Zijuan Lai. "Infrared: Theory." ChemWiki. University of California, Davis. n.d. Web. 13 Jan 2014. http://chemwiki.ucdavis.edu/Physical_Chemistry/Spectroscopy/Vibrational_Spectroscopy/Infrared_Spectroscopy/Infrared%3A_Theory
13. Barnet C. Introduction to infrared radiative transfer. JCSDA Summer Colloquium on Data Assimilation. Joint Center for Satellite Data Assimilation. 10 July 2007. http://www.jcsda.noaa.gov/documents/meetings/2009summercoll/Barnet2_InfraRadTran.pdf, Lecture. 13 Jan 2014
14. Satzer B. Thermal infrared radiation and carbon dioxide in the atmosphere. The Mathematics of Climate Change Seminar. University of Minnesota: 3M. 12 Feb 2013. Lecture. 13 Jan 2014. http://www.math.umn.edu/~mcgehee/Seminars/ClimateChange/presentations/2013-1Spring/20130212ThermalIRandCarbonDioxideintheAtmosphere.pdf
15. USDOE. Annual global fossil-fuel carbon emissions – graphics. Carbon Dioxide Information Analysis Center. Oak Ridge: US Department of Energy. 26 Sept 2012. Web. 13 Jan 2014. http://cdiac.esd.ornl.gov/trends/emis/glo_2010.html
16. USDOE. Global annual temperature anomalies from land meteorological stations, 1880–2012. Carbon Dioxide Information Analysis Center. Oak Ridge: US Department of Energy. 26 Aug 2013. Web. 13 Jan 2014. http://cdiac.esd.ornl.gov/trends/temp/hansen/land_trend.html
17. IPCC. IPCC fourth assessment report: climate change 2007. AR4 WGI Chapter 3: Observations: surface and atmospheric climate change. Intergovernmental Panel on Climate Change. n.d. Web. 13 Jan 2014. http://www.ipcc.ch/publications_and_data/ar4/wg1/en/figure-3-1.html
18. Robert M. Climate change science. Engineering for Professionals. Johns Hopkins University. 2010. Lecture. 13 Jan 2014. http://media.ep.jhu.edu/EPmedia/play/?file=epmedia/courses/online/575_720/Module12/Module12Part2.mp4
19. USEPA. Air pollution and the clean air Ac. United States Environmental Protection Agency. 15 Aug 2013. Web. 13 Jan 2014. http://www.epa.gov/air/caa/
20. Bernstein L, et al. Observed changes in climate and their effects. Climate Change 2007: Synthesis Report Summary for Policymakers. Valencia: Intergovernmental Panel on Climate Change 12–17 Nov 2007 p3 Web 13 Jan 2014. http://www.ipcc.ch/pdf/assessment-report/ar4/syr/ar4_syr_spm.pdf
21. Fagre D. USGS repeat photography project documents retreating glaciers in Glacier National Park. Northern Rocky Mountain Science Center. United States Geological Survey. May 2013. Web. 13 Jan 2014. http://nrmsc.usgs.gov/repeatphoto/
22. USGS. USGS repeat photography project documents retreating glaciers in Glacier National Park. Northern Rocky Mountain Science Center. United States Geological Survey. May 2013. Web. 13 Jan 2014. http://nrmsc.usgs.gov/research/glacier_retreat.htm
23. NOAA. Use earth system models to study the impact of climate change on ecosystems. Geophysical Fluid Dynamics Laboratory. National Oceanic and Atmospheric Administration. 18 June 2013. Web. 13 Jan 2014. http://www.gfdl.noaa.gov/climate-and-ecosystems-projections-of-future-climate-changes
24. Lee J. Massive manatee is spotted in hudson river. The New York Times. N.p. 7 Aug. 2006. nytimes.com Web. 13 Jan 2014. http://www.nytimes.com/2006/08/07/nyregion/07manatee.html
25. NOAA. Stern review: the economics of climate change. Coral Reef Conservation Program. National Oceanic and Atmospheric Administration. N.d. Web. 13 Jan 2014. http://coralreef.

noaa.gov/aboutcrcp/strategy/reprioritization/wgroups/resources/climate/resources/stern_execsum.pdf
26. Wang MHS, Wang LK (2016) Glossary of land and energy resources engineering. In: Wang LK, Wang MHS, Hung YT, Shammas NK (eds) Natural resources and control processes. Springer, Cham, USA, pp 493–623
27. Wang MHS, Wang LK (2015) Environmental water engineering glossary. In: Yang CT, Wang LK (eds) Advances in water resources engineering. Springer, Cham, USA, pp 471–556
28. Wang MHS, Wang LK, Shammas NK (2019) Glossary of climate change, global warming, and ozone layer protection. In: Hung YT, Wang LK, Shammas NK (eds) Handbook of environment and waste management, Acid rain and greenhouse gas control, vol 3. World Scientific, Singapore
29. Berwyn, Bob. Austria's Treasured National Resource, Its Glaciers, are Melting Fast. https://insideclimatenews.org. April 28 2016
30. Berwyn B. Global warming is pushing pacific salmon to the brink, Federal Scientists warn. https://insideclimatenews.org. July 29 2019
31. Pulland N. Oceans are melting glaciers from below much faster than predicted, study finds. https://insideclimatenews.org. July 25 2019
32. Princess Cruise Lines, Ltd. Glacier Bay: rangers' top ten questions. www.nps.gov. July 2019
33. Berwyn B. Dying orchards, missing fish as climate change fueled Europe's Record Heat. https://insideclimatenews.org. August 3 2019
34. Crozier L, Burke B, Scheuerell M, Buhle E, Zabel R, Axel G, Sandford B, Sanderson B, Faulkner J, Fullerton A, Jorgensen J, Gosselin J, Sullivan K, Dorfmeier E (2019) Assessing climate impacts to pacific Salmon. Northwest Fisheries Science Center (NWFSC), Seattle
35. Crozier L (2014) Impacts of Climate Change on Columbia River Salmon: a review of the scientific literature published in 2013. Pages D1-D50 in Endangered Species Act Section 7(a)(2) supplemental biological opinion: consultation on remand for operation of the Federal Columbia River Power System. U.S. National Marine Fisheries Service, Northwest Region
36. McClure MM, Alexander M, Borggaard D, Boughton D, Crozier L, Griffis R, Jorgensen JC, Lindley ST, Nye J, Rowland MJ, Seney EE, Snover A, Toole C, Van Houtan K (2013) Incorporating climate science in applications of the U.S. Endangered Species Act for aquatic species. Conserv Biol 27(6):1222–1233
37. Zimmermann D (2019) Chinook Salmon – Life blood of the Yukon River. Yukon North of Ordinary, Government of Yukon, Canada pp. 98–100
38. USDC. Homing, straying, and colonization. U.S. Dept Commerce/NOAA/NMFS/NWFSC/Publications. Archived from the original on 20 November 2018. Retrieved 11 August 2015
39. Scholz AT, Horrall RM, Cooper JC, Hasler AD (1976) Imprinting to chemical cues: the basis for home stream selection in salmon. Science 192(4245):1247–1249. https://doi.org/10.1126/science.1273590. PMID 1273590
40. Ueda H (2011) Physiological mechanism of homing migration in Pacific salmon from behavioral to molecular biological approaches. Gen Compar Endocrinol 170(2):222–232. https://doi.org/10.1016/j.ygcen.2010.02.003.hdl.2115/44787. PMID 20144612
41. Wong JO, Wang N-Y, Wang LK, Wang, M-HS (2019) Scientific studies of global warming, climate change, glacier melting and salmon protection. In: Wang LK, Tsao H-P (eds) Evolutionary progress in science, technology, engineering, arts, and mathematics (STEAM), vol 1(9). Lenox Institute Press, Newtonville, 110 pages. 12128–0405, USA. No. STEAM-VOL1-NUM9-SEP2019; ISBN 978-0-9890870-3-2
42. Hung Y-T, Wang LK, Shammas NK (eds) (2020) Handbook of environment and waste management, vol 3. World Scientific, Singapore

Chapter 2
Village-Driven Latrines with "Engineers Without Borders USA"

Joshua Knight, Melissa Montgomery, Debbie Heuckeroth, Eugene Lendzemo, and David Sacco

Contents

1 Sanitation 39
 1.1 Sanitation Introduction 39
 1.2 Lessons Learned from Past Projects 39
2 Engineers Without Borders USA 40
 2.1 Overview, Vision, and Offices of EWB-USA 40
 2.2 Project Variables of EWB-USA 42
3 Latrine Technologies 42
 3.1 Introduction to Latrine Design 42
 3.2 Latrine Design Checklist 43
 3.3 Existing Latrine Technologies 47
4 Planning, Monitoring, Evaluation, and Learning of Latrine Projects 49
 4.1 Mission and Framework 49
 4.2 Methodology 50
 4.3 Defining Success 51
 4.4 Lessons Learned: Factors Impacting Success 51
5 Latrine Case Study: Kikoo, Cameroon with Yale Ewb-Usa Chapter 54
 5.1 Case Study Introduction 54
 5.2 Case Study Implementation 56

J. Knight (✉)
Global Hope Network International, Orlando, FL, USA
e-mail: joshua.knight@ghni.org

M. Montgomery
Engineers Without Borders USA (EWB-USA), Denver, CO, USA
e-mail: melissa.montgomery@ewb-usa.org

D. Heuckeroth
Sanitation Standing Content Committee, EWB-USA Philadelphia Professional Chapter, Philadelphia, PA, USA

E. Lendzemo
Greenery Association, Kumbo, Cameroon

D. Sacco
EWB-USA Yale University Student Chapter, New Haven, CT, USA
e-mail: dsacco@tpadesigngroup.com

L. K. Wang, M. -H. S. Wang, Y. -T. Hung, N. K. Shammas (eds.), *Integrated Natural Resources Management*, Handbook of Environmental Engineering 20,
https://doi.org/10.1007/978-3-030-55172-8_2

5.3 Case Study Current Status 57
5.4 Case Study Lessons Learned 58
5.5 Case Study Conclusions 59
6 Conclusions and Next Steps 59
7 Author Introduction 60
Glossary 60
References 61

Abstract This article provides an introduction to proper latrine project design in rural communities. Latrines are usually the most appropriate sanitation technology for rural villages. However, without an adequate understanding of proper design standards and the village context, a latrine project can be a complete failure, with community members returning to previous, unhealthy methods for sanitation. Engineers Without Borders USA (EWB-USA) has provided humanitarian engineering services since 2001 in rural communities on infrastructure projects and has collected many lessons learned to assist those in international development to complete sustainable latrine projects which will effectively decrease waterborne and water-related illnesses in communities. These design standards and best practices are included here for the benefit of rural villages and those who serve them. Additionally, a case study is provided with specific lessons learned from an EWB-USA sanitation project in Kumbo, Cameroon, Western Africa. Both the overall and case study lessons learned point to the need for early, thorough, and respectful village engagement as well as thorough technical investigation of the water, soil, and current sanitation practices in the community.

Keywords Engineers Without Borders USA · Humanitarian engineering service · Rural communities · Villages · Latrine technologies · Design · Calculations · Sanitation · Ventilated improved pit latrine · Pour-flush latrine · Composting latrine · Project categories · Lessons learned · Nongovernmental organization

Acronyms

EWB-USA	Engineers Without Borders USA
NGOs	Nongovernmental organizations
O&M	Operation and maintenance
SDG	Sustainable Development Goals
UN	United Nations
UNDP	United Nations Development Program
VIP	Ventilated improved pit
WASH	Water, sanitation, and hygiene

1 Sanitation

1.1 Sanitation Introduction

In rural villages without access to adequate sanitation, latrines are typically the most appropriate technology that can be introduced at a relatively low cost to improve sanitation and public health. Engineers Without Borders USA (EWB-USA) works with rural communities to improve their basic infrastructure, for example, assisting to develop reliable access to clean drinking water and to design and construct facilities such as latrines for sanitation. EWB-USA also assists communities with behavior change such as proper handwashing and learning to use and maintain latrines instead of open defecation. These measures are aligned with the United Nations Sustainable Development Goal (SDG) 6, "Ensure availability and sustainable management of water and sanitation for all."The combination of clean water, sanitation (such as latrines), and handwashing is the best way to reduce the incidence of diarrheal diseases, which "account for 1 in 9 child deaths worldwide, making diarrhea the second leading cause of death among children under the age of 5" [1].

In 2014 in India, a case of rape and murder of two young girls brought to the public discourse another risk of open defecation – the personal security risk of not having access to sanitation in the home. However, open defecation is the cultural norm in many poor communities globally. Research in five northern Indian states found that "19 percent of women with access to a latrine still preferred to defecate in the open. . . The research is clear that 'building toilets without addressing common norms, attitudes and beliefs around latrine use is unlikely to reduce open defecation in rural India'" [2].

1.2 Lessons Learned from Past Projects

Clearly there is still much work to be done in the area of water, sanitation, and hygiene (WASH), and there are many lessons that can be learned from past projects. According to the Report of the UN Secretary-General, *Special edition: progress towards the Sustainable Development Goals* [3]:

1. "Globally, the proportion of population using safely managed drinking water services increased from 61 to 71 percent between 2000 and 2015 and remained unchanged in 2017. An additional 19 percent of the global population used basic drinking water services. This means that 785 million people still lacked even a basic drinking water service.
2. . . . Between 2000 and 2017, the proportion lacking even a basic sanitation service decreased from 44 to 27 per cent, yet 701 million people still practiced open defecation in 2017.
3. In 2017, some 60 percent of people worldwide and only 38 per cent in the least developed countries had a basic handwashing facility with soap and water at

home, leaving an estimated 3 billion people without basic handwashing facilities at home."

Numerous organizations have been working on these water and sanitation challenges globally. Learning from the project experiences of these organizations can assist in improving the chances of success of future projects. The importance of village-driven development cannot be overstated to improve the success rate of water, sanitation, and related behavior change initiatives. "Access to water and sanitation are recognized by the United Nations as human rights, reflecting the fundamental nature of these basics in every person's life. Lack of access to safe, sufficient and affordable water, sanitation and hygiene facilities has a devastating effect on the health, dignity and prosperity of billions of people, and has significant consequences for the realization of other human rights" (https://www.unwater.org/water-facts/human-rights/).

EWB-USA uses a community-based or village-driven approach to ensure that infrastructure improvements make use of technologies that the community needs, *wants*, and has the capacity to operate, maintain, repair, and reconstruct as needed. These technologies are described as "appropriate" for the community. Selection of an appropriate technology is essential to the sustainable success of any village-driven infrastructure project. When a project proves to be sustainable over time, this provides an indication that the technology selected was appropriate for the community. As an example, on one EWB-USA latrine project in Nicaragua, EWB-USA returned years later to evaluate the success of the project. Happily, EWB-USA discovered that the community had rebuilt the latrines on their own after they were destroyed by a hurricane, demonstrating that the village-driven approach had been a success and the project was sustainable.

Use of a village-driven approach enables the community itself to develop, thereby improving the capacity of the community to further improve its conditions with subsequent projects and initiatives. For additional background on the importance and use of a village-driven approach, refer to "Engineering for Sustainable Human Development, A Guide to Successful Small-Scale Community Projects," by EWB-USA's founder, Bernard Amadei, Ph.D., NAE [4]. For an earlier attempt of the organization's efforts at learning from sanitation projects, please see the article completed by Heuckeroth, Knight, Montgomery, and Walker [5].

2 Engineers Without Borders USA

2.1 Overview, Vision, and Offices of EWB-USA

The vision of Engineers Without Borders USA (EWB-USA) is a world in which the communities we serve have the capacity to sustainably meet their basic human needs and that in turn our members and volunteers develop enriched global perspectives. EWB-USA collaborates on more than 390 active projects in over 40 countries driven

Fig. 2.1 Map of EWB-USA project locations

by approximately 200 chapters across the United States, partnering directly with communities in the United States and around the world to meet their self-identified needs (Fig. 2.1). Of these types of projects, approximately 11 percent are of the sanitation discipline, primarily of the latrine type. The most common type of engineering needs identified by local community partners relate to water supply, which is part and parcel of any sanitation project. With our long-term commitments, EWB-USA is perfectly poised to address sanitation at a holistic level.

EWB-USA currently has four country offices, including the United States, Guatemala, Nicaragua, and Uganda, and is transitioning to a field office model to increase overall impact. These field offices have enabled increased support directly to local partners and allowed completion of long-term monitoring and evaluation of our work. EWB-USA programming has expanded to address underserved communities here in our own backyard, through the Community Engineering Corps program. Visit the EWB-USA website for more information about our work.

Our approach to development is based on long-term partnerships with communities and local partners including nongovernmental organizations (NGOs) and local governments who help support the communities and the projects with which EWB-USA assists. We work together to develop solutions to improve the health of communities through safe collection of human excreta, proper wastewater disposal, hygiene education, and waste management training.

Our mission is not to provide access to sanitation facilities as quickly as possible. Rather, it is to make sure that those implemented projects have greater sustainability, so that we move away from the long history of failed development projects. Therefore, it is essential to acknowledge that there is no one-size-fits-all solution for a community's water, sanitation, and other basic needs. The geographic, cultural, and political fabric of each community we partner with is unique, which is why our engineering solutions are unique, too.

2.2 *Project Variables of EWB-USA*

We take into consideration several specific variables to ensure the project is built to last. These are:

1. *Program Scope*. The community's request might be for latrines, but there could be other sources of waste and pollution or hygiene issues to address, such as gray water, groundwater protection, trash, lack of handwashing, etc.
2. *Education and Training*. EWB-USA chapters are encouraged to work with public health specialists, such as health delegates from the community or volunteers from the United States, to collect data on the diseases present in the community related to poor sanitation and hygiene. This information is used to create appropriate health education training programs and resources, as well as guide the design of an appropriate infrastructure solution.
3. *Gender Approach*. We consider the safety of young women in the design of sanitation facilities, such as distance from homes or schools and privacy. Women could be at risk if they must travel long distances in isolated pathways by themselves, even during the day. Women are included in design discussions and operation and maintenance (O&M) trainings. Women will usually be the ones to clean the sanitation facilities and will be directly affected by design decisions.
4. *Financing Strategies for On-Site Household Systems*. Different strategies can be used to help families pay for their sanitation facilities. In certain regions, microfinancing institutions might offer loans for community members looking to build or improve their sanitation systems. A revolving fund might help build more latrines in large communities. The best strategy needs to be discussed and agreed upon with the community before implementation.

3 Latrine Technologies

3.1 *Introduction to Latrine Design*

Many rural low-cost latrine designs exist from respectable organizations like the United Nations (UN) Development Program [6], the World Bank [7], the Sphere Project [8], the RedR [9], and the World Health Organization (WHO) [10]. This basic introduction doesn't look to add one more latrine design to the library but rather to include a basic introduction of latrine design. Below is a list of information that a team would look to compile to design a low-cost, effective, sustainable latrine in a rural setting. These are minimum recommendations and specific projects may include the need for additional requirements. For instance, to meet wastewater treatment requirements for a specific country or region, additional design details beyond what is explained here would be necessary.

3.2 Latrine Design Checklist

The following is a minimum list of set criteria for a latrine project, using the EWB-USA Latrine Design Checklist [11], where the information is decided in collaboration with the village driving the project.

1. *Problem Definition*. Give a concise description of the problem that the latrine is meant to solve. For example, a family of six currently is using a neighbor's latrine and would like to have their own latrine.
2. *Design Standards and Criteria*. Decide which country-specific or worldwide standard is to be used (local or national government standard, World Health Organization, United Nations, World Bank, etc.)
3. *Current System*. Note the community's current methodology for sanitation. Include types, pictures, sketches, sizes, condition (new, not in use, dilapidated, almost full, etc.), volume (original volume and current level of liquid and solids), number of people served, year installed, and maintenance level (flies visible, cleanliness, emptying schedule, etc.).
4. *Solution Comparison*. Give a short summary of different methods that could solve the given problem, and explain why the chosen solution is the best one. Potential rationales include depth to water table during the rainy season, soil infiltration rate, current accepted sanitation practices, etc.
5. *Intended Use*. Provide a description of how the latrine is to be used. For example, the proposed system will provide 100% of the latrine needs for the family of 6 for 20 years, with yearly emptying necessary.
6. *Land Ownership*. Provide confirmation and ideally documentation that the proposed land for the latrine is owned or allowed to be used by the proposed user.
7. *Soils Analysis*. Give a soil description of the site with potential issues for drainage, erosion, and excavation. For example, knowing the soil stability in the dry season helps understand the safety of excavation; knowing the saturated soil condition helps to understand the need for pit wall stabilization.
8. *Water Table*. Provide the water table depth, ideally for 1 month into the dry season and 1 month into the rainy season, when historical highs and lows are most likely to occur. Note dates measured and methodology of measurement.
9. *Siting Latrine Locations*. Consider the source(s) of drinking water for the community, and use this information to determine a suitable downstream or otherwise hydrologically segregated location for the latrine to reduce the potential for contamination of drinking water sources. Consider flooding risks based on area flood maps, community historical knowledge, seasonal rainfall, proximity of streams, etc., to avoid siting the latrines in an area that could become flooded. This is an important consideration to support climate change resiliency.
10. *Calculations*. Ensure that your calculations are consistent with your latrine drawings.

(a) *Site Selection*: List distances from drinking water sources (should be uphill and $\geq$ 30 m away from water sources and $\geq$ 2 m above the rainy season water table) and community dwellings (should be $\geq$6 m and $\leq$ 50 m from dwellings for avoidance of disease transmission but ease of use). Note, more space is necessary in coarse or fissured ground – find local hydrogeological expertise [12].
(b) *Latrine Demand*
    (i) *Latrine Unit Demand.* State the unit demand of the system (i.e., X liter per person per day) and references (preferably multiple sources, like literature, on-site percolation tests, and current latrine pit rate use).
    (ii) *Time to Fill Pit.* State design time to fill pit, which is usually between 1 and 5 years (i.e., designed to fill in 3 years).
(c) *Pit Size.* Use a conservative estimate of pit use per person, and give pit size given the unit demand and number of people to use pit for the given time to fill pit.
(d) *Pit Emptying.* State how the pit will be emptied or closed with the slab moved, and provide references to current practices near or in the community.
(e) *Geotechnical Analysis.* Give the soil bearing pressure used for the design, and state any geological considerations that are considered in the design with references.
(f) *Structural Calculations.* Use calculations to show that the pit and superstructure can resist lateral water loads, any anticipated roof loading, and lateral loads from earthquakes or wind. Give calculations for the live and dead loads experienced by the superstructure.
(g) *Pit Lining.* Include the pit lining material and provide calculations and/or justification that the lining can handle foreseen seismic and rain events.
(h) *Hydraulic Analysis.* Calculate pipe sizes and slopes for all connections from appurtenances to the pit and urine diversion system if present.
(i) *Latrine Closure.* Explain how the latrine pit will be closed when finished to minimize pathogen destruction, including emptying method if necessary.
(j) *Ventilation.* Provide calculations to show that foul smells will not remain in superstructure.
(k) *Composting Latrines Only*:
    (i) *C:N Ratio.* State carbon to nitrogen ratio and reference (i.e., 30:1).
    (ii) *Temperature.* Provide heap temperature and reference (40–50 ° C).
    (iii) *Ventilation.* Calculate ventilation of water vapor and $CO2$ out and $O2$ in.
    (iv) *Moisture.* Provide moisture content evaluation (45–70%, Eawag Tech Compendium [13]).
    (v) *Urine Diversion.* Explain methodology to divert and use urine, if applicable.

(vi) *Bulking Material*. Refer to available materials that will meet your C/N ratio, how they have been tested, and comparison to current practices.
(vii) *pH Control*. Provide optimal pH and how that will be maintained.
(viii) *Unacceptable Materials*. Explain what is unacceptable for the latrine and how those will be kept out (i.e., corn cobs, trash, etc.).
(ix) *Aeration*. Explain how aeration will be provided for aerobic digestion and whether compost will be turned manually and whether bulking material will be necessary.
(x) *Sun-Drying*. Calculate storage volume and space to ensure the family can make it to the next dry season if this method is to be used which greatly reduces pathogens.
(xi) *Pathogen-Free Compost*. Show how this will be provided and checked, like by volume decrease, lack of odor, etc.

(l) *Pour Flush Latrines Only: Water Source*. Explain how owners will get water and ensure that unsanitary conditions will not occur with a temporary water shortage.
(m) *Ventilated Improved Pit (VIP) Latrines Only*:

(i) *Darkness*. Low light should be maintained in the latrine so the brightest light is through the ventilation pipe, driving flies up the pipe rather than through the seat.
(ii) *Ventilation Rate*. For air to come into the stall and out through the ventilation pipe, there needs to be a relatively open front toward the prevailing wind with an open seat. Provide for this in the design.

11. *Design Drawings*. Provide drawings in one set which are consistent with the calculations.

(a) *Site Plan*. Plan drawing is at an appropriate scale and includes at least the following: the outline of the building served by the latrine, the pit, area topography, property lines, nearby manmade features (nearby latrines, buildings, fences, areas for special use like cooking or farming, roadways, major walkways, etc.), and nearby natural features (trees, drainage areas, steep slopes). Site grading is included. Note distances from important items like dwellings and water sources as listed above.
(b) *Pit Foundation*. Provide foundation dimensions, reinforcement, formwork, and cement mix ratio.
(c) *Pit Structure*. This includes detailed structural drawings at appropriate scale.

(i) *Walls (if necessary)*. With a concrete or masonry block pit, include the foundation and wall configuration, as well as reinforcing bars and corner bars and dowels. If present, include details of roof hatch with consideration for practicality of removing hatch. Include securable access points and screened vents, as applicable.

(ii) *Water Tightness (if necessary)*. For masonry and cementitious pits, include water tightness provisions for the roof and sidewall.

(d) *Superstructure*. Include drawings necessary to build the superstructure with details like the door frame, hinges, roof, etc.
(e) *Plumbing*. Provide details for piping connections, pit connection, and urine diversion if necessary.
(f) *Appurtenance*. Ensure that the appurtenance follows local practices, customs, and religion for the specific household.
(g) *Additional Plumbing*. Give plumbing details of pipe lengths between appurtenance and pit. Provide trench details for underground pipelines.

12. *Material and Cost List*

(a) *Material List*. List materials to be used, all of which must be found during the assessment phase at the local hardware store, which may be quite different from those available in western hardware stores.
(b) *Costs*:

(i) *Capital Costs*. Include material and paid labor.
(ii) *Operation and Maintenance (O&M) Costs*. Include costs for emptying the pit or spreading sludge if necessary, as well as other costs such as cleaning and hygiene supplies.
(iii) *In-Kind Costs*. Quantify labor, materials, food for workers, translation, etc.

13. *Operation and Maintenance*. The following should be provided to the community, with consideration for local customs and religions, in the local language or in pictures (the more pictures the better).

(a) *Latrine Cleaning*. Consider the difficulty of cleaning all parts of the system, and include a method for cleaning and note who will be responsible, if necessary. For ventilated pits, the pipe opening must be screened and include a cleaning and replacement schedule and responsibility (start out more frequently). For composting latrines, a mixing schedule of the compost must be calculated which maintains an aerobic environment and reduces odor. A source for the bulking material and a schedule should be included.
(b) *Personnel*. For households, the homeowner is responsible for maintenance, but this is not so clear for a communal building (market, community center, etc.). This should be discussed with the community and documented in the community agreement. If paid personnel are necessary (pumping or compost handling), providers should be researched and documented, as well as the method to pay for it.

14. *Education*. Water, sanitation, and hygiene (WASH) considerations must be included to truly reduce waterborne and water-related diseases, as a latrine alone is not enough. Monitoring (soap and water availability, latrine cleanliness, clean water sources), education, and future project plans should be included, as appropriate.

3.3 Existing Latrine Technologies

Several latrine technologies are widely used today. Selection of the technology depends on water and material availability, soil analysis, community practices, and estimated population growth, as explained in the previous section.

The three most common types of latrines are:

1. Ventilated improved pit latrine
2. Pour-flush latrine
3. Composting latrine

3.3.1 Ventilated Improved Pit (VIP)

The VIP system (Fig. 2.2) produces a continuous airflow through the ventilation pipe. The airflow vents away odors, assists waste breakdown by drying action, and

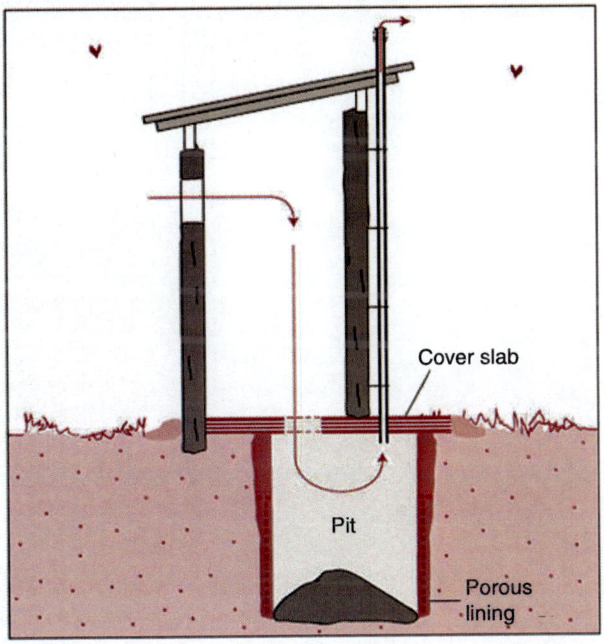

Fig. 2.2 Ventilated improved pit latrine

acts as an effective fly control mechanism. Despite their simplicity, well-designed VIPs can be completely smell-free and more pleasant to use than some other more complicated technologies if designed, used, and maintained correctly.

3.3.2 Pour-Flush Latrine

A pour-flush latrine (Fig. 2.3) is similar to a regular cistern flush toilet, except that the water is poured in by the user instead of coming from the cistern above. Pour-flush latrine pits often have unsealed bottoms allowing liquid to percolate down into the soil. In areas of high groundwater, these latrine pits have sealed bottoms. The additional water from flushing causes sealed pits to fill up faster. Full pour-flush pits are then emptied using the vacuum hose of a sludge pumping truck or allowed to settle before being emptied manually using a shovel. To ensure proper sanitation is maintained, the destination of the pumped or shoveled sludge must be carefully considered. For example, if a truck is allowed to discharge the untreated sludge to a nearby waterway, there will be a significant negative impact to the environmental health of the waterway and in turn to human health if the waterway is used as a water supply for drinking, bathing, or fishing.

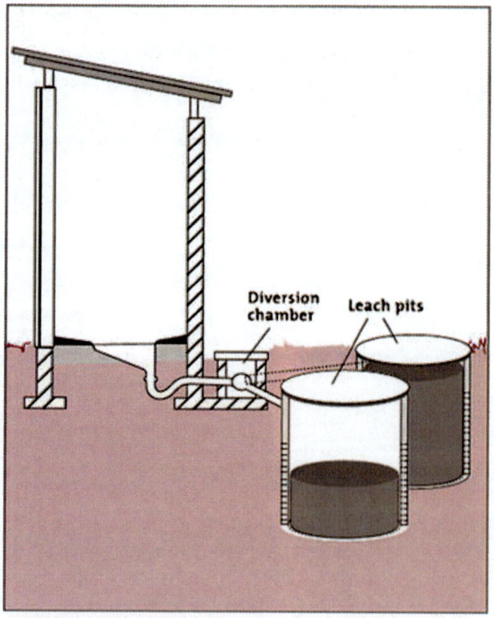

Fig. 2.3 Pour-flush latrine

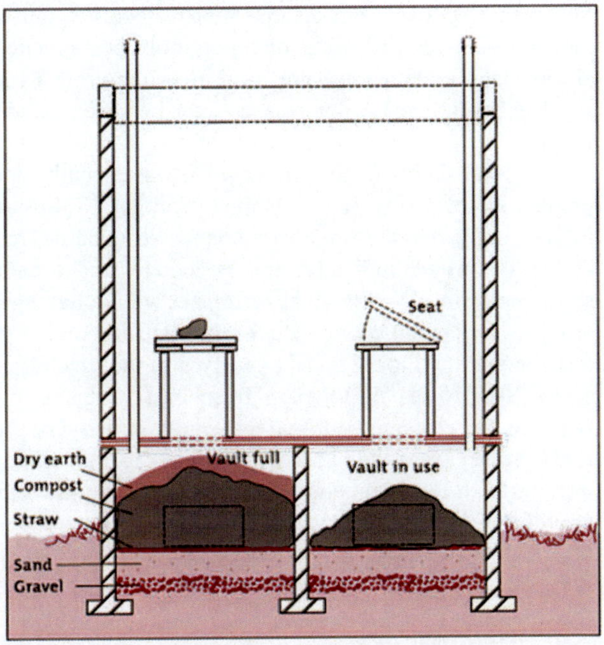

Fig. 2.4 Composting latrine

3.3.3 Composting Latrine

Composting toilets (Fig. 2.4) use the natural processes of decomposition and evaporation to recycle human waste. Excreta entering the toilets are over 90 percent water, which is evaporated and carried back to the atmosphere through the vent system. The small amount of remaining solid material is converted into useful fertilizing soil by natural decomposition. The quality of the compost must be considered to ensure full destruction of pathogens, worm eggs, etc., prior to its use for growing any plants for edible consumption.

4 Planning, Monitoring, Evaluation, and Learning of Latrine Projects

4.1 Mission and Framework

EWB-USA's mission is to build a better world through engineering projects that empower communities to meet their basic human needs. As such, an integral part of all EWB-USA programs is the planning, monitoring, evaluation, and learning (PMEL) framework. The PMEL framework provides the feedback we need to maximize our effectiveness and enable leaders in their fields to solve the world's most pressing problems. This framework is embedded in the project delivery process to provide feedback and inform improvements to our overall process and delivery model. EWB-USA provides planning tools, best practice guidance, and resources to aid in planning. We measure more than the project functionality. Our framework assesses quality-of-life changes from the perspectives of both successful projects and failures.

The EWB-USA PMEL framework was recently updated to align our standard project indicators with the United Nations 17 Sustainable Development Goals (SDGs) defined in the "2030 Agenda for Sustainable Development" (*United Nations 2015*). The project indicators are assigned by project category. This shift allows us to report uniformly on the level of impact we are having across the organization and also provides reliable evidence for the contribution our efforts make toward both our mission and the global goals expressed in the SDGs.

EWB-USA strives to learn from past mistakes, big and small, to constantly improve our efficiency and effectiveness in delivering community-driven projects. Each project team is asked to contribute to this effort by writing helpful lessons learned, both technical and nontechnical, for future teams to utilize. In addition, it provides insight to improve our project process and identify gaps.

4.2 Methodology

EWB-USA tracks projects on three high-level metrics, each with specific indicators. These performance and quality indicators look broadly at the project design and determine if the project is meeting the relevant local and global standards, as well as best practices for quality.

1. *Functionality*. These indicators will document that the project functions as designed.
2. *Maintenance*. These indicators will document that the project is being maintained in a way to foster a sustainable improvement into the future.
3. *Community Capacity*. These indicators will document that the community can sustain the project into the future without outside assistance and that the community members have the knowledge, financial capacity, and interest to keep the system operational. As of this writing, there are a total of 37 active latrine projects, from which this research is drawn.

Table 2.1. summarizes the five sanitation project categories that EWB-USA delivers and the number of active (not closed) projects in each category.

Baseline data are documented during the initial assessment. Monitoring on all indicators is done on every trip into the community throughout the entire project process. A final evaluation trip is conducted at least 1 year after implementation has been completed. These data are reported within a database platform and are used to provide transparency to our work, share lessons learned through our membership and staff, and provide regular reporting on our organizational impact and overall contributions to the SDGs.

Table 2.1. EWB-USA sanitation project categories

Category	Description	Number of active EWB-USA projects
Latrine and handwashing (school)	Any latrine project at a school	12
Latrine and handwashing (hospital)	Any latrine project at a clinic or hospital	1
Latrine and handwashing (community/household)	Any latrine project not at school or clinic	24
Stand-alone handwashing Facility	Handwashing station not built as part of the latrine	0
Other sanitation	Wastewater treatment, solid waste management, medical waste management, black water management, stand-alone hygiene education, showers	6

4.3 Defining Success

The established indicators that speak to the higher metrics of performance and quality, maintenance and functionality, and community capacity are used to determine the overall success of an EWB-USA project. Success is dependent not only on a quality design but also on the community's capacity to operate and maintain the project well into the future, after the partnership has ended. The alignment of indicators across all EWB-USA projects allows for high-level reporting across all our work on an annual basis.

4.4 Lessons Learned: Factors Impacting Success

As a critical component of the learning aspect within the PMEL framework, lessons learned provide useful data to help teams and the communities they serve to set themselves up for success. These reported lessons also inform potential gaps in resources. EWB-USA has created a dashboard tool (Fig. 2.5) to explore the lessons learned that have been reported for a given region, project type, phase, or country.

Some of the most common reported lessons learned factors are community relations, construction materials, and finances. Figure 2.6 notes general lessons learned from the database applicable to latrines. These common factors among all project types speak to some of the bigger challenges within all of our projects, including latrine projects, incorporating:

1. *Importance of Community Engagement and Buy-In*. Regardless of how well-designed a project may be, the real success factor is whether the community takes ownership and responsibility. This includes cultural acceptance and the behavior/

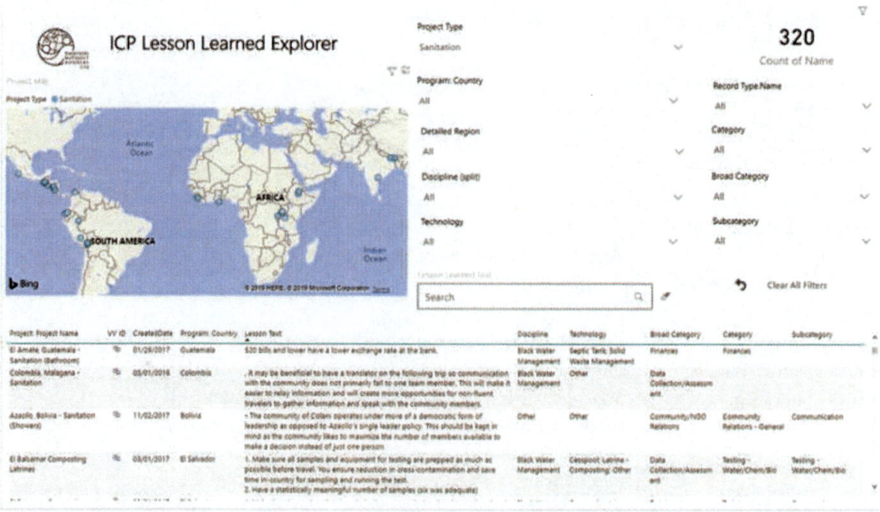

Fig. 2.5 Dashboard

system adjustment of the latrine usage and maintenance. As part of the collaborative mindset of the project process, opportunities can also arise for community leadership and technical skill-building.

2. *Construction Materials*. Sourcing local materials and local labor for construction increases the likelihood that the community will take on ownership and long-term maintenance.
3. *Finances*. The financial components of any engineering project are important for timely project delivery to fulfill the commitments we have made to our community partners, as well as being responsible stewards of donor funding.

Figure 2.6 describes more examples of the lessons learned that can be found using the dashboard tool.

As noted previously, the high-level metrics that are monitored for performance and quality include functionality, maintenance, and community capacity. Figure 2.7 shows the current reporting on 37 active EWB-USA sanitation projects on these three metrics. This reporting is completed by project teams at least 1 year after implementation is completed. Highlights of the graphs include:

1. Approximately 79 percent of sanitation projects are reported to be functioning as intended.
2. Approximately 91 percent of sanitation projects are reported as being maintained by the community.
3. Approximately 96 percent of sanitation projects are reported as the community having the financial and technical capacity to take on long-term operation and maintenance of the system.

The following are two recent lessons learned quotes from project teams:

Examples of General Lessons Learned as Reported from the Dashboard Tool		
Community/Nongovernmental Organization (NGO) relations		
Ø Certain communities are democratic in nature; thus, choosing large groups of people to make decisions related to a project, rather than a select few, may be beneficial. Ø Respecting local culture may be key to the success of the project. Ø Define expectations and capacity of cash contributions in the assessment phase.	Ø Importance of involving local officials (mayor, president, chief, etc.) to help gain acceptance and credibility of EWB-USA members by the community as well as provide maintenance assistance. Ø Effective community interviews/surveys in the early phase (assessment) of the project to confirm needs, develop a data profile, and facilitate trust and a long-term working relationship.	Ø Importance of communicating EWB-USA's level of involvement to the community and integrating the community into the project. Ø Expectations of the NGO to coordinate, including when EWB-USA chapters are not present in the country.
Technical		
Ø Understanding and accounting for existing field conditions and environment risks, such as topography, groundwater, climate change. Ø Be prepared for the unexpected, such as ground material inconsistency, rain affecting construction progress, etc.	Ø Understanding availability of local materials, and compromises of material strength properties in those materials. Ø Men in the community may tend to migrate to the inner cities for work, resulting in less manual laborers available locally.	Ø Ensure technology is appropriate for community. Ø Long-term operation and maintenance plan. Ø Consider incorporating hygiene training in conjunction with the technical project, where applicable.
Logistics/Management		
Ø Be constantly aware of construction safety (e.g. around excavations) and personal protective equipment (PPE).	Ø Schedule planning, including general delays and delivery of materials/equipment from local manufacturer or contractor.	Ø Look up common building practices in the region. Ø Ensure security of materials.

Fig. 2.6 Examples of general lessons learned as reported from the dashboard tool

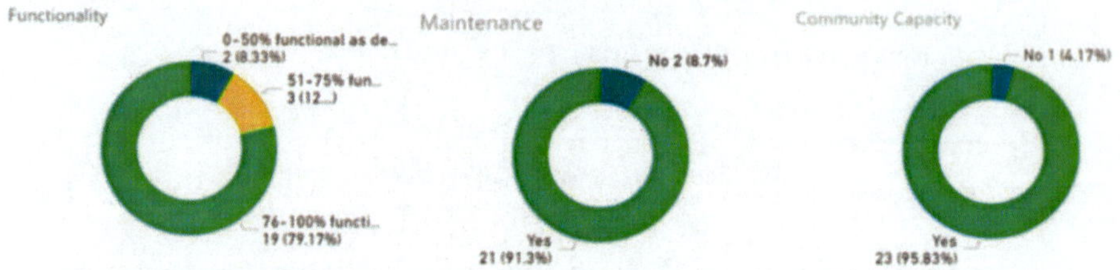

Fig. 2.7 Reporting metrics

- "Our monitoring team found a huge link between the construction quality of the superstructure and the overall cleanliness of implemented latrines. Well-built superstructures usually housed remarkably clean latrines. Nice superstructures also protect the space better from the elements and offer recipient families more privacy and a greater sense of pride for the latrines. A few of the latrines from the second phase of implementation did not receive enough attention with regard to the superstructure, especially due to the lack of labor that came in the coffee season after our chapter's departure. Moving forward, our chapter would like to explore requiring families to acquire all the materials for the superstructure before constructing the latrine – that way, the superstructure can be well-constructed as soon as the latrine is." (2017, Guatemala)
- "We learned a lot from the focus groups and community members about what the designs for the ... latrine should be and what the community will find the most helpful. Through focus groups and meeting with a group of school girls, we learned that the biggest issue for girls menstruating during school is that they don't have a place to clean or store menstrual pads. We are incorporating a room with drying racks and cubbies into our design for the latrine as well.... These design constraints are, as a result, integral elements of our final implemented designs." (2018, Nepal)

5 Latrine Case Study: Kikoo, Cameroon with Yale Ewb-Usa Chapter

To provide a better sense of what a latrine project is like to other practitioners, governments, and rural communities, the following is a brief case study of a latrine project completed in Kikoo, Cameroon, western Central Africa.

5.1 Case Study Introduction

In 2006, the community members of Kikoo, Cameroon, along with Joshua Knight, then missionary engineer with the Kumbo Catholic Diocese, applied for a project

Fig. 2.8 Traditional toilet: traditional toilet enclosed with a palm front fence with wooden crossbars and squatting hole

with EWB-USA. The Yale University EWB-USA chapter accepted the request which began with a successful water project. After the water project was completed, it was followed by a latrine project for which the Yale chapter provided 90% of the cash needs for three latrines, while the community provided 10% cash and in-kind contributions (labor, food for the visiting engineers). The EWB-USA Yale teams identified open pit toilets common in the community as one of the sanitary problems affecting the livelihood of the people.

Traditionally, pit toilets were dug 3–4 m deep and the top surface closed with selected wooden cross bars constituting the floor, and a squatting hole is then fashioned at the center of the floor in the form of a slit in the wood. The walls in most cases were absent or made of palm front fence, with toilet roofs literally absent (Fig. 2.8). The open toilets provided access to flies to gain access to feces deposited in the toilets. The absence of a roof and walls provided a meager user-friendly environment whenever it rained and deprived users of a sense of pride and privacy while performing nature's call. Wood used as crossbars was not treated with preservative substances to ensure sustainability for a longer period of time consequently; they degraded and became weak and unsafe for users.

Gender sequestration was hardly taken into consideration when locating and constructing the toilets. Children and women more often were accompanied during the nights to use the toilets because they are far from the houses. In extreme cases, children wet the beds at night. The odor from the open pit toilets attracted flies and other organisms like chameleons that come to trap flies for their food, bees attracted by urine dropped on the floor, and snakes hunting for chameleons, rodents, and birds that perch on the walls of the toilets. These toilets provided a unique ecological system where snake bites were common within the homes. Accumulation of water in the toilets especially during the rainy seasons provided excellent mosquito breeding ponds.

A recommendation was made for the construction of prototype "ventilated improved pit" (VIP) toilets, using available local materials and hands-on training of the community technician to replicate in the future with principal focus on improving the sanitary welfare of the people.

The VIP toilets were designed to make a 3–4 m deep pit into the ground, and on the surface, a solid foundation is provided on which concrete slabs with appropriate cement, aggregate, sand, and mixture interwoven with iron rods. In the process, squatting holes are opened at specific locations as well as provision for the ventilation pipes. The slabs are made at appropriate sizes that allow transportation to another site when the toilet is full. The walls are constructed without slits on the walls and a roof provided. The ventilation pipe is then installed into the slab rising above the roof hosting a fine wire mesh trap at the top surface with an elbow facing sunrise direction. The doors are provided with shutters ensuring that the toilets are kept shut when not in use to ensure a near dark interior of the toilet. Due to the temperature difference between the cool air in the underground pit and the warmer air at the top of the vent, an updraft is created and air in the vent pipe rises, which evacuates odors from the interior of the toilet and dissipates them into the atmosphere. Once a fly ventures into the toilet, it will gain access to the interior through the squatting hole, and the vent pipe will be the only source of light guiding it to go out; the wire mesh will prevent the fly from leaving and it will die, thus preventing transmission of disease from the fecal matter in the pit.

5.2 *Case Study Implementation*

Three toilets were constructed in public places with access to a majority of the population: government primary school, Catholic primary school, and community center. The chapter did not undertake a master plan exercise or even review the proposed locations with the schools because when the chapter went to assess for the project, they found 6-meter-deep pits already dug for the latrines, with planks being the only protection against children falling in. For that reason, the chapter felt that the appropriate thing to do was address the sanitation issue by building VIP latrines as planned, but to build them over the existing trenches rather than over new pits. They expected pushback to filling in the existing trenches and wanted to address the safety hazard quickly, so using the pits seemed like the best thing to do under the circumstances.

The location of government primary school latrine was not well reflected upon at the time; with the improvement of school infrastructure, it was abandoned and allowed to crumble on its own; meanwhile, the slabs are still intact. A similar toilet was constructed (Fig. 2.2d) though not respecting the same technology by a contractor employed by the government.

The walls and roof of the toilet at the community center were constructed with old and torn corrugated iron sheets (Fig. 2.9b) that allowed light to penetrate into the toilets.

At the Catholic primary school, the toilet was well-located behind the school buildings (Fig. 2.9a). The walls were partly made of cement and mud blocks, with poor finishing at the level of the roof with multiple openings. The doors were void of shutters.

Fig. 2.9 Kikoo project toilets: (**a**) Catholic primary school toilet, (**b**) community center toilet, (**c**) men's weekly cultural meeting hall, and (**d**) new toilet at the government primary school

5.3 Case Study Current Status

Two of the three toilets are still in use in spite of the numerous construction shortcomings and are serving the community. Multiple slits on the roofs and walls (Fig. 2.10b) together with the absence of shutters provide light in the toilets aiding flies to escape. At the Catholic primary school, the toilet and school has been abandoned (Fig. 2.10a) because of the current northwest and southwest Anglophone crisis affecting Cameroon that has kept schools closed for the past 3 years. The external walls of the toilets are still intact (Fig. 2.10c, d), but the roofs have multiple holes allowing light rays (Fig. 2.10c, d) to penetrate into the toilet. In spite of the external wall and the roof that are degrading, the interior is good though the vicinity of the squatting holes is apparently degrading (Fig. 2.10c), perhaps due to acid from urine. The walls were partly made of cement and mud blocks (Fig. 2.10c, d) which are still in good state. Rehabilitation can be easy requiring clearing around the overgrown bush, mending of the roof, and provision of shutters on the door.

At the community center, the external walls are already obsolete because of rusted and torn corrugated iron sheets and more openings on the walls (Fig. 2.3b), and the roof of the toilet now illuminates the toilet and allows flies to escape without passing through the vent pipe. The concrete slab is still intact and safe to use.

Fig. 2.10 Kikoo project toilet details: (**a**) Catholic primary school toilet, (**b**) community center toilet with torn wall and degraded external enclosure, (**c**), (**d**) internal view of the Catholic primary school toilet with intact walls, sunlight rays on the floor, squatting hole, and ventilation pipe installed on concrete floor

5.4 *Case Study Lessons Learned*

The VIP toilet technology has been a welcome relief to the Kikoo community, with the following lessons:

1. Floor Sealing. Toilet floors have to be sealed completely to prevent vectors such as snakes and flies from having access to feces. This was a success story of the project since most toilet floors in the community either with wooden cross bars are sealed with soil or slabs which sealed the toilets completely.
2. Build Close to Home. Another success of the project was to show that with good ventilation, a VIP toilet can be located close to the home rather than far in the woods or farm. This is very secure for children and women especially at night and gives them pride and dignity.
3. Ventilated Pipe. The ventilated pipe for latrines has been replicated in two community houses since the project was realized.
4. Maintenance Requirement. As visible from Fig. 2.10c, d, the two remaining VIP latrines have considerable light infiltration. While it is not necessary to have a completely dark latrine, a dim latrine is necessary for flies not to escape. Ongoing maintenance is critical to ensure that walls, doors, and roofs are kept in good repair. Thankfully, neither latrine has a current need to be emptied, but with the current design, the latrine slab can be moved to a new location when necessary.
5. Location and Master Planning. The government school latrine was removed when the school was improved. No master planning was performed with the community. This may have shown that the best place for the school to expand was

into the proposed latrine site. It would behoove future teams to walk through a master planning exercise before locating sites for their project.

5.5 *Case Study Conclusions*

The future generation of the Kikoo community will likely embrace the VIP technology. Primary school pupils acknowledge that their school toilets do not smell, and they have a greater sense of pride and privacy each time they attain nature's call and look forward to building similar toilets in their own homes.

In spite of the VIP difficulties encountered in the project, they are much better than the traditional toilets. The lessons were taken forward to build latrines with concrete slabs and ventilation, but ensuring ongoing maintenance and forward thinking will help sanitation in the community of Kikoo going forward.

6 Conclusions and Next Steps

As seen in the lessons learned figures in Section 4, as well as the case study in Section 5, there continues to be room for improvement on the delivery of projects with our partners. The overall latrine project data shows that approximately 21 percent of implemented projects are not functioning as designed. While it is not unexpected with development work to have some setbacks, EWB-USA believes that we have a responsibility to improve upon this indicator. The data also report that nearly 10 percent of sanitation projects are not being maintained. The latrine lessons learned from general EWB-USA projects and the case study show the need for engaging the community early on in the project cycle, ensuring they are fully invested through early planning, design, implementation, and maintenance and ensuring that all customs and future plans are incorporated. Also, extensive technical investigations of the water, soil, and current sanitation practices in the community are necessary for long-term project success. Long-term operation and maintenance is a key indicator that is used to monitor the success of a project. With the new EWB-USA country office model, there is greater opportunity to improve upon the metrics that are used to evaluate the success of a project with EWB-USA staff on the ground to provide comprehensive and long-term monitoring of the success of our work. Also, if there is an observed lack of maintenance, the country offices can engage with the community and/or NGO partner to reeducate or repair as necessary. As a learning organization, EWB-USA embraces continuous improvement and has established a robust system to enable the organization to identify lessons learned that will inform organizational learning and continuous improvement to better assist the rural villages and communities we serve.

7 Author Introduction

Joshua Knight, PE, PhD, is the Program Engineer for Global Hope Network International and a member of the EWB-USA Sanitation Standing Content Committee and Application Review Committee and may be reached at joshua.knight@ghni.org.

Melissa Montgomery, PE, is the Chief Design and Build Programs Officer at Engineers Without Borders USA and may be reached at melissa.montgomery@ewb-usa.org.

Eugene Lendzemo is the Project Geologist and Technical Manager at Greenery Association, Kumbo, Cameroon, and may be reached at euglendz@yahoo.com.

Debbie Heuckeroth, PE, ENV SP, is a member of the EWB-USA Philadelphia Professional Chapter and Sanitation Standing Content Committee and may be reached at debbieheuckeroth@yahoo.com.

David Sacco, PE, is a mentor for the EWB-USA Yale University Student Chapter and may be reached at dsacco@tpadesigngroup.com.

Glossary

Compost	Decayed organic material used as a plant fertilizer.
Composting latrine	Type of dry (not using water) latrine which treats human excreta using a biological process called composting, which decomposes excreta and organic matter into a compost-like material but does not destroy all pathogens.
Engineers Without Borders USA	A nonprofit organization established to partner with developing communities in order to improve their quality of life.
Excreta	Waste matter discharged from the body, especially feces and urine.
Latrine	A receptacle (like a toilet or something as simple as a pit in the ground) for use as a sanitation system which functions to receive human excreta.
NGO or nongovernmental organization	Any nonprofit, voluntary citizens' organization which can be organized at a local, national, or international level.
Pathogen	A bacterium, virus, or other microorganism that can cause disease.
Pit latrine (also simple pit latrine)	The cheapest and most basic form of improved sanitation which consists of a pit dug into the ground with some kind of sturdy cover with a hole in it through which excreta go into the pit.
Pour-flush latrine	Latrine where a small quantity of water is poured in to flush excreta through a water seal into a pit.
Sanitation	The development and application of measures for the sake of cleanliness and protecting health, especially the disposal of excreta and solid waste.

Solid waste Garbage or refuse which is typically not disposed of in a sewer or into a latrine.

Ventilated improved pit latrine (VIP) A pit latrine with a black vent pipe going from the pit to above the superstructure with a fly screen on top. VIP latrines have the advantage over simple pit latrines of removing flies and odors if designed, built, and used properly.

WASH An acronym that means water, sanitation, and Hygiene. Access to WASH usually includes safe water, appropriate sanitation, and hygiene education. Adequate access to WASH can greatly reduce water-borne diseases in a community.

References

1. Diarrhea WHO (2013) Common illness, global killer. World Health Organization, US
2. McCarthy J (2014) How a lack of toilets puts India's women at risk of assault. NPR News. http://www.npr.org/sections/parallels/2014/06/09/319529037/indias-rape-uproar-ignitesdemandto-end-open-defecation
3. United Nations (2019) Special edition: progress towards the Sustainable Development Goals. Report of the Secretary-General 2019 session. 26 July 2018–24 July 2019. Agenda items 5 (a) and 6. https://undocs.org/E/2019/68
4. Amadei B (2014) Engineering for sustainable human development: a guide to successful small-scale community projects. American Society of Civil Engineers
5. Montgomery M, Knight J, Heuckeroth D, Walker W (2020) Community-based latrine development with "Engineers Without Borders USA". Evolutionary Progress in Science, Technology, Engineering, Arts, and Mathematics (STEAM), pp 1–27
6. Ryan BA, Mara DD (1983) Ventilated improved pit latrines: vent pipe design guidelines. https://www.ircwash.org/sites/default/files/321.4-6326.pdf
7. Mara DD (1985) The design of pour-flush latrines. World Bank, Washington, DC. https://sswm.info/sites/default/files/reference_attachments/MARA%201985%20The%20Design%20of%20Pour%20Flush%20Latrines_0.pdf
8. Russell S, Standard CH, Komino TP (2018) The sphere handbook. https://spherestandards.org/wp-content/uploads/Sphere-Handbook-2018-EN.pdf
9. Davis J, Lambert R (2002) Engineering in emergencies. A practical guide for relief workers. Practical Action Publishing
10. Franceys R, Pickford J, Reed R, World Health Organization (1992) A guide to the development of on-site sanitation. https://www.who.int/water_sanitation_health/hygiene/envsan/onsitesan.pdf
11. Engineers Without Borders USA (2017). Latrine checklist. 13 June 2017 (Article)
12. Harvey P (2007) Excreta disposal in emergencies: a field manual. © WEDC, Loughborough University. https://www.unicef.org/cholera/Chapter_9_community/19_Interagency-Excreta_disposal_in_emergencies.pdf
13. Gensch R, Jennings A, Renggli S, Reymond P (2018) Compendium of sanitation technologies in emergencies—. https://www.bmz.de/de/zentrales_downloadarchiv/Presse/emergency-sanitation-compendium.pdf

Chapter 3
Surface Water Quality and Analysis

Hamidi Abdul Aziz, Salem S. Abu Amr, and Yung-Tse Hung

Contents

1 Introduction 65
1.1 Water Quality Monitoring and Analysis 65
1.2 Classification of Surface Water 68
1.3 Water Quality Indices (WQIs) 72
2 Water Quality Module Formation 89
2.1 Module Subdivision and Classification 92
3 Oxygen Consumption and Replenishment in Receiving Water 94
3.1 Effect of Physical Characteristics of Water on Dissolved Oxygen 95
3.2 Biochemical Decay of Organic Waste Materials 96
3.3 Reaeration Portion of the DO Mass Balance 99
3.4 Oxygen Percent Saturation and Temperature 100
3.5 Quantifying the DO Mass Balance 100
3.6 Nitrification 101
4 Microbiological Water Quality 102
4.1 Indicator of Microbiological Water Quality 103
4.2 Monitoring of Microbiological Water Quality 104
4.3 Fecal Coliform Bacterial Die-Off in Urban Watershed 107
Glossary 110
References 111

H. A. Aziz
School of Civil Engineering, Engineering Campus, Universiti Sains Malaysia, Nibong Tebal, Penang, Malaysia

Solid Waste Management Cluster, Science & Engineering Research Centre (SERC) Engineering Campus, Universiti Sains Malaysia, Nibong Tebal, Penang, Malaysia
e-mail: cehamidi@usm.my

S. S. Abu Amr
Faculty of Engineering, Karabuk University, Demir Campus, Karabuk, Turkey

Y.-T. Hung (✉)
Department of Civil and Environmental Engineering, Cleveland State University, Strongsville, OH, USA

L. K. Wang, M. -H. S. Wang, Y. -T. Hung, N. K. Shammas (eds.), *Integrated Natural Resources Management*, Handbook of Environmental Engineering 20,
https://doi.org/10.1007/978-3-030-55172-8_3

Abstract The fundamental of water quality management and analysis is to adopt several monitoring actions to save water bodies and protect users. Evaluation of physiochemical and biological parameters of surface water is an important aspect for water quality model formation and adopted suitable management actions. This chapter focused on evaluation, monitoring, and analysis of surface water quality. Recently, Water Quality Indices (WQIs) was introduced as new methods for water quality management. Dissolved oxygen in surface water is necessary for aquatic life. Microbial quality of surface water should be safe and should not have adverse effect on human health. In this chapter, WQIs formation, definitions, and limitations were discussed, surface water modeling, standards and indicators of chemical and microbial quality of surface water were presented and discussed.

Keywords Water quality · Module · Physicochemical · Biological · Evaluation · Standards

Acronyms

BOD	Biochemical oxygen demand
DO	Dissolved oxygen
NH_4^+	Ammonia
NO_3^-	Nitrate
NO_2	Nitrite
EC	Electrical Conductivity
ANOVA	Analysis of variance
NSF	National Sanitation Foundation
PCA/FA	Principal component analysis/factor analysis
SO_4^{2-}	Sulfate
SWQM	Surface Water Quality Monitoring
WQIs	Water Quality Indices

Nomenclature

J	is the "diffusion flux" [(amount of substance) per unit area per unit time], for example $\frac{\text{mol}}{\text{m}^2.\text{s}}$, J measures the amount of substance that will flow through a small area during a small time interval.
D	is the diffusion coefficient or diffusivity in dimensions of [length2 time−1], for example $\frac{\text{m}^2}{\text{s}}$
Φ	(for ideal mixtures) is the concentration in dimensions of [amount of substance per unit volume], for example $\frac{\text{mol}}{\text{m}^3}$
L	is the position [length], for example m
dM	is the change of storage mass in the system (kg)
ρ	is the density (kg/m^3)
v	is the speed (m/s)

A	is the area (m^2)
dt	is an increment of time (s)
t	is time [s]
$D_{(in\ BOD)}$	is the saturation deficit, which can be derived from the dissolved oxygen concentration at saturation minus the actual dissolved oxygen concentration ($D = DO_{sat} - DO$). D has the dimensions $\left[\frac{g}{m^3}\right]$.
K_1	is the deoxygenation rate, usually in d^{-1}.
K_2	is the reaeration rate, usually in d^{-1}.
L_a	is the initial oxygen demand of organic matter in the water, also called the ultimate BOD (BOD at time t = infinity). The unit of is $\left[\frac{g}{m^3}\right]$.
L_t	is the oxygen demand remaining at time t.
$[O_2]$	is the observed oxygen concentration
[Osat]	is the saturated concentration of oxygen at the local temperature (and possibly altitude, barometric pressure, and salinity or conductivity).
Da	is the initial oxygen deficit $\left[\frac{g}{m^3}\right]$.
t	is the elapsed time, usually [d].

1 Introduction

1.1 Water Quality Monitoring and Analysis

Surface water is water collecting on the ground or in a stream, river, lake, wetland, or ocean; it is related to water collecting as groundwater or atmospheric water. Surface water is naturally replenished by precipitation and naturally lost through discharge to evaporation and sub-surface seepage into the ground.

Testing water quality data for trend over a period of time has received considerable attention recently. The interest in methods of water quality trend arises for two reasons. The first is the intrinsic interest in the question of changing water quality arising out of the environmental concern and activity. The second reason is that only recently has there been a substantial amount of data that is amenable to such an analysis. Recently, several researchers reported different methods and techniques for water quality evaluation and analysis. Naddeo et al. [1] focused on 13 rivers of southern Italy in order to evaluate and optimize the monitoring procedure of surface water. The study recommends minimizing the sampling frequencies in order to reduce the cost of samples analysis. Boyacıoğlu et al. [2] investigated the priorities in surface water quality management based on correlations and variations of different organic and inorganic parameters. Wang et al. [3] used multivariate statistical techniques, such as cluster analysis (CA) and principal component analysis/factor analysis (PCA/FA), to assess the surface water quality and identification of the source of water pollution in the Songhua River in Harbin region, China. Data on 15 parameters, including organic, inorganic, physical, chemical, heavy metals, and hazardous material through the period 2005–2009 were used. This study will provide useful information to managers to know better about how to improve water quality. Selle et al. [4] utilized spatial–temporal patterns of scores technique for surface water, springs, and deep groundwater from the wells in order to understand the processes governing water quality at catchment scales. The study demonstrated the potential analysis to identify dominant processes at catchment scales. Chung and Yoo [5] designed a wireless sensor network (WSN) with deployed field servers to detect water pollution in streams, rivers, and coastal areas. The proposed system can be efficiently performed to monitor the variation of the water quality data in streams, rivers, and coastal areas in real time. Hatvani et al. [6] used Dynamic Factor Analysis method to determine the driving background factors of a river located in an agricultural watershed to separate the role of the diffuse and point source nutrient loads. In this study, Dynamic Factor Analysis was applied to the time series (1978–2006) of 21 response parameters measured in its watershed. The study concluded that, with the aid of Dynamic Factor Analysis, the superimposed effects of the socio-economic changes which began in the mid-1980s, and the introduction of advanced wastewater treatment in the river catchment in the early 1990s, could be separated and their relative importance assessed. Chen et al. [7] prepared a comparative study of surface water quality for the major rivers and lakes in china. Data from 33,612 observations for the major rivers and lakes between 2012 and 2018 was used to evaluate the performance of ten learning models (seven traditional and three ensemble models) to explore the potential key water parameters for future model prediction. Busico et al. [8] utilized a multivariate statistical analysis to investigate a novel hybrid method for the effect of anthropogenic pollutions on groundwater in Italy. Jahin et al. [9] developed irrigation water quality index for surface water in Egypt by using multivariate analysis. Weerasinghe and Handapangoda [10] investigated the analysis of physiochemical parameters of surface water in Sri Lanka. In this study, two-way ANOVA, followed by Tukey's pairwise comparison, were used to assess the spatial and temporal variability. Carstens and Amer [11] perform spatio-temporal analysis study of urban changes and surface water quality in southeast Louisiana. The study reported that the high levels of fecal coliform were consistent with increased urbanization in water bodies. Khan et al. [12] investigated the effect of chemical and microbiological quality of sea water on reverse osmosis membrane and on fouling of RO membrane modification in RO sea water desalination plant in Saudi Arabia. Won et al. [13] evaluated the microbiological quality of tow irrigation canals and four surface reservoirs located in Ohio, USA. The study reported that the level of *Escherichia coli* in irrigation canals was higher than that in reservoirs and increased during heavy rain season. According to Texas commotion on environmental quality [14, 15], the Surface Water Quality Monitoring (SWQM) Program reported that around 1800 samples were collected from different surface water sites statewide to characterize physical, chemical, and biological parameters in order to identify emerging problems and evaluate the effectiveness and trends of water quality program. Standard values and criteria of surface water quality and monitoring has been adopted by Colorado department of public health and are presented in Tables 3.1, 3.2, and 3.3.

Table 3.1 Standard values for terrestrial surface water bodies (rivers, lakes)

Class	Standard values						
	Hydrogen ion concentration index (pH)	**Dissolved oxygen (DO) (mg/l)**	**Biological oxygen demand (BOD) (mg/l)**	**Suspended solids (SS) (mg/l)**	***E. coli* (CFU/100 ml)**	**Ammonia nitrogen (NH_3-N) (mg/l)**	**Total phosphorus (TP) (mg/l)**
A	6.5–8.5	Over 6.5	Less than 1	Less than 25	Less than 50	Less than 0.1	Less than 0.02
B	6.0–9.0	Over 5.5	Less than 2	Less than 25	Less than 5000	Less than 0.3	Less than 0.05
C	6.0–9.0	Over 4.5	Less than 4	Less than 40	Less than 10,000	Less than 0.3	–
D	6.0–9.0	Over 3	–	Less than 100	–	–	–
E	6.0–9.0	Over 2	–	No floating material and no oil	–	–	–

Marine surface water bodies

Class	Standard values			
	Hydrogen ion concentration index (pH)	**Dissolved oxygen (DO) (mg/l)**	**Biological oxygen demand (BOD) (mg/l)**	***E. coli* (CFU/100 ml)**
A	7.5–8.5	Over 5.0	Less than 2	Less than 1000
B	7.5–8.5	Over 5.0	Less than 3	–
C	7.0–8.5	Over 2.0	Less than 6	–

Notes

The following units are used for standard values in the foregoing standards:

6. pH: pH value
7. *E. coli*: colony forming units per 100 ml water sample
8. Others: milligram/liter

Table 3.2 Environmental standards for protecting human health

Water quality item		Standard value (units: mg/l)
Heavy metals	Cadmium	0.01
	Lead	0.1
	Hexavalent chromium	0.05
	Arsenic	0.05
	Mercury	0.002
	Selenium	0.05
	Copper	0.03
	Zinc	0.5
	Manganese	0.05
	Silver	0.05
Agricultural chemicals	Total organophosphates (parathion, Diazinon, Methamidophos, Monocrotophos,EPN, Chlorpyrifos) and carbamates (Fenobucarb, Carbofuran, Methomyl)	0.1
	Endrin	0.0002
	Lindane	0.004
	Toxaphene	0.005
	Endosulfan	0.003
	Heptachlor and derivatives (heptachlor, heptachlor epoxide)	0.001
	DDT and derivatives (DDT, DDD, DDE)	0.001
	Aldrin, Dieldrin	0.003
	Pentachlorophenol and salts	0.005
	Herbicides (Butachlor, Paraquat, 2,4-D)	0.1

Notes

I. The above standards provide specific values for substances posing a cumulative hazard to human health

II. These values are the maximum allowable volumes

III. These values are uniformly applicable to all public water areas

IV. The central competent authority shall add and officially announce maximum allowable volumes for other agricultural chemicals harmful to water quality

1.2 *Classification of Surface Water*

Calcification of water quality has been reported by Water framework directive (WFD) in 2012 [16]. WFD specified the quality elements that are used to assess the ecological and chemical status of a water body. Quality elements are generally biological (e.g. fish, invertebrates, macrophytes) or chemical (e.g., heavy metals, pesticides, nutrients). Classifications indicate where the quality of the environment is good, where it may need improvement, and what may need to be improved. They can also be used, over the years, to plan improvements, show trends, and to monitor success. Based on WFD, there are two status classifications which are commonly reported: ecological and chemical.

Table 3.3 Value standards (concentrations in μg/l unless noted) [8]

Parameter[a]	Table value standards[b, c]
Ammonia[d]	Cold water $\text{acute} = \frac{0.275}{1+10^{7.204-\text{pH}}} + \frac{39.0}{1+10^{\text{pH}-7.204}}$ $\text{chronic} = \left(\frac{0.0577}{1+10^{7.688-\text{pH}}} + \frac{2.487}{1+10^{\text{pH}-7.688}}\right) * \text{MIN}\left(2.85, 1.45 * 10^{0.028(25-T)}\right)$ Warm water $\text{acute} = \frac{0.411}{1+10^{7.204-\text{pH}}} + \frac{58.4}{1+10^{\text{pH}-7.204}}$ $\text{chronic (Apr1 - Aug31)} = \left(\frac{0.0577}{1+10^{7.688-\text{pH}}} + \frac{2.487}{1+10^{\text{pH}-7.688}}\right) * \text{MIN}\left(2.85, 1.45 * 10^{0.028(25-T)}\right)$ $\text{chronic (Sep1 - Mar31)} = \left(\frac{0.0577}{1+10^{7.688-\text{pH}}} + \frac{2.487}{1+10^{\text{pH}-7.688}}\right) * 1.45 * 10^{0.028*(25-\text{MAX}(T,7))}$
NH3 = old TVS	Cold water acute = 0.43/FT/FPH/2[e] in mg/l (N) Warm water acute = 0.62/FT/FPH/2[e] in mg/(N)
Cadmium	$\text{Acute} = (1.136672-[\ln(\text{hardness}) \times (0.041838)]) \times e^{0.9151[\ln(\text{hardness})]-3.1485}$ $\text{Acute(trout)} = (1.136672-[\ln(\text{hardness}) \times (0.041838)]) \times e^{0.9151[\ln(\text{hardness})]-3.6236}$ $\text{Chronic} = (1.101672-[\ln(\text{hardness}) \times (0.041838)] \times e^{0.7998[\ln(\text{hardness})]-4.4451}$
Chromium III[f]	$\text{Acute} = e^{(0.819[\ln(\text{hardness})] + 2.5736)}$ $\text{Chronic} = e^{(0.819[\ln(\text{hardness})] + 0.5340)}$
Chromium VI[f]	Acute = 16 Chronic = 11
Copper	$\text{Acute} = e^{(0.9422[\ln(\text{hardness})]-1.7408)}$ $\text{Chronic} = e^{(0.8545[\ln(\text{hardness})]-1.7428)}$
Lead	$\text{Acute} = (1.46203-[\ln(\text{hardness})*(0.145712)]) * e^{(1.273[\ln(\text{hardness})]-1.46)}$ $\text{Chronic} = (1.46203-[\ln(\text{hardness})*(0.145712)]) * e^{(1.273[\ln(\text{hardness})]-4.705)}$
Manganese	$\text{Acute} = e^{(0.3331[\ln(\text{hardness})] + 6.4676)}$ $\text{Chronic} = e^{(0.3331\ [\ln(\text{hardness})] + 5.8743)}$
Nickel	$\text{Acute} = e^{(0.846[\ln(\text{hardness})] + 2.253)}$ $\text{Chronic} = e^{(0.846[\ln(\text{hardness})] + 0.0554)}$

(continued)

Table 3.3 (continued)

Parameter[a]	Table value standards[b, c]
Selenium[g]	Acute = 18.4 Chronic = 4.6
Silver	Acute = $\frac{1}{2}e^{(1.72[\ln(\text{hardness})] - 6.52)}$ Chronic = $e^{(1.72[\ln(\text{hardness})]-9.06)}$ Chronic(trout) = $e^{(1.72[\ln(\text{hardness})]-10.51)}$
Uranium	Acute = $e^{(1.1021[\ln(\text{hardness})] + 2.7088)}$ Chronic = $e^{(1.1021[\ln(\text{hardness})] + 2.2382)}$
Zinc	Acute = $0.978\ e^{(0.8525[\ln(\text{hardness})] + 1.0617)}$ Chronic = $0.986\ e^{(0.8525[\ln(\text{hardness})] + 0.9109)}$

[a]Metals are stated as dissolved unless otherwise specified

[b]Hardness values to be used in equations are in mg/l as calcium carbonate and shall be no greater than 400 mg/L. The hardness values used in calculating the appropriate metal standard should be based on the lower 95% confidence limit of the mean hardness value at the periodic low flow criteria as determined from a regression analysis of site-specific data. Where insufficient site-specific data exists to define the mean hardness value at the periodic low flow criteria, representative regional data shall be used to perform the regression analysis. Where a regression analysis is not appropriate, a site-specific method should be used. In calculating a hardness value, regression analyses should not be extrapolated past the point that data exist

[c]Both acute and chronic numbers adopted as stream standards are levels not to be exceeded more than once every three years on an average

[d]FT = 100.03(20–TCAP)

Where TCAP is $\leq T \leq 30$

FT = 100.03(20-T)

Where 0 is $\leq T \leq$ TCAP

TCAP = 20 °C cold water aquatic life species present

TCAP = 25 °C cold water aquatic life species absent

FPH = 1; Where $8 < pH \leq 9$ 6

FPH = 1 + 10(7.4-pH)

1.3 Where $6.5 \leq pH \leq 8$

FPH means the acute pH adjustment factor, defined by the above formulas

FT means the acute temperature adjustment factor, defined by the above formulas

T means temperature measured in degrees

TCAP means temperature CAP; the maximum temperature which affects the toxicity of ammonia to salmonid and non-salmonid fish groups

NOTE: If the calculated acute value is less than the calculated chronic value, then the calculated chronic value shall be used as the acute standard

[e]For acute conditions the default assumption is that salmonids could be present in cold water segments and should be protected, and that salmonids do not need to be protected in warm water segments. For chronic conditions, the default assumptions are that early life stages could be present all year in cold water segments and should be protected. In warm water segments, the default assumption is that early life stages are present and should be protected only from April 1 through August 31. These assumptions can be modified by the commission on a site-specific basis where appropriate evidence is submitted

[f]Unless the stability of the chromium valence state in receiving waters can be clearly demonstrated, the standard for chromium should be in terms of chromium VI. In no case can the sum of the instream levels of Hexavalent and Trivalent Chromium exceed the water supply standard of 50 ug/l total chromium in those waters classified for domestic water use

[g]Selenium is a bioaccumulative metal and subject to a range of toxicity values depending upon numerous site-specific variables

Chemical status is assessed from compliance with environmental standards for chemicals that are priority substances and/or priority hazardous substances. Table 3.4 presents the list of priority substances for chemical status adapted from Environmental Quality Standards Directive (2008/105/EC). Chemical status is recorded as 'good' or 'fail'. Chemical status for a water body is determined by the worst scoring chemical (one-out-all-out approach). The priority substances were monitored only in water bodies where there are known discharges of these pollutants. Water bodies without discharges of priority substances are reported as being at good chemical status.

Ecological status classification consists of four different types of assessments:

1. An assessment of status indicated by a biological quality element, such as fish, invertebrates, or algae
2. An assessment of compliance with environmental standards for supporting physico-chemical conditions, such as dissolved oxygen, phosphorus, and ammonia
3. An assessment of compliance with environmental standards for concentrations of specific pollutants, such as zinc, cypermethrin, or arsenic and in determining high status only
4. A series of tests to make sure that hydromorphology is largely undisturbed and Tables 3.5 and 3.6 illustrate the biological and chemical quality elements, respectively.

Based on WFD, ecological status is recorded as high, good, moderate, poor, or bad. 'High' represents "largely undisturbed conditions". Figure 3.1 shows a diagram for the summary of classification.

1.3 Water Quality Indices (WQIs)

Water quality assessment can be defined as the evaluation of the physical, chemical, and biological nature of water in relation to natural quality, human effects, and intended uses. Water quality indices (WQIs) can be defined as the translation of numerical values of several water quality characteristics of a sample into a single value used in the monitoring, comparison, and control of water quality [16]. WQIs reduce a great amount of parameters to a simpler expression to enable easier interpretation of the monitoring data.

The concept of WQIs has been reported in Germany in 1848 (the presence or absence of certain organisms in water was used as the indicator) [17]. A wide range of WQIs has been developed and applied to classify the quality of water in different regions. Traditionally, most of these WQIs were introduced to deal with data from different manual and automated sampling networks and monitoring programs.

WQIs can be classified into three categories: hysic-chemical, biological, and hydro-morphological WQIs, as shown in Table 3.2. A large number of WQIs has

Table 3.4 EQS Directive priority substances

CAS number[a]	EU number[b]	Name of priority substance[c]	Identified as priority hazardous substance
15,972-60-8	240-110-8	Alachlor	
120-12-7	204-371-1	Anthracene	X
1912-24-9	217-617-8	Atrazine	
71-43-2	200-753-7	Benzene	
Not applicable	Not applicable	Brominated diphenylether[d]	X
32,534-81-9	Not applicable	Pentabromodiphenylether (congener numbers 28, 47, 99, 100, 153 and 154)e	
7440-43-9	231-152-8	Cadmium and its compounds	X
85,535-84-8	287-476-5	Chloroalkanes, C10-13[d]	X
470-90-6	207-432-0	Chlorfenvinphos	
2921-88-2	220-864-4	Chlorpyrifos	
(Chlorpyrifos-ethyl)			
107-06-2	203-458-1	1,2-Dichloroethane	
75-09-2	200-838-9	Dichloromethane	
117-81-7	204-211-0	Di(2-ethylhexyl)phthalate (DEHP)	
330-54-1	206-354-4	Diuron	
115-29-7	204-079-4	Endosulfan	X
206-44-0	205-912-4	Fluoranthene[f]	
118-74-1	204-273-9	Hexachlorobenzene	X
87-68-3	201-765-5	Hexachlorobutadiene	X
608-73-1	210-158-9	Hexachlorocyclohexane	X
34,123-59-6	251-835-4	Isoproturon	
7439-92-1	231-100-4	Lead and its compounds	
7439-97-6	231-106-7	Mercury and its compounds	X
91-20-3	202-049-5	Naphthalene	
7440-02-0	231-111-4	Nickel and its compounds	
25,154-52-3	246-672-0	Nonylphenols	X
104-40-5	203-199-4	(4-nonylphenol)	X
1806-26-4	217-302-5	Octylphenols	
140-66-9	Not applicable	(4-(1,1′,3,3′-tetramethylbutyl)-phenol)	
608-93-5	210-172-5	Pentachlorobenzene	X
87-86-5	201-778-6	Pentachlorophenol	
Not applicable	Not applicable	Polyaromatic hydrocarbons	X
50-32-8	200-028-5	(Benzo(a)pyrene)	X
205-99-2	205-911-9	(Benzo(b)fluoranthene)	X
191-24-2	205-883-8	(Benzo(g,h,i)perylene)	X
207-08-9	205-916-6	(Benzo(k)fluoranthene)	X
193-39-5	205-893-2	(Indeno(1,2,3-cd)pyrene)	X
122-34-9	204-535-2	Simazine	

(continued)

Table 3.4 (continued)

CAS number[a]	EU number[b]	Name of priority substance[c]	Identified as priority hazardous substance
Not applicable	Not applicable	Tributyltin compounds	X
36,643-28-4	Not applicable	(Tributyltin-cation)	X
12,002-48-1	234-413-4	Trichlorobenzenes	
67-66-3	200-663-8	Trichloromethane (chloroform)	
1582-09-8	216-428-8	Trifluralin	

[a]CAS: Chemical Abstracts Service
[b]EU number: European Inventory of Existing Commercial Substances (EINECS) or European List of Notified Chemical Substances (ELINCS)
[c]Where groups of substances have been selected, typical individual representatives are listed as indicative parameters (in brackets and without number). For these groups of substances, the indicative parameter must be defined through the analytical method
[d]These groups of substances normally include a considerable number of individual compounds. At present, appropriate indicative parameters cannot be given
[e]Only Pentabromobiphenylether (CAS number 32534 81 9)
[f]Fluoranthene is on the list as an indicator of other, more dangerous polyaromatic hydrocarbons

Table 3.5 Biological quality elements monitored for each water category [9]

Category	Quality element	Description
Rivers	Macrophytes and phytobenthos – diatoms	Microscopic diatoms (algae) found on rocks and plants
	Macrophytes and phytobenthos – macrophytes	Water plants visible to the naked eye, growing in the river
	Macroinvertebrates	Insects, worms, molluscs, crustacea, etc., living on the river bed
	Fish	Including eel
Lakes	Phytoplankton	Free-floating microscopic plants
	Macrophytes and phytobenthos – diatoms	Microscopic diatoms (algae) found on rocks and plants
	Macrophytes and phytobenthos – macrophytes	Water plants visible to the naked eye, growing in the lake
	Macroinvertebrates	Insect larvae, worms, molluscs, crustacean, etc., living on the lake bed.
TraC	Phytoplankton	Free-floating microscopic plants
	Macroalgae	Seaweeds visible to the naked eye
	Angiosperms	Sea grasses and saltmarsh plants
	Benthic invertebrates	Worms, molluscs, and crustacean, etc., living in or on the bed of the estuary or sea
	Fish (transitional only)	Fish which spend all or part of their life in transitional waters

Table 3.6 Physico-chemical quality elements monitored for each water category [9]

Quality element	Rivers	Lakes	TraC
pH	✓	✓	
Ammonia (total as N)	✓	✓	
Phosphate	✓	✓	
Dissolved inorganic nitrogen			✓
Dissolved oxygen	✓	✓	✓
Specific pollutants (annex VIII)	✓	✓	✓
Acid neutralizing capacity		✓	
Temperature	✓		

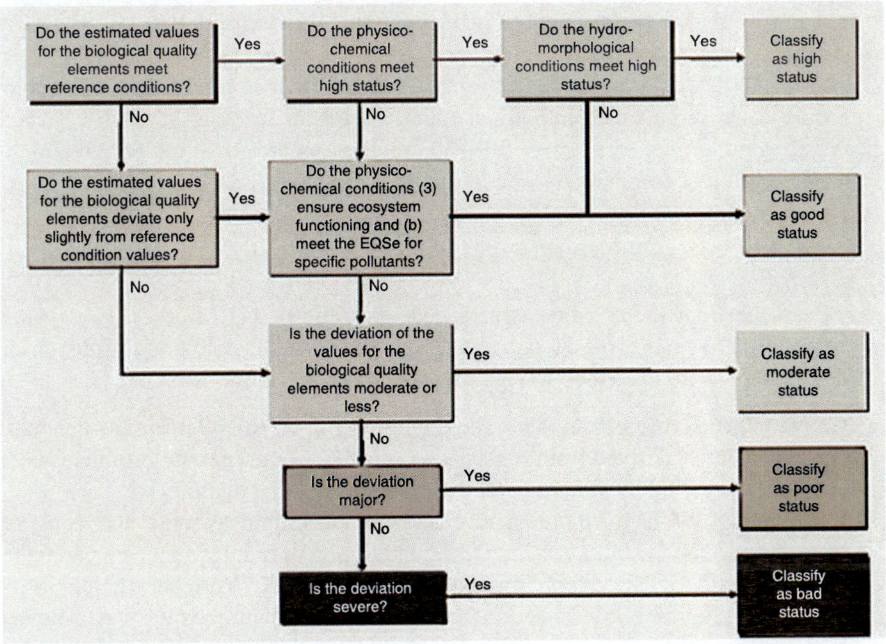

Fig. 3.1 Decision-tree illustrating the criteria determining the different ecological status classes

been designed for hysic-chemical parameters. The development of a WQI is normally through different steps [18–20] and are as follows:

1. Selection of the optimum set of parameters that together reflect the overall quality of the water body with respect to a given end use
2. Transformation of the required parameters of different units and dimensions into a common scale based on their impact on the health and ecosystem
3. Determination of weights of selected parameters
4. Calculation of the index score is obtained by aggregating the respective sub-indices into a common function

Suitable WQIs should have clear objectives, good synthetic capacities, and be able to achieve a reasonable balance between the simplification of reality and the complexity of the environment. They should include variables that are normally and continuously monitored and that have a clear effect on water quality (e.g., potentially affecting aquatic life, bathing, public supply, irrigation, and recreational uses). WQI is most useful for comparative purposes and for general questions. Site-specific questions that should be addressed by an analysis of the original data. It is limited in that while a certain site may receive a good score, it may still be impaired or degraded based on a parameter not included in the index calculation. Also, aggregation of data may either mask or over-emphasize short-term (acute) water quality problems. Table 3.4 summarizes developments and practical WQIs' applications. The WQIs summarized below are a version of a WQI that was adapted from work conducted by the National Sanitation Foundation (NSF) in the 1970s. Concentrations of nine parameters (dissolved oxygen, fecal coliform, pH, biochemical oxygen demand (BOD), total nitrates and phosphates, total solids, temperature, and turbidity) were each assigned an individual rating based on existing standards or best professional judgment on a scale of 0–100. Each rating was then multiplied and the root of the product computed to obtain the final rating (Eq. 3.1)

$$\mathrm{WQI} = (\mathrm{Pi} * \mathrm{Pi2} * \mathrm{Pi3} * \ldots * \mathrm{Pi} + n)^{1/n} \tag{3.1}$$

Where the final index value is assigned as follows: 0–20 poor, 20–40 below average, 40–60 average, 60–80 above average and from 80–100 good (Table 3.7).

Three steps were described to calculate the WQI as follow:

1. Converted each result to an index score ranging from 1 to 100 using the quadratic equation (Eq. 2) derived from regression curve data. The specific formula used at each station varied by stream class or ecoregion for that station.
2. Aggregating WQIs by month and calculating a simple average and applying penalty factors if necessary to reduce the likelihood of one low-scoring parameter being masked by the averaging process. The overall WQI per station is the average of the three lowest-scoring months. A similar procedure was followed to determine a WQI for each parameter.
3. Moderation of low scores that could be attributed to natural variance.

$$\mathrm{WQI} = a + b_1(\mathrm{Parameter}) + b_2(\mathrm{Parameter})^2 \tag{3.2}$$

Table 3.8 summarized indicators of water quality and the reasons for including in the WQI.

Table 3.7 Description of water quality indices

References	Purpose of WQI	Parameters	WQI Development
[15]	Simplify WQ reporting to the public.	DO as percent saturation, fecal coliform, pH, BOD_5, total nitrates, total phosphates, TS, temperature, and turbidity	Standardize parameter concentration to sub-index score. Final WQI computed using an unweighted product equation of sub-indices.
[16]	Used as a tool to summarize and report routine stream monitoring data to the public. Indicates whether WQ is less than expected to support designated uses.	DO as concentration, fecal coliform, pH, total nitrogen, total phosphorus, TSS, and turbidity.	Quadratic equation to convert results to sub-index score. Apply weightings and other rules to account for strongly correlated parameters and to avoid double weighting. Final WQI is average of three lowest sub-index scores.
[17]	Evaluate WQ data in a way that is scientifically valid and easily understood by professionals and the public.	Variable. Based on recorded biological response of indicator organism (e.g., fish or benthos) by ecoregion. As an example, they used field measurements of DO, pH, temperature, and conductance.	Standardization of parameter concentration to biological response and narrative category. Assign numerical rank to narrative category. Final WQI is average of all worse case ranks over monitoring period.
[18]	Same as above, but they wanted to develop a new WQI with fewer parameters that is simpler than most other WQIs.	DO, fecal coliform, turbidity, total phosphorus, and specific conductance.	No standardization of variables to sub-index score. Rank and weight parameters based on significance. Conversion to logarithmic scale to keep final WQI a small number.
[19]	Used as an aid to describe WQ variables related to a watershed study exploring relationships between urban non-point pollution, land use (imperviousness), habitat and benthos across an urban-forest gradient in New Haven, Connecticut.	TSS, TDS, fecal coliform, nitrate, phosphate, the chloride to sulfate ratio, and the nitrate to total nitrogen ratio.	No standardization of variables to a sub-index score. Tuned a common WQI to quantify non-point source pollution levels in urbanized watersheds. Equation was described as a normalized average of seven parameters.
[20]	Intended to develop a WQI that would not hide or "eclipse" vital information.	Varied base on designated use. DO, pH, TSS, turbidity, actual temperature, temperature above natural, BOD_5, ammonia, and fecal coliform.	Panel of experts selected parameters and produced graphs to standardize results to sub-index scale and aggregated sub-index scores for final WQI using the lowest sub-index score or "minimum operator" function.

Table 3.8 List of parameters or indicators and reasons for including them in the WQI

Parameters/indicators of water quality	Include as part of WQI?	Reasoning
Specific conductance (μS/cm)	Yes	Compared to reference conditions, it indicates dissolved pollutants and potential upstream watershed disturbances.
Dissolved oxygen (%)	Yes	A standard water quality parameter. It indicates oxygen demand as well as supersaturation (evidence of excessive alga and nutrient enrichment).
pH (s.u.)	Yes	A standard water quality parameter. Values outside of desired range may indicate a water quality problem.
Water temperature (°C)	Maybe	If monitoring is conducted in summer then yes; otherwise, no. project dependent (Mtns vs. coast)
Nitrite-nitrate nitrogen (mg/L)	Yes	Compared to reference conditions, it provides evidence of nutrient enrichment, disturbance, and pollution sources.
Total Kjeldahl nitrogen (mg/L)	Yes	Same as above.
Ammonia nitrogen (mg/L)	Yes	Same as above.
Total phosphorus (mg/L)	Yes	Same as above.
Total suspended residue (mg/L)	No	Turbidity and bank stability metrics below will capture potential sediment pollution. During baseflow, this parameter may not adequately reflect the potential for sediment pollution.
Turbidity (NTU)	Depends	Indicates potential sediment or particulate pollution. If stream bank stability metric below is included, turbidity may not need to be included. However, it could provide evidence of sources if turbidity is present during baseflow.
Sodium (mg/L)	No	Specific conductance will capture problem levels.
Chloride (mg/L)	No	Same as above.
Metals (Cu, Zn, Pb, etc.)	No	Typically levels are very low. Uncertainty remains in interpreting total recoverable vs. dissolved data. Turbidity and other sediment indicators will capture potential problem.
Fecal coliform (cols/100 mL)	Yes	Indicates potential sewer line leaks, failing septic systems and upstream livestock operations.
Organics	No	Too expensive. Other parameters could be a surrogate for organic pollutants.
Livestock access	Yes	We know that livestock represent a stressor to both water quality and the riparian environment.
Light penetration (shading)	No	Riparian vegetation would capture this parameter.
Bank stability	Yes	This metric captures a source of sediment pollution during storm flow and may be the sole metric to represent potential sediment pollution during storm flow. Others are listed below.

(continued)

Table 3.8 (continued)

Parameters/indicators of water quality	Include as part of WQI?	Reasoning
Bank height ratio (BHR)	Maybe	This estimate may a more robust measure of potential sediment pollution during storm flow than simply bank stability. However, this metric and estimates of bank stability (a sort of pseudo BEHI) would be a powerful parameter combination indicating sediment pollution.
Riparian zone width/ vegetation	Yes	We know that riparian conditions (width and type of vegetation) affect water quality functions. Conventional chemistry parameters alone do not adequately address this metric.
Bottom substrate	Maybe	This would address sediment pollution in terms of bed load and may not be needed if one of the other sediment related metrics are included.

1.3.1 WQIs for Data Generated from Automated Networks

Automated sampling networks generate a limited range of physico-chemical parameters that are measured continuously in specific locations at high temporal frequency. This process creates a significant volume of data that is eventually stored in data tables, which are not translated into intelligible information describing the status of the water body. As not all the existing indices are suitable to deal with data of this nature, we need to select an index. Terrado et al. [17] proposed different suitable WQIs to deal with data generated from automated sampling networks, and Table 3.4 summarizes the main characteristics of proposed indices (Tables 3.9 and 3.10).

Figure 3.2 shows the proposed criteria that can be used to compare the different WQIs [18]. Five proposed indices were selected depending on whether they fulfill these particular criteria in a good, a fair, or a bad way. Significant parameters, such as pH, conductivity, turbidity, dissolved oxygen, water temperature, ammonia, nitrates, chlorides, and phosphates, were considered for proposed WQIs. Different objectives and a flexible index that allowed use of different parameters were established depending on various water uses. A higher value on simplicity in programming, tolerance to missing and erroneous data, and the possibility of the index working with non-synchronized data were performed. Accordingly, Canadian Council of Ministers of the Environment (CCME) was selected as the most suitable tool for categorizing water bodies using data generated by automated sampling stations. Figures 3.3 and 3.4 illustrate chart for developing WQIs.

For example, British Columbia Ministry of Environment [21] developed formula for CCME WQI. The index number ranges between 0 (poor water quality) and 100 (excellent water quality), divided into five descriptive categories: Poor: (0–44), Marginal: (44.1–64), Fair: (64.1–79), Good: (79.1–94) and Excellent: (CCME 94.1–100). Table 3.11, summarizes descriptive index criteria and Table 3.7 illustrates the advantages and disadvantages of CCME WQI index.

There are three elements that can be used for calculation and modification of the range of index; F_1 (scope), F_2 (frequency) and F_3 (amplitude). F_1 represents the

Table 3.9 Classification of water quality indices [11]

	Indices for general	Horton Index
Physico-chemical indices	Water quality	National Sanitation Foundation Water Quality Index
		Prati's Implicit Index of Pollution
		McDuffie and Haney's River Pollution Index
		Diniu's Water Quality Index
		British Columbia Water Quality Index
		Oregon Water Quality Index
		Florida Stream Water Quality Index
		Overall Index of Pollution
		Pesce and Wunderlin's Water Quality Index
		Water Quality Index of Central Pollution Control Board
		River Pollution Index
		Universal Water Quality Index
		Canadian Water Quality Index
		Simplified Water Quality Index
		Said et al.'s Water Quality Index
	Indices for specific water uses	O'Connor Indices: Fish and Wildlife Index and Public Water
		Supply Index
		Deininger and Landwehr Index for Public Water Supply
		Walski and Parker's Index for Recreation
		Stoner's Index for Dual Uses (PWS and Irrigation)
		Nemerow and Sumitomo's Pollution Index for Three Uses
		Smith's Index for Four Water Uses: 1) General, 2) Regular
		Public Bathing, 3) Water Supply, and 4) Fish Spawning
		Viet and Bhargava's Index
		Gekov et al.'s Index
		Haire et al.'s Nutrient Loading Index and Eutrophication Index
		Li's Regional Water Resource Quality Assessment Index
	Indices for planning	Truett et al.'s Prevalence Duration Intensity Index
		Truett et al.'s National Planning Priorities Index
		Truett et al.'s Priority Action Index
		Dee et al.'s Environmental Evaluation System
		Inhaber's Canadian National Index
		Zoeteman's Pollution Potential Index
		Johansson and Johnson Pollution Index

(continued)

Table 3.9 (continued)

	Statistical approaches	Shoji et al.'s Composite Pollution Index
		Joung et al.'s Index of Partial Nutrients (Factor Analysis)
		Joung et al.'s Index of Total Nutrients (Factor Analysis)
		Coughlin et al.'s Principal Component Index (Principal
		Component Analysis)
		Shin and Lam (Principal Component Analysis)
		Parinet et al.'s (Principal Component Analysis)
		Harkins's Index (Kendall Ranking Approach, 1975) (Non-
		Parametric Classification)
		Schaeffer and Janardan's Beta Function Index
		Kung et al.'_s Fuzzy Clustering
Biological indices	Macroinvertebrates	Biomonitoring Working Party
		Biological Families Index
	Fish	Index of Biological Integrity
		Extended Biological Index (Adapted from Woodiwis (1978)
		Biological Index)
	Diatoms	Index of Sensitivity to Pollution (CEMAGREF, 1982)
		Biological Index of Diatoms
	Macrophytes	Macrophytes Index
		Index of Macroscopic Aquatic Vegetation
Hydro-morphological indices	Connectivity Habitat	Fluvial Connectivity Index
		Fluvial Habitat Index
		Qualitative Habitat Evaluation Index
	Vegetation	Fluvial Vegetation Index
		Bank Vegetation Quality Index

percentage of variables that do not meet their objectives at least once during the time period under consideration (failed variables) in relation to the total number of variables measured; F_2 represents the percentage of individual tests that do not meet objectives (failed tests) and; F_3 represents the amount by which failed test values do not meet their objectives.

$$F_1 = \left(\frac{\text{Number of failed variable}}{\text{Total nomber of variables}}\right) \times 100 \tag{3.3}$$

Table 3.10 Parameters, sub-index and aggregation functions, weights, and quality classification procedure of the different proposed physico-chemical indices for the evaluation of water quality

Index	Parameter	Application limits	Sub-index function	Weight	Aggregation function	Water-quality classification
ISQA	Temperature (_C)	≤20	IT = 1 -	$ISQA = I_T(I_{TOC} + I_{SS} + I_{DO} + I_{COND})$		0 (minimum quality) – 100 (maximum quality)
		>20	IT = 1–(T–20)*0.0125			
	Total organic carbon	≤5	ITOC = 30–TOC			
	(mg/L)	12 ≥ TOC > 5	ITOC = 21–(0.35*TOC)			
		>12	ITOC = 0			
	Suspended solids	≤100	ISS = 25–(0.15*SS)			
	(mg/L)	250 > SS > 100	ISS = 17–(0.07*SS)			
		>250	ISS = 0			
	Dissolved oxygen	<10	IDO = 2.5*DO			
	(mg/L)	≥10	IDO = 25			
	Conductivity (lS/cm)	≤4000	ICOND = (3.6-log COND)*15.4			
		>4000	ICOND = 0			
Pesce and Wunderlin	Dissolved oxygen (mg/L)	≤1 2 ≥ DO>1	IDO = 0 –IDO = 10	$WQI_{min} \frac{1}{3} \cdot \sum_{i=1}^{3} I_i$		0 (minimum quality) – 100 (maximum quality)
		3 ≥ DO>2	IDO = 20			
		3.5 ≥ DO>3	IDO = 30			
		4 ≥ DO>3.5	IDO = 40			
		5 ≥ DO>4	IDO = 50			
		6 ≥ DO>5	IDO = 60			
		6.5 ≥ DO>6	IDO = 70			
		7 ≥ DO>6.5	IDO = 80			

		7.5 > DO>7	IDO = 90		
		≥7.5	IDO = 100		
	Conductivity (lS/cm)	>12,000	ICOND = 0		
		12,000 ≥ COND ≥ 8000	ICOND = 10		
		8000 > COND ≥ 5000	ICOND = 20		
		5000 > COND ≥ 3000	ICOND = 30		
		3000 > COND ≥ 2500	ICOND = 40		
		2500 > COND ≥ 2000	ICOND = 50		
		2000 > COND ≥ 1500	ICOND = 60		
		1500 > COND ≥ 1250	ICOND = 70		
		1250 > COND ≥ 1000	ICOND = 80		
		1000 > COND ≥ 750	ICOND = 90		
		<750	ICOND = 100		
	Turbidity (NTU)	>100	ITURB = 0		
		100 ≥ TURB ≥ 80	ITURB = 10		
		80 > TURB ≥ 60	ITURB = 20		
		60 > TURB ≥ 40	ITURB = 30		
		40 > TURB ≥ 30	ITURB = 40		
		30 > TURB ≥ 25	ITURB = 50		
		25 > TURB ≥ 20	ITURB = 60		
		20 > TURB ≥ 15	ITURB = 70		
		15 > TURB ≥ 10	ITURB = 80		
		10 > TURB ≥ 5	ITURB = 90		
		<5	ITURB = 100		

(continued)

Table 3.10 (continued)

Index	Parameter	Application limits	Sub-index function	Weight	Aggregation function	Water-quality classification
Liou	Dissolved oxygen (mg/L)	>6.5 IDO = 1 4.6–6.5 IDO = 3	–	$IWQI = \frac{1}{4} \cdot \sum_{i=1}^{4} I_i$		<2 = good 2–3 = slightly polluted 3.1–6 = moderately polluted >6 = very polluted
		2–4.5 IDO = 6				
		<2 IDO = 10				
	BOD5 (mg/L)	<3 IBOD = 1				
		3–4.9 IBOD = 3				
		5–15 IBOD = 6				
		>15 IBOD = 10				
	Suspended solids	<20 ISS = 1				
	(mg/L)	20–49 ISS = 3				
		50–100 ISS = 6				
		>100 ISS = 10				
	Ammonia nitrogen	<0.5 INH3-N = 1				
	(mg/L)	0.5–0.99 INH3-N = 3				
		1–3 INH3-N = 6				
		>3 INH3-N = 10				

NSF WQI	Dissolved oxygen (% saturation)	Values extracted from the quality graphic elaborated for each parameter[a]	0.17	$\mathrm{NSWFQI}^{(3)} = \frac{\sum_{i=1}^{n} W_i I_i}{\sum W_i}$	0–25 = very bad 26–50 = bad 51–70 = regular 71–90 = good 91–100 = excellent
	Fecal coliforms[b]		0.16		
	(1b/100 ml)				
	pH		0.11		
	BOD5 (mg/L)		0.11		
	Nitrates (mg/L)		0.10		
	Phosphates (mg/L)		0.10		
	Temperature (_C)		0.10		
	Turbidity (NTU)		0.08		
	Total solids (mg/L)		0.07		
CCME WQI	No specific parameters defined.	F1: Scope (% of variables that do not meet their objectives at least once)	–	CCME WQI = 100-$\left(\frac{\sqrt{F_1^2 + F_2^2 + F_3^2}}{1.732}\right)$	0–44 = poor 45–64 = marinal 65–79 = regular 80–94 = good 95–100 = excellent
	It is a flexible index used with several different parameters.	F2: Frequency (% of individual tests that do not meet their objectives)			
	For each parameter, a maximum and a minimum water quality objective need to be	F3: amplitude (amount by which failed tests do not meet their objectives)			
		** sub-indices F1, F2 and F3			

[a]Graphics for each of the parameters are available on the http://www.water-research.net/watrqualindex/waterqualityindex.htm web page.
[b]Fecal coliforms is not a parameter usually obtained from the automated sampling networks. However, this is not a problem, since some of the variables can be omitted in calculating the NSF WQI.

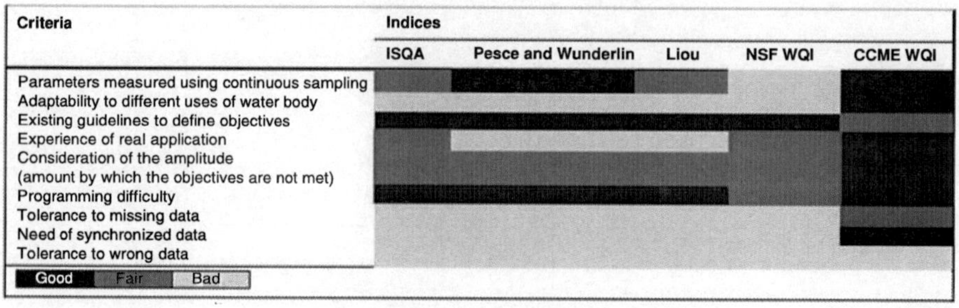

Fig. 3.2 Comparison of WQI with different criteria [11]

$$F_2 = \left(\frac{\text{Number of failed tests}}{\text{Total number of tests}}\right) \times 100 \tag{3.4}$$

F_3 can be calculated as follows: the number of times by which an individual concentration is greater than (or less than, when the objective is minimum). The objective is termed "excursion", then; when the test value not exceed the objective, excursion can be influenced by Eq. (3.5), while when test value not fall below the objective, Eq. (3.6) can be performed:

$$\text{excursion}_i = \left(\frac{\text{Failed test value}_i}{\text{Objective}_i}\right) - 1 \tag{3.5}$$

$$\text{excursion}_i = \left(\frac{\text{Objective}_i}{\text{Failed test value}_i}\right) - 1 \tag{3.6}$$

The collective amount by which individual tests are out of compliance can be calculated by summing the excursions of individual tests from their objectives and dividing the total number of tests (both those meeting objectives and those not meeting objectives). This variable, referred to as the normalized sum of excursions, or nse, can be calculated as:

$$\text{nse} = \frac{\sum_{i=1}^{n} \text{excursion}_i}{\text{total number of tests}} \tag{3.7}$$

F_3 is then calculated by an asymptotic function that scales the normalized sum of the excursions from

objectives (nse) to yield a range between 0 and 100.

$$F_3 = \left(\frac{\text{nse}}{0.01\text{nse} + 0.01}\right) \tag{3.8}$$

Fig. 3.3 Flowchart of progressive tasks to be completed for development of a WQI for surface water

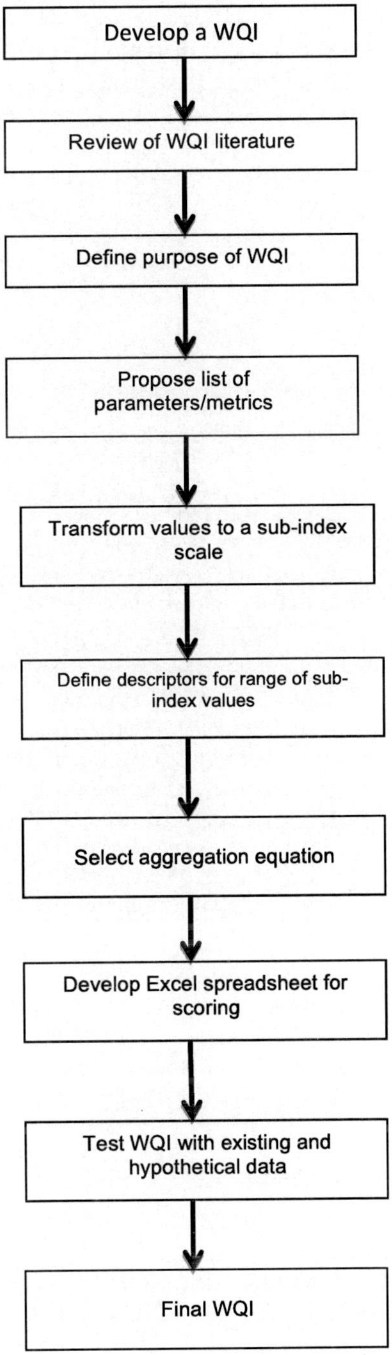

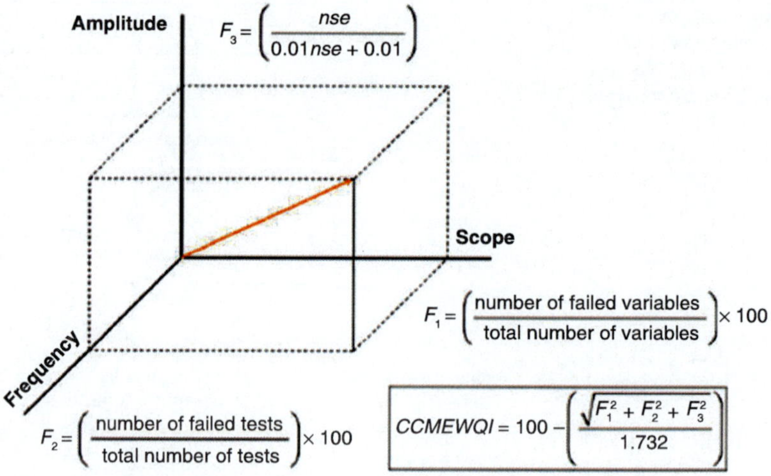

Fig. 3.4 Graphical representation of water quality indices (WQI) calculated in a three dimensional space by summing three factors (F_1, F_2, and F_3) as vectors

Table 3.11 Description of CCME WQI index criteria [11, 20]

Poor	CCME WQI value 0–44: water quality is almost always threatened or impaired conditions usually depart from natural or desirable levels
Marginal	CCME WQI value 44.1–64: Water quality is frequently threatened or impaired conditions often depart from natural or desirable levels
Fair	CCME WQI value 64.1–79: Water quality is usually protected but occasionally threatened or impaired conditions sometimes depart from natural or desirable levels
Good	CCME WQI value 79.1–94: Water quality is protected with only a minor degree of threat or impairment conditions rarely depart from natural or desirable levels
Excellent	CCME WQI value 94.1–100: Water quality is protected with a virtual absence of threat or impairment – Conditions very close to natural or pristine levels.

Fig. 3.5 Mass balance conservation

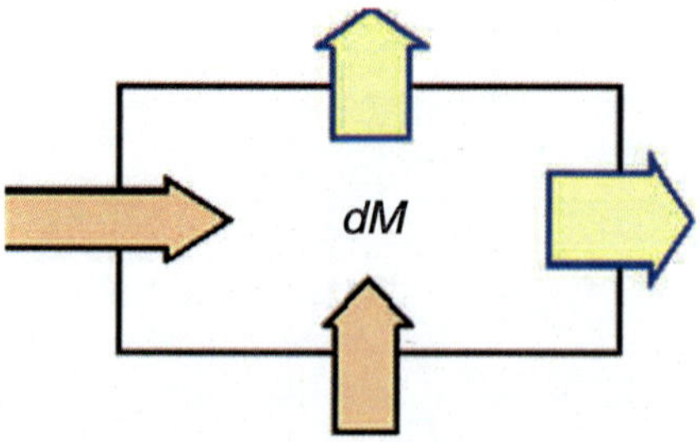

Then the index, as a three-dimensional space, can be calculated (Eq. 3.7) by summing the three factors as shown in Fig. 3.2. The sum of the squares of each factor is therefore equal to the square of the index. This approach treats (Fig. 3.5, Table 3.12)

Table 3.12 Advantages and disadvantages of CCME WQI [11]

Advantages		Disadvantages	
1.	Flexibility in the selection of input parameters and objectives	1.	Missing guidelines about the variables to be used for the index calculation
2.	Adaptability to different legal requirements and different water uses	2.	Missing guidelines about the objectives specific to each location and particular water use
3.	Statistical simplification of complex multivariate data		
4.	Clear and intelligible diagnostic for managers and the general public	3.	Easy to manipulate (biased)
		4.	The same importance is given to all variables
5.	Suitable tool for water quality evaluation in a specific location	5.	No combination with other indicators or biological data
6.	Easy to calculate	6.	Only partial diagnostic of the water quality
7.	Tolerance to missing data	7.	F1 not working appropriately when too few variables are considered or when too much covariance exists among them. The n factor has too much weight in calculating the index
8.	Suitable for analysis of data coming from automated sampling		
9.	Experience in implementation		
10.	Considers amplitude (of difference from the objective)		

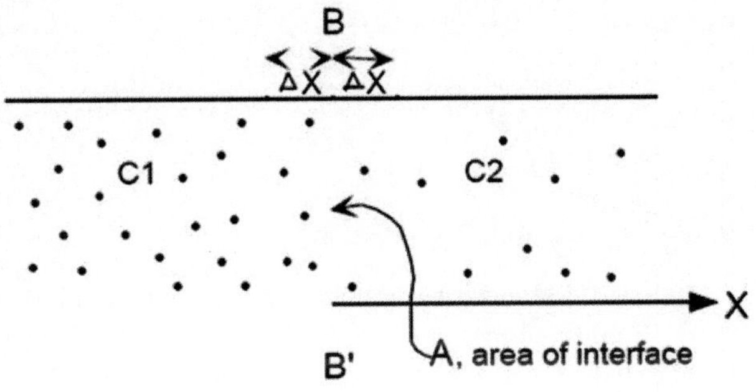

Fig. 3.6 Conceptual model for diffusion

$$\text{CCME WQI} = 100 - \left(\frac{\sqrt{F_1^2 + F_2^2 + F_3^2}}{1.732}\right) \tag{3.9}$$

2 Water Quality Module Formation

A general water quality module with diffusion transportation can be readily derived based on the mass conservation and *Fick's* diffusion principle (Eq. 3.10).

$$J = -D\frac{\partial \Phi}{\partial x} \tag{3.10}$$

where

J is the "diffusion flux" [(amount of substance) per unit area per unit time], for example $\frac{mol}{m^2.s}$, *J* measures the amount of substance that will flow through a small area during a small time interval.

D is the diffusion coefficient or diffusivity in dimensions of [length2 time−1], example $\frac{m^2}{s}$

Φ (for ideal mixtures) is the concentration in dimensions of [amount of substance per unit volume], example $\frac{mol}{m^3}$

x is the position [length], example m

The Law of Conservation of Mass states that mass can neither be created nor destroyed. The inflows, outflows, and change in storage of mass in a system must be in balance. The mass flow in and out of a control volume (through a physical or virtual boundary) can, for a limited increment of time, be expressed as (Fig. 3.6):

$$dM = \rho i\ vi\ Ai\ dt - \rho o\ vo\ Ao\ dt \tag{3.11}$$

where

dM = change of storage mass in the system (kg)
ρ = density (kg/m^3)
v = speed (m/s)
A = area (m^2)
dt = an increment of time (s)

If the outflow is higher than the inflow, the change of mass dM is negative and the mass of the system decreases, while the mass in a system increases if the inflow is higher than the outflow. The Law of Mass Conservation is a fundament in fluid mechanics and a basis for the Equation of Continuity and the Bernoulli Equation.

Diffusion is defined as the net transport due to random motion. *Fick's second law* (Eq. 3.12) predicts how diffusion causes the concentration to change with time.

$$\frac{\partial \Phi}{\partial t} = D\frac{\partial^2 \Phi}{\partial x^2} \tag{3.12}$$

Where

Φ is the concentration in dimensions of [(amount of substance) length^{-3}], example $\frac{mol}{m^3}$

t is time [s]

D is the diffusion coefficient in dimensions of [length2 time^{-1}], example $\frac{m^2}{s}$

x is the position [length], example m.

The use of a particular model depends on the systems to be modeled and on the legislation in place in that country. Water Quality Models are usually classified into subdivision categories based on:

1. Identifying the environment modeled
2. Purpose of the model
3. Consideration of the number of 'dimensions'
4. Description of the main process
5. The data used are discrete observed measurements or statistical distributions, and
6. Consideration of temporal variability.

A model for diffusive flux can be constructed from the following example. Consider a one-dimensional system with motion in the X direction only (Fig. 3.7). An interface B-B′ separates two regions of different concentration, C1 and C2 = particles/volume on the left and right side of the interface, respectively. The motion of each particle is a one-dimensional random walk. In each time interval, Δt, each particle will move a distance $\pm\, \Delta X$, moving right (+ ΔX) or left (− ΔX) with equal probability.

Within each time step, any particle within a distance ΔX of the interface B-B′ has a 50% probability of crossing over that interface. The number of particles with the potential to cross B-B′ from left to right (positive mass flux) is (C1 $\Delta X\, A$), where A is the area of interface B-B′. On average, half of these take a positive step and cross the interface in time Δt such that the flux left to right is (0.5 C1 $\Delta X\, A$). Similarly, the number of particles crossing right to left in Δt (negative mass flux) will be (0.5 C2 $\Delta X\, A$). The resulting mass flux, q_X, is

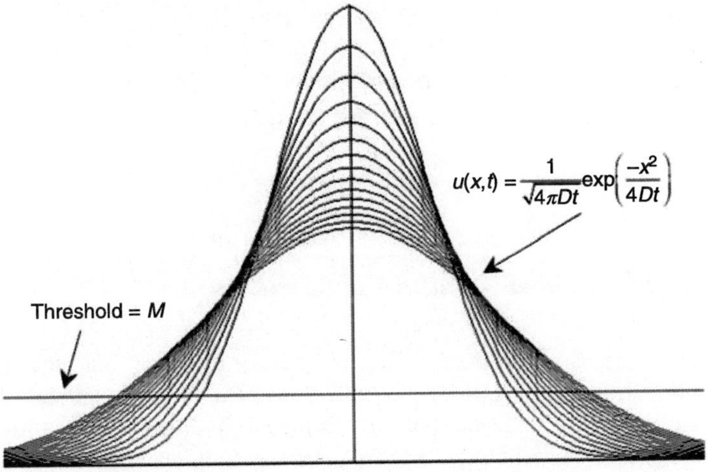

Fig. 3.7 The formula for distance within which the pheromone is sensed is given by $\sqrt{-2\, D\, t.\mathrm{Ln}\left(4M^2\pi\, D.t\right)}$

$$q_x = \frac{0.5\ \Delta X\ A\ (C1 - C2)}{\Delta t} \tag{3.13}$$

If $C(x)$ is continuous, then C2 ≈ C1 + $\Delta X\ \partial C/\partial x$, and Eq. 3.12 becomes

$$q_x = -\left\lfloor \frac{\Delta X^2}{2\Delta t} \right\rfloor A\ \frac{\partial C}{\partial x} = -D\,A\frac{\partial C}{\partial x}\left[\frac{\text{mass}}{\text{time}}\right] \tag{3.14}$$

Which is the mathematical expression of Fick's equation.

The coefficient of diffusion, D~(1/2)ΔX2/Δt, has units of [length^2 time^{-1}]. The diffusivity of a chemical molecule in a given fluid depends on the ease with which the molecule can move, specifically, how far, ΔX, the molecule can move in a given time interval. The ease of molecular motion, and thus the diffusivity of a particular chemical, will depend on the molecule size and polarity, the type of fluid, and the temperature.

Diffusion from a point source

If there is a unit mass at $x = 0$ and $t = 0$, then the concentration of the diffusing material is given by the following formula.

$$U(x,t) = \frac{1}{\sqrt{4\pi Dt}}\ \exp\left(\frac{-x^2}{4Dt}\right) \tag{3.15}$$

The graph below shows how the concentration changes with time. Diffusion from a point source can be used to model diffusion of an insect pheromone. In this case, there would be a threshold below which the pheromone would not be detected. The horizontal line in the graph indicates this threshold level. Wherever the concentration is above the line, it would be sensed, and wherever below the line, it would not be sensed (Fig. 3.8).

The duration of the effect of the pheromone will be $t_{\max} = \frac{1}{4M^2\pi.D}$. The greatest distance at which it can be detected during its duration will be at time $t_0 = \frac{1}{4M^2\pi.D.e}$ and the distance will be $X_{\max} = \sqrt{\frac{\log(e)}{2\ M^2\pi.e}}$

2.1 *Module Subdivision and Classification*

The subdivisions based on purpose can be rather subjective, but they do provide useful information with regard to the limitations of a particular model. A mixing-zone model will only represent that proportion of the system that is immediately downstream of, or adjacent to, a discharge into the main water body, and a time-of-travel model provides the user with the time of arrival of pollutants downstream of an 'incident' and so is only used to simulate simple pollution incidents. Time-of-travel models do not generally include anything other than a conservative description of solute movements, but are essentially simple instream water quality models. The

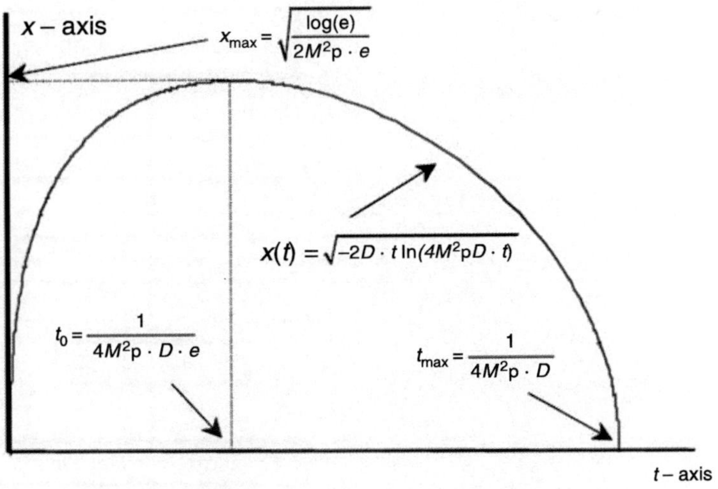

Fig. 3.8 Plot of range of pheromone

division is used here because instream models are not used routinely in pollution incident assessments unless they have been calibrated specifically for that purpose. The dimensions simulated by a particular model will provide information on both the complexity of a model and also on its suitability to specific applications. A zero-dimensional (0D) model does not represent the processes of dispersion of contaminants in any direction, but simply represents the volumes and concentrations assuming that the water body is completely and instantaneously mixed. A one-dimensional (1D) model represents the water flow and the advection and dispersion of solutes in just one direction (i.e., downstream in a river model) and so the stream is assumed to be completely (and instantaneously) mixed across its width and depth. Following from this, a two dimensional (2D) model will either simulate dispersion across the width or the depth of the stream, but not both. A width-averaged model is often used in simulating thermal stratification of deepwater bodies or when there may be layers of salt and freshwater at different depths in estuaries. Depth-averaged models are useful when the river is broad and shallow such that stratification is limited, but dispersion across the width of the river is slow. Three-dimensional (3D) models account for the water flows and solute transport in all directions. These models are highly sophisticated, and 3D water quality models are usually reserved for large (i.e., deep and wide) estuaries where the mixing patterns are complex (Fig. 3.9).

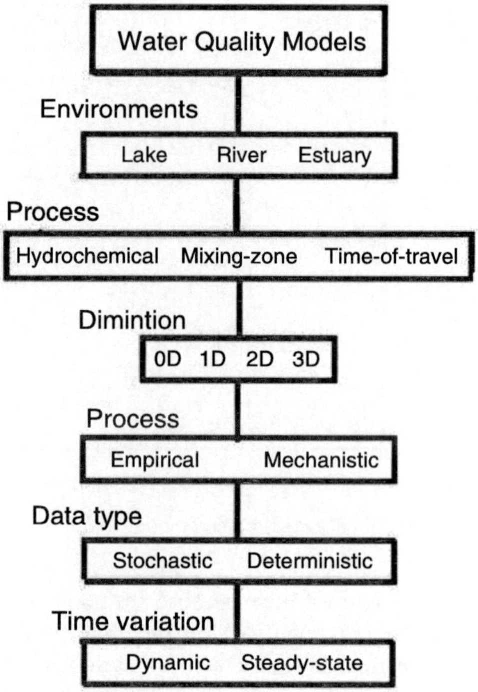

Fig. 3.9 Subdivisions of water quality models in common use

3 Oxygen Consumption and Replenishment in Receiving Water

Dissolved oxygen analysis measures the amount of gaseous oxygen (O_2) dissolved in an aqueous solution. Dissolved oxygen is one of the most important parameters in aquatic systems. This gas is an absolute requirement for the metabolism of aerobic organisms and also influences inorganic chemical reactions. Therefore, knowledge of the solubility and dynamics of oxygen distribution is essential to interpreting both biological and chemical processes within water bodies. Oxygen gets into water by diffusion from the surrounding air, by aeration (rapid movement) and as a waste product of photosynthesis. Adequate dissolved oxygen is necessary for good water quality. Oxygen is a necessary element to all forms of life. Natural stream purification processes require adequate oxygen levels in order to provide for aerobic life forms. Moreover, oxygen affects a vast number of other water indicators, not only biochemical but esthetic ones, like odor, clarity, and taste. Consequently, oxygen is perhaps the most well-established indicator of water quality. Some pollutants, such as acid mine drainage, produce direct chemical demands on oxygen in the water. Dissolved oxygen is consumed in the oxidation-reduction reactions of introduced chemical compounds, such as nitrate (NO_3^-) and ammonia (NH_4^+), sulfate (SO_4^{2-}) and iron ions. In this section, these factors are discussed.

3.1 *Effect of Physical Characteristics of Water on Dissolved Oxygen*

Oxygen is typically a limiting factor in aquatic ecosystems. Dissolved oxygen (DO) concentrations are expressed as milligrams of oxygen per liter of water (mg/L). DO also regulates the availability of certain nutrients in the water. Many physical factors affect the amount of dissolved oxygen in a stream. The physical factors that influence DO are temperature, altitude, salinity, and stream structure. Temperature inversely controls the solubility of oxygen in water; as temperature increases, oxygen is less soluble. In contrast, there is a direct relationship between atmospheric pressure and DO; as the pressure increases due to weather or elevation changes, oxygen solubility increases. Salinity also reduces the solubility of oxygen in water. Table 3.13 presents the relationship between amounts of dissolved oxygen in the water and temperature and salinity of water. Cold water can hold more dissolved oxygen than warm water and fresh water can hold more dissolved oxygen than salt water. So the warmer and saltier the water, the less dissolved oxygen there can be. The maximum amount of dissolved oxygen that the water can hold is called

Table 3.13 Maximum dissolved oxygen concentrates vary with temperature [14]

Temperature (°C)	DO (mg/l)	Temperature (°C)	DO (mg/l)
0	14.60	23	8.56
1	14.19	24	8.40
2	13.81	25	8.24
3	13.44	26	8.09
4	13.09	27	7.95
5	12.75	28	7.81
6	12.43	29	7.67
7	12.12	30	7.54
8	11.83	31	7.41
9	11.55	32	7.28
10	11.27	33	7.16
11	11.01	34	7.16
12	10.76	35	6.93
13	10.52	36	6.82
14	10.29	37	6.71
15	10.07	38	6.61
16	9.85	39	6.51
17	9.65	40	6.41
18	9.45	41	6.41
19	9.26	42	6.22
20	9.07	43	6.13
21	8.90	44	6.04
22	8.72	45	5.95

the saturation value. Dissolved oxygen measurements are given as a percent of saturation (%) or in units of milligrams per liter (mg/l) [22]

The oxygen content of natural waters can vary widely depending on the physical, chemical, and biological processes in the water body as already indicated. Running waters generally have higher oxygen content than stagnant waters. As explained, the major source of oxygen in natural waters is the atmospheric air. Oxygen enters water by the process of diffusion and higher the mixing of air and water due to surface-water agitation, such as by wave action and turbulence in running water or due to artificial process of aeration by compressed air diffusers and use of agitators, in aquaria and ponds, the higher the oxygen saturation. Apart from diffusion from the air, water bodies gain oxygen through photosynthesis of chlorophyll-bearing aquatic organisms, i.e., higher plants, phytoplankton, and photosynthetic bacteria. Oxygen is lost from the water body by the respiration of animals and plants and also by decomposition of organic matter. Excess oxygen will also be released from water to air by diffusion.

The four commonly recognized profiles described are indicated in Fig. 3.10. Stratification of lakes with reference to temperature and physico-chemical conditions, though mainly recognized in temperate conditions, are also seen in tropical water bodies in a modified form; these are of special importance for aquaculturists working to stock open waters, which is also part of the present course.

3.2 Biochemical Decay of Organic Waste Materials

Natural organic detritus and organic waste from waste water treatment plants, failing septic systems, and agricultural and urban runoff, act as food sources for water-borne bacteria. Bacteria decompose these organic materials using dissolved oxygen. Biochemical oxygen demand (BOD) is a measure of the amount of oxygen that bacteria will consume while decomposing organic matter under aerobic conditions. The main focus of wastewater treatment plants is to reduce the BOD in the effluent discharged to natural waters. Wastewater treatment plants are designed to function as bacteria farms, where bacteria are fed oxygen and organic waste [23].

The BOD test can be used to investigate the controls on consumption of dissolved oxygen in the water column and to derive decay rates for water quality models.

$$\mathrm{BOD}_t = \mathrm{BOD}_u \left(1 - e^{-kt}\right) \tag{3.16}$$

Where, BOD_t is the exerted biochemical oxygen demand (mg/L of O_2) at time t, BOD_u is the ultimate BOD (mg/L of O_2), k is the first-order decay rate (1/day), and t is the time (days).

The Streeter-Phelps equation is used in water pollution as a water quality modeling tool. The model describes how dissolved oxygen (DO) decreases in a river or stream along a certain distance by degradation of biochemical oxygen demand

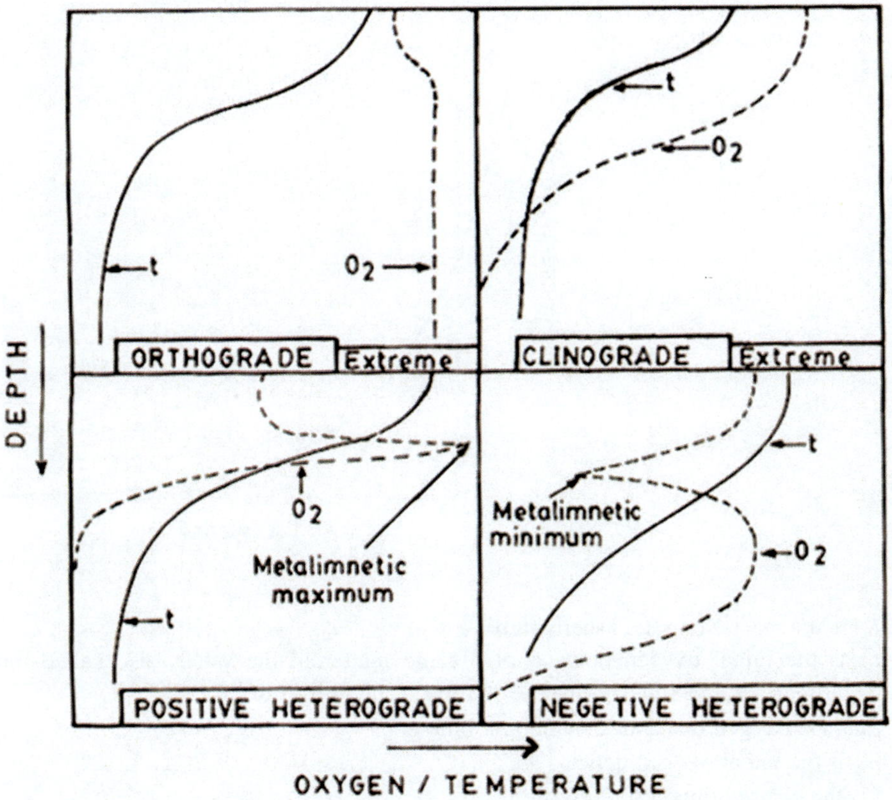

Fig. 3.10 Types of vertical-oxygen profiles in lakes typically of a stratified condition [15]

(BOD). The equation was derived by Streeter and Phelps in 1925, based on field data from the Ohio River. The equation is also known as the DO sag equation. It determines the relation between the dissolved oxygen concentration and the biological oxygen demand over time and is a solution to the linear first order differential equation [24]. This differential equation states that the total change in oxygen deficit (*D*) is equal to the difference between the two rates of deoxygenating and reaeration at any time (Fig. 3.11).

$$D = \frac{K_1 L_1}{K_2 - L_2} \left(e^{-k1t} - e^{-k2t}\right) + D_a e^{-k2t} \tag{3.17}$$

Where

D is the saturation deficit, which can be derived from the dissolved oxygen concentration at saturation minus the actual dissolved oxygen concentration ($D = DO_{\text{sat}} - DO$). *D* has the dimensions $\left[\frac{g}{m^3}\right]$.

K_1 is the deoxygenation rate, usually in d^{-1}.

Fig. 3.11 Streeter-Phelps DO sag curve and BOD development

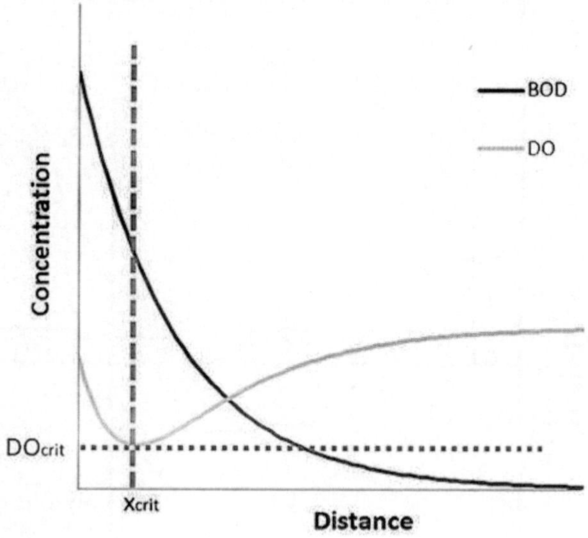

K_2 is the reaeration rate, usually in d^{-1}.
L_a is the initial oxygen demand of organic matter in the water, also called the ultimate BOD (BOD at time t = infinity). The unit of is $\left[\frac{g}{m^3}\right]$.
L_t is the oxygen demand remaining at time t.
D_a is the initial oxygen deficit $\left[\frac{g}{m^3}\right]$.
t is the elapsed time, usually [d].

The relationship for the change in oxygen concentration due to oxidation of organics should be developed first. The rate that oxygen is used will be proportional to the rate that substrate (or biochemical oxygen demand) is oxidized. The rate of substrate utilization by bacteria is given by the Monod relationship

$$\frac{dL}{dt} = \frac{-kLX}{K_s + L} \tag{3.18}$$

where L is substrate concentration expressed as oxygen demand or BODL [mg/L], k is the maximum specific substrate utilization rate, K_s is the half velocity constant, and X is the concentration of bacteria. However, the concentration of bacteria is a function of the substrate concentration and thus application of the Monod equation to a polluted river is not trivial. Often the bacterial concentration remains relatively constant. If the half velocity concentration is large relative to the concentration of substrate we obtain

$$\frac{dL}{dt} = \frac{-kXL}{K_s + L} \cong \left[\frac{-kX}{K_s}\right] L \cong -k_{\text{ox}} L \tag{3.19}$$

where k_{ox} is a first-order oxidation rate constant that includes both the approximation that the bacteria concentration is roughly constant and that the substrate concentration is smaller than the half velocity constant.

Separate variables and integrate

$$\int_{L_o}^{L} \frac{dL}{L} = \int_{0}^{t} (-k_{\text{ox}}) dt \tag{3.20}$$

$$L = L_o e^{-k_{\text{ox}} t} \tag{3.21}$$

The rate of oxygen utilization is equal to the rate of substrate utilization (when measured as oxygen demand) and thus we have

$$\frac{\partial C_{\text{oxidation}}}{\partial t} = \frac{dL}{dt} = -k_{\text{ox}} L \tag{3.22}$$

where C is the dissolved oxygen concentration [mg/L]. Now we can substitute for L in Eq. 3.22 using Eq. 3.21 to obtain.

3.3 Reaeration Portion of the DO Mass Balance

The driving force for reaeration is the dissolved oxygen deficit, D (mgO_2/L), a representation of how 'hungry' the water is for oxygen. The deficit is defined as the departure from saturation or the difference between the maximum amount of oxygen that the water can hold and the actual oxygen concentration,

$$D = DO_{\text{sat}} - DO_{\text{act}} \tag{3.23}$$

The maximum amount of oxygen that the water can hold is termed the saturation concentration and varies with temperature according to Henry's Law,

$$DO_{\text{sat}} = K_{\text{H}} \cdot P_{O_2} \tag{3.24}$$

The value for Henry's Law constant for oxygen (K_{H}) decreases as temperature increases and thus the value of DO_{sat} is lower at higher temperatures.

3.4 Oxygen Percent Saturation and Temperature

Oxygen saturation levels indicate how much biological processes have affected the water recently. Community respiration, mainly by bacteria, reduces oxygen concentrations. The high water temperature and decomposable organic matter and other bacterial substrates in the water are the main factors for dissolved oxygen consumption and reduction. Phytoplankton photosynthesis, plus some contribution from photosynthesis of submersed plants and benthic algae around the shallow edges of a lake or pond, can increase oxygen concentrations above saturated levels, but only during the day when photosynthesis is occurring. Oxygen percent saturation compares an observed oxygen concentration to the absolute solubility of oxygen at a particular water temperature. This index often takes into account barometric pressure and salinity effects at the measurement site, but usually ignores effects of water pressure at depths below a lake or stream surface. Solubility of gases increases by an amount equal to the surface saturation concentration for about every 10 m increase in depth.

$$\% = ([O_2]/[Osat]) \times 100 \tag{3.25}$$

where

$\%$ = percent saturation
$[O_2]$ = observed oxygen concentration, and
$[Osat]$ = saturated concentration of oxygen at the local temperature (and possibly altitude, barometric pressure, and salinity or conductivity).

Reaeration is a first-order reaction based on the reaeration rate coefficient, K_a (d^{-1}), and a driving force as represented by the dissolved oxygen deficit,

$$\frac{dO_2}{dt} = k_a \cdot D \tag{3.26}$$

The magnitude of the reaeration rate coefficient varies with temperature (increases with increasing temperature) and the turbulence of the stream (increases with increasing turbulence). Consider the effects of temperature and turbulence on the DO sag curve.

3.5 Quantifying the DO Mass Balance

All of the inputs to the DO mass balance can be quantified following:

$$\frac{dO_2}{dt} = k_a \cdot D - k_L \cdot L \tag{3.27}$$

In practice, the mass balance is written in terms of *D* because it is really *D* not DO that drives the reaeration term,

$$\frac{dD}{dt} = k_L \cdot L - k_a \cdot D \tag{3.28}$$

In order to apply this equation in modeling the oxygen resources of a river, the equation must be integrated,

$$D_t = \frac{k_1 \cdot L_0}{(k_2 - k_1)} \cdot \left(e^{-k_1 \cdot t} - e^{-k_2 \cdot t}\right) + D_0 \cdot e^{-k_2 \cdot t} \tag{3.29}$$

3.6 Nitrification

Nitrification is a microbial process that involves the transformation of ammonia to nitrate in the presence of organic nitrogen and ammonia in surface water. Nitrification is a critical step in the biochemical nitrogen (N) cycle. This process is very significant for water quality of fluvial ecosystems (mainly polluted ones) and the organisms occupying them, because it is an integral part of the biological mechanism of river self-purification and takes part in the nitrogen cycle [25]. Nitrification can have adverse impacts of increasing nitrite and nitrate levels, reducing alkalinity, pH, dissolved oxygen, and chloramine residuals, and promoting bacterial regrowth [26]. Summary of water quality problems associated with nitrification were presented in Table 3.14. When unbalanced by anthropogenic activities, rapid nitrification overwhelms denitrification in the N cycle, leading to the accumulation of nitrate and resulting in the contamination of ground waters and eutrophication of lakes. The oxidation of ammonia into nitrite is performed by two groups of organisms, ammonia-oxidizing bacteria, *Nitrosomonas*, which convert ammonia to nitrite (NO_2), and *Nitrobacter*, which convert nitrite to nitrate (NO_3^-) [27–30].

$$2\,NH_3 + 3\,O_2 \rightarrow 2\,NO_2^- + 2\,H_2O + 2\,H^+ \text{ (Nitrosomonas)} \tag{3.30}$$

$$2\,NO_2^- + 1\,O_2 \rightarrow 2\,NO_3^- \text{ (Nitrobacter, Nitrospina)} \tag{3.31}$$

$$NH_3 + O_2 \rightarrow NO_2^- + 3H^+ + 2e^- \tag{3.32}$$

$$NO_2^- + H_2O \rightarrow NO_3^- + 2H^+ + 2e^- \tag{3.33}$$

Nitrifying bacteria are very sensitive to pH (Fig. 3.12). *Nitrosomonas* has an optimal pH between approximately 7.0 and 8.0, and the optimum pH range for

Table 3.14 Summary of water quality problems associated with nitrification

Chemical issues	Biological issues
Disinfectant depletion	HPC increase
Nitrite/nitrate formation	Ammonia oxidizing bacteria (AOB) increase
Dissolved oxygen depletion	Nitrite oxidizing bacteria (NOB) increase
Reduction in pH and alkalinity	
DBP formation due to mitigation techniques	

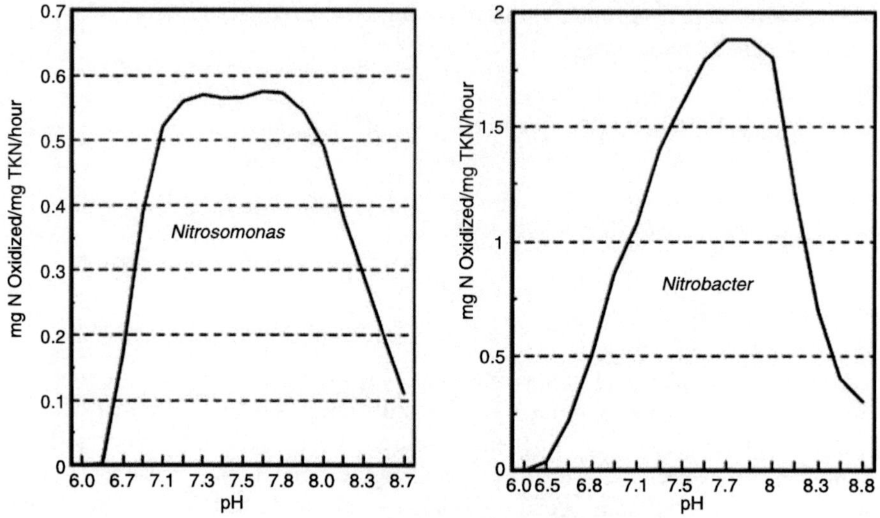

Fig. 3.12 Effects of pH on *Nitrosomonas* and *Nitrobacter* enrichment cultures [33]

Nitrobacter is approximately 7.5 to 8.0. Some utilities have reported that an increase in pH (to greater than 9)—prepared by American Water Works Association (AWWA) with assistance from Economic and Engineering Services, Inc.—10 can be used to reduce the occurrence of nitrification [31]. However, many other factors contribute to the viability of nitrifying bacteria and, as a result, nitrification episodes have been observed at pH levels ranging from 6.6 to 9.7 [32].

4 Microbiological Water Quality

Microbial water quality is regularly monitored and used to identify changes in a watershed that are useful for regulatory compliance and future watershed planning. Monitoring can also identify locations where there are problems related to land use practices. Surface water quality affects human and animal health and microbial water quality is especially important. The evaluation of microbial water quality for a

Table 3.15 Categories of water quality based on the River Bacterial Index formula [35]

96–100	Excellent – guidelines almost always met; best quality
81–95	Good – guidelines occasionally exceeded, but usually by small amounts; threat to quality is minimal
66–80	Fair – guidelines sometimes exceeded by moderate amounts; quality occasionally departs from desirable levels
46–65	Marginal – guidelines often exceeded, sometimes by large amounts; quality is threatened, often departing from desirable levels
0–45	Poor – guidelines almost always exceeded by large amounts; quality is impaired and well below desirable levels; worst quality

specific water body is typically based on the major uses for that source. Regions of extensive agriculture and sewage outfalls have the potential to introduce pathogens into surface waters which must be removed by water treatment before direct consumption. Bacterial water quality is an indication of the potential presence of various organisms such as *Escherichia coli, Giardia*, and *Cryptosporidium*. Fecal contamination by *E. coli* is widely used as an indicator of related pathogenic organisms such as *Salmonella, Shigella, Campylobacter, and Yersinia* [34]. *E. coli* occurs naturally in human and animal bodies, but some strains contain toxic cell wall components, some of which can be fatal as shown by the Walkerton tragedy. The guidelines for water quality stipulated by Alberta Environment indicate acceptable levels of indicator organisms that can be present to minimize the possibility of waterborne disease [35]. Livestock can be important sources of these microorganisms, especially during periods of high runoff. Agricultural inputs are generally considered to be non-point sources because of their wide geographic distribution and are difficult to control. Point sources, such as sewage outfalls, are much easier to regulate and treat. Based on Alberta formula for surface water quality guidelines [35], the river index bacteria formula was used, it includes the number of times bacterial density in these samples exceeds a guideline (frequency) and the extent to which the guideline is exceeded (amplitude). The result is a number between 0 and 100, where 100 represents the excellent guality of the river water and zero represents the poor quality. Table 3.15 presents the five categories of water quality based on the River Bacterial Index formula.

4.1 Indicator of Microbiological Water Quality

Indicator bacteria are types of bacteria used to detect and estimate the level of fecal contamination of water.

Indicator bacteria are not themselves dangerous to health but are used to indicate the presence of a health risk. These bacteria may include species of pathogenic bacteria, such as *Salmonella* or *Campylobacter*, associated with gastroenteritis. In addition, feces may contain pathogenic viruses, protozoa, and parasites. Fecal

Table 3.16 Definitions for indicator and index micro-organisms of public health concern [36]

Group	Definition
Process indicator	A group of organisms that demonstrates the efficacy of a process, such as total heterotrophic bacteria or total coliforms for chlorine disinfection.
Fecal indicator	A group of organisms that indicates the presence of fecal contamination, such as the bacterial group's thermotolerant coliforms or *E. coli*. Hence, they only infer that pathogens may be present.
Index and model organisms	A group/or species indicative of pathogen presence and behavior, respectively, such as *E. coli* as an index for *Salmonella* and F-RNA coliphages as models of human enteric viruses.

material can enter the environment from many sources including waste water treatment plants, livestock or poultry manure, sanitary landfills, septic systems, sewage sludge, pets, and wildlife. If sufficient quantities are ingested, fecal pathogens can cause disease. Table 3.16 provides definitions of indicators of microorganisms [36]. The variety and often low concentrations of pathogens in environmental waters makes them difficult to test for individually. Public agencies therefore use the presence of other more abundant and more easily detected fecal bacteria as indicators of the presence of fecal contamination. Table 3.17 presents Summary of Water Quality Criteria for Microbiological Indicators [37].

4.2 *Monitoring of Microbiological Water Quality*

Fecal pollution of surface water can lead to health problems because of the presence of infectious microorganisms. These may be derived from human sewage or animal sources (Tables 3.18 and 3.19).

Sanitary inspection and assessment of microbiological water quality are very important aspects in the monitoring of surface water bodies. Kostyla et al. [38] reviewed 22 studies in developing countries that used fecal contamination as an indicator for surface water quality. Figure 3.13 shows classification criteria for microbiological water quality. This classification is very important to grade the recreational areas and provide on-site safety guidance and identification and promotion of effective management interventions.

Microbial contamination of surface water quality may be strongly influenced by factors such as rainfall leading to relatively short periods of elevated fecal pollution. Classification of recreational water quality based on sanitary inspection and microbial quality assessment with prevention of exposure can be designed as shown in Fig. 3.14. Users can be shown to be effectively discouraged from entering the water following occasional and predictable water quality deteriorations (linked to, for example, rainfall), the area may be upgraded to reflect the water quality that users are exposed to, but only with the accompanying explanatory material.

Table 3.17 Summary of water quality criteria for microbiological indicators [37]

Water use	*Escherichia coli*	Enterococci	*Pseudomonas aeruginosa*	Fecal coliforms
Raw drinking water No treatment	0/100 mL	0/100 mL	0/100 mL	0/100 mL
Raw drinking water Disinfection only	Less than or equal to 10/100 mL 90th percentile	Less than or equal to 3/100 mL 90th percentile	None applicable	Less than or equal to 10/100 mL 90th percentile
Raw drinking water Partial treatment	Less than or equal to 100/100 mL 90th percentile	Less than or equal to 25/100 mL 90th percentile	None applicable	Less than or equal to 100/100 mL 90th percentile
Raw drinking water Complete treatment	None applicable	None applicable	None applicable	None applicable
Aquatic life Shellfish harvesting	Less than or equal to 43/100 mL 90th percentile	Less than or equal to 11/100 mL 90th percentile	None applicable	Less than or equal to 43/100 mL 90th percentile
Aquatic life Shellfish harvesting	Less than or equal to 14/100 mL median	Less than or equal to 4/100 mL median	None applicable	Less than or equal to 14/100 mL median
Wildlife	None applicable	None applicable	None applicable	None applicable
Livestock Free range animals	None applicable	None applicable	None applicable	None applicable
Livestock General livestock use	200/100 mL maximum	50/100 mL maximum	None applicable	200/100 mL maximum
Livestock Closely confined (no treatment)	0/100 mL maximum	0/100 mL maximum	None applicable	0/100 mL maximum
Livestock Closely confined (disinfection only)	Less than or equal to 10/100 mL 90th percentile	Less than or equal to 3/100 mL 90th percentile	None applicable	Less than or equal to 10/100 mL 90th percentile
Livestock Closely confined (partial treatment)	Less than or equal to 100/100 mL 90th percentile	Less than or equal to 25/100 mL 90th percentile	None applicable	Less than or equal to 100/100 mL 90th percentile
Livestock Closely confined (complete treatment)	None applicable	None applicable	None applicable	None applicable

(continued)

Table 3.17 (continued)

Water use	*Escherichia coli*	Enterococci	*Pseudomonas aeruginosa*	Fecal coliforms
Irrigation Crops eaten raw	Less than or equal to 77/100 mL geometric mean	Less than or equal to 20/100 mL geometric mean	None applicable	Less than or equal to 200/100 mL geometric mean
Irrigation Public access Livestock access	Less than or equal to 385/100 mL geometric mean	Less than or equal to 100/100 mL geometric mean	Less than or equal to 10/100 mL 75th percentile	None applicable
Irrigation General irrigation	Less than or equal to 1000/100 mL geometric mean	Less than or equal to 250/100 mL geometric mean	None applicable	Less than or equal to 1000/100 mL geometric mean
Recreation Aesthetics Non-contact	None applicable	None applicable	None applicable	None applicable
Recreation Secondary contact Crustacean harvesting	Less than or equal to 385/100 mL geometric mean	Less than or equal to 100/100 mL geometric mean	Less than or equal to 10/100 mL 75th percentile	None applicable
Recreation Primary contact	Less than or equal to 77/100 mL geometric mean	Less than or equal to 20/100 mL geometric mean	Less than or equal to 2/100 mL 75th percentile	Less than or equal to 200/100 mL geometric mean
Industrial water (dairy, food processing) No treatment	0/100 mL	0/100 mL	None applicable	0/100 mL
Industrial water (dairy, food processing) Disinfection only	Less than or equal to 10/100 mL 90th percentile	Less than or equal to 3/100 mL 90th percentile	None applicable	Less than or equal to 10/100 mL 90th percentile
Industrial water (dairy, food processing) Partial treatment	Less than or equal to 100/100 mL 90th percentile	Less than or equal to 25/100 mL 90th percentile	None applicable	Less than or equal to 100/100 mL 90th percentile
Industrial water (dairy, food processing) Complete treatment	None applicable	None applicable	None applicable	None applicable
Industrial water Other industries	Less than or equal to 385/100 mL geometric mean	Less than or equal to 100/100 mL geometric mean	Less than or equal to 10/100 mL 75th percentile	None applicable

Table 3.18 Examples of Pathogens and index organism concentrations in raw sewage [38]

Pathogen/index organism	Disease/role	Numbers per 100 ml
Bacteria		
Campylobacter spp.	Gastroenteritis	10^4–10^5
Clostridium perfringens spores	Index organism	6×10^4–8×10^4
Escherichia coli	Index organism (except specific strains)	106–107
Fecal streptococci/intestinal enterococci	Index organism	4.7×10^3–4×10^5
Salmonella spp.	Gastroenteritis	0.2–8000
Shigella spp.	Bacillary dysentery	0.1–1000
Viruses		
Polioviruses	Index organism (vaccine strains), Poliomyelitis	180–500,000
Rotaviruses	Diarrhea, vomiting	400–85,000
Adenoviruses	Respiratory disease, gastroenteritis	Not enumerated
Norwalk viruses	Diarrhea, vomiting	Not enumerated
Hepatitis A	Hepatitis	Not enumerated
Parasitic protozoa		
Cryptosporidium parvum oocysts	Diarrhea	0.1–39
Entamoeba histolytica	Amoebic dysentery	0.4
Giardia lamblia cysts	Diarrhea	12.5–20,000
Helminths (ova)		
Ascaris spp.	Ascariasis	0.5–11
Ancylostoma spp. and Necator sp.	Anemia	0.6–19
Trichuris spp.	Diarrhea	1–4

4.3 Fecal Coliform Bacterial Die-Off in Urban Watershed

Fecal coliform bacteria are microscopic organisms that live in the intestines of warm-blooded animals. They also live in the waste material or feces excreted from the intestinal tract. When fecal coliform bacteria are present in high numbers in a water sample, it means that the water may have received fecal matter from one source or another. Unlike the other conventional water quality parameters, fecal coliform bacteria are living organisms. They multiply quickly when conditions are favorable for growth and die in large numbers when they are not. Because bacterial concentrations are dependent upon specific conditions for growth and these conditions change quickly, fecal coliform bacteria counts are not easy to predict. For example, although winter rains may wash more fecal matter from urban areas into a lake, cool water temperatures may cause many of the organisms to die. Direct exposure to sunlight is also lethal to bacteria, so die-off may be high even in the warmer water of summer time. Based on bacterial die-off modeled (Fig. 3.15), about 90% of bacteria will disappear from fresh water within two to five days [39], while the die-off rate is

Table 3.19 Microbial indicators (average numbers per gram wet weight) excreted in the feces of warm-blooded animals [37]

Group	Thermotolerant coliforms	Fecal streptococci	*Clostridium perfringens*	F-RNA coliphage	Excretion (g/day)
Farm animals					
Chicken	1,300,000	3,400,000	250	1867	182 (71.6)
Cow	230,000	1,300,000	200	84	23,600 (83.3)
Duck	33,000,000	54,000,000	–	13.1	336 (61.0)
Horse	12,600	6,300,000	<1	950	20,000
Pig	3,300,000	84,000,000	3980	4136	2700 (66.7)
Sheep	16,000,000	38,000,000	199,000	1.5	1130 (74.4)
Turkey	290,000	2,800,000	–	–	448 (62.0)
Domestic pets					
Cat	7,900,000	27,000,000	25,100,000		
Dog	23,000,000	980,000,000	251,000,000	2.1	413
Human	13,000,000	3,000,000	1580a	<1.0–6.25	150 (77.0)
Ratios in raw	50	5	0.3	1	–
Sewage					

Fig. 3.13 Sampled classification matrix of microbial water quality [38]

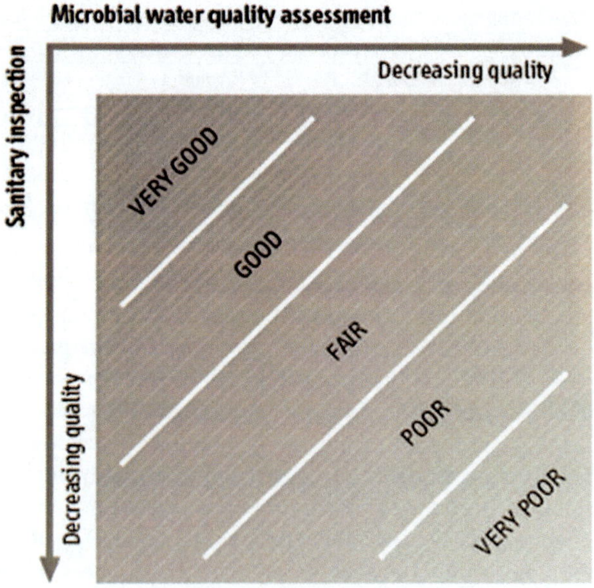

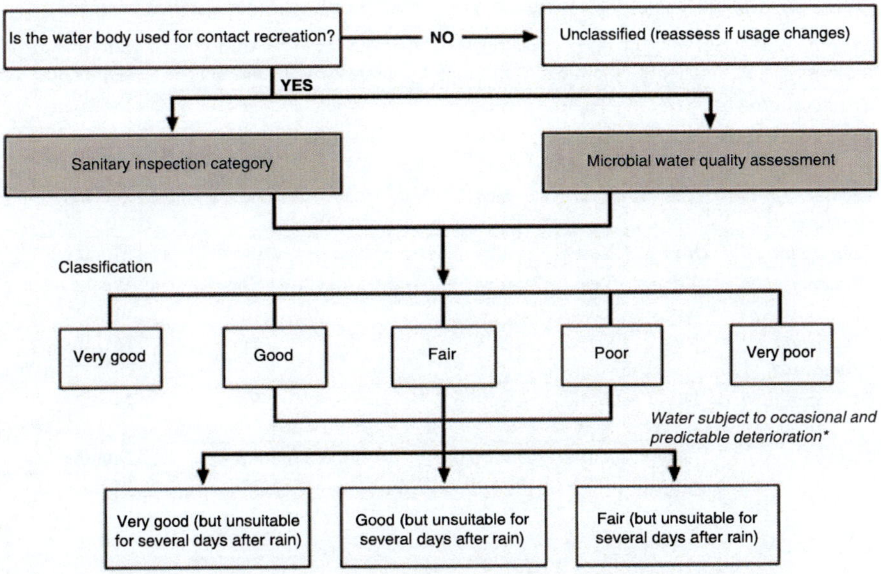

Fig. 3.14 Simplified framework for assessing surface (recreational) water environments

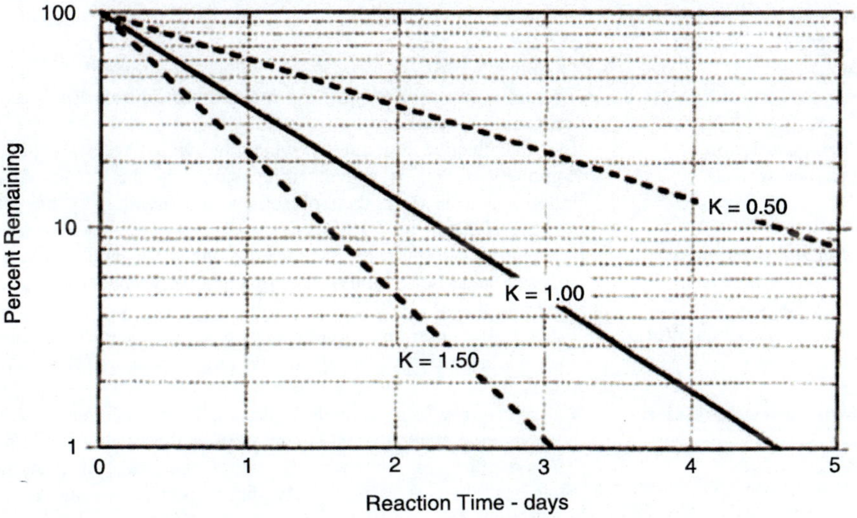

Fig. 3.15 Effect of different die-off rates (k) on bacteria mortality [39]

generally much faster in marine and estuarine waters than freshwater [40, 41]. Exposure of bacteria to sunlight is one of the most important factors causing bacteria die-off, while different treatment processes of surface water can significantly affect die-off of bacteria. Table 3.20 summarizes comparison of die-off rates and treatment effectiveness for different microbes.

Table 3.20 Comparison of die-off rates and treatment effectiveness for different microbes

Microbial indicator	Light?	Settling?	Surface filtration?	Die-off rates (k)	Ability to multiply	Survival in sediments?
Total coliforms	Yes	Yes	Yes	1/day	Yes	Moderate
Fecal coliforms	Yes	Yes	Yes	0.7–1.0/ day	Yes	Days
Fecal streptococci	Yes	Yes	Yes	1/day	Low	Weeks
Escherichia coli	Yes	Yes	Yes	1/day	Low	Months
Salmonella spp.	Yes	Yes	Yes	1.5/day	Yes	Weeks to months
Psuedonomas aeruginosa	Yes	Partial	Yes	?	Yes	Months
Crytospoidium spp.	No	Partial	Partial	1.5/day	No	Months
Giardia spp.	No	Partial	Partial	1.5/day	No	Months

Glossary

American Water Works Association (AWWA) An international non-profit, scientific, and educational association founded to improve water quality and supply

Canadian Council of Ministers of the Environment (CCME) Is the primary minister-led intergovernmental forum for collective action on environmental issues of national and international concern

DO_{sat} Dissolved oxygen in water at saturation concentration

K_H The value for the Henry's Law constant for oxygen decreases as temperature increases

Normalized Sum of Excursions (*nse*) Is the collective amount by which individual tests are out of compliance. This is calculated by summing the excursions of individual tests from their objectives and dividing by the total number of tests

Principal component analysis/factor analysis (PCA/FA) are multivariate statistical methods that analyze several variables to reduce a large dimension of data to a relatively smaller number of dimensions.

Surface Water Quality Monitoring (SWQM) Program that monitors and evaluates physical, chemical, and biological characteristics of aquatic systems as a basis for effective policy.

Water framework directive (WFD) EU directive which commits European Union member states to achieve good qualitative and quantitative status of all water bodies

Water quality The condition of the water, including chemical, physical, and biological characteristics, usually with respect to its suitability for a particular purpose, such as drinking or swimming.

Water Quality Indices (WQIs) A tool that aggregates results of several types of physical, chemical, and biological measurements into a single indicator of water quality conditions for streams and lakes.

Exerted biochemical oxygen demand (mg/L of O_2) at time t (BOD_t) Is the increases of the amount of dissolved oxygen needed (i.e., demanded) by aerobic biological organisms to break down organic material present in a given water sample by a time

Ultimate BOD (mg/L of O_2) (BOD_u)	The total amount of oxygen consumed when the biochemical reaction is allowed to proceed to completion. First-order decay rate (1/day) (K): An equation relating the rate constant k to the initial concentration [A]0 and the concentration [A]t present after any given time t.

References

1. Naddeo V, Scannapieco D, Zarra T, Belgiorno V (2013) River water quality assessment: implementation of non-parametric tests for sampling frequency optimization. Land Use Policy 30(1):197–205
2. Boyacıoğlu H, Gündogdu V, Boyacıoğlu H (2013) Investigation of priorities in water quality management based on correlations and variations. Mar Pollut Bull 69(1–2):48–54
3. Wang Y, Wang P, Bai Y, Tian Z, Li J, Shao X, Mustavich LF, Li B (2013) Assessment of surface water quality via multivariate statistical techniques: a case study of the Songhua River Harbin region, China. J Hydro Environ Res 7(1):30–40
4. Selle B, Schwientek M, Lischeid G (2013) Understanding processes governing water quality in catchments using principal component scores. J Hydrol 486:31–38
5. Wan-Young Chung W, Yoo J (2015) Remote water quality monitoring in wide area. Sensors Actuators B Chem B217:51–57
6. Hatvani IG, Kovács J, Márkus L, Clement A, Hoffmann R, Korponai J (2015) Assessing the relationship of background factors governing the water quality of an agricultural watershed with changes in catchment property (W-Hungary). J Hydrol 521:460–469
7. Chen K, Chen H, Zhou C, Huang Y, Qi X, Shen R, Liu F, Zuo M, Zou X, Wang J, Zhang Y, Chen D, Chen X, Deng Y, Ren H (2020) Comparative analysis of surface water quality prediction performance and identification of key water parameters using different machine learning models based on big data. Water Res 171(15):115454
8. Busico G, Kazakis N, Cuoco E, Colombani N, Tedesco D, Voudouris K, Mastrocicco M (2020) A novel hybrid method of specific vulnerability to anthropogenic pollution using multivariate statistical and regression analyses. Water Res 171(15):115386
9. Jahin HS, Abuzaid AS, Abdellatif AD (2020) Using multivariate analysis to develop irrigation water quality index for surface water in Kafr El-Sheikh governorate, Egypt. Environ Technol Innov 17:100532
10. Weerasinghe VPA, Handapangoda K (2019) Surface water quality analysis of an urban lake, East Beira, Colombo, Sri Lanka. Environ Nanotechnol Monitor Manag 12:100249
11. Carstens D, Amer R (2019) Spatio-temporal analysis of urban changes and surface water quality. J Hydrol 569:720–734
12. Khan MT, Manes CO, Aubry C, Croué J (2013) Source water quality shaping different fouling scenarios in a full-scale desalination plant at the Red Sea. Water Res 47(2):558–568
13. Won G, Kline TR, LeJeune JT (2013) Spatial-temporal variations of microbial water quality in Marta used for irrigation. Agric Water Manag 116:73–78
14. Texas commotion on environmental quality, surface water quality monitoring (2013) Assessed on 22 of Jun 2013 from http://www.tceq.texas.gov/waterquality/monitoring
15. Colorado Department of Public Health and Environment, Water Quality Control Commission (2013) 5 ccr 1002-32 regulation no. 32 classifications and numeric standards for Arkansas river basin, http://www.colorado.gov/cs/Satellite/CDPHE-Main/CBON/1251595703337
16. Environmental Agency (2012) Method statement for the classification of surface water bodies, Water framework directive, Monitoring Strategy, v3
17. Terrado M, Borrell E, de Campos S, Barceló D, Tauler R (2010) Surface reservoirs and canals surface-water-quality indices for the analysis of data generated by automated sampling networks. Trends Anal Chem 29(1):40–52

18. Ott WR (1978) Water quality indices: a survey of indices used in the United States, EPA-600/4-78-005. Monitoring Technology Division, Environmental Protection Agency, Washington, DC
19. Abbasi SA (2002) Water quality indices, state of the art report, INCOH/ SAR-25/2002. National Institute of Hydrology, Roorkee, Uttarakhand
20. Abbasi SA, Arya DS (2000) Environmental impact assessment. Discovery Publishing House, New Delhi
21. B.C. Ministry of Environment, Land and Parks (1996) British Columbia water quality status report. 181 pp
22. American Public Health Association, American Water Works Association, and Water Environment Federation (1998) Standard methods for the examination of water and wastewater. American Public Health Association, Washington, DC
23. Streeter HW, Phelps EB (1925) A study of the pollution and natural purification of the Ohio river. III. Factors concerned in the phenomena of oxidation and reaeration, public health bulletin no. 146, reprinted by U.S. Department of Health, Education and Welfare, Public Health Service, 1958, ISBN B001BP4GZI, http://dspace.udel.edu:8080/dspace/bitstream/handle/19716/1590/C%26EE148.pdf?sequence=2
24. Polak J (2004) Nitrification in the surface water of the Włocławek dam reservoir. The process contribution to biochemical oxygen demand (N-BOD). Pol J Environ Stud 13(4):415–424
25. Wilczak A, Jacangelo JG, Marcinko JP, Odell LH, Kirmeyer GJ, Wolfe RL (1996) Occurrence of nitrification in chloraminated distribution systems. J AWWA 88(7):74–85
26. Treusch AH, Leininger S, Kletzin A, Schuster SC, Klenk HP, Schleper C (2005) Novel genes for nitrite reductase and Amo-related proteins indicate a role of uncultivated mesophilic crenarchaeota in nitrogen cycling. Environ Microbiol 7:1985–1995
27. Purkhold U, Pommerening-Roser A, Juretschko S, Schmid MC, Koops H-P, Wagner M (2000) Phylogeny of all recognized species of ammonia oxidizers based on comparative 16S rRNA and amoA sequence analysis: implications for molecular diversity surveys. Appl Environ Microbiol 66:5368–5382
28. Könneke M, Bernhard AE, de la Torre JR, Walker CM, Waterbury JB, Stahl DA (2005) Isolation of a mesophilic autotrophic ammonia-oxidizing marine archaeon. Nature 437:543–546
29. Martens-Habbena W, Berube PM, Urakawa H, de la Torre JR, Stahl DA (2009) Ammonia oxidation kinetics determine niche separation of nitrifying archaea and Bacteria. Nature 461:976–981
30. Skadsen J, Sanford L (1996) The effectiveness of high pH for control of nitrification and the impact of ozone on nitrification control. In proc. 1996 AWWA water quality technology conference. AWWA, Boston
31. Odell LH, Kirmeyer GJ, Wilczak A, Jacangelo JG, Marcinko JP, Wolfe RL (1996) Controlling nitrification in Chloraminated systems. J AWWA 88(7):86–98
32. Grady CPL Jr, Lim HC (1980) Biological wastewater treatment. Marcel Dekker, New York
33. United States Environmental Protection Agency, Drinking Water Containments. Basic information about E. coli.0157:H7 in Drinking Water. 26 November 2006. http://www.epa.gov/safewater/contaminants/ecoli.html. 29 June 2013
34. Alberta Environment (2009) Alberta river quality index. http://environment.alberta.ca/1777.html. 29 June 2013
35. Ashbolt NJ, Grabow WOK, Snozzi M (2001) Indicators of microbial water quality, chapter in book, World Health Organization (WHO). Water Quality: Guidelines, Standards and Health, London edn. IWA Publishing
36. Warrington PD (2001) Water quality criteria for microbiological indicators, Overview Report, Resource Quality Section, Ministry of Environment and Parks
37. Fecal Pollution of Water Quality (2013) Chapter in book, Guidelines for Safe Recreational Water Environments, World Health Organization (WHO), http://wwwwhoint/water_sanitation_health/bathing/srwe1-chap4pdf, 30 June 2013

38. Kostyla C, Bain R, Cronk R, Bartram J (2015) Seasonal variation of fecal contamination in drinking water sources in developing countries: a systematic review. Sci Total Environ 514:333–343
39. Hydroqual I (1996) Design criteria report, Kensico watershed Stormwater best management facilities, appendix C. Dept. of Environmental Protection, New York City, USA, 240 pp
40. Mancini J (1978) Numerical estimates of coliform mortality rates under various conditions. J WPCF 50(11):2477–2484
41. Thoman R, Mueller P (1987) Principles of surface water modeling and control. Chapter 5. Harper Collins, New York, USA, 450 pp

Chapter 4
Treatment of Electrical and Electronic Component Manufacturing Wastes

Omotayo Sarafadeen Amuda, Yung-Tse Hung, and Lawrence K. Wang

Contents

1 Introduction 117
1.1 Overview of Electrical and Electronic Component Industry 117
1.2 General Classification 119
2 Subcategory Classification 119
2.1 Dry Products Subcategory 119
2.2 Dielectric Materials Subcategory 131
2.3 Electric Lamp Subcategory 135
2.4 Electron Tube Subcategory 141
2.5 Semiconductor Subcategory 153
2.6 Capacitor Subcategory 160
2.7 Printed Wiring Boards 163
3 Electrical and Electronic Industry and Waste Generation 171
3.1 Waste Generation 171
3.2 Waste Generation Prevention 178
4 Electrical and Electronic Component Industry Point Source Discharge Effluent Limitations, Performance Standards, and Pretreatment Standards 178
4.1 US Environmental Regulations for Subcategory of Luminescent Materials 186
4.2 Cathode Ray Tube Subcategory 187
4.3 Semiconductor Subcategory 187
4.4 Electronic Crystal Subcategory 189
5 Pollution Prevention Case Studies 189
5.1 Texas Instruments 190
5.2 Tri-Star Technologies 192
6 Summary 194
References 195

O. S. Amuda (✉)
Analytical/Environmental Chemistry Unit, Department of Pure and Applied Chemistry, Ladoke Akintola University of Technology, Ogbomoso, Nigeria
e-mail: osamuda@lautech.edu.ng

Y.-T. Hung
Civil and Environmental Engineering, Cleveland State University, Cleveland, OH, USA

L. K. Wang
Lenox Institute of Water Technology, Newtonville, NY, USA

L. K. Wang, M. -H. S. Wang, Y. -T. Hung, N. K. Shammas (eds.), *Integrated Natural Resources Management*, Handbook of Environmental Engineering 20,
https://doi.org/10.1007/978-3-030-55172-8_4

Abstract The electrical and electronic components industry is generally classified as standard industrial classification (SIC) with the code 36. The SIC is a Statistical Classification Standard used for all US-based establishments of Federal Economic Statistics which is under the Department of Commerce. Pollution prevention techniques and processes applicable to the electrical and electronic industry can be, broadly, divided into process or equipment modification, raw materials substitution or elimination, waste segregation (separation or preparation), and, finally, recycling. This chapter discusses industrial description and production processes, waste generation and pollution prevention, water use in the industry, wastewater treatment processes, treatment technology, and cost. Furthermore, US regulations such as best applicable technology (BAT), best conventional pollution control technology (BCT), best practicable control treatment (BPT), point source discharge limitation, performance standards for new source (PSNS), and pretreatment standards for existing sources (PSES) for subcategories in electrical and electronic components manufacturing industry are also presented. A case study of Texas Instruments is also included.

Keywords Electrical/electronic industry · Semiconductor · Cathode ray tube · Luminescent material · Pollution generation · Pollution prevention · BAT · BPT · BCT · PSNS · NSPS

Nomenclature

BAT	Best available treatment
BCT	Best conventional pollution control technology
BPT	Best practicable pollution control technology
CFC	Chlorofluorocarbons
CRT	Cathode ray tube
DSP	Digital signal processing solutions
ICs	Integrated circuits
LCD	Liquid crystal display
LED	Light-emitting diode
NSPS	New source performance standards
PFC	Power factor correction
POTW	Public owned treatment works
PSES	Pretreatment standards for existing sources
PSNS	Pretreatment standards for new sources
PWB	Printed wiring board
SIC	Standard industrial classification
TCA	Trichloroethane
TRI	Toxic release inventory

1 Introduction

1.1 Overview of Electrical and Electronic Component Industry

The electronic industry came into being about 1883, a decade after Thomas Edison invented the light bulb. His preliminary work on vacuum tube was complimented by Lee De Forest, who carried out a detailed research on Edison's discovery and patented the work. The industry earlier assumed slow growth with the production of few components that have less practical applications in electronics. The advent of World War I (WW I) spurred the industry's pace of improvement. Discoveries like piezoelectricity, in 1808, were effectively used to produce underwater acoustic wave in submarine-detecting sonar during WW I, before being used as control devices in radios. This was commercialized after WW I for broadcasting and public usage between 1920s and 1930s. The industry expands in research and output during the period. The arrival of integrated circuits (ICs) and other advanced semiconductors shortly after the mid-twentieth century gave the industry a geometrical expansion, and so the demand for electrical components sowed high. Between 1960s and 1970s, introduction of some equipment like computers, electronic scientific apparatus, and those used in space and communication sectors opened a wider demand for more electronic components like rectifiers, piezoelectric devices, and host of others essential to those equipments. The industry's revenue also climbed along expansion such that toward the end of 1970s, it was more than eight times before WW I.

The introduction of personal computer and its accessories, telecommunications equipment, and electronics into the stream of industrial and consumer products, during the 1980s, led to a boost in electronic industry. This growth in research and production in the electrical and electronic industry soared geometrically through the 1990s, and till today it is now one of the largest and the fastest growing domains in the manufacturing industry [1], and this is expected to be on the increase annually through the mid-2000s [2]. Countries like Malaysia, Canada, Taiwan, Mexico, and Japan are top on the list in terms of importation of electrical components from the USA. Japan dominated the USA. Miscellaneous components imports with a staggering 53% share pioneer companies in the electrical and electronic world like the British Marconi Company, the National Electric Signaling Company of the US General Electric Company (formed by Edison), and the Westinghouse and the Radio Corporation of America. Today most of the high-tech companies combine computer with electronics and some of such include Hewlett-Packard Co, Toshiba Corp, NEC Corp, and others. Table 4.1 shows the top ten worldwide electronics/ computer industries surveyed in 1992.

By the year 2001, the profile of the leading electrical and electronic component industry is given in Table 4.2.

The electrical and electronic components' industries worldwide are valued at 1 trillon US dollars and grow on the average of 7–8% per year [1]. Worldwide

Table 4.1 Top ten worldwide electronics/computer industry companies

Company name	1992 electronic sales in millions of dollars
IBM	$53,600
Matsushita Electric	$48,668
Toshiba	$29,232
NEC	$28,375
Fujitsu	$25,879
Philips	$25,747
Hitachi	$25,107
Siemens	$24,550
Sony	$22,959
Alcatel-Alsthom	$20,892

Source: [20]

Table 4.2 2001 Profile of the leading electrical and electronic component industries

Company name	Location	No of employee	Revenue ($)
Hughes Electronics	El Segundo, California	13,700	8.9 billion
Northrop Grumman Corp	Baltimore	7,300	6.6 billion
Agere Systems Inc	Allentown, Pennsylvania	10,700	2.2 million

Source: [3]

Table 4.3 Geographic distribution of electronics/computer facilities

State	Percentage
California	34
Texas	6.5
Massachusetts	6.4
New York	4.5
Illinois	4.4
Pennsylvania	4.0

Source: [4]

workforce in the electrical and electronic component industry (including computer) is estimated to be four million, and the USA has the largest share with about two and half million [3]. The workforce of the electronic industry in most Asian nations like Japan, the Republic of Korea, Taiwan, Malaysia, and others is also experiencing rapid growth. The geographic distribution of the electrical and electronic industry facilities in the USA is uneven. Some have concentration of many facilities while some have large plants for a particular electrical and electronic component. The distribution of 60% of the electronics/computer industry facilities is located in six states of the USA (Table 4.3).

1.2 General Classification

The electrical and electronic component industry is generally classified as standard industrial classification with the code 36. The SIC is a Statistical Classification Standard used for all US-based establishments of Federal Economic Statistics which is under the Department of Commerce [4]. The wide range of electrical and electronic components are broadly sourced from the product areas which include carbon and graphite, electric and electronic components, electron tubes, fluorescent lamps, cathode and television tubes, capacities, incandescent lamps, insulators (mica, plastic, and laminates), semiconductors, resistance heaters, switchgears and fuses, and wet transformers.

To indicate main category and subcategory, the electrical and electronic components are assigned four digits under the SIC. This is summarized in Table 4.4, based on the number of subcategories and discharges.

The invention of computer and its accessories has led to overlapping in the classification of electrical and electronic components. The recent document addressed the industry as computer and electronic component industry [5]. However, the computer and electronic group is classified under the SIC with the code 357 and with the following subcategories (Table 4.5).

2 Subcategory Classification

The electrical and electronic component industry would be discussed, for simplification purpose, under seven subcategories. They are dry products – carbon and graphite products, electric materials, electric lamps, electric tubes, semiconductors, and capacitors [6]. These subcategories shall be discussed under the subheadings: description and production process, waste characterization, pollutant resolvability, and treatment.

2.1 Dry Products Subcategory

This subcategory is named dry products because it includes electrical and electronic components which are produced in plants which do not use manufacturing process water. These products include switchgears and fuses, resistance, heaters, ferrite electronic parts, motors and generators, fuel cells, alternators, insulated wire, and cables (nonferrous). In actual fact, accessories employed in this subcategory are products of metal finishing industries, and associated wastes are studied under the metal finishing industries.

Table 4.4 Standard industrial classification of electrical and electronic component industries

Codes	Component category
SIC 361	
SIC 3612[a]	Power, distribution, and specialty transformer
SIC 3613[a]	Switchgear and switchboard apparatus
SIC 362	Electrical industrial apparatus
SIC 3621[a]	Motors and generators
SIC 3622[a]	Industrial control
SIC 3623[a]	Electric welding apparatus
SIC 3624[a]	Carbon and graphite products
SIC 3625[b]	Relays and industrial controls
SIC 3629[a]	Electrical industrial apparatus (nec)
SIC 363[a]	Household appliances
SIC 3631[a]	Household cooking equipment
SIC 3632[a]	Household refrigerators (home and farm freezers)
SIC 3633[a]	Household laundry equipment
SIC 3634[a]	Electric housewares and fans
SIC 3635[a]	Household vacuum cleaners
SIC 3639[a]	Household appliances (nec)
SIC 364[b]	Electrical industrial lighting and wiring equipment
SIC 3641[b]	Electric lamp bulbs and tubes
SIC 3643[a]	Current-carrying wiring devices
SIC 3644[a]	Noncurrent-carrying wiring devices
SIC 3645[a]	Residential electric lighting fixtures
SIC 3646[a]	Commercial, industrial, and institutional electric lighting fixtures
SIC 3647[a]	Vehicular lighting equipment
SIC 3648[a]	Electrical industrial, lighting equipment (nec)
SIC 365	
SIC 3651[a]	Radio and television receiving sets, except communication types
SIC 3652[a]	Phonograph records and prerecorded magnetic tapes
SIC 366	Communication equipment
SIC 3661[a]	Telephone and telegraph apparatus
SIC 3663[b]	Radio and television broadcasting and communication equipment
SIC 3669	Communication equipment (nec)
SIC 367[b]	Electronic components and accessories
SIC 3671[b]	Electron tubes
SIC 3672[b]	Printed circuit boards
SIC 3673[a]	Transmitting, industrial, and special purpose electron tubes
SIC 3674[a]	Semiconductors and related devices
SIC 3675[a]	Electronic resistors
SIC 3676[a]	Electronic coils, transformers, and other inductors
SIC 3677[a]	Electronic connectors
SIC 3678[b]	Electronic components (nec)
SIC 3679	Radiographic X-ray, fluoroscopic X-ray, therapeutic X-ray, and other X-ray apparatus and tubes

(continued)

Table 4.4 (continued)

Codes	Component category
SIC 3693[a]	Electromedical and electrotherapeutic apparatus Electrical equipment for internal combustion engine
SIC 36941[a]	Electrical machinery, equipment, and supplies (nec)
SIC 3699[a]	

Source: [2]
nec not elsewhere classified
[a]Early classification
[b]As reclassified

Table 4.5 SIC code for computer and electronic components

SIC code	Component category
SIC 3571	Electronic computers
SIC 3572	Computer storage devices
SIC 3575	Computer terminals
SIC 3577	Other computer equipment
SIC 3578	Calculating and accounting machines except electronic computers
SIC 3579	Office machines (nec)

Source: [2]
nec not elsewhere classified

2.1.1 Carbon and Graphite Subcategory

Description and Production Process

The major products of this subcategory are electrodes and other carbon-and-graphite products besides electrodes. The electrodes are principal parts of electronic furnace as well as electrolytic cell which are used in metal industry for smelting, refining, and electroplating purposes. The electrodes are employed individually or combined depending on the industry and the desired products. Carbon electrodes are used singularly in the aluminum industries, while the graphite electrodes are used in the metallurgical industries as well as inorganic chemical industries for the production of caustic soda, chlorine, potash, and sodium chlorate. Combined carbon and graphite electrode are used for structural, refractory, nuclear, and other non-electrical applications. They are also used in the production of brushes, brush stock contacts, jigs and fixtures, battery carbons, seals and rings, as well as rods for electric arc lighting, welding, and metal cutting. Both types of electrodes are produced from conglomerate of small particles, which are further heat-treated at temperature above 1000 °C. Graphite requires special treatment at a higher temperature in order to give its special anisotropic crystalline structure. Graphite electrodes are produced from the baked carbon in an electric furnace. The baked carbon is packed with metallurgical coke, graphite, or charcoal which helps to conduct current at the start of heating and at an increased temperature. Similar to carbon electrodes, production of graphite last about 2 weeks; however, in order to meet standard, peak temperature is monitored and controlled to improve its physical and electrical properties. After this, the finished graphite is unpacked and air-cooled at room temperature and some of its properties, like density, are improved by impregnating with coal tar pitch or petroleum oil. Though, this may equally be done before the production of graphite. Handling and machinability, particularly, with conventional woodworking or metalworking tools, are also improved by quenching the impregnated graphite with water.

Required quantities of calcined carbon filler, binders, and additives are weighed and mixed together in a heated mixer, which turns the mixture to a "green" aggregate with time. This aggregate is the compressed mold or extrude and may be air-cooled or water quenched to solid form. The solidified green aggregate is often packed in sand, coke granules, or charcoal where it is held to distribute heat uniformly, absorb excess far or pitch binder, and prevent possible oxidation of the mixture during the early stage of baking process. The baking process is carried out in a furnace, which may be gas or oil fired, at a temperature of about 1000 °C; however, the process continues for week after the furnace temperature has been reduced to 400 °C. The baked carbon products are collected and processed further through cleaning, shaping, and surface grinding with diamond wheel to give a good finishing.

Wastewater Characterization

Generally, water generated in the production plants for carbon and graphite can be categorized as contact water, used predominantly in the post-extrusion quench, post-impregnation quench, machining, and scrubbing. Water is used for cooling, dust control, and wet grinding in the carbon electrode plants, while it is used for wet-scrubbing in the graphite production plant. Depending on the size of the manufacturing plant of this subcategory, water use in some verified plants ranges, approximately, from 302,000 to 2,650,000 L/day and is mainly employed as contact water for scrubbing. Major pollutants generated in this subcategory can be classified as classical pollutants (suspended solids, oil and grease, etc) and toxic pollutants (metals, inorganics, organics, and pesticides). In the study of wastewater characterization of plant 36173 used in the carbon and graphite manufacturing subcategory, Table 4.6 presents wastewater characterization in the plant.

Elsewhere [6] the results of the quantification of wastewater generated in the extrusion quench, impregnation quench, machining, and scrubbing streams are given in Table 4.7. Some pollutants that were not detected in the wastewater streams are equally listed in Table 4.8.

Table 4.6 Wastewater characterization for plant 36173, carbon and graphite subcategory

Pollutants	Extrusion quench flow = 7.67 L/s			Impregnation quench flow = 2.32 L/s			Machining flow = 1.64 L/s		
	Influent	Effluent	Percent removal	Influent	Effluent	Percent removal	Influent	Effluent	Percent removal
TSS	5.0	6.0	NM	19	5.0	74	53	24	55
TOC	1.0	2.0	NM	<10	4.0	NM	7,200	2.0	>99
BOD	2.0	2.0	0	19	7.0	61	6.0	6.0	0
Oil and grease	14	11	21	44	6.0	86	7.0	19	NM
Phenols pH, pH Units	0.02	0.02	0	0.019	0.014	26	0.12	0.01	92
Toxic pollutants, μg/L									
Metals and inorganics									
Antimony	<1	1	NM	<1	<1	NM	1	1	0
Arsenic	<1	1	NM	1	1	0	4	<1	88
Beryllium	<5	<1	50	<5	1	30	1	<1	50
Cadmium	<5	<15	NM	<5	5	NM	5	<5	50
Chromium	<15	<15	NM	<15	<13	NM	31	43	NM
Copper	<13	<13	NM	<15	<13	NM	1300	160	84
Cyanide	<10	30	NM	10	10	NM	65	<10	92
Lead	<50	<50	NM	<50	<50	NM	170	440	NM
Mercury	<1	<1	NM	<1	<1	NM	41	<1	NM
Nickel	<13	<13	NM	<13	<13	NM	70	<13	91
Selenium	2	16	NM	6	2	66	3	13	NM
Silver	<1	<1	NM	<1	<1	NM	7	1	86
Thallium	<1	<1	NM	<1	<1	NM	4	1	75
Zinc	23	140	NM	34	41	NM	240	27	89

(continued)

Table 4.6 (continued)

Pollutants	Extrusion quench flow = 7.67 L/s			Impregnation quench flow = 2.32 L/s			Machining flow = 1.64 L/s		
	Influent	Effluent	Percent removal	Influent	Effluent	Percent removal	Influent	Effluent	Percent removal
Toxic organics									
Acenaphthene	<10	<10	NM	<10	<10	NM	<10	ND	>99
Benzene	<10	ND	>99	<10	<10	NM	ND	<10	134
Chloroform	<10	<10	NM	<10	<10	NM	<10	ND	>99
2-Chlorophenol				<10	ND	>99	<10	<10	NM
1,2-Diphenylhydrazine	<10	<10	NM	<10	<10	NM			
Fluoranthene	<10	<10	NM	<10	<10	NM	<10	ND	>99
Methylene chloride	62	90	NM	30	23	23	<10	59	NM
Dichlorobromomethane							<10	ND	>99
Naphthalene	<10	<10	NM	ND	<10	NM	<10	ND	>99
2-Nitrophenol	–	–	–	–	–	–	<10	ND	>99
N-Nitrosodiphenylamine	<10	<10	NM	<10	<10	NM	<10	<10	NM
Phenol	<10	<10	NM	<10	ND	>99	39	ND	>99
Bis (2-ethylhexyl)-phthalate	<11	<10	9	<10	<10	NM	<10	<10	NM
Butylbenzyl-phthalate	–	–	–	<10	<10	NM	<10	<10	NM
Di-n-butyl-phthalate	<10	<10	NM	<10	<10	NM	<10	<10	NM
Diethyl phthalate	<10	<10	NM	<10	<10	NM	<10	<10	NM
Dimethyl phthalate	<10	<10	NM	<10	<10	NM	ND	<10	NM
1-2-Benzanthracene	11	<10	55	19	10	47	–	–	–
3,4-Benzopyrene	13	<10	62	<10	<10	NM	–	–	–
11,12-Benzo(k)-fluoranthene	11	<10	55	<10	<10	NM	–	–	–
Chrysene	11	<10	55						
Anthracene	<10	<10	NM	<10	<10	NM	<10	<10	NM
1,12-Benzo(ghi) perylene	<10	<10	NM	<10	<10	NM	–	–	–

Fluorene	<10	<10	NM	<10	<10	NM	<10	<10	NM
Dibenzo(ah) anthracene	<10	<10	NM	<10	<10	NM	<10	<10	NM
Indeno(1,2,3-cd)-pyrene	<10	<10	NM	<10	<10	NM			
Toluene	<10	<10	NM	<10	<10	NM	<10	<10	NM
Trichloroethylene		<10	<10	NM	<10				
Pesticides									
4,4-DDE							<10	ND	>99
Alpha-endosulfan				<10	ND	>99			
Alpha-BHC							<10	ND	>99

Source: [6]
Blanks indicate data not available
ND not detected, *NM* not meaningful

Table 4.7 Summary of classical and toxic pollutants in carbon and graphite subcategory raw wastewater by individual process units

Pollutant	Number of samples/number of detection	Range of detections	Mean of detections	Median of detections	Number of samples/number of detection	Range of detections
Machining, flow = 0.32 L/S						
Classical pollutants mg/L						
TSS	3/3	53–5,000	1,600	1,600	1/1	630
TOC	3/3	190–7,200	2,700	680	1/1	120
BOD	3/2	6–23	16		1/1	[a]
Oil and grease	3/3	7–70	32	20	1/1	9.5
Phenols	3/3	0.01–0.12	0.05	0.20	1/1	<0.005
pH, pH units					1/1	7.1
Calcium	3/3	8.5–38	20	14	–	–
Magnesium	3/3	1.2–8.3	3.8	1.8	–	–
Sodium	3/3	11–20	15	13	–	–
Aluminum	3/3	0.23–2.2	0.94	0.42	–	–
Magnesium	3/3	0.006–0.071	0.048	0.066	–	–
Vanadium	3/3	0.023–0.12	0.075	0.082	–	–
Boron	3/3	0.041–0.042	0.19	0.11	–	–
Barium	3/3	0.085–0.88	0.48	0.48	–	–
Molybdenum	3/3	0. 2–0.44	0.16	0.02	–	–
Tin	3/3	0.013–0.21	0.086	0.035	–	–
Yttrium	3/3	<0.004–0.007	<0.005	0.004	–	–
Cobalt	3/3	0.013–<0.051	<0.038	<0.051	–	–
Iron	3/3	0.57–6.7	3.3	2.6	–	–
Titanium	3/3	0.047–0.34	0.14	0.05	–	–

Toxic pollutants μg/L						
Metals and inorganics						
Antimony	3/3	1–3	<2	<2	1/1	<5
Arsenic	3/3	1–4	<2.3	<2	1/1	<10
Beryllium	3/3	<1–<5	<2.3	1	1/1	<10
Cadmium	3/3	3–5	<4.3	<5	1/1	19
Chromium	3/3	21–54	15	11	1/1	5
Copper	3/3	39–5,300	2,200	1,300	1/1	<10
Cyanide	3/2	<10–65	<38		1/1	10
Lead	3/3	62–1,700	640	170	1/1	<5
Mercury	3/3	<1–2	<1.3	<1	1/1	<5
Nickel	3/3	20–150	79	70	1/1	<10
Machining, flow= 0.32 L/S						
Selenium	3/3	<3–80	<29	5	1/1	<10
Silver	3/3	<1–7	<3.3	<2	1/1	<10
Thallium	3/3	<1–<20	<8.3	4	1/1	<2
Zinc	3/3	61–240	120	64	1/1	38
Toxic organics						
Acenaphthene	3/2	<10–<10	<10	–	1/0	–
Benzene	3/2	<10–<10	<10	–	1/0	–
1,1,1-Trichloroethane	3/1	68	–	–	1/0	
Chloroform	3/1	<10	–	–	1/0	
2-Chlorophenol	3/1	<10–<10	<10		1/0	
1,2-Diphenylhydrazine	3/2	<10			1/0	
Fluoranthene	3/1	<10–<10	<10		1/0	
Methylene chloride	3/2	<10–49	<23	<10	1/0	
Naphthalene	3/3	<10–<10	<10	<10	1/0	

(continued)

Table 4.7 (continued)

Pollutant	Number of samples/number of detection	Range of detections	Mean of detections	Median of detections	Number of samples/number of detection	Range of detections
N-Nitrosodiphenylamine	3/2	<10–<10	<10		1/0	
Phenol	3/2	<10–39	<24		1/0	
2-Nitropheno1	3/2	<10–<10	<10		1/0	
Dichlorobromomethane	3/1	<10			1/0	
Bis(2-ethylhexyl) phthalate	3/3	<10–<10	<10	<10	1/0	
Butylbenzy1 phthalate	3/3	<10–<19	13	<10	1/0	
Di-n-buty1 phthalate	3/3	<10–<10	<10	<10	1/0	
Di-n-octy1 phthalate	3/3	<10			1/0	
Benzo(a) pyrene	3/1	<10–<10	<10	<10	1/0	
Benzo(b) fluoranthene	3/2	<10			1/0	
Anthracene	3/3	<10			1/0	
Fluorine	3/3	<10–<10	<10	<10	1/0	
Pyrene	3/1	<10–<10	<10	<10	1/0	
Tetrachloroethylene	3/1	<10			1/0	
Toluene	3/3	<10			1/0	
Trichloroethylene	3/3	<10–<10	<10	<10	1/0	
Machining, flow= 0.32 L/S						
Pesticides						
4,4-DDO	3/1	<10			1/0	
Heptachlor	3/1	<10			1/0	
Alpha-BHC	3/2	<10–<10		<10	1/0	

Source: [6]
[a]Interference present

Table 4.8 Pollutants not detected in carbon and graphite subcategory raw wastewater

Toxic organics	
Acrolein	2,4-Dinitrophenol
Acrylonitrile	4,6-Dinitro-o-cresol
Benzidine	N-Nitrosodimethylamine
Carbon tetrachloride	N-Nitrosodi-n-propylamine
1,2,4-Trichlorobezene	Pentachlorophenol
Hexachlorobenzene	Di-n-octyl phthalate
1,2-Dichloroethane	3,4-Benzofluoranthene
Hexachloroethane	Acenaphthylene
1,1-Dichloroethane	Phenanthrene
1,1,2,2-Tetrachloroethane	Vinyl chloride
Chloroethane	
Bis(choromethyl) ether	*Pesticides*
Bis(2-choroethyl) ether	Aldrin
2-Chloroethyl vinyl ether (mixed)	Dieldrin
2-Chloronaphthalene	Chlordane
2,4,6-Trichlorobenzene	4,4^{1}-DDT
Parachlorometa cresol	4,4^{1}-DDE
1,2,-Dichlorobenzene	Toxaphene
1,3,-Dichlorobenzene	Beta-endosulfan
1,4-Dichlorobenzene	Endosulfan sulfate
3,3,-Dichlorobenzidine	Endrin
1,1,-Dichloroethylene	Endrin aldehyde
1,2,-Tans-dichloroethylene	Heptachlor epoxide
2,4-Dichlorophenol	Beta-BHC
1,2-Dichloroethylene	Delta-BHC
1,2-Dichloropropylene	
1,3-Dichloropropene	*Polychlorinated biphenyls*
2,4-Dimethylphenol	Aroclor 1242
2,4-Dinitrotoluene	Aroclor 1254
2,6-Dinitrotoluene	Aroclor 1221
4-Chlorophenyl phenyl ether	Aroclor 1232
4-Bromophenyl phenyl ether	Aroclor 1248
Bis(2-chloroisopropyl) ether	Aroclor 1260
Bis(2-chloroethoxy) methane	Aroclor 1016
Methyl chloride	*Miscellaneous*
Methyl bromide	2,3,7,8-Tetrachlorodibenzo-p-dioxin
Trichlorofluoromethane	(TCDD)
Dichlorodifluoromethane	
Hexachlorocyclopentadiene	
Isophorone	
Nitrophenol	
2-Nitrophenol	
4-Nitrophenol	

Source: [6]

Table 4.9 Oil separation performance in carbon and graphite subcategory

Pollutant	Raw waste	Treated waste	Percent removal
Extrusion quench oil separation			
Toxic pollutants, µg/L			
Methylene chloride	62[a]	90	NM
1,2-Benzanthracene	11	<10	>9
3,4-Benzopyrene	13	<10	>23
11,12-Benzofluranthene	11	<10	>9
Chrysene	11	<10	>9
Selenium	2	18[a]	9
Zinc	23	140	NM
Classical pollutants, mg/L			
Tss	5	6	NM
TOC	1	2	NM
BOD	2	2	0
Oil and grease	14	11	21
Phenols	0.022	0.020	9
Impregnation quench oil separation			
Toxic pollutants µg/L			
Methylene chloride	[a]	23	
1,2-Benzanthracene	19	10	47
Arsenic	1	1	0
Zinc	34	41	NM
Classical pollutants, mg/L			
TSS	19	5	74
TOC	<1	4	NM
BOD	19	7	63
Oil and grease	44	6	86
Phenols	0.019	0.014	26

Source: [6]
NM not meaningful
[a]Interference present

Wastewater Treatment Process

Wastewater generated in the manufacturing of carbon and graphite subcategory is treated using two approaches. The first approach involves skimming and contract hauling of sludge with provision for settleable solid removal that may be used for extrusion and impregnation quenches within the industry.

Table 4.9 shows result of a typical pollutant removal using this approach.

The second approach, which is aimed at recycling the water for reuse within the plant, is carried out mostly on the machining, grinding, and scrubber effluent. It consists of clarifier for settling of the pollutant and oil skimming simultaneously. Associated problem with this approach is the likelihood of increased oil and solid pollutant buildup in the streams; however, addition of polymer can correct this particularly solid removal. This approach has not proved effective for use for the treatment of extrusion and impregnation quench streams. Results of effectiveness of a typical clarifier used in this approach are given in Table 4.10.

Treatment Technologies and Cost

Different treatment technologies are employed in the two approaches involved in the treatment of pollutants in the carbon and graphite subcategory. For the first approach (level 1), oil skimming is used for extrusion and impregnation quench streams, while chemical coagulation and sedimentation with oil skimming are recommended for the machining, grinding, and scrubber wastewater streams. Because, the treatment of wastewater in the second approach (level 2) is for recycling purpose, a more effective pollutant reduction technology is needed to avoid buildup. Therefore, polymer is added to the technologies involved in the first approach, to improve the wastewater quality that is fit for recycling into the plant.

The estimate total annual costs for the two approaches of treatment in this subcategory are shown in Table 4.11.

2.2 *Dielectric Materials Subcategory*

2.2.1 Description and Production Process

The major components of this subcategory are dielectric-containing devices, such as oil-filled transformers, capacitors, electric products like mica paper, phenolics, laminates, fiberglass, polyesters, and rubber (excluding porcelain and glass). Transformers are broadly classified into oil-filled (wet) transformers and non-oil-filled (dry) transformers. Oil-filled (wet) transformers are often large in size; they operate at high voltage and are designed for power distribution. However, some small high-voltage transformers, such as automotive ignition coils, are available. The dry transformers are low-voltage transformers and their small types are widely use as fluorescent lamp ballasts, doorbell transformers, machine troll controls, power regulation, instrument transformers, and other general industrial and commercial applications.

Oil-filled capacitors are classified into large- and small-sized capacitors based on sizes. The larger-sized capacitors are used for power factor correction (PFC), while smaller-sized capacitors are used for electric motor control and in fluorescent lamps. Dielectric products, particularly, mica paper, are important component of capacitors. In fact, a capacitor may be defined as layered structures of conductive and dielectric (insulating) surfaces. Mica paper is predominantly used in the fixed types of capacitor which are in the form of rigid plates, thin sheets, or rolled flexible materials. Other types of dielectric are electrical insulators made from rubber and

Table 4.10 Machining setting data in carbon and graphite subcategory

Pollutant	Grinding effluent	Clarifier effluent	Percent removal
Toxic pollutants mg/L			
Methylene chloride	[a]	59	>47
Butyl benzyl phthalate	19	<10	>47
Antimony	3	1	67
Arsenic	1	<1	NM
Chromium	54	43	20
Copper	5,300	160	97
Lead	1,700	440	74
Mercury	2	<1	>50
Nickel	20	<13	>35
Selenium	[a]	13	
Zinc	61	27	56
Classical pollutants, mg/L			
TSS	3,000	24	99
TOC	190	2	99
BOD	25	6	75
Oil and grease	20	19	5
Phenols	0.014	0.01	28

Source: [6]
NM not meaningful
[a]Interference present

Table 4.11 Total annual unit costs for two levels of treatment for dielectric materials subcategory

Level of treatment	Flow[a]	Cost[a]	Flow	Cost	Flow	Cost
Extrusion and impregnation quenches						
Level 1	43	0.15	615	0.02	1880	0.01
Level 2	43	0.31	615	0.06	1880	0.03
Machining, grinding, and scrubber effluent						
Level 1	4	2.2	31	0.42	63	0.26
Level 2	4	2.3	31	0.48	63	0.34

Source: [6]
[a]Flow rate, liter/min, and cost, c/liter, in August 1979 dollars

plastics of different types through various machine-forming operations like molding, cutting, stamping, extrusion, punching, and others. The processes involved in the production of the above identified electrical components under this subcategory can be described as wet, unlike in the carbon and graphite subcategory.

Manufacturing of a power transformer is passed on the manufacture of its parts, particularly the steel core, winding of the coil, and general assembly on frame. Wet transformers have additional accessories such as tank that holds the dielectric fluid; there are radiators designed to air-cool the dielectric fluid when the transformer is in use. Copper wire with 12–18 wraps of paper insulator is wound round the coil to form a cylindrical pressboard (or phenolic plastics). The windings are separated with series of paper or materials of similar properties, and at the end of the winding process, the windings are axially compressed to required size, heated, and vacuum dried to remove residual moisture. Finally, the impregnated oil in the coil is compressed out. The transformer core is made by stacking strips of steel together to form the core which is then placed over legs of the core. The assembly, well insulated with Kraft paper and/or phenolic spacers, has its top yoke dried in a chamber and later saturated manually with transformer oil. Other processes to improve the qualities of the transformer are done in different industries.

Large oil-filled capacitor containers are produced manually from steel or aluminum. Since they are used in environments susceptible to corrosion, the surfaces of these containers are treated with phosphate or chromate before being painted. Special ceramic insulators are often used to separate the cover from the terminals in order to avoid discharging. Small oil-filled capacitor containers are made of same materials as large oil-filled capacitor containers but are fabricated on a dry automatic impact extruder. The manufacturing processes further involve connection of insulated materials to the interior wiring, soldering of the electrical leads to the terminal connection on the cover, welding of the cover to the body to form a sealed capacitor, and oil-filling operations.

Mica paper is produced from mica slurry which is first heated in a kiln before grinding and re-slurried with water. This is further screened with double serene separator to separate the undersized and oversized particles. The fine screen slurry is thoroughly mixed in a mixing pit before being sent to a vortex and then papermaking machine where remaining moisture content is evaporated. The machine turns out the mica matrix in a sheet form and this is further cured in an oven before being wound on rollers having met standard specifications.

2.2.2 Description of Industrial Plant

Different plants producing some of the components under this subcategory are found in the USA. Plant 30082 is one of the largest oil-filled capacitor plants in the USA. Plant 43055 is another plant with a large mica paper dielectric manufacturing facility using 3,300,000 L of water per day.

2.2.3 Water Uses

Less than 5% of the approximated 199 million L/day of gross water used in the manufacture of wet and dry transformers is processed water employed for non-contact cooling. In a typical manufacturing plant, for oil-filled capacitor, water use is mainly for detergent wash process and phosphatizing while solvent is used in many other processes, thus reducing the volume of water use. However, paper mica production has the largest water use in this subcategory; its production process is water dependent, particularly to saturate, separate, flake, and convey the processed mica from one unit to the other. Water usage in this section ranges between 31,000 and 3,600,000 L/day (minimum to maximum).

2.2.4 Waste Characterization

Investigation of wastewater produced from the above identified plants shows a wide disparity from one another. An average of 190,000 L/day of wastewater discharged from plant 30082 in the production of oil-filled capacitors is first sent to an effluent pool before being discharge. From the pool, the wastewater is filtered through double media filter system in parallel arrangement before being sent into a pair of carbon adsorbers and diatomaceous earth filter before being discharged into the environment. Table 4.12 summarizes the characterization data of the plant.

The mica paper plant, plant 43055, uses large volume of water and so wastewater that is well laden with mica particles is produced. To settle out the mica particles, the raw waste is made to pass through numbers of settling pond with a calculated settling period to achieve effective removal. The characterization data for the plant is presented in Table 4.13.

2.2.5 Wastewater Treatment Processes

Adsorption technology of pollutant removal from industrial wastewater has been employed by several workers [7–9]. Treatment processes for the removal of pollutants identified in some plants under this subcategory are filtration, adsorption, and settling or sedimentation. The two recommended primary approaches (levels 1) of oil skimming are applied to the segments involved in the manufacturing of wet transformer and oil-filled capacitor, while double-stage sedimentation is recommended for mica paper dielectric segment. The secondary approach (level 2) is designed for effective removal of pollutant from the subsector and consists of, in addition to oil skimming, multimedia filtration, carbon adsorption, and diatomaceous earth filtration. Result of treated effluent showing the effectiveness of the two approaches is presented in Table 4.14.

2.2.6 Treatment Technologies and Cost

The best practice control technology (BPT) for the subcategory in the treatment of wastewater includes two-stage sedimentation for mica paper dielectric and oil skimming for the wet transformers and oil-filled capacitor. Suggested best available technology (BAT) includes those of the BPT and, in addition, multimedia filtration, carbon adsorption, and diatomaceous earth filtration. Total annual unit costs for the two approaches are summarized in the Table 4.15 below.

For the primary approach, the total annual cost is given as 0.002 4/liter for 2,630 liters/min system excluding expenses like lining of ponds and land cost.

Table 4.12 Wastewater characterization for plant 30062, dielectric subcategory

Pollutant	Lagoon effluent	Final effluent	Percent removal
Toxic pollutants, µg/L			
Material and inorganics			
Antimony	1.7	<1.7	50
Arsenic	2.3	<1.3	72
Cadmium	<5	<5	NM
Chromium	<20	<20	NM
Copper	67	<13	90
Lead	<50	<3.7	NM
Nickel	<13	<50	NM
Silver	<1	3.7	NM
Zinc	640	50	92
Toxic organic			
1,2,4-Trichlorobenzene	78	ND	>99
1,1,1-Trichloroethane	1,400	1.7	>99
1,1-Dichloroethane	7	2	71
1,1-Dichloroethylene	53	ND	>99
1,2-Trans-dichloroethylene	620	ND	92
Methylene chloride	15	47	27
Naphthalene	140	40	71
Bis(2-ethylhexyl) phthalate	52	13	75
Di-n-butyl phthalate	7.6	1.3	83
Diethyl phthalate	6.6	5.7	14
Toluene	37	5.3	86
Trichloroethylene	1,700	5.3	>99

Source: [6]
NM not meaningful, *ND* not detected

2.3 Electric Lamp Subcategory

2.3.1 Description Production Process

Incandescent, fluorescent, electric discharge lamps (besides conventional fluorescent) and tungsten filaments used in the production of electric lamps are major electrical and electronic components that are manufactured under this subcategory. As lamps, they are used for illuminator, lighting, photoelectric, and photochemical effect purposes. Some of them are important components of electric circuits for demonstrative, household, and industrial uses. The production of components of this subcategory is meant for commercial purpose and acceptability, optimized light generation, illumination efficiency, as well as lamplife. Table 4.16 summarizes the types of artificial light production and related products. The flow chart of lamp phosphor process is presented in Fig. 4.1.

Table 4.13 Mica paper manufacturing plant-specific verification date for the dielectric subcategory, plant 43055

Pollutant	Raw waste	Treated waste	Percent removal
Toxic pollutant, μg/L			
Toxic organics			
Benzene	<10	NA	
Chlorobenzene	<10	NA	
1, 1, l-Trichloroethane	<10	NA	
Chloroform	180	NA	
Ethylbenzene	<10	NA	
Methylene chloride	<10	NA	
Bis(2-ethylhexyl) phthalate	29	NA	
Butyl benzyl phthalate	<10	NA	
Di-n-butyl phthalate	<10	NA	
Tetrachloroethylene	<10	NA	
Toluene	<10	NA	
Trichloroethylene		<10	NA
Toxic metals			
Antimony	<5	<5	NM
Arsenic	<3	<3	NM
Beryllium	<1	<1	NM
Cadmium	1	1	0
Chromium	4	4	0
Copper	11	10	9
Lead	13	16	NM
Mercury	<1	<1	NM
Nickel	1	1	NM
Selenium	23	39	NM
Silver	<3	<3	NM
Thallium	<25	<25	NM
Zinc	17	16	6
Classical pollutant, mg/L			
Flow, L/s	40.6	40.6	58
Aluminum	0.26	0.11	31
Barium	0.013	0.009	15
Boron	0.052	0.044	0
Calcium	9.7	9.7	NM
Cobalt	<0.001	<0.001	NM
Iron	0.11	0.001	65
Magnesium	3.3	3.3	0
Manganese	0.004	0.002	50
Molybdenum	0.002	0.001	50
Sodium	1.6	1.5	6
Tin	0.027	0.034	NM
Titanium	0.002	0.001	50

(continued)

Table 4.13 (continued)

Pollutant	Raw waste	Treated waste	Percent removal
Vanadium	0.028	0.024	14
pH	7.0	6.9	
Yttrium	0.001	0.003	NM
Cyanide, total	0	0	0
Oil and grease	0.4	1.0	NM
TOC	1.9	1.7	11
Classical pollutant, mg/L			
BOD	2.0	1.2	40
TSS	10.0	7.2	28
Phenols	0.008	0.006	25

Source: [6]
NA not analyzed, *NM* not meaningful

Table 4.14 Treated effluent concentrations in the dialectic materials subcategory

	Dielectric materials excluding mica paper	
Pollutant	Level 1	Level 2
Toxic pollutants, g/L		
Carbon tetrachloride	ND	ND
1,2-Trans-dichloroethylene	ND	50
Methylene chloride	930	10
Trichloroethylene	ND	5
PCB 1254	<10	ND
Copper	100	10
Lead	<40	48
Zinc	100	700
Classical pollutant, mg/L		
TSS	11	47
Oil and grease	3.5	3.9
Total organic carbon	16	8.5
Mica paper dielectric manufacture		
Classical pollutant, mg/L		
TSS	7.2	

Source: [6]
ND not detected

2.3.2 Description of Industrial Plant

In plant 28077, incandescent lamps, mercury vapor lamps, and coiled tungsten filaments are manufactured solely for the production of various types of incandescent and fluorescent lamps.

Table 4.15 Total annual unit costs for two levels of treatment for dielectric materials subcategory

Level of treatment	Flow[a]	Cost[a]	Flow	Cost	Flow	Cost
Extrusion and impregnation quenches						
Level 1	43	0.15	615	0.02	1880	0.01
Level 2	43	0.31	615	0.06	1880	0.03
Machining, grinding, and scrubber effluent						
Level 1	4	2.2	31	0.42	63	0.26
Level 2	4	2.3	31	0.48	63	0.34

Source: [6]
[a]Flow rate, liter/min, and cost, c/liter, in August 1979 dollars

Table 4.16 Summaries of the types of artificial light production and related products

Types	Products
Incandescence	Tungsten filament lamp filament manufacture
Luminescence	Electric discharge lamps (other than fluorescent)
	Fluorescent lamps
	Filament manufacture

Source: [20]

2.3.3 Water Use

Less volume of water is employed in this subcategory and is used mainly in the manufacturing process for dissolution of filament mandrel and acid cleaning wastes.

2.3.4 Wastewater Characterization

Wastewater generated in a typical plant, like plant 28077, is settled in tank particularly for pH adjustment before discharge. Results of sample of wastewater verified before and after pH adjustment in the above plant are given in Table 4.17. Wastewater from the electric lamp subcategory manufacturing plants, where wet process is mainly employed for filament mandrel dissolution generally, consists of pollutants as listed in Table 4.18.

2.3.5 Wastewater Treatment Processes

General treatment processes to remove pollutants in the electric lamps subcategory are achievable through sedimentation, chemical precipitation, and dewatering; pH adjustment of the wastewater is carried out for recycling and discharge purposes. Any or all the steps of treatment processes are applicable to any segment of the subcategory involved with wet processes. Two steps of treatment are commonly applied to the process water generated in the manufacture of fluorescent, to arrive at

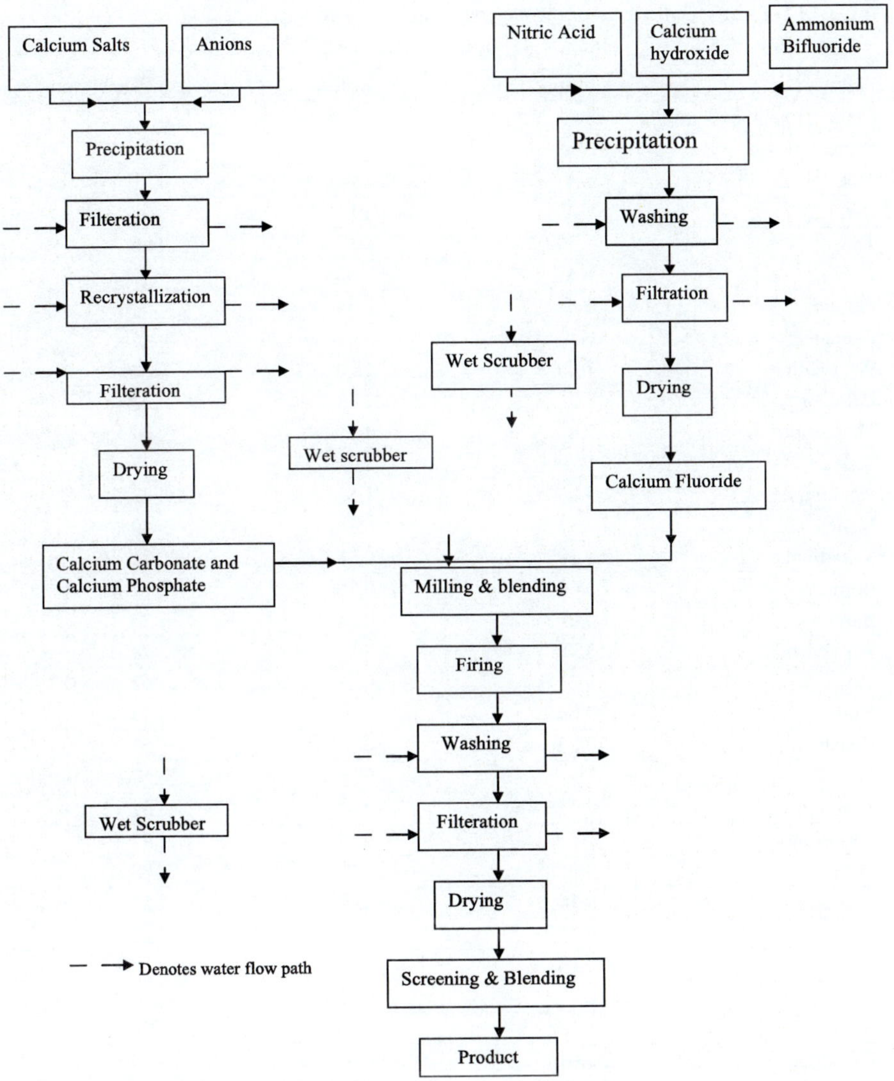

Fig. 4.1 Lamp phosphor process. (Source: [6])

water suitable for recycling into the plant. The results of the effectiveness of such treatment process are given in Table 4.19. All the three steps of treatment are applicable to wastewater generated in the manufacture of tungsten filaments resulting from the dissolution of the filament mandrel. Steps involving collection and contract removal have been found suitable for the treatment of wastewater resulting from the acid cleaning process of quartz mercury vapor lamps during its production. General treatment technologies involved in each step for each segment are summarized in Table 4.20.

Table 4.17 Plant-specific data for the electric lamp subcategory, plant 28077

	Flow = 0.40 L/s		
Pollutant	Influent	Effluent	Percent removal
Classical pollutants, mg/L			
Oil and grease	NA	0	
TOC	NA	13	
BOD	NA	30	
TSS	1,500	60	96
Phenols	NA	0.008	
Fluoride	0	0.54	NM
Cyanide	NA	0.01	
PH, pH units	10.0	12.0	
Calcium	8.9	24	NM
Magnesium	1.8	4	NM
Sodium	14,000	3,200	77
Aluminum	9.8	1.8	82
Manganese	0.53	0.27	49
Vanadium	0.28	0.08	71
Boron	0.46	0.38	17
Barium	0.053	0.04	25
Molybdenum	350	76	78
Tin	0.21	0.08	62
Yttrium	0.045	0.03	33
Cobalt	0.23	0.1	57
Iron	50	59	NM
Titanium	0.026	0.01	62
Potassium	76	6	92
Gallium	0.068	0.01	85
Germanium	0.14	0.06	57
Rubidium	0.019	0.007	63
Strontium	0.14	0.16	NM
Zirconium	0.7	0.025	96
Niobium	1	0.045	96
Palladium	0.014	0.03	NM
Indium	3.7	1.6	57
Tellurium	0.27	0.1	63
Tungsten	37	5.3	86
Osmium	0.28	0.045	84
Platinum	0.18	0.045	75
Gold	1.7	ND	>99
Bismuth	0.14	0.02	86
Uranium	1.4	0.035	98
Antimony	4.2	0.5	88
Arsenic	17	2.5	85

(continued)

Table 4.17 (continued)

Pollutant	Flow = 0.40 L/s Influent	Effluent	Percent removal
Beryllium	5	5	0
Cadmium	68	30	56
Chromium	31,000	11,000	65
Copper	19,000	3,600	81
Lead	1,400	290	79
Mercury	0.5	0.5	0
Nickel	1,900	500	74
Selenium	3.9	2.5	36
Silver	260	59	77
Thallium	440	170	61

Source: [6]
ND not detected, *NM* not meaningful, *NA* not analyzed

2.3.6 Treatment Technology and Cost

The BPT varies from one segment of the subcategory to the other. Those achievable, under step 1, for the fluorescent lamps include sedimentation of phosphor waste stream, use of lime, coagulant aid and electrolyte for precipitation and sedimentation in a clarifier, use of vacuum filter for sludge dewatering, and neutralization of the effluent before discharge. Some of these technologies particularly equalization, chemical precipitation, and sedimentation are the BPTs recommended under step 1 for the tungsten filament segments, and they have been confirmed to be effective in the treatment of industrial wastewater [10–19]. Since low volume of wastewater results from the manufacture of quartz mercury vapor lamps, the recommended technology is restricted to BPT and includes collection and contract removal. Suggested BAT, in addition to the recommended BPT in the step 1 above, includes total recycling of the wet air scrubber waste and phosphor waste stream after sedimentation for the fluorescent filament segment. The BAT consists of multimedia filtration, chemical precipitation, clarification, neutralization, pressure filtration, and reverse osmosis. The total annual unit costs for the fluorescent and tungsten filament lamp treatment technology involving all the steps are given in Table 4.21.

2.4 *Electron Tube Subcategory*

2.4.1 Description and Production Process

Early subclassification under the SIC identified this subcategory as electron tube, which includes receiving-type electron tubes, television picture tube, and transmitting-type electron tubes. However, in recent reclassification [20], the subsector is addressed as cathode ray tube subcategory where tube glass, color picture tubes,

Table 4.18 Summary of classical and toxic pollutants for the electric lamp subcategory wastewater by individual process units

Pollutant	Filament mandrel dissolution Flow = 0.32 L/h Number of samples/ number of detection	Range of detection	Mean of detection	Median of detection
Classical pollutant, mg/L				
Oil and grease	1/0			
TOC	1/1	190		
TSS	4/4	1–4,600	1,200	58
Phenol	1/1	0.025		
pH, pH unit	4/4	<2–12	<6.1	5.4
Calcium	4/4	0.66–29	15	16
Magnesium	4/4	0.22–6	4	5
Sodium	3/3	1,400–19,000	7,600	2,700
Aluminum	3/3	0.36–14	4.9	0.62
Manganese	3/3	0.03–0.75	0.4	0.43
Vanadium	3/3	0.067–0.36	0.17	0.09
Boron	3/3	0.23–0.51	0.36	0.34
Barium	3/3	0.003–0.06	0.039	0.05
Molybdenum	4/4	1.9–52,000	13,000	270
Tin	3/3	0.051–0.28	0.13	0.06
Yttrium	3/3	0.01–0.06	0.032	0.026
Cobalt	3/3	0.01–0.33	0.13	0.061
Iron	3/3	0.7–72	29	13
Titanium	4/4	0.005–0.23	0.074	0.03
Potassium	3/3	1.3–110	38	<1.4
Calcium	3/3	<0.006–<0.19	<0.071	<0.016
Geranium	3/3	<0.013–<0.4	<0.16	<0.06
Rubidium	3/3	<0.005–<0.05	<0.022	<0.01
Strontium	3/3	0.082–0.15	0.12	0.13
Zirconium	3/3	<0.012–<2	<0.68	<0.02
Niobium	3/3	<0.010–<3	<0.10	0.04
Palladium	3/3	<0.016–<0.2	<0.04	<0.02
Indium	3/3	1.8–4.5	2.7	1.6
Tellurium	3/3	<0.11–<0.7	<0.37	<0.2
Tungsten	3/3	0.66–51	19	6.4
Osmium	3/3	<0.017–<0.8	<0.29	<0.04
Platinum	3/3	<0.016–<0.5	<0.19	<0.05

(continued)

Table 4.18 (continued)

Pollutant	Filament mandrel dissolution Flow = 0.32 L/h Number of samples/ number of detection	Range of detection	Mean of detection	Median of detection
Gold	3/2	0.23–2.4	1.3	
Bismuth	3/3	0.010–<0.2	<0.08	<0.03
Uranium	3/3	<0.02–<2	<0.68	<0.03
Antimony	4/4	<1–<9	<5	<5
Arsenic	4/4	<5–64	<30	26
Beryllium	4/4	<1–<10	<6	6.5
Cadmium	4/4	<3–130	<39	11
Chromium	4/4	75–100,000	25,000	760
Copper	4/4	70–27,000	6,800	180
Lead	4/4	59–1,900	570	150
Mercury	4/4	<1–<1	<1	<1
Nickel	4/4	240–2,500	870	370
Selenium	4/4	<3–7	<3	<5
Silver	4/4	10–360	100	26
Thallium	4/4	<30–590	<200	88
Zinc	4/4	18–560	280	270
Toxic pollutant, μg/L				
Toxic organics				
Acrylonitrile	1/1	<10		
1,1,1-Trichloroethane	1/1	<10		
Chloroform	1/1	36		
Methylene chloride	1/1	21		
Dichlorobromomethane	1/1	10		
Chlorodibromoethane	1/1	<10		
Bis (2-Ethylhexyl) phthalate	1/1	<10		
Di-n-butyl phthalate	1/1	<10		
Di-ethyl phthalate	1/1	<10		
Toluene	1/1	<10		

Source: [6]

single phosphor tubes, television sets, and computer displays are produced. Electron tubes are devices with vacuum or ionized gas-filled tubes made of glass, quartz, ceramic, or metal, which facilitate the conduction of electrons or ions between the electrodes. In a given electron tube, electrons are emitted by certain elements or compounds when the energy of the surface atoms is raised by the addition of heat or increase in energy of bombarding particles, which excites the screen. The receiving-type electron tubes are important components that are manufactured under this subcategory and are widely used as low-voltage and low-power appliances such as radio and television receivers, computers, and sensitive control and measuring

Table 4.19 Treated effluent data for fluorescent lamp manufacture, electric lamp subcategory

Pollutant, mg/L	Level 1			
	Raw waste[a] flow weighted mean concentration	Daily maximum	30-day average	Long-term average
Antimony	0.460	0.15	0.07	0.05
Cadmium	0.310	0.04	0.02	0.012
Lead	0.029	0.15	0.07	0.05
Tin	22	0.37	0.16	0.13
TSS	140	52	23	18

Source: [6]

[a]This is a developed process raw waste stream. It does not include phosphor coating equipment and floor wash-downs

Table 4.20 General treatment technologies involved in each step for each segment under the electric lamp subcategory

Segment	Step 1	Step 2	Step 3
Fluorescent lamps	Settling of phosphor wastes Chemical precipitation and sedimentation of scrubber waste in a clarifier Sludge dewatering pH adjustment Collection and removal of silicone coating waste Chemical precipitation and sedimentation in a clarifier	Recycling of treated waste for the scrubber Chlorine coating Wastewater recycling for cleaning of sponge mix, rack, etc.	Not recommended
Tungsten filaments	Sludge dewatering pH adjustment	Multimedia filtration of suspended solids	Iron precipitation with NaOH and pH adjustment Iron oxide precipitation through pressure filtration Waste concentration through reverse osmosis products
Quartz mercury vapor lamps	Wastewater collection and contract removal.		

Source: [6]

NB. Recommended steps under tungsten filament lack available data and may indicate that most of them are not in use

equipment. The electrodes in these components are usually made of nickel, filled on the electron gun, and are classified based on the number of electrode they contain. The television picture tubes consist of large glass envelopes that provide electrical insulation for high voltage. The number of modified high-velocity electrons striking a luminescent surface corresponds to the transmitted signals in the picture tubes and is influenced by electrical impulses within a given time. Also, there are three electron

Table 4.21 Total annual unit cost for treatment of fluorescent and tungsten filament lamp wastewater

Level of treatment	Flow[a]	Cost[a]	Flow	Cost	Flow	Cost
Fluorescent lamps						
Level 1	115	0.30	127	0.27		
Level 2	115	0.30	127	0.21		
Tungsten filaments						
Level 1	54	0.45	132	0.22	263	0.15
Level 2	54	0.52	132	0.27	0.18	
Level 3	54	0.74	132	0.42	263	0.39

Source: [6]
[a]Flow rate, liter/min, and cost, c/liter, August 1979 dollars

guns representing the three primary colors (red, green, and blue) which produce individual color via a high stream of high velocity electrons that are modified by static and dynamic convergence mechanisms and an electromagnetic system. The incident beam exited the phosphor on the aperture mask behind the face of the screen to produce images.

The transmitting-type electron tubes are completely vacuumed tubes operating over a wide range of frequency. The stream of electrons within the tube is under the influence of both electrostatic and electromagnetic fields that are usually applied externally. A good example of this is the klystron tube. Production of the components under this subcategory shares, virtually, the same procedure which primarily includes preparation of glass panel and shadow mask, coating of the glass panel interior, installation of electron shield, preparation and connection of the glass funnel to panel or shadow mask assembly, the electron gun installation, and finally putting finishing processes in place [4]. In the preparation of the panel and shadow mask, the shadow mask, constructed from a thin layer of aluminum steel, is molded to match the contour of the interior surface of the glass panel. It is then welded to a blackened metal frame before being cleaned with degreasing solvents and caustics. All other accessories like springs to fix the glass panel and shadow mask assembly in position, in relation to each other, and for onward processing are put in place. This step is followed by coating of interior surface of the panel mask through the screening process which makes the panel mask to create image. After usual washing processes, the glass panel is subjected to carbon stripe process where it is coated with photoresist which contains chromate, dilute acids, oxidizers, and others like binding agents, before spurning to activate the photoresist and then dried. The shadow mask is later reinserted in the glass panel where it is subjected to photolithography process involving exposures to ultraviolet light. It is removed after the exposure and water-washed before being recoated and developed in order to improve its photoresist exposure pattern. However for color tubes, the assembly undergoes phosphor stripe process to fix the three primary colors (green, blue, and red); necessary cleaning and finishing are done before the installation of the electron shield. The shield, which prevents stray electrons from reacting outside the screen area, is made up of thin aluminum and is welded to the shadow mask assembly before being connected to the funnel alongside the shadow mask. The washed funnel is coated with black graphite which is a good electrical conductor and a non-reflective coating. The panel mask assembly is then connected to the funnel and fixed to position. Thereafter, the assembly is oven heated to fuse the frit joint between the panel and funnel at the seal edges. The electron guns, which give the color used in the screen, are combination of several assembled metal components embedded in glass parts. The gun assembly is then fused to the neck of the funnel using high temperature flame. After this, the tube is sealed and all the gases in it are removed using vacuum pumps. Furthermore, the tube is coated with carbon black paint and metal band is glued to the outside of the panel to protect the tube when subjected to voltage testing and other specifications. Figure 4.2 summarizes the manufacturing process of cathode ray tube.

2.4.2 Water Use

According to SIC code 36, plants 11114 and 30172 which are involved in the manufacturing of color television picture tubes in different locations produce wastewater, approximately as large as 280,000 and 150,000 L/day, respectively. Large volume of water is used in the manufacturing of the components under this subcategory for washing, rinsing, and solution preparation. In the manufacturing of receiving and transmitting tubes, water is used for backwashing, tower cooling, and steam boiling. Water is employed in processes leading to electroplating of anodes, cathodes, and grids and also for washing and rinsing of basic solution and other solvents involved in the processes. Water is equally used for bath dumps and subsequent rinsing during processes accompanying finishing.

2.4.3 Wastewater Characterization

Large amount of wastewater is generated from the production process of the television picture tubes and other cathode ray tubes; these resulted from solution and rinses in the electroplating process and washing off of grease, solvents, and particles on the surface of the panel, funnels, and glass tubes. Both alkaline cleaning and chemical etching processes and subsequent rinsing contribute to the wastewater stream in most of the plants.

The results of the wastewater characterized in plant 11114 and plant 30172 are presented in Tables 4.22 and 4.23, respectively.

2.4.4 Wastewater Treatment Process

Wastewater treatment technologies in this subcategory target the segments that manufacture television picture tubes as well as steel aperture masks. These are applied in three steps, with one complementing the other. The treatment processes

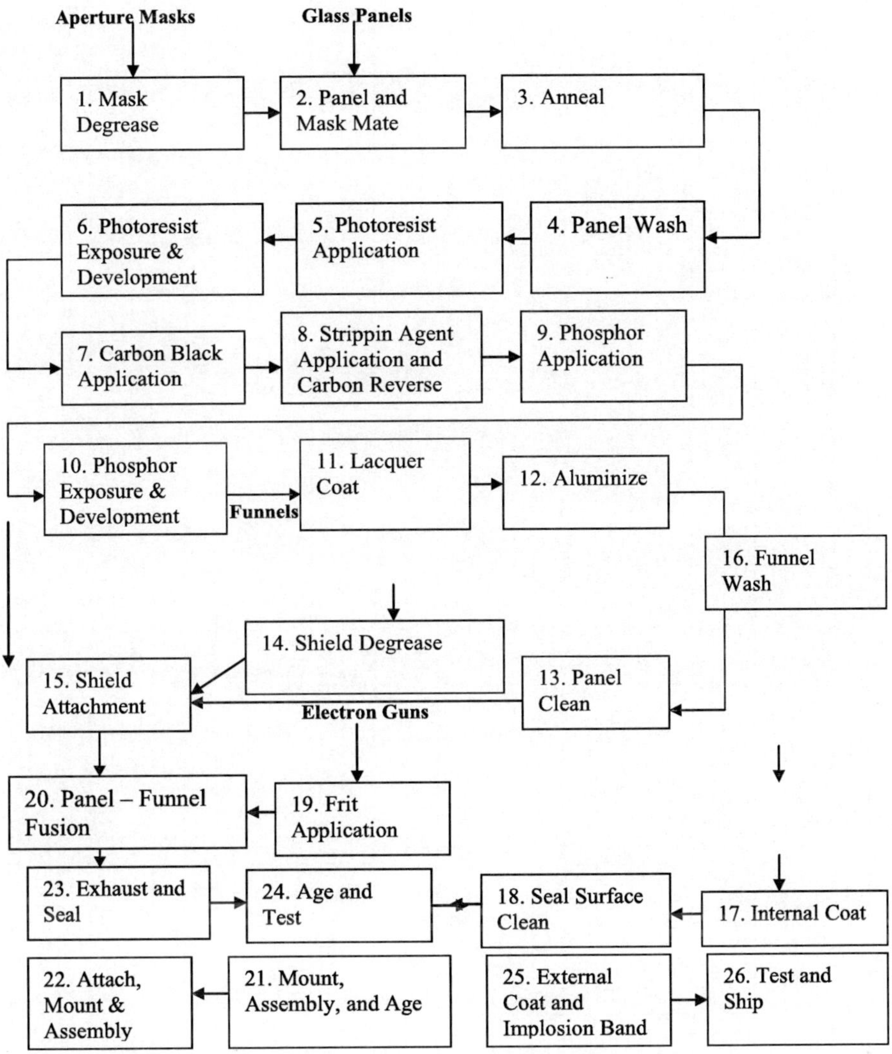

Fig. 4.2 Color cathode ray manufacturing process. (Source: [20])

fall under chemical precipitation, sedimentation, clarification, sludge dewatering, and pH adjustment before discharge.

Step 1 is the treatment of the picture tube segment which involves collection and removal of solvent, use of chemical coagulant and coagulant aid for precipitation and sedimentation of metal waste in the stream, removal of fluoride through precipitation, reduction of chromium, sludge dewatering, and pH adjustment of the wastewater before discharge. Multimedia filtration is the main technology used in step 2 treatment system in addition to treatments in step 1. In step 3, all the above-stated treatments are repeated with the objective of treating the wastewater for reuse in the

Table 4.22 Summary of the influent and effluent data in the electron tube subcategory, plant 11114

	System 1 flow = 6.19 L/s			System 2 flow = 8.52 L/s	System 3 flow = 1.42 L/s		
Pollutants	Influent	Effluent	% Removal	Effluent	Influent	Effluent	% Removal
Classical pollutant, mg/L							
Oil and grease	23	51	NM	10			
TOC	84	89	NM	8			
TSS	78	80	NM	140	2,300	26	99
Phenols	0.002	0.034	NM	40			
Fluoride	2400	1100	54	700			
Cyanide	0.21	0.52	NM	0.52			
pH, pH units	5.6	6.1		7.3	4.9		
Calcium	100,000	8,300	92	15,000	2,000	510	74
Magnesium	41,000	7,700	81	5,700	440	71	84
Sodium	2,600,000	1,200,000	54	1,000,000	740,000	16,000	98
Aluminum	53,000	7,600	86	5,100	540	860	NM
Manganese	710	48	93	200	0.5		
Vanadium	61	0.5	99	2	58	0.5	99
Boron	2,300	19,000	NM	11,000	1,000	200	80
Barium	45,000	500	99	230	330	360	NM
Molybdenum	150	18	88	37	62	18	71
Tin	270	12	96	12	280	12	96
Yttrium	20,000	49	99	81	440,000	880	99
Cobalt	410	25	94	25	1,700200	88	
Iron	540	2,000	NM	57,000	39	15	62
Titanium	480	120	75	110	14	2.7	81
Toxic pollutant, mg/L							
Metals and inorganic							
Antimony	440	61	86	79	0.67	1.7	NM

Arsenic	1,200	64	95	62	5.3	1	81
Beryllium	2.5	2.5	0	2.5	2.5	2.5	0
Cadmium	11,000	370	97	6	62,000	3,900	94
Chromium	2,700	300	89	3,800	4,400	2,900	34
Copper	790	30	96	100	84	6.5	92
Lead	620,000	14,000	98	320	25	25	0
Mercury	0.5	0.5	0	0.5	0.5	0.5	0
Nickel	2,700	110	96	97	6.5	6.5	0
Selenium	22	1	96	5	5	7	NM
Silver	83	2	98	0.5	120	3.2	97
Thallium	1.7	1	41	0.5	0.5	0.5	0
Zinc	1,200,000	33,000	97	320	1,200,000	17,000	99

Source: [6]

Table 4.23 Summary of influent and effluent data in the electron tube subcategory, plant 30172

Pollutants	Flow = 3.58 L/s		
	Influent	Effluent	Percent removal
Classical pollutant, mg/L			
Oil and grease	31	21	32
TOC	640	40	94
BOD	7.4	2.2	70
TSS	1.0	3.1	NM
Phenols	0.01	0.01	0
Fluoride	16	33	NM
Cyanide	0.002	0.02	NM
pH, pH units	4.4	77	NM
Calcium	16	310	NM
Magnesium	4	7.8	NM
Sodium	74	140	NM
Aluminum	1	0.3	70
Manganese	0.060	0.007	88
Vanadium	0.027	0.001	96
Boron	32	2	94
Barium	19	0.140	99
Molybdenum	0.27	0.024	91
Tin	0.37	0.066	82
Yttrium	1.7	0.00195	99
Cobalt	0.32	0.025	92
Iron	180	0.12	99
Titanium	0.033	0.001	97
Palladium	0.031	0.0025	92
Tellurium	0.004	0.002	50
Platinum	0.019	0.0025	87
Gold	0.001	0.001	0
Toxic pollutant, mg/L			
Antimony	14	120	NM
Arsenic	27	9.3	65
Beryllium	1.3	0.5	62
Cadmium	100	1	99
Chromium	81,000	210	99
Copper	37	14	62
Lead	83,000	163	99
Mercury	0.5	0.5	0
Nickel	1,700	15	99
Selenium	4.0	2	50
Silver	6.2	0.8	87
Thallium	16	1	94
Zinc	140,000	75	99

(continued)

Table 4.23 (continued)

Pollutants	Flow = 3.58 L/s		
	Influent	Effluent	Percent removal
Toxic organics			
Benzene	5	5	0
1,1,l-Trichloroethane	58	5	91
Fluoranthene	5		
Methylene chloride	490	400	18
Naphthalene	5	5	0
Bis(2-ethylhexyl) phthalate	460	5	99
Butyl benzyl phthalate	10	5	50
Anthracene	5	5	0
Phenanthrene	5	5	0
Pyrene	5	ND	>99
Toluene	29	5	83
Trichloroethylene	10	5	50
Acenaphthene	5	NM	
Chloroform	5	NM	
Di-n-butyl phthalate	5	NM	
Diethyl phthalate	5	NM	

Source: [6]
Blanks indicate pollutants not analyzed
ND not detected, *NM* not meaningful

plant. About 60% and 100% reuses of the treated wastewater are recommended for the television picture tube and aperture mask segments, respectively. Summary of effectiveness of all these recommended technologies applied to the television picture tubes is presented in Table 4.24.

2.4.5 Treatment Technology and Cost

The best practicable control treatment (BPT) for this subcategory, particularly for the television picture tube and aperture mask waste under the step 1 treatment process, is achievable in three separate treatment systems summarized in Table 4.25.

The BAT treatment level involves all the treatments recommended for BPT and, in addition, multimedia filtration, to improve the standard of the effluent before neutralization and discharge into the environment. To maximize the use of water, particularly reuse, in the manufacturing of television picture tubes and aperture masks, it is recommended that wastewater resulting from the plants should be treated to 50% and 100%, respectively, before recycling. The total annual unit cost covering all the treatment systems involved in all the steps stated above for the electronic tube subcategory is summarized in Table 4.26.

Table 4.24 Treated effluent date for electron tube subcategory

Pollutant	Raw waste flow weighted mean concentration	Level 1			Level 2		
		Daily maximum	30-day average	Long-term average	Daily maximum	30-day average	Long-term average
Total toxic organics[a]	0.290	NA	NA	0.290	NA	NA	0.290
Antimony	0.11	0.15	0.07	0.05	NA	NA	NA
Arsenic	0.08	0.15	0.07	0.05	NA	NA	NA
Cadmium	3.2	0.04	0.02	0.012	0.03	0.015	0.01
Chromium	1.14	2.2	0.80	0.57	1.24	0.45	0.32
Lead	4.6	0.15	0.07	0.050	0.10	0.04	0.034
Zinc	38	1.6	0.72	0.55	0.74	0.32	0.25
Boron	13	5.7	2.6	2.0	5.2	2.4	1.8
Barium	0.48	0.5	0.22	0.16	0.42	0.19	0.14
Iron	4.9	2.3	1.0	0.8	0.75	0.33	0.26
Yttrium	11	0.017	0.008	0.006	0.009	0.004	0.003
Oil and grease	62	34	16	12	21	9.2	7.8
Total suspended solids	140	52	23	18	37	16	13
Fluoride	840	44	20	15	14	6.2	4.8

Source: [6]

NA performance data not available

[a]Included in total organic carbons are 1,1,1-trichloroethane, methylene chloride, toluene, and trichloroethylene

Table 4.25 The best practicable control treatment (bpt) under the step 1 treatment process in three separate treatment systems

System	Recommended system
First system	Equalization, chemical precipitation, and sedimentation of lead waste (involving lime, sodium carbonate coagulant aid, and polyelectrolyte). Chemical precipitation of the treated waste (using lime and calcium chloride)
Second system	Chemical reduction of chromium roasts (using sulfuric acid and sodium bisulfite). Recovery of phosphors (through separate sedimentation of red phosphor, blue phosphor, and green phosphor). Final chemical precipitation and sedimentation of combined chromium and phosphor from the above processes
Third system	Equalization (of aperture mask formation waste). Chemical precipitation (using lime) and sedimentation

Source: [22]

NB. Finally, the effluents resulting from the above recommended systems are neutralized before being discharge while sludge is dewatered by a vacuum filter

Table 4.26 Total annual units costs for treatment of picture tube and aperture mask wastewater

Level of treatment	Flow[a]	Cost[a]	Flow	Cost	Flow	Cost
Picture tube						
Level 1			889	0.19	3380	0.07
Level 2			889	0.21	3380	0.08
Level 3			889	0.22	3380	0.08
Aperture mask						
Level 1			631	0.09		
Level 2			631	0.11		
Level 3			631	0.12		

Source: [6]
[a]Flow rate, liter/min, and cost, c/liter in August 1979 dollars

2.5 *Semiconductor Subcategory*

2.5.1 Description and Production Process

Semiconductors are solid electrical devices that conduct, by guiding or moving, an electrical current and act as insulator, by preventing the passage of heat or electricity. Simple diodes and integrated circuits (ICs) are variety of semicondutors manufactured in the electrical and electronic industries. The diodes are mono-purpose, mainly affecting the flow of electrical current. ICs combine multiple of such diodes (>200) and as such they are multipurpose and effective. ICs are assembled on chips with sizes as small as a thumbnail of a mature man. Several of the functions they perform include information processing and display, power handy, data storage, signal conditioning, and conversion between light energy and electrical energy. They are widely used in consumer electronic products, telecommunication equipment, industrial machinery, transportation equipment, and military hardware; however, their usage in computers account for about 40% of the semiconductor available in the market [15]. Semiconductors are made purely of silicon or with other crystalline materials like gallium and germanium; they are classified according to the type of material used in their production. Common semiconductor products available include silicon-based integrated circuits, gallium arsenide, gallium phosphide glass water devices, silicon, and germanium. They are used in forming different devices like bipolar, light-emitting diodes (LED), diodes, transistors, liquid crystal display, etc.

Production process involved in the manufacturing of semiconductors is complex, and as such some of the processes are repeated to complete the production. Distinctly, these processes may be divided into five steps [20], which are design, crystal processing, water fabrication, final layering and cleaning, and assembly.

Production of Silicon-Based ICs

Depending on types, purpose, and specifications, the intended semiconductors to be used are first designed using all available engineering design aids ranging from manual to software. This gives the viability of the semiconductors when tested with computer simulation. Following fabrication of pattering "masks," manufacturing equipment is selected and operating conditions are set [20]. In the crystal processing steps; silicon ingots are chemically doped through diffusion or ion implantation process to produce silicon crystals with improved electrical conductivity and other properties of semiconductors. Some of the doping materials (in either element or compound forms) commonly used include antimony, arsenic, phosphorus, boron, aluminum, gallium, germanium, tin, tellurium, and beryllium. Wafers are then produced by slicing the ingots into cylindrical shapes of about 7.6×10^{-4} m thickness and are etched mechanically and chemically. Etching generally facilitates cutting into or imprinting on the surface material, which could be wet etching using acid solution or dry etching involving reactive bases (halogenated or non-halogenated gaseous compounds). Wafer fabrication involves oxidation step in which a film of silicon dioxide is formed on the exterior surface of the silicon wafer. Patterns are then imprinted onto the substrate through photolithography and etching processes. Photolithography sets incorrect patterns affecting the electrical function of the semiconductor. The process is summarized in Fig. 4.3. The photolithographic component is then treated with chemical developers to remove unnecessary coatings or resist materials adhering to the substrate. This is then followed by etching in acid solution to create depressions or patterns on selected portion of the

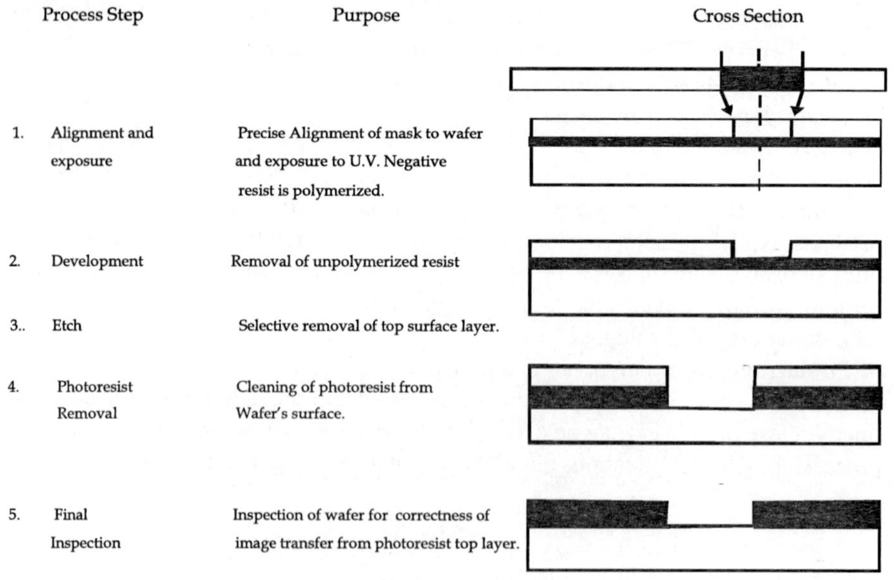

Fig. 4.3 Photolithography process. (Source: [20])

oxide layers. Thereafter, the wafer is rinsed by dropping in stripping solution in order to remove unwanted photoresist before being dried.

Through diffusion or implantation process, dopants are applied to the surface of the patterned wafer and additional layers of silicon may be applied to the wafer using epitaxial growth or chemical vapor deposition techniques. Other materials applicable in this process include silane, silicon tetrachloride, ammonia, nitrogen, and hydrogen. These steps may be repeated more than once in order to meet specifications. The fourth general step involved in the production of semiconductors is layering and cleaning; in this step, the surface of the patterned wafer is coated with thin layers of metal through metallization process involving spluttering. High-vacuum evaporation metals like aluminum, platinum, nickel, chromium, silver, and others can be used in this process, and their layers facilitate circuit/conduction properties within the finished semiconductor. As applied in the steps above, the metalized wafers are photolitographed and etched using halogenated solvents or acid solutions to remove unnecessary metal. Final layers of oxide of varying thickness is then applied over the surface of the processed wafer to provide a protective coat which protects the semiconductor from exterior influences like metal contacts. The semiconductor may range in thickness through the passivation process. The surface of the layers is rinsed in deionized water, and back of the wafer is mechanically ground to remove unnecessary materials. In order to improve connection of lead to the bonding pads during future processes and use, a film of gold is often applied to the back of the wafer. In order to determine if the produced chips meet the designed specifications, they are tested with alcohol compounds and cleaned with solvents like deionized water, acetone, methanol, and isopropyl alcohol. In assembly stage, the processed and certified chips are mounted unto a metal frame which connects them to lead and enclose them against mechanical shock and defects liable from the surrounding. They may be packaged for use using plastic or ceramic, though, the plastic packaging is the most widely available. Each package contains the chip, the lead frame (of the package), the die-attach pad, the wire bond, and the molded plastic housing [5]. Figure 4.4 shows a typical plastic packaged chip. The chip is fixed to the attach

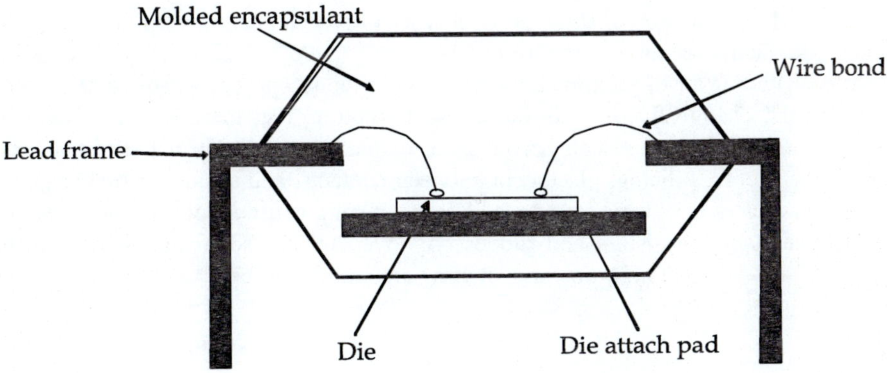

Fig. 4.4 Plastic package components of semiconductor. (Source: [20])

pad with epoxy material, and its parts are bonded to the leads of the package with tiny gold or aluminum wires before being placed into a molding press that encapsulates the setup in plastic [20].

The semiconductor is further cured, excess material is removed through chemical or mechanical deflash process, and finally, the leads are trimmed and formed. Many of the semiconductor manufacturing companies in the USA do not dice finished their wafers but export them to other facilities overseas where dicing and assembly process relatively cost less.

Gallium Arsenide and Gallium Phosphide Wafers

Often times the starting material for these product types of semiconductor is the wafer which is usually procured from crystal growers. The procured wafers are first placed in a furnace to grow a silicon nitride layer on it. Following this, the wafer is coated with a layer of photoresist and then exposed to ultraviolet light through glass photomasks that allow the light to strike only selected areas. It is, however, developed with a xylene-based developer. The resist is then stripped off after the wafer has been chemically etched and rinsed with deionized water. Thereafter, its surface is doped with a selected dopant, covered with metal oxide and photoresist. After this, it is masked, chemically etched with cyanide-containing chemical, and then rinsed in deionized water. During the lapping process, the desired thickness is achieved, and the pack is further sputtered with a layer of metal, usually gold, to provide contacts for conductivity before the finishing, testing, and assembly. Semiconductors of this type are predominantly used for the production of light-emitting diodes (LEDs).

Silicon and Germanium Semiconductors

The silicon and germanium semiconductors are discrete devices of small semiconductors of pure crystal placed between two and three contacts. The crystal materials used for the production of these product types of semiconductor are first cleaned in acid (or alkali) solution and thoroughly rinsed in deionized water before being coated with a layer of photoresist. As in the production processes above, the wafer is exposed chemically to hydrofluoric acid solution, and thereafter, the wafer is rinsed, dried and doped with boron, gold, or phosphorus in diffusion furnace. Its finishing involves dicing, placement between contacts, and sealing in rubber glass, plastics, or ceramic; wires are attached before testing certification for use. They are mostly used in transistors and diodes employed in less sophisticated equipment designs as well as in high-power switching and amplification.

Liquid Crystal Display (LCD)

Production of LED starts with cutting of flat glass into a square of 10 by 10 cm dimension which is then cleaned in ammonium hydroxide solution. It is later immersed in a stripping solution of alkaline and then rinsed in deionized water, spun dried, and sent for photolithographic process where it is masked with liquid, exposed, and developed. Thereafter, it is thoroughly rinsed with deionized water, dried before being chemically etched with acid, re-rinsed in deionized water, and then solvent dry. After this, the square is stripped in alkaline solution, rinsed with deionized, and spun dried. As in the production of silicon-based semiconductor, the square is subjected to passivation process. Liquid silicon dioxide is further deposited on the glass layer; however, silicon and oxygen gas may be used with phosphine gas as a dopant. Sodium ions on the glass surface are removed from spots to improve the electronic characteristic of the LED. The glass is dipped in an ammonium fluoride bath, rinsed with deionized water, and dry spinned. The glass is then reprocessed through the passivation procedures and screen printed with devitrified liquid glass in matrix. The glass is further vitrified through baking and sawed into patterns indicated by the vitrified glass boundaries. This is followed by cleaning the glass in alkaline solution, rinsing in deionized water and a layer of silicon oxide is evaporated onto its surface. Two mirror-image pieces of vitrified glass are bonded with a space between the two pieces of glass by heating them in a furnace. The glass is transferred into a vacuum chamber, where air in between the two glass pieces is removed and liquid crystal of biphenyl compound is injected, sealed with epoxy, and finished by vapor degreasing using solvent and finally shaped for assessment and assemblage. This device is widely used in machines that need to display responses.

2.5.2 Water Use

Water is significantly employed in virtually all steps involved in the manufacturing of all the components identified under this subcategory as contact water. Large volume of deionized water is consumed in the semiconductor subcategory, essentially, to rinse the body of the wafer after each production step, which includes cutting, etching, sputtering, electroplating, etc. Water is further employed for cooling and lubrication of some machines like the diamond saws and grinding machines used for slicing, lapping, and dicing wafer. An estimate of 628 million L/day of water is consumed in the entire semiconductor industry and very small amount is recycled for reuse.

2.5.3 Wastewater Characterization

Water is used to wash off deposited materials, chemical solvent, and grease in the production steps of semiconductor manufacturing plants, which invariably result to generation of wastewater in the plant. The large volume of the wastewater is

Table 4.27 Process steps and wastewater generated in semiconductor subcategory plant

Process step	Process wastewater generated
Crystal preparation	Spent deionized water, spent solvent, spent alkaline cleaning solution, spent acids, spent resist material
Wafer fabrication	Spent solvents, spent acid, spent etchant solution, spent aqueous developing solution
Layering and clearing	Spent deionized water, spent solvents, spent acids, spent etchants, spent aqueous developing solutions, spent, cleaning solution.
Assembly	Spent cleaning solution, spent solvents, spent aqueous developing solution

Source: [20]

generated after processes like cleaning operation to remove particles and spent solvents and acids. It is also generated during the developing, etching, and resist slipping processes. This is presented in Table 4.27.

Sampling efforts to verify the types and quantities of toxic pollutants present in semiconductor plants were restricted to scrubber wastewater because of their consistency throughout the plants.

2.5.4 Wastewater Treatment Process

The recommended treatment processes and technology for the semiconductor subcategory are primarily designed to achieve effective control of the pollutants after characterization of the wastewater from the associated plants. In step 1 treatment approach, mostly, the wastewater from the scrubber is targeted for recycle, and its waste content is concentrated and bled off to waste treatment section for further processing, while water suitable for recycling is engaged in the plant. The excess solvents resulting from use and spill can be collected and removed. The wastewater stream is also treated with chemical coagulants like lime in order to precipitate the dilute acid and to remove fluoride. All the treated waste streams are sent to the clarifier, sedimentation, and clarification tanks. Finally, the sludge generated is dewatered and discharged into the channels. Treatment process in step 2 focuses more on reclamation and recycling; thus in addition to treatment processes in step 1, dilute acid wastewater and waste solvents are collected and recovered for use. In step 3, after the use of the processes involved in step 2, adsorption process, involving carbon-packed columns, is employed to treat the wastewater resulting from the sludge dewatering process before being discharged into the environment. Table 4.28 shows the effluent data for the treatment processes identified in all the steps above.

2.5.5 Treatment Technology and Cost

Some of the best practicable control treatments (BPT) achievable in the semiconductor subcategory include solvent collection, chemical precipitation, and

Table 4.28 Treated effluent data for the semiconductor subcategory

Pollution, mg/L	Raw wastes flow weighted worn concentration	Daily maximum	30-day average	Long-term average	Daily maximum	30-day average	Long-term average	Daily maximum	30-day	Long-term average
Total toxic organics[a]	3.4	2.1	10A	0.82	2.1	10A	0.02	0.12	0.05	0.04
Arsenic	0.021	[b]	[b]	[b]	NA	NA	NA	NA	NA	NA
Chromium	0.16	(o)	(o)	(o)	(o)	(o)	(o)	(o)	(o)	(o)
Copper	0.06	2.6	1.1	0.81	2.6	1.1	0.81	0.15	0.07	0.05
Lead	0.098	0.15	0.07	0.05	0.15	0.07	0.05	2.7	1.2	0.94
Nickel	1.0	2.7	1.2	0.98	2.7	1.2	0.98	[b]	[b]	(B)
Oil and grease	4.4	34	16	12	34	10	12	52	23	18
TSS	48	52	23	18	52	23	18	44	20	15
Fluoride	57	44	20	15	44	20	15			

NA not available

[a]Total toxic organics includes 1,2,4-trichlorobenzene, 1,1,1-trichloroethane, chloroform, 1,2-dichlorobenzene, 1,3-dichlorobenzene, 1,4-toluene, and trichloroethylene

[b]No change in performance from previous level or raw waste

Table 4.29 Total annual unit costs for treatment of semiconductor subcategory wastewater

Level of treatment	Flow[a]	Cost[a]	Flow	Cost	Flow	Cost
Level 1	107	0.03	1700	0.10	7330	0.07
Level 2	27	0.83	426	0.23	1830	0.17
Level 3	27	1.5	426	0.56	1830	0.46

[a]Flow rate, liter/min, and cost, c/liter in August 1979 dollars

Table 4.30 The application, uses, and examples of fixed capacitors

Application	Purposes (uses)	Example
Energy storage devices	Accumulation of electrical energy at a rate and discharging the energy at a faster rate	Motor starting capacitor, fluorescent lamp ballasts, automotive ignition condensers
Protective devices	In combination with resistors, are used for reducing radio interference caused by arcing	Electrical instrumentation, television components, audio equipment, specialized instrumentation
Bypass devices	Prevention of flow of direct current from impeding the flow of alternate current. Attenuation of low-frequency current while allowing the bypass of higher-frequency current	Television components, audio equipment, specialized instrumentation

Source: [20]

sedimentation in a clarifier for dilute acid wastes and vacuum filtration of finely dewatered sludge that is generated from the sedimentation process.

The following treatment processes are the BAT treatment recommended for the semiconductor subcategory and involve those suggested under BPT above. They are recycling off dilute acid wastewater to the water production unit, without any treatment. Also, the solvents are separated and recovered for reuse. This is followed by treatment of the filtrate resulting from sludge dewatering process in carbon-packed adsorber before discharge. The summary of estimate of the total annual cost for all the treatment systems identified in the semiconductor category is shown in Table 4.29.

2.6 Capacitor Subcategory

2.6.1 Description Production Process

Two distinct components that are manufactured in the capacitor industry are variable and fixed capacitors. The fixed capacitors consist of conductive and dielectric surfaces in a well-arranged structure [21]. The layered structure is in two forms: rigid plates and thin flexible sheets. The type of conducting material, dielectric material, and encapsulating material distinguishes one type of fixed capacitor from the other. The application, uses, and examples of fixed capacitors are summarized in Table 4.30.

The common type of fixed and variable capacitors under this subcategory include paper dielectric, film dielectric, metalized dielectric, dual dielectric, tantalum dry slug and wire, tantalum foil, tantalum wet slug, aluminum electrolytic, mica dielectric, ceramic and glass encapsulated capacitors. Other types of capacitors like the oil-filled capacitors are classified under the SIC 3629 category. Some of the capacitors under the SIC 3675 have the same manufacturing processes, particularly, in relation to the use of process water.

2.6.2 Description of Industrial Plant

A typical plant used for the production of capacitor is plant 09062. This one-process line plant produces dry tantalum slug capacitors, and the water use in this plant is less compared to those engaged for cooling tower and boiler, blowdown, and water deionizer backwash. Table 4.31 shows the plant's specific verification data.

2.6.3 Water Use

Large volume of water used in the manufacturing process of capacitors is employed as process water. An estimate of 0.95 million L/day of process water usage has been recorded for the entire capacitor industry and is discharged as wastewater. Product type, production rate, location, and facility size are the factors that influence the flow rate of water leaving a particular plant. About 40% of this water has been subjected to treatment before being discharged into the environment.

2.6.4 Wastewater Characterization

Besides water use in the manufacturing process steps, which is relatively less, large volume of wastewater is generated in some capacitor manufacturing ancillary operations like cooling tower, blowdown, boiler blowdown, and water deionizer backwash. Information on the specific raw wastewater data is not available; however, pollutants that were detected in raw waste that was verified in some plants in the capacitor subcategory are presented in Table 4.32.

2.6.5 Wastewater Treatment Process

Wastewater treatment technologies applicable in the capacitor manufacturing facilities are subdivided into three steps. Step 1 treatment for capacitor manufacturing facilities is primarily pH adjustment of the wastewater before discharge. The pH of the ball mill discharge is also adjusted and then subjected to settling process before discharge. Step 2 treatment systems include chemical precipitation and sedimentation to remove solids and other metals in the wastewater stream. Others are

Table 4.31 Plant-specific data for the capacitor subcategory, plant 09062

Pollutant	Influent	Effluent	Percent removal
Classical pollutants, mg/L	0	0	0
Oil and grease	120	110	8
TOC	120	110	8
BOD	75	62	17
TSS	56	1.0	98
Phenols	<0.001	<0.001	0
Fluoride	1.3	4.7	NM
Cyanide	<0.005	<0.005	0
pH, pH units	7.0	7.7	NM
Aluminum	0.51	0.97	NM
Barium	0.01	0.02	NM
Boron	0.09	0.08	NM
Calcium	4.2	11	NM
Cobalt	0.01	0.005	50
Iron	2.4	0.68	72
Magnesium	0.43	1.3	NM
Manganese	4.1	4.3	NM
Molybdenum	0.03	0.04	NM
Sodium	19	120	NM
Tin	0.03	0.02	33
Titanium	0.02	0.01	50
Vanadium	0.01	0.03	NM
Yttrium	0.005	0.03	NM
Toxic pollutants, μg/L			
Toxic metals			
Antimony	0.05	0.05	0
Arsenic	2.5	2.5	0
Beryllium	5	5	0
Cadmium	5	5	0
Chromium	40	50	NM
Copper	850	60	93
Lead	30	50	NM
Mercury	0.5	0.5	0
Nickel	80	150	NM
Selenium	2.5	2.5	0
Silver	9	3	67
Thallium	2.5	2.5	0
Zinc	170	290	NM
Toxic organics			
Benzene		5	
Chloroform		5	
Methylene chloride		1,300	

(continued)

Table 4.31 (continued)

Pollutant	Influent	Effluent	Percent removal
Methyl chloride		24	
Di-n-Butyl phthalate		5	
Di-Ethyl phthalate		11	
Toluene		21	
Trichloroethylene		5	

Source: [6]
Blanks indicate pollutant not analyzed
NM not meaningful

neutralization, vacuum filtration, and treatment of the dewatered sludge through contract hauling. Oil skimming is also employed in the treatment of wastewater from other parts of the facilities besides those generated from the ball milling. Step 3 treatment processes are applied to compliment the treatment processes in step 2. The processes specifically remove the total suspended solids by using polishing filter after the neutralization step. Efficiency of step 1 treatment systems observed in a capacitor manufacturing industry is summarized in Table 4.33.

2.6.6 Treatment Technology and Cost

All the treatment technologies including neutralization which is applicable to both the capacitor waste stream and ball mill wastes are the best practicable control treatment (BPT) applicable to the capacitor subcategory. Furthermore, the ball milling waste is always subjected to sedimentation process after neutralization in order to effectively treat the effluent to standard specification. The pretreated wastewater from the treatment process is further subjected to neutralization before discharge. Vacuum filter is used to dewater the sludge. The best available technologies (BAT) include all the treatment system recommended in BPT in addition to oil skimming, chemical precipitation (using lime and coagulants), and sedimentation (ball milling waste streams). It is further recommended under the BAT that the resulting effluent be passed through a multimedia filter to free it of total suspended solids. Table 4.34 shows the total annual unit costs for the steps of treatments applicable to the capacitor subcategory of the electrical and electronic component industry.

2.7 *Printed Wiring Boards*

2.7.1 Description and Production Process

In the reclassification of the SIC 36 category of the electrical and electronic component industry [2], the printed wiring boards are coded as SIC 3672 and as such an important subcategory of the industry. They are hard structure on which electronic

Table 4.32 Pollutants not detected in the capacitor subcategory raw wastewater

Toxic organics	Pesticides
Acenaphthene	Chlordane
Acrolein	4, 4^{1}-DDT
Acrylonitrile	4, 4^{1}-DDE
Hexachlorobenzene	4, 4^{1}-DDD
1, 2-Dichloroethane	Alpha-endosulfan
Hexachloroethane	Beta-endosulfan
1,1,2,2,-Tetrachloroethane	Endosulfan sulfate
Chloroethane	Endrin
Bis(chloromethyl)ether	Endrin aldehyde
Bis(2-chloroethyl) ether	Heptachlor
2-Chloroethyl vinyl ether (mixed)	Toxaphene
2-Chloronaphthalene	
Parachlorometa cresol	*Polychlorinated biphenyls*
3, 3^{1}-Dichlorobenzidine	Aroclor 1242
1, 2-Trans-dichloroethylene	Aroclor 1254
1, 2-Dichloropropane	Aroclor 1221
1, 2-Dichloropropylene	Aroclor 1232
(1, 3-Dichloropropene)	Aroclor 1248
2, 4-Dimethylphenol	Aroclor 1260
2, 4-Dinitrotoluene	Aroclor 1016
2, 4-Dinitrotoluene	
4-Chlorophenyl phenyl ether	
4-Bromophenyl phenyl ether	*Miscellaneous*
Bis (2-chloroisopropyl) ether	2,3,7,8-Tetrachlorodibenzo-p-dioxin
Bis(2-chloroethoxy) methane	(TCDD)
Methyl chloride	
Methyl bromide	
Bromoform	
Dichlorodifluoromethane	
Hexachlorobutadiene	
Hexachlorocyclopentadiene	
2, 4-Dinitrophenol	
4, 6-Dinitro-o-cresol	
N-Nitrosodimethylamine	
Toxic organics	
N-Nitrosodi-n-propylamine	
Benzo(a)anthracene	
Benzo(a)pyrene	
Benzo(b)fluoranthene	
Benzo(k)fluoranthene	
Chrysene	
Acenaphthylene	

(continued)

Table 4.32 (continued)

Toxic organics	Pesticides
Benzo (ghi) perylene	
Fluorine	
Dibenzo(a,h)anthracene	
Indeno(1,2,3-cd)pyrene	
Vinyl chloride	

Source: [6]

Table 4.33 Treated effluent data for level 1 treatment in the capacitor subcategory

	Flow weighted mean concentration	Range	Mean
Treated capacitor (excluding ball mill wastes) effluent			
Toxic pollutants, µg/L			
Copper	53	49–60	54
Lead	50	50–50	
Zinc	13	67–290	180
Classical pollutants, mg/L			
Aluminum	0.42	0.23–0.97	0.6
Barium	0.84	0.02 –1.1	0.57
Iron	0.40	0.03 –0.68	0.49
Manganese	1.1	0.01–4.3	2.1
TSS	15	1.0–20	10
Oil and grease	0	0	
pH, pH units	7.0–7.7	7.4	
Treated ball milling raw waste			
Toxic pollutants, Ng/L			
Chromium			7
Lead			11
Zinc			36
Classical pollutants, mg/L			
Barium			160
TSS			210

Source: [6]

components like semiconductors, capacitors, and others are mounted. They appear as single-sided, double-sided, multilayer, and flexible boards. Production of printed wiring boards can be achieved through three methods: additive, subtractive, or semi-additive technology. However, the subtractive process is the most widely used method by many manufacturing industries. Furthermore, the actual steps and material use in the manufacturing process depend on the customers' requirement and manufacturer's desires.

Under the board preparation step, insulating layer is introduced in between a two-sided-etched copper dielectric boards consisting of fiberglass and epoxy resin, which are bounded (laminated) together by heat and pressure. Patterns are imaged on the

Table 4.34 Total annual unit cost for treatment of capacitor and ball milling wastes

Level of treatment	Flow rate (liter/min) Flow[a]	Cost[a]	Flow	Cost	Flow	Cost
Capacitor						
Level 1	1	7.7	13	0.08	55	0.31
Level 2	1	816	13	0.03	55	0.39
Level 3	1	9.9	13	1.12	55	0.47
Ball milling						
Level 1	<1	149	9	0.72	19	0.37
Level 2	<1	219	9	1.9	19	1
Level 3	<1	244	9	2.1	19	1.2

Source: [6]
[a]Flow rate, liter/min, and cost, c/liter in August 1979 dollars

board using photographic tools; holes are then drilled on the boards to create an electric path between the layer and position to mount components. In this process, the holes are subjected to etch-back process to remove smeared epoxy resin and other contaminants using sulfuric or hydrochloric acids and potassium permanganate. Then, the holes are chemically plated with copper or graphite carbon and dried for hours to prevent the board from rusting. The board is mechanically cleaned to remove copper contaminants; it is chemically cleaned with alkaline acid and rinsed water; the board is then set for electroless plating. The general steps involved in the subtractive manufacturing process of Printed Wire Boards are board preparation, application of conductive coating, soldering, fabrication, and assemblage.

Common materials used in the plating process are listed in Table 4.35.

The next step of the PWB production is imaging which involves photolithographic process. In this process, part of the board, where the circuit pattern will not be set, is covered with photoresist; thereafter, it is developed after exposure to light, to remove the resist layer on the unwanted areas. This is followed by light etching using ammoniacal etchants to remove rust initiator and other metal, which is usually copper. The board is then subjected to stencil printing process which is followed by etching process, and thereafter, the photoresist is removed with photoresist stripper. The next important step in the production of PWB is electroplating which is employed to improve the electrical conductivity of the board and protection against corrosion. Copper laminate is improved through photolithography and stenciling processes to create the circuit pattern. Essentially, the PWB plating process involves deposition of metals, particularly copper and tin-lead, in the plating bath to a designed thickness; this is followed by etching or stripping processes. After this, the board is thoroughly rinsed with water to end the chemical reactions and to remove metal deposits and other impurities. The tin-lead layer is removed and the panel conductivity is tested.

Soldering coating is another step in the manufacturing of PWB in which panel is dipped into molten solder consisting of 60% tin and 40% lead. The required thickness of the solder coat is between 50 and 1,200 microinches. Therefore, after

Table 4.35 Materials used in copper and tin-lead electro- and electroless plating process

Type of plating	Electroplating chemicals	Electroless chemical
Copper	Copper pyrophosphate Orthophosphate Pyrophosphate Nitrates Acid copper Copper sulfate Sulfuric acid	Hydrochloric acid Palladium chloride Stannous chloride Metallic tin pellets Sodium hydroxide Copper sulfate Formaldehyde
Tin-lead	Tin-lead Fluoroboric acid Boric acid Peptone	Tin chloride Sodium hypophosphite Sodium citrate

dipping, the excess solder is removed through solder leveling process involving the use of hot air or hot oil. The next processing steps are electrical and mechanical testing of the board. Examination of the plated holes based on electrical test, dimensional inspection, visual inspection, and quality audits on a cross section cut from a sample panel are done in these steps. After these have been certified to meet standard or consumer's specification, the PWBs are then packed and labeled for onward processing or shipping to other manufacturing companies where it serves as a vital raw material. Prominent step after this is the assembly process where electrical and electronic components are attached to the board through wave, dip, or drag soldering methods.

2.7.2 Water Use

General production steps for the production of PWB can be described as wet process, involving the use of water and solvents. Water is predominantly used for washing and rinsing of the PWB surfaces, to remove scrap-based materials as well to wash the solvent and greases. In other parts of the factory, water is employed in plant wash-downs to dissolve the used solvent and other chemicals. In recent times, there has been reduction in the use of non-environmentally friendly chemicals like chlorofluorocarbons (CFCs) and trichloroethane (TCA); deionized water is widely used as alternative.

2.7.3 Wastewater Characterization

Large volume of wastewater is generated during the production of the PWBs. The wastewater contains used solvents, dissolved and undissolved metals, and other particles like photographic resist. Table 4.36 presents output of printed wiring board pollution. The results of releases verified in some printed wiring board manufacturing facilities are summarized in Table 4.37.

Table 4.36 Printed wiring board pollution outputs

Process	Air emissions	Process wastes (liquids/ wastewaters)	Other wastes (solids/rcra)
Board preparation	Particulates, acid fumes, and VOCs	Spent acids and spent alkaline solutions	Sludge and scrap board material
Electroless plating		Spent electroless copper baths, spent catalyst solutions	Waste rinse water and sludges from wastewater treatment
Imaging	Organic vapors and acid fumes	Spent developing solution , spent resist material, spent acid solutions, and aqueous metals	F001-5, depending on concentration and mixture of solvents. Sludges from wastewater treatment
Electroplating	Acid fumes, ammonia fumes, and VOCs	D008 (lead), D002, D003, spent etchants, spent acid solutions, spent developing solutions, spent plating baths	F006, F007, and F008
Solder coating	VOCs and CFCs		
PWD assembly and soldering	VOCs and CFCs	Metals (nickel, silver, and copper), D008 (lead), flux residue, spent deionized water; spent solvents	Solder dross, scrap boards, filters, gloves, rags, wastewater treatment sludge

2.7.4 Wastewater Treatment Process

Wastewater generated in the PWB production processes is treated at each step; however, wastewater generated from similar processes is pooled to a tank for treatment. The wastewater treatment methods employed in this subcategory are primarily for water reuse in the plant. The pH of the wastewater is often adjusted in tanks before discharge or recycles. The use of reactive chemicals for rinsing and multiple drag-out baths will further reduce the pollution level of the wastewater emanating from the plant. The combined electrolysis and chemical precipitation is a very good process of removing and recovering lead and tin from the waste stream. Sludge dewatering and sludge processing for the recovery of metal are part of the processes applicable to the wastewater treatment.

2.7.5 Treatment Technologies and Cost

All the treatment technologies identified under this subcategory are classified both as best practicable control treatment (BPT) and best available technologies (BAT). The costs involving best practicable control treatment (BPT) and best available technologies (BAT) are stated in Table 4.38.

Table 4.37 Releases for printed wiring board manufacturing facilities (SIS 3672) in TRI, by number (releases reported in pounds/year)

Chemical name	# Facilities reporting chemical	Fugitive air	Point air	Water discharges	Underground injection	Land disposal	Total releases	Average release per facility
Sulfuric acid	208	25640	98477	0	0	250	124367	598
Ammonia	117	80332	480081	28029	0	0	588442	5029
Copper	89	1345	1840	27	0	8500	11732	132
Copper compound	73	6830	7532	1831	0	9739	25932	355
Hydrochloric acid	70	13268	40342	32189	0	27	85826	1226
Nitric acid	59	7572	12750	0	0	0	20322	344
Glycol ethers	25	82099	132118	23057	0	0	237274	9491
Formaldehyde	22	3225	14912	255	0	0	18392	836
Chlorine	16	1545	5992	50	0	0	7587	747
Lead	12	250	750	5	0	3500	4505	375
Acetone	10	117974	70711	0	0	0	188685	18869
Freon 113	9	83258	37550	0	0	0	120808	13423
Lead compounds	7	760	1260	252	0	0	2272	325
Ammonium sulfate (solution)	6	0	0	100000	0	0	100000	16667
Methyl ethyl ketone	6	13770	25023	0	0	0	38793	6466
Phosphoric acid	6	510	505	0	0	0	1015	169
Methanol	5	62978	7394	0	0	0	70372	14074
Dichloromethane	4	51269	125288	5	0	0	176562	44141
1,1,1-Trichloroethane	3	24930	8310	0	0	0	33240	11080
2-Methoxyethanol	3	5000	40960	0	0	0	45960	15320
Hydrogen fluoride	2	0	250	0	0	0	250	125
Nickel	2	0	0	0	0	0	0	0
Toluene	2	29425	14125	0	0	0	43550	21775
Zinc compounds	2	750	0	0	0	0	750	375

(continued)

Table 4.37 (continued)

Chemical name	# Facilities reporting chemical	Fugitive air	Point air	Water discharges	Underground injection	Land disposal	Total releases	Average release per facility
Ammonium nitrate (solution)	1	0	0	0	0	0	0	0
Barium compounds	1	250	0	0	0	0	250	250
Ethylbenzene	1	250	2600	0	0	0	2850	2850
Ethylene glycol	1	600	1200	0	0	0	1800	1800
Isopropyl alcohol (manufacturing)	1	0	0	0	0	0	0	0
Methylenebis (phenylisocyanate)	1	0	0	0	0	0	0	0
Phenol	1	750	750	250	0	0	1750	1750
Silver	1	0	0	0	0	0	0	0
Tetrachloroethylene	1	12900	22300	0	0	0	35200	35200
Trichloroethylene	1	14920	26000	0	0	0	40920	40920
Xylene (mixed isomers)	1	1000	16560	0	0	0	17560	17560
1,2-Dichlorobenzene	1	1800	2130	0	0	0	3930	3930
Totals:	—	645,200	1.197,730	185,950	0	22,016	2,050,896	—

Table 4.38 Costs involved in the practices of BAT and BPT

Technology	Cost	
	Capital investment	Savings
Sludge pretreatment process	–	10,000 in disposal cost
Drag-out reduction with drip bars	100/tank	600/year
Reducing drag-out generation with drain boards	25/tank	450/year
Drag-out reduction with racking	–	600/year
Use of drag-out recovery tank	500/tank	4,700/year
Electrodialysis reversal process for metal salt in wastewater	–	40,100/year

Source: [20]
NB. All numerical values are in US dollars ($)

3 Electrical and Electronic Industry and Waste Generation

3.1 Waste Generation

The pollutants generated in the wastes of electrical and electronic industry are predominantly solvents, acids, and heavy metals. Acetone, xylene, and methanol are some of the commonly released solvents, while sulfuric, hydrochloric, and nitric acids are common acids released in the manufacturing processes. Ammonia is also released in large quantities in some of the subcategories of the electrical and electronic industry. This section focuses on the Toxics Release Inventory (TRI) of the semiconductor, printed wiring boards (PWBs), and cathode ray tube (CRT) subcategories as representatives to study the pollution releases of the industry. Releases are defined as on-site discharge of toxic chemical to the environment, namely, air, surface water, land, and underground water while transfer is regarded as transfer of chemicals in waste to a facility that is geographically or physically separated from the facility reporting under the Toxics Release Inventory [21].

In the semiconductor subcategory, common acids released during etching and cleaning processes are sulfuric acid and hydrochloric acid. Processes like photolithography and cleaning involve the use of solvents like acetone, glycol, ethers, xylene, and freon 113.

Tables 4.39 and 4.40 summarize the chemical releases and transfers for the semiconductor subcategory.

Common acids released in the printed wiring board facilities are similar to those of semiconductors, but are released during cleaning, electroless plating, and electroplating and etching processes. Lead and copper are common metals released in PWB production during electroplating, etching, and soldering processes. Resulting ammonium sulfate solution from PWB manufacturing facilities is transferred to POTWs. Table 4.41 presents the pollutant transfers from PWB facilities.

Table 4.39 Releases for semiconductor manufacturing facilities (SIC 3674) in TRI, by number of facilities (releases reported in pounds/year)

Chemical name	# Facilities reporting chemical	Fugitive air	Point air	Water discharges	Under-ground injection	Land disposal	Total releases	Average releases per facility
Sulfuric acid	125	13544	88209	17	250	139	102259	818
Hydrochloric acid	78	8262	69429	3	0	10	77704	996
Hydrogen fluoride	71	4940	55479	9902	0	5	70326	991
Phosphoric acid	69	4039	25674	0	0	5	29718	431
Nitric acid	57	5403	47628	23	0	5	53059	931
Acetone	53	121794	890290	1460	659	5	1014208	19136
Ammonia	42	42770	101717	42082	17805	8600	212974	5071
Glycol ethers	27	41317	212900	500	0	82000	336717	12471
Xylene (mixed isomers)	25	9952	252661	0	139	0	262752	10510
Ethylene glycol	16	1688	9316	1600	0	0	12604	788
Methanol	16	31049	135566	0	129	0	166744	10422
Freon 113	10	41211	73335	0	0	0	114546	11455
1,1,1-Trichloroethane	8	1691	82366	0	1	0	84058	10507
Methyl ethyl ketone	6	1332	128250	0	0	5	129587	21598
Tetrachloroethylene	4	514	55034	1	0	0	55549	13887
Ammonium nitrate (solution)	3	0	0	0	0	0	0	0
Ammonium sulfate (solution)	3	250	0	0	0	0	250	83
Lead	3	0	0	0	0	0	0	0
Phenol	3	50	2745	0	0	0	2795	932
Toluene	3	25170	33580	0	0	0	58750	19583
Trichloroethylene	3	14009	21896	0	0	0	35905	11968
Copper	2	0	0	12	0	0	12	6
Ethylbenzene	2	175	1300	0	0	0	1475	738

Methyl isobutyl Ketone	2	750	9325	0	0	0	10075	5038
1,2-Dichlorobenzene	2	200	49234	0	0	0	49434	24717
1,2,4-Trichlorobenzene	2	0	6519	0	0	0	6519	3260
Antimony Compounds	1	18	5	1	0	0	24	24
Chlorine dioxide	1	5	5	0	0	0	10	10
Cobalt compounds	1	5	2	0	0	0	7	7
Isopropyl alcohol (manufacturing)	1	0	0	0	0	0	0	0
Lead compounds	1	0	0	0	0	0	0	0
N-Butyl alcohol	1	21	84	0	0	0	105	105
Nickel compounds	1	0	0	0	0	0	0	0
Nitrilotriacetic acid	1	5	5	0	0	0	10	10
P-Xylene	1	0	430	0	0	0	430	430
Totals	—	370,264	2,352,984	55,601	18.983	90,774	2,888,606	—

Source: [20]

Table 4.40 Transfers for semiconductor manufacturing facilities (SIC 3674) in TRI, by number of facilities (transfers reported in pounds/year)

Chemical name	# Reporting chemical	POTW discharges	Disposal	Recycling	Treatment	Energy recovery	Total transfers	Average transfer per facility
Sulfuric acid	125	147449	500380	1039071	169372	0	1856272	14850
Hydrochloric acid	78	236415	29599	21664	84745	5	372428	4775
Hydrogen Fluoride	71	11733	198630	525	151929	0	362817	5110
Phosphoric acid	69	1103	269124	200000	33594	0	503821	7302
Nitric acid	57	56177	99817	20910	62904	0	239808	4207
Acetone	53	104090	1582	136987	116610	1075656	1442137	27210
Ammonia	42	944298	52771	650	10806	0	1008525	24013
Glycol ethers	27	30889	3345	139100	56330	1049440	1279104	47374
Xylene (mixed isomers)	25	3891	824	31304	127501	728688	892208	35688
Ethylene glycol	16	458412	2027	15194	623	102016	578272	36142
Methanol	16	14474	0	27715	64502	716413	823104	51444
Freon 113	10	25	592	36937	2435	5660	45649	4565
1,1,1-Trichloroethane	8	263	5	75267	18264	8000	101799	12725
Methyl ethyl ketone	6	869	750	0	2105	276109	279833	46639
Tetrachloroethylene	4	0	0	10215	59628	53000	122843	30711
Ammonium Nitrate (solution)	3	224302	0	0	0	0	224302	74767
Ammonium Sulfate (solution)	3	1488462	0	122000	0	0	1610462	536821
Lead	3	0	1500	59125	13961	0	74586	24862
Phenol	3	2331	0	0	27	94679	97037	32346
Toluene	3	0	0	0	17000	5970	22970	7657
Trichloroethylene	3	0	0	59736	0	0	59736	19912
Copper	2	0	18	0	166	0	184	92
Ethylbenzene	2	0	146	0	190	16800	17136	8568

Methyl isobutyl ketone	2	0	0	0	9300	12190	21490	10745
1,2-Dichlorobenzene	2	10	0	0	2157	93600	95767	47884
1,2,4-Trichlorobenzene	2	1413	0	0	32273	0	33686	16843
Antimony compounds	1	0	18100	0	0	0	18100	18100
Chlorine dioxide	1	0	0	0	0	0	0	0
Cobalt compounds	1	0	3780	0	0	0	3780	3780
Isopropyl alcohol (manufacturing)	1	5	0	10165	0	0	10170	10170
Lead compounds	1	0	6630	0	0	0	6630	6630
N-Butyl alcohol	1	10430	0	0	0	1433	11863	11863
Nickel compounds	1	381	0	3574	0	0	3955	3955
Nitrilotriacetic acid	0	0	0	0	0	0	0	0
P-Xylene	1	0	0	0	10380	0	10380	10380
Zinc compounds	1	0	267300	0	0	0	267300	267300
Totals	—	3,737,422	1,456,920	2,010,139	1,046,802	4,239,659	12,498,154	—

Table 4.41 Transfer for printed wiring board manufacturing facilities (SIC 3672) in TRL by number of facilities (transfers reported in pounds/year)

Chemical name	# Reporting chemical	POTW discharges	Disposal	Recycling	Treatment	Energy recovery	Total transfers	Average transfers per facility
Sulfuric acid	208	34956	15558	85488	456242	28400	620284	2982
Ammonia	117	412348	2513	6102550	212950	0	6730361	57524
Copper	89	18527	77880	5159806	104791	0	5361004	60236
Copper compounds	73	31441	101998	7949551	263240	0	8346230	114332
Hydrochloric acid	70	1317	750	1056064	1453601	3100	2514832	35926
Nitric acid	59	265	8500	169722	202665	0	381152	6460
Glycol ethers	25	475285	1350	6974	240182	21792	745583	29823
Formaldehyde	22	64501	0	0	2500	0	67001	3046
Chlorine	16	655	0	94152	111000	0	205807	12863
Lead	12	1025	13297	268496	4231	40	287089	23924
Acetone	10	2100	45	3000	1600	188153	194898	19490
Freon 113	9	250	0	77460	1700	5	79415	8824
Lead compounds	7	1559	14454	92233	5125	0	113371	16196
Ammonium sulfate (solution)	6	338933	0	0	0	0	338933	56489
Methyl ethyl ketone	6	0	250	0	750	397048	398048	66341
Phosphoric acid	6	250	0	0	460	0	710	118
Methanol	5	41902	170	0	10746	0	52818	10564
Dichloromethane	4	253	0	71940	2526	38970	113689	28422
1,1,1-Trichloroethane	3	0	0	115750	1410	8180	125340	41780
2-Methoxyethanol	3	0	0	0	0	12250	12250	4083
Hydrogen fluoride	2	0	0	0	5600	0	5600	2800
Nickel	2	251	0	381	0	0	632	316
Toluene	2	8905	0	0	0	121600	130505	65253

Zinc compounds	2	4334	10876	0	1087	0	16297	8149
Ammonium nitrate (solutions)	1	73000	0	0	0	0	73000	73000
Barium compounds	1	0	500	0	0	0	500	500

Source: [20]

In the cathode ray tube subcategory, acetone, methyl ethyl ketone, toluene, and methanol are the major solvents that are usually released during cleaning and degreasing processes. Lead and zinc are the respective metals that are released during the frit sealing process and the phosphor strip process. Other releases are presented in Table 4.42 and the subcategory transfers are summarized in Table 4.43.

In comparison with other industries, the Toxics Release Inventory of the electrical and electronic component industry (SIC 36) is relatively the least. Moreover, the average transfer per facility in this industry is relatively better than some industrial sectors like fabricated metal industry. However, with the growing rate in this sector, among the manufacturing industry, this position might have shifted to the higher side. The Toxics Release Inventory data for selected industries is presented in Table 4.44.

3.2 Waste Generation Prevention

Waste generation or pollution prevention in manufacturing industry simply entails the use of raw materials and production processes that minimize or eradicate the creation of waste or pollutions at the source. This can be achieved through materials input reduction, reengineering processes to reuse by-products, improving management practices, and the use of alternatives to toxic chemicals. Many environmental agencies, particularly the US Environmental Protection Agency, are vanguard in the promotion of pollution prevention because of its cost-effectiveness as against the cost of reducing pollutants that pose risk to the environment and human health. Pollution prevention techniques and processes applicable to the electrical and electronic industry can be broadly divided into process or equipment modification, raw materials substitution or elimination, waste segregation (separation or preparation), as well as recycling. Examples of some of these techniques are summarized in Table 4.45.

4 Electrical and Electronic Component Industry Point Source Discharge Effluent Limitations, Performance Standards, and Pretreatment Standards

Limitations and standards for discharge of effluent primarily target the selection of best method, technology, and equipment as well as the public environmental concern that can be put in place by the manufacturing industries to reduce the discharge of pollutants into the environment. Some of these are discussed below under subcategories of the Electrical and Electronic Component Industry Category.

Table 4.42 Releases for cathode ray tube manufacturing facilities (SIC 3671) in TRI< by number of facilities (releases reported in pounds/year)

Chemical name	#Facilities reporting chemical	Fugitive air	Point air	Water discharge	Underground injection	Land disposal	Total releases	Average releases per facilities
Hydrochloric acid	9	359	589	0	0	0	948	105
Acetone	8	121559	102405	0	0	0	223964	27996
Nitric acid	8	2767	77073	0	0	0	79840	9980
Lead compounds	7	99	2637	435	0	0	3171	453
Sulfuric acid	7	1580	152	1550	0	0	1732	247
Methanol	6	41906	35307	0	0	0	78763	13127
Trichloroethylene	6	151543	393048	476	0	0	544591	90765
Barium compounds	5	6	5	0	0	0	487	97
Hydrogen fluoride	5	328	4175	1681	0	0	5935	1187
Toluene	5	38856	480286	164	0	0	520823	104165
Zinc compounds	4	205	5017	65	0	0	5386	1347
Copper	3	10	255	3103	0	0	330	110
Ammonia	2	1069	8411	2	0	0	12583	6292
Arsenic compounds	2	0	0	0	0	0	2	1
Freon 113	2	34718	5227	0	0	0	39945	19973
Methyl ethyl ketone	2	72778	54045	5	0	0	126823	63412
1,1,1, Trichloroethane	2	1484	35983	146	0	0	37472	18736
Chromium compounds	1	0	0	5	0	0	146	146
Copper compounds	1	10	200	0	0	0	215	215
Methyl isobutyl ketone	1	139	13777	0	0	01	13916	13916
Methylenebis (phenyl isocyanate)	1	0	0	0	0	0	0	0
Nickel	1	5	5	50	0	0	10	10

(continued)

Table 4.42 (continued)

Chemical name	#Facilities reporting chemical	Fugitive air	Point air	Water discharge	Underground injection	Land disposal	Total releases	Average releases per facilities
Nickel compounds	1	0	0	50	0	0	50	50
Tetrachloroethylene	1	0	0	0	0	0	0	0
Xylene (mixed isomers)	1	70418	0	0	0	70488	70488	
Total	—	**470,923**	**1,289,015**	**7,682**	**0**	**1**	**1,767,620**	—

Source: [20]

Table 4.43 Transfers for cathode ray tube manufacturing facilities (SIC 3671) in TRI, by number of facilities (transfers reported in pounds/year)

Chemical name transfers per facilities	#Facilities	POTW reporting	Disposal discharges	Recycling	Treatment	Energy	Total recovery	Average transfers
Hydrochloric acid	9	250	0	0	250	0	500	56
Acetone	8	173	0	21712	60	38674	60619	7577
Nitric acid	8	0	0	0	333274	0	33274	4165659
Lead compounds	7	1175	1924617	487010	137506	0	2550308	364330
Sulfuric acid	7	0	0	250	20639	0	20889	2984
Methanol	6	202029	0	64240	5000	5820	277089	46182
Trichloroethylene	6	250	0	151155	150000	0	301405	50234
Barium compounds	5	255	295228	138785	1850	0	436118	87224
Hydrogen Fluoride	5	39347	0	0	215536	106983	254883	50977
Toluene	5	81	0	626179	277	0	733520	146704
Zinc compounds	4	1397	56654	212504	59710	0	330265	82566
Copper	3	61	279	80492	0	0	80832	26944
Ammonia	2	0	0	0	0	0	0	0
Arsenic compounds	2	0	7388	7579	0	0	14967	7484
Freon 113	2	0	0	7170	0	0	7170	3585
Methyl ethyl ketone	0	0	0	0	15549	15549	7775	
1,1,1-Trichloroethane	2	7	0	10845	0	0	10852	5426
Chromium compounds	1	0	162	2	0	0	164	164
Copper compounds	1	45	0	68700	0	0	68745	68745
Methyl isobutyl ketone	1	0	4192	0	0	0	4192	4192
Nickel	1	63	0	24146	0	0	24209	24209
Nickel compounds	1	0	36	40260	0	0	40296	40296
Tetrachloroethylene	1	0	0	0	20600	0	20600	20600
Xylene (mixed isomers)	1	0	0	0	0	0	0	0
Totals	—	**245,133**	**2,288,556**	**1,941,029**	**944,702**	**168,748**	**5,588,168**	—

Source: [20]

Table 4.44 Comparative TRI (Toxics Release Inventory) data for selected industries

Industry sector	SIC range	#TRI facilities	Total releases (10^6 pounds)	Average transfer per facility (pounds)	1993 total (10^6pounds) (pounds)	Average transfer per facility (pounds)	Total releases + transfers	Average releases + transfers per (10^6 pounds) facility (pounds)
Stone, clay, and concrete	32	634	26.6	41.895	2.2	3,500	28.2	46,000
Lumber and wood products	24	491	8.4	17,036	3.5	7,228	11.9	24,000
Furniture and fixtures	25	313	42.2	134,883	4.2	13,455	46.4	148,000
Printing	2711–2789	318	36.5	115,000	10.2	732,000	46.7	147,000
Electronics/ computers	36	406	6.7	16,520	47.1	115,917	53.7	133,000
Rubber and misc. plastics	30	1,579	118.4	74,986	45.0	28,537	163.4	104,000
Motor vehicle bodies, parts, and accessories	371	609	79.3	130,158	145.5	238,938	224.8	369,000
Pulp and paper	2611–2631	309	169.7	549,000	48.4	157,080	218.1	706,000
Inorganic chemical mfg.	2812–2819	555	179.6	324,000	70.0	126,000	218.1	706,000
Petroleum refining	2911	156	64.3	412,000	417.5	2,676,000	481.9	3,088,000
Fabricated metals	34	2,363	72.0	30,476	195.7	82,802	267.7	123,000
Iron and steel	3312–3313 3321–3325	381	85.8	225,000	609.5	1,600,000	695.3	1,825,000

Nonferrous	333,334	208	182.5	877,269	98.2	472,335	280.7	1,349,000
Metal organic chemical mfg.	2861–2869	417	151.6	364,000	286.7	688,000	438.4	1,052,000
Metal mining	10			Industry sector not subject to TRI reporting				
Nonmetal mining	14			Industry sector not subject to TRI reporting				
Dry cleaning	7215, 7216, 7218			Industry sector not subject to TRI reporting				

Source: [20]

Table 4.45 Pollution prevention techniques applicable to the electrical and electronic component industry (semiconductor printed wiring board, and cathodes ray tube subcategory)

Industry	Operations	Process or equipment modification	Raw material substitution/ elimination	Waste segregation/ separation/ preparation	Recycling
General electrical and electronic industry	General electroplating	Modification of rinsing methods to control drag-out Water conservation through improved rinsing systems. Use of drip bars and drain boards to reduce drag-out Regeneration of plating bath by activated carbon filtration to remove buildup organic contaminants Installation of pH controller to reduce the alkaline and acid concentrations in tanks	Substitution of cyanide plating solutions with alkaline zinc, acid zinc, acid sulfate copper, alkaline copper, electroless nickel, etc. Use of sodium bisulfite and sulfuric acid for ferrous sulfate to oxidize chronic acid waste. Replacing cyanide with non-cyanide baths Substitution of non-cyanide for a sodium cyanide solution used in copper plating baths		Sending drag-out waste to another company for waste exchange Reuse drag-out waste back into process tank General recycling of rinse water Sludge slagging to extract and recycle metals Removal and recovery of lead and tin from boards by electrolysis of chemical precipitation Recovery of dissolved copper from dray-out rinses Recovery of chromatic acid drag-out and recycle it into chromic acid etch line. Conversion of sludge to smelter feed
	General etching	Substitution of sodium persulfate etchant with hydrogen peroxide/sulfuric acid	Replacement of chlorinated solvent baths with nonhazardous products to reduce and later replace use of chlorinated solvents		Recovery of copper by electrolytic process

Semiconductor manufacturing	General operation	Reduction of pressure to vaporize water at cooler temperature and concentrate and precipitate solutes Installation of systems to facilitate chrome waste generation	Substation requires photoresist for TCA and methylene chloride during board manufacturing Substitution of CFC 113 and TCA with non-clean fluxes, aqueous clean fluxes, semi-aqueous cleaning agents and other solvents		
Printed wiring boards	General operation	Modification of sludge pretreatment processes. Redesigning of PWB during board assembly Reduction of pressure to vaporize water at cooler temperature and concentrate and precipitate solutes		Segregation of wastewater sludge for subsequent metal recovery	

Source: [20]

4.1 US Environmental Regulations for Subcategory of Luminescent Materials [16]

New source of this subcategory that introduces pollutants into a Publicly Owned Treatment Works (POTWs) must achieve the pretreatment standards for new sources (PSNS) as stated in Table 4.46. The flowchart of wastewater treatment in luminescent materials subcategory is presented in Fig. 4.5.

Furthermore, the requirement for the new source performance standard (NSPS) for the luminescent material subcategory is listed in Table 4.47.

Table 4.46 Pretreatment standard for new sources for luminescent material subcategory

Pollutant	One day (mg/L)	Monthly average (mg/L)
Cadmium	0.55	0.26
Antimony	0.10	0.04
Zinc	1.64	0.67
Fluoride	35.0	18.0

Source: [26]

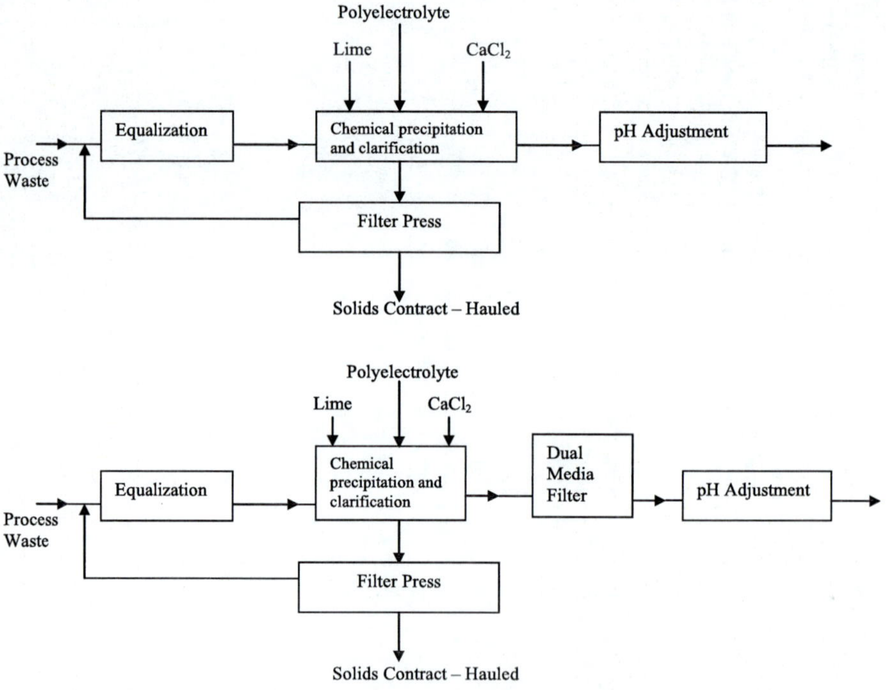

Fig. 4.5 Wastewater treatment for luminescent material subcategory. (Source: [6])

Table 4.47 New source performance standard (NSPS) for luminescent material subcategory

Pollutant	One day (mg/L)	Monthly average (mg/L)
pH	6.00–9.00	6.00–9.00
Cadmium	0.55	0.26
Antimony	0.10	0.04
Zinc	1.64	0.67
Fluoride	35.0	18.0
TSS	60.0	31.0

Source: [26]

Table 4.48 Pretreatment standards for existing sources (PSES) for cathode ray tube subcategory

Pollutant	One day (mg/L)	Monthly average (mg/L)
Total toxic organic	1.58	–
Cadmium	0.06	0.03
Chromium	0.65	0.30
Lead	1.12	0.41
Zinc	1.38	0.56
Fluoride	35.0	18.0

Source: [26]

4.2 Cathode Ray Tube Subcategory

Table 4.48 shows pretreatment standards which any existing source under the cathode ray tube subcategory that introduces pollutants into a Publicly Owned Treatment Works (POTWs) must achieve.

Under the regulations for new source performance standards (NSPS), a new source subject to the cathode ray tube subcategory must achieve the following standard as presented in Table 4.49.

Any new source of the cathode ray tube subcategory that introduces pollutants into a POTWs must achieve the under listed pretreatment standards for new sources (PSNS) in Table 4.50.

4.3 Semiconductor Subcategory

The degree of effluent reduction achievable by the application of the best practicable control technology (BPT) currently available for the semiconductor subcategory is presented in Table 4.51.

The new source performance standards for the semiconductor subcategory are given in Table 4.52 and any new source must achieve the standard stated therein.

In the application of the best conventional pollution control technology (BCT), the degree of effluent reduction attainable is documented in Table 4.53, and any existing point source subjected to the semiconductor subcategory must achieve this standard.

Table 4.49 New source performance standards (NSPS) for cathode ray tube subcategory

Pollutant	One day (mg/L)	Monthly average (mg/L)
pH	6.00–9.00	6.00–9.00
Total toxic organics	1.58	–
Cadmium	0.06	0.03
Chromium	0.56	0.26
Lead	0.72	0.27
Zinc	0.80	0.33
Fluoride	35.00	18.00
TSS	46.00	24.00

Source: [26]

Table 4.50 Pretreatment standards for new source for cathode ray tube subcategory

Pollutant	One day (mg/L)	Monthly average (mg/L)
Total toxic organics	1.58	–
Cadmium	0.06	0.03
Chromium	0.56	0.26
Lead	0.72	0.27
Zinc	0.80	0.33
Fluoride	35.00	18.00

Source: [26]

Table 4.51 Semiconductor BPT effluent limitations

Pollutant	One day (mg/L)	Monthly average (mg/L)
Total toxic organics	1.37	Not applicable
pH	6.00–9.00	6.00–9.00

Source: [26]

Table 4.52 Semiconductor NSPS effluent limitations

Pollutant	One day (mg/L)	Monthly average (mg/L)
Total toxic organics	1.37	Not applicable
Fluoride	32.00	17.40
pH	Not applicable	6.00–9.00

Source: [26]

Table 4.53 Semiconductor BCT effluent limitations

Pollutant	One day (mg/L)	Monthly average (mg/L)
pH	6.00–9.00	6.00–9.00

Source: [26]

4.4 *Electronic Crystal Subcategory*

Table 4.54 presents the degree of effluent reduction attainable by the application of the best available technology (BAT) economically available for the electronic crystal subcategory which any existing point source must achieve.

The effluent limitations for the electronic crystal subcategory representing the degree of effluent reduction attainable by the application of the best practicable control technology currently available (BPT) are listed in Table 4.55.

Any existing source of the electronic crystal subcategory that introduces pollutant into a Publicly Owned Treatment Work (POTWs) must base its treatment on the pretreatment standards for existing sources (PSES) as stated in Table 4.56.

The new source performance standard (NSPS) for the electronic crystal subcategory is documented in Table 4.57, which any new source must comply with.

The degree of effluent reduction attainable by the application of the best conventional pollution control technology (BCT) currently available for the electronic crystal subcategory is given in Table 4.58.

5 Pollution Prevention Case Studies

Pollution prevention programs that are most successful involve looking for opportunities "beyond the barrels."

Table 4.54 Electronic crystals BAT effluent limitations

Pollutant	One day (mg/L)	Monthly average (mg/L)
Total toxic organic	1.37	Not applicable
Arsenic[a] (T)	2.09	0.83
Fluoride (T)	32.00	17.40

Source: [26]
[a]The arsenic limitation only applies to manufacturers of gallium or indium arsenide crystals

Table 4.55 Electronic crystals BPT effluent limitations

Pollutant	One day (mg/L)	Monthly average (mg/L)
Total toxic organic	1.37	Not applicable
Arsenic[a] (T)	2.09	0.83
Fluoride (T)	32.00	17.40
Total suspended solids, TSS	61.00	28.00
pH	6.00–9.00	6.00–9.00

Source: [26]

Table 4.56 Electronic crystals PSES effluent limitations

Pollutant	One day (mg/L)	Monthly average (mg/L)
Total toxic organics	1.37	Not applicable
Arsenic[a] (T)	2.09	0.83

Source: [26]

[a]The arsenic limitation only applies to manufacturers of gallium or indium arsenide crystals

Table 4.57 Electronic crystals NSPS effluent limitations

Pollutant	One day (mg/L)	Monthly average (mg/L)
Total toxic organics	1.37	Not Applicable
Arsenic[a] (T)	2.09	0.83
Fluoride (T)	32.00	17.40
Total suspended solids	61.00	23.00
pH	6.00–9.00	6.00–9.00

Source: [26]

[a]The arsenic limitation only applies to manufacturers of gallium or indium arsenide crystals

Table 4.58 Electronic crystals BCT effluent limitations

Pollutant	One day (mg/L)	Monthly average (mg/L)
Total toxic organics	61.0	23.0
pH	6.00–9.00	6.00–9.00

Source: [26]

5.1 Texas Instruments

One of the top electrical and electronic industries in the USA is Texas Instruments, which is the fifth ranking in the manufacture of semiconductors [22]. With headquarter in Dallas and worldwide manufacturing facilities, Texas Instruments workforce is approximately 36,000. The company is leading in the design and supply of Digital Signal Processing Solutions (DSPs) and others like materials and controls, educational and productivity solutions, and digital imaging. Some other manufacturing industries particularly inorganic chemical manufacturers operate within the facility in the company to facilitate supply of resources needed for production [23]. Prominent among the Texas Instrument facilities is the Texas Instruments Sherman located near Dallas, which produces chips. The 1997 verification of industrial waste generated in the facility waste stream is illustrated in Table 4.59.

Facility, primarily for waste minimization as well as cost benefit. By 1996, following the technology and compliance with various environmental acts, the Sherman plant was able to reduce the quantities of hazardous wastes it generated to 60% [24]. Some of these approaches are summarized in Table 4.60.

Table 4.59 Texas Instruments Incorporated waste generation data, 1997

Number of waste stream							
List only	Listed and characteristic	Characteristic only	Total	Physical form	Waste type	Annual waste quantity (tons)	Average quantity per waste (tons)
2	3	9	14	Non-wastewater	Hazardous	216	15
0	0	2	2	Wastewater	Hazardous	2,759	1,380
N/A	N/A	N/A	15	Non-wastewater	Nonhazardous	234	16
Total waste stream			31	Total waste quantity		3,209	

Source: [24]

Table 4.60 Sherman's facility pollution prevention initiatives

Company	Technology	Investment cost ($)
Raytheon	Simultaneous clarification and filtration treatment plant	600,000–800,000
Texas Instruments	Phosphate removal system	50,000
Instruments	Recycling of treated wastewater to cool towers and scrubbers	250,000–300,000

Source: [24]

5.2 *Tri-Star Technologies*

Tri-Star Technologies, Inc. is a manufacturer of double-sided and multilayer printed wiring boards, specializing in products for the electronics industry. With 220 employees at their 120,000 ft^2 facility, they produce 1,000,000 surface ft^2 annually [25].

Broadening of pollution prevention by Tri-Star Technologies, Inc. in Methuen, Massachusetts, has led to reductions in energy and water use that have resulted in significant cost savings. During implementation of pollution prevention program, Tri-Star discovered that demonstrating cost savings is the key to a successful pollution prevention program.

With several successes to their credit (both from an economic and a waste reduction standpoint), Tri-Star's pollution prevention team was able to obtain funding for projects that are capital intensive. These projects offered increased cost savings in the long term. For instance, the facility currently installed a cupric chloride regeneration system to recycle their inner-layer etchant. Such a system requires a capital investment of about $150,000, but the payback is expected to be less than 18 months through significant reductions in spent (hazardous) etchant and virgin chemical purchases and by selling the recovered copper by-product.

Tri-Star engaged cost saving opportunities in energy and water conservation by looking beyond regulated materials. Such opportunities are a "cheap, and easy way to reduce a facility's environmental impact while saving money," says Ed Gomes, Environmental Health and Safety Manager for the facility. On a global level pollution prevention can be successfully achieved by reductions of energy that can lead to reductions in the by-products of energy use, which can subsequently cause global warming, acid rain, and smog. Though, Tri-Star projects are directly not transferable to every facility, however, other manufacturers could use the experienced gained from Tri-Star to examine their own energy and water use.

Tri-Star has implemented several energy conservation projects. The two projects described in this case study involved collaborative efforts with the electric and/or gas companies. Together, these two projects have resulted in savings of thousands of dollars per month or about $51,800 annually.

In order to save energy, Tri-Star was able to balance its airflow rate at the flow-rate air make up units. With several pieces of equipment exhausting air, such as wet scrubbers and an electrostatic precipitator, Tri-Star has been using two gas-fired air makeup units, each with a capacity of 40,000 cubic feet per minute (cpm) to balance the airflow. Since these operated at a fixed rate, the facility is continuously operating, even during non-production hours.

Aside the operating energy being consumed by this facility, flow-rate air makeup units also required the air conditioning or heating systems to work overtime such that the units blow hot, humid air into the facility during summer and heat up the air blown in during winter, thus resulting in a concentrated heat source that causes nonuniform high temperature in the facility, in which thermostats are shut down in some areas in the facility and making the cold areas colder.

Tri-Star uses state-of-the-art variable speed controllers, which could be retrofitted to the makeup air units, because of its alliance with the gas and electric companies. Using the variable speed controllers, the flow rate is determined based on the air exhaust rate from the exhausting equipment. Through this, significant savings in electric and gas bills were realized. An outstanding feature of this project is that it requires no capital investment; the new equipment was supplied by the electric and gas company.

The annual savings on gas and electricity by the facility are estimated to be $22,900 and $15,600, respectively. Also, energy savings of 31,000 therms and 192,800 kilowatt-hours (KWH) through the project caused reduction of air pollution, which translates to annual reductions of CO_2 (global warming) of 212,100 pounds, SO_2 (acid rain) of 1,700 pounds, and NOx (acid rain and smog) of 600 pounds.

Tri-Star examined compressed air in its energy conservation project. The industry had been using two 100 HP and two 50 HP compressors to provide the facility with compressed air. However, these units did not meet the compressed air demand. Investigation of the facility's compressor by "Energy Initiative" program conducted by the electric company (Mass Electric) resolved into addition of a reserve air tank, replacement of the four compressors of 300 HP combined capacity with three 50 HP energy efficient compressors (150 HP combined capacity), and setting up of units to cycle, based on the compressed air demand in the facility.

Based on these changes Tri-Star was able to save energy and eliminate the problems of meeting the facility demand for compressed air. Annual energy cost savings from this project are estimated to be $13,300, based on a 164,800 KWH reduction in electricity use. This translates to annual reductions of CO_2 (global warming) of 181,300 pounds, SO_2 (acid rain) of 1,500 pounds, and NOx (acid rain and smog) of 500 pounds. These energy use reduction projects have improved plant operations, saved money, and reduced virgin oil consumption, waste oil (hazardous waste) generation, and air pollution to the community. Tri-Star's pollution prevention project that goes beyond regulated materials has led to water conservation in the industry.

Tri-Star expanded their fabrication business to add assembly operations and the facility considered the different types of systems that could be used to clean flux residue from the wave solder unit. The facility discovered that hot deionized (DI) water could equally clean the boards just as effectively as the chemical-based cleaning systems. Thus, the facility purchased a closed-loop DI water generation system that delivers 5–7 gallons per minute (gpm). The system generates DI water and recycles it through the cleaning process in a closed-loop system.

Tri-Star initially experienced problem that solder paste was degrading the DI water generation system's resin columns. The problem was overcome when the facility installed a sink on the side cleaning unit where the operator could manually clean the reject boards with the hot DI water instead of placing them on the system conveyor. The effluent from the sink was made to pass through a filter and directly to waste treatment. This simple installation did not allow the passage of the solder mask from the rejected boards to the closed-loop system, thus, extending the life of the system's resin columns.

The advantages of this closed-loop system cannot be compared to other technologies such as flux residue cleaning systems; some of these advantages are that it conserves water, it does not use solvents, and it does not involve the use of chemicals.

In its attempt to further conserve water, Tri-Star put some "smart rinsing" techniques on their electroless copper line, installed flow controls on rinses and increased counterflow rinsing. With all these changes on the deposition line, the facility was able to cut the incoming water sources from seven to four and reduced its water usage on the line from 19 to 4 gpm. To make these changes in any process, Tri-Star put up the following cross contamination criteria, such as do not contaminate:

- Electroless copper bath with palladium from the catalyst
- Catalyst with cleaner/conditioner
- Accelerator with electroless copper
- Electroless copper with microetch or acids
- Microetch with cleaner/conditioner

Overall, the water usage at Tri-Star was reduced by 2.5 million gallons per year, resulting in cost savings of approximately $15,000. This was due to the operating rate of the electroless copper line for 12–16 hours per day and 5 days per week. The total fees for water and sewer are $6.26/1,000 gal. These water conservation techniques may be applied to other facilities that use multiple rinses.

Another merit of these changes was that the amount of chemicals needed to maintain the process baths was reduced by 25% for the affected baths (glass etch, microetch, sulfuric acid dip, and accelerator), thus making savings on chemical purchases.

6 Summary

Waste generation or pollution prevention in manufacturing industry simply entails the use of raw materials and production processes that minimize or eradicate the creation of waste or pollutions at the source. This can be achieved through materials input reduction, reengineering processes to reuse by-products, improving management practices, and the use of alternatives to toxic chemicals [26–29]. Many environmental agencies, particularly the US Environmental Protection Agency, are vanguard in the promotion of pollution prevention because of its cost-effectiveness as against the cost of reducing pollutants that pose risk to the environment and human health. Pollution prevention techniques and processes applicable to the electrical and electronic industry can be, broadly, divided into process or equipment modification, raw materials substitution or elimination, waste segregation (separation or preparation) and, finally, recycling. Technical glossary terms can be found from the literature [30–33].

References

1. Gurauskiene I, Varizinskas V (2006) Eco-design methodology for electrical and electronic equipment industry. Env Res Eng Manage 3(37):43–51
2. US Department of Labour, Bureau of Labour Statistics, Economic and Employment Projections (2004)
3. US industry and trade outlook: New York McGraw Hill (2000)
4. US Department of Commerce, Bureau of the Census (1995)
5. Department of Trade and Economic Department and Electronic Computer Industry Group (1998) A review of key location factors Saint Paul, Minnesota
6. USEPA Office of Compliance Sector Notebook Project Profile of the Electronics and Electrical Industry (1981) U.S. Environmental Protection Agency 401 M St., SW (MC 2221-A) Washington, DC 20460
7. Amuda OS, Ibrahim AO (2006) Industrial wastewater treatment for chemical oxygen demand (COD) using natural material as adsorbent. Afr J Biotechnol **5**(16):1483–1487
8. Amuda OS, Ojo IO, Edewor TI (2007) Biosorption of Lead in Industrial wastewater using shell of seed of *Chrysophyllum albidum*. Biorem J 11(4):183–194
9. Amuda OS, Giwa AA, Bello IA (2007) Removal of heavy metal from industrial wastewater using modified activated coconut shell carbon. J Biochem Eng **36**:174–181
10. Amuda OS, Alade A (2006) Coagulation/flocculation process in the treatment of abattoir wastewater. Desalination **196**:22–31
11. Amuda OS, Amoo IA, Ipinmoroti KO, Ajayi OO (2006) Coagulation /flocculation process in the removal of trace metals present in industrial wastewater. J Appl Sci Environ Manag **10** (3):159–162
12. Amoo IA, Amuda OS, Ipinmoroti KO, Ajayi OO (2006) Performance optimization of some chemical coagulants/flocculants in the treatment of wastewater from food industry. Sci Focus **11** (1):38–45
13. Amuda OS, Amoo IA, Ajayi OO (2006) Performance optimization of some coagulants/flocculants in the treatment of a beverage industrial wastewater. J Hazard Mater B **129**:69–72
14. Amuda OS, Amoo IA (2007) Coagulation/flocculation process and sludge conditioning in beverage industrial wastewater treatment. J Hazard Mater **141**:778–783
15. Hargreaves AJ, Vale P, Whelan J, Alibardi L, Constantino C, Dotro G, Cartmell E, Campo P (2018) Coagulation–flocculation process with metal salts, synthetic polymers and biopolymers for the removal of trace metals (Cu, Pb, Ni, Zn) from municipal wastewater. Clean Techn Environ Policy 20:393–402
16. Ghorpade A, Ahammed MM (2017) Water treatment sludge for removal of heavy metals from electroplating wastewater. Environ Eng Res 23(1):92–98
17. Lugo LMA, Diaz J, Pérez-Flórez A, Celis C (2020) Implementation of modified acacia tannin by Mannich reaction for removal of heavy metals (Cu, Cr and Hg). Water 12:352–363

18. Song X, Li L, Zhou L, Chen P (2018) Magnetic thiolated/quaternized-chitosan composites design and application for various heavy metal ions removal, including cation and anion. Chem Eng Res Des 136:581–592
19. Tiruneh AT, Debessai TY, Bwembya GC, Nkambul SJ (2018) Combined clay adsorption-coagulation process for the removal of some heavy metals from water and wastewater. Am J Environ Eng 8(2):25–35
20. USEPA Office of Compliance Sector Notebook Project Profile of the Electronics and Electrical Industry (1995) Office of Compliance Office of Enforcement and Compliance Assurance U.S. Environmental Protection Agency 401 M St., SW (MC 2221-A) Washington, DC, USA, 20460 September
21. USEPA (1993) Toxic release inventory database, USA
22. USEPA (1999, September) Office of enforcement and compliance assurance, profile of the electronics and industry, USA, p 9–11
23. Baker DJ (ed) (2003) Wards Business directory of US private and public companies. Thomson Gale, Detroit, MI, USA
24. Texas Instrument Inc (1997) Annual waste summary, Dallas, TX, USA
25. USEPA Office of Solid Waste (1999) Hazardous waste identification rule (HWIR) industry case studies. USEPA, Washington, D.C., USA
26. Code of Federal Regulations (2003, July 1) The U.S. Government Printing Office via GPO Access PART 469 electrical and electronic components point source category title 40, Volume 27
27. DfE Project (2013, February 1) Pollution prevention information clearinghouse (PPIC) U.S. EPA. 1200 Pennsylvania Ave. NW (7407-T) Washington, DC, USA, 20460-0001. http://www.ipc.org
28. Wang LK (2010) Recycling and disposal of hazardous solid wastes containing heavy metals and other toxic substances. In: Wang LK, Hung YT, Shammas NK (eds) Handbook of advanced industrial and hazardous wastes treatment. CRC Press, Boca Raton, FL, USA, pp 1213–1232
29. Lee D (2011) Water and wastewater treatment solutions for electrical & electronic, food & beverage, and power industries in APAC. Frost Sullivan. http://www.slideshare.net/FrostandSullivan/water-and-wastewater-treatment-solutions-for-electrical-electronic-food-beverage-and-power-industries-in-apac
30. Nood L (2012, May) Research and markets: Water and wastewater treatment solutions in electrical and electronic industry. http://www.businesswire.com/news/home/20120531005747/en/Research-Markets-Water-Wastewater-Treatment-Solutions-Electrical
31. Jimlynch J (2013, May 7) Snapshot of worldwide electronics recycling 2013. http://ifixit.org/category/e-waste-2/
32. Wang MHS, Wang LK (2015) Environmental water engineering glossary. In: Yang CT, Wang LK (eds) Advances in water resources engineering. Springer Science + Business Media, New York, USA, pp 471–556
33. Wang MHS, Wang LK (2016) Glossary of land and energy resources engineering. In: Wang LK, MHS W, Hung YT, Shammas NK (eds) Natural resources and control processes. Springer Science + Business Media, New York, USA

Chapter 5
Geographic Information Systems and Remote Sensing Applications in Environmental and Water Resources

Samuel O. Darkwah, Michael D. Scoville, and Lawrence K. Wang

Contents

1 Introduction 198
2 Fundamentals of Remote Sensing 199
2.1 Characteristics of Various Sensor Data 199
2.2 Satellite Data Processing 201
3 Fundamentals of GIS 201
3.1 GIS Subsystems 203
3.2 Data Models 204
4 Applications to Water Resources 204
4.1 Mapping and Monitoring 205
4.2 Modeling 211
5 US Environmental Protection Agency Programs 216
Glossary 217
References 233

Abstract The fundamentals of both remote sensing (RS) and geographic information systems (GIS) are introduced in detail. For RS, the characteristics of various sensor data and the satellite data processing technology are explained. For GIS, the authors describe various GIS subsystems and data models. There are many applications to water resources, such as mapping, monitoring, and modeling. The US Environmental Protection Agency (USEPA) has established many useful programs to support the activities of American water resources engineers and planners. These important USEPA support programs are presented.

S. O. Darkwah (✉)
Lenox Institute of Water Technology, Lenox, MA, USA

M. D. Scoville
Blasland, Bouck & Lee, Inc., Syracuse, NY, USA

L. K. Wang
Lenox Institute of Water Technology, Newtonville, NY, USA

L. K. Wang, M. -H. S. Wang, Y. -T. Hung, N. K. Shammas (eds.), *Integrated Natural Resources Management*, Handbook of Environmental Engineering 20,
https://doi.org/10.1007/978-3-030-55172-8_5

Keywords Geographic information systems.GIS · Remote sensing · RS. Sensor data · Satellite data processing. GID data models. Mapping. Monitoring. GIS application. · Water resources. USEPA programs. Glossary

Nomenclature

DEM	Digital elevation model
ETM	Enhanced Thematic Mapper
GIS	Geographic information systems
Landsat	US satellite
MSS	Multispectral Scanner carried on Landsat satellite
P	Panchromatic sensor on SPOT satellite
SAR	Synthetic aperture radar
SPOT	Satellite Pour l'Observation de la Terre, a French satellite
TIN	Triangulated irregular network
TM	Thematic Mapper
USEPA	US Environmental Protection Agency
XS	Multispectal sensor on SPOT satellite

1 Introduction

Conflicting interests and parameters often characterize the decision-making for water resources projects. Systematic and exhaustive analysis of these issues would depend on the quality and quantity of information available on both natural and socioeconomic resources. Remote sensing technology has demonstrated its capabilities to generate information in spatial and temporal domain, which is very crucial for successful water resources monitoring, model analysis, prediction, and validation. However, the use of remote sensing technology involves large amount of spatial data management and requires an efficient system to handle such data. Geographic information systems (GIS) are excellent tools available to store, retrieve, and analyze different types of data for management of water resources [1]. Remote sensing and GIS techniques can be used to analyze large or small areas, integrate numerous variables into the evaluation process, and easily update database information. These two technologies thus add a spatial dimension to water resources analysis process and provide a more effective tool for a quick and transparent decision-making. Remote sensing and GIS also play an important role in evolving alternate scenarios for water resources investigation.

The current domain of GIS includes all aspects of spatial data analysis from data acquisition and manipulation through problem solving to the presentation of results. GIS needs large quantities of accurate data to be useful. Remote sensing data have a number of features which make them ideal GIS data source. Satellites generate inexpensive, reliable, and timely information on various natural resources such as surface water, groundwater, land use/cover, soil, and environmental hazards. For most water resources investigations, GIS provides an extremely useful technology for interaction between remote sensing data and a multitude of geographic data such as topography, rainfall, evaporation, vegetation, and geomorphology. This chapter is intended to provide the background information necessary to successfully use GIS technology in conjunction with remotely sensed data to plan and manage water resources projects. The chapter also provides case studies that demonstrate the proper application of GIS and remote sensing principles.

A GIS references real-world spatial data elements to a coordinate system. Spatial data elements are also known as graphic or feature data elements. The feature data elements can be separated into different layers or map themes. A layer is a thematic set of spatial data described and stored in a database or map library. Layers organize a database or map library by subject matter such as soils, rivers, and wells. A GIS can also store attribute data, which is a descriptive information of the map features. GIS can combine satellite data and other types of geographic data to generate maps and reports, enabling users to collect, manage, and interpret location-based information in a planned and systematic way. A satellite is the vehicle or platform which carries a sensor. Some important satellite platforms are Satellite Pour l'Observation de la Terre (SPOT), Landsat, Ikonos, and Indian Remote Sensing satellite (IRS-1A and IRS-1C). The HRV (High Resolution Visible) sensor carried on SPOT satellite operates into two modes: the XS or multispectral mode and P or Panchromatic mode. The three sensors carried on Landsat satellites are MSS (MultiSpectral Scanner), TM (Thematic Mapper), and ETM (Enhanced Thematic Mapper). The readers are referred to the Glossary Section of this book chapter or the USEPA web site for the glossary details.

2 Fundamentals of Remote Sensing

Remote sensing may be broadly defined as the collection of information about an object without being in physical contact with the object. Since 1972, satellites have provided repetitive, synoptic, global coverage of high-resolution imageries of the earth surface that can be used to interpret numerous phenomena. Satellites measure and record the spectral reflectance (the portion of the incident energy that is reflected) of the earth's surface in several different wavelength ranges, from microwave to visible light. The electromagnetic spectrum illustrated in Fig. 5.1 is the basis of all environmental remote sensing. Ultraviolet, visible, and short infrared radiations (<3 μm wavelength) are mainly reflected solar energy while mid-infrared, thermal infrared, and microwave radiations (>3 μm wavelength) are mostly emitted from the earth's surface. For more in-depth information on basic remote sensing, the reader is referred to remote sensing books by Lillesand and Keifer [2] and Barret and Curtis [3].

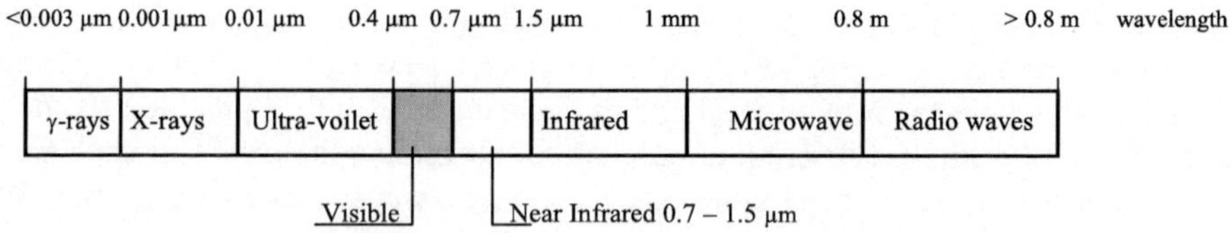

Fig. 5.1 The electromagnetic spectrum

2.1 Characteristics of Various Sensor Data

Aircraft and satellites are the common platforms from which remote sensing observations are made. Environmental satellites carry two types of sensors, electro-optical sensors and synthetic aperture radar (SAR) for gathering data from the earth's surface. Most of the electro-optical sensors are passive instruments that capture reflected energy from the sun and therefore only operate during the daytime and in cloud-free environments. Active scanners such as SAR supply their own energy source that bounces off the target and is captured by the instrument. SAR is therefore used at night and in areas such as the tropics that are cloud covered for most of the year. An example of an active sensor is RADARSAT, a Canadian radar sensor.

Most remote sensing images are from electro-optical sensors such as aerial photography and multispectral scanners. The commonly used electro-optical imageries are described below.

Panchromatic Imagery Panchromatic imagery is represented as black and white and acquired by a digital sensor that measures reflectance in one wide portion of the electromagnetic spectrum. These wavelength portions of the spectrum are often called bands. For most current panchromatic sensors, this single band usually spans the visible to near-infrared part of the spectrum. Panchromatic images are reliable for identification of land forms, erosional and depositional features, disturbed land, many kinds of man-made features, and land/water interface.

Multispectral Imagery Multispectral imagery is acquired by a digital sensor that measures reflectance in many wavelengths or bands. These multiple reflectance values are combined to create color images. The nature of multispectral data makes them attractive for land use/cover and surface character analysis.

Hyperspectral Imagery Hyperspectral imagery refers to a spectral sensor capable of measuring reflectance in many individual bands (up to 250) of the electromagnetic spectrum. The theory behind hyperspectral sensing is that measurement of reflectance in numerous narrow portions of the spectrum can detect very subtle characteristics and differences among surface features.

Some of the most commonly used data from passive sensors comes from the Landsat MSS, Landsat TM, Landsat ETM, SPOT P, SPOT XS, and IRS1-C/D sensors. One of the newest sensors to recently go into orbit is Space Imaging IKONOS sensor. Launched on September 24, 1999, the Ikonos-2 satellite provides 1 m panchromatic data and 4 m multispectral data, paving the way for highly

Table 5.1 Commonly used satellite data for water resources application

Satellite	Sensor	Spatial resolution (m)	Number of bands	First launched	Potential application
Landsat	TM	30	7	1984	1,2,3,4,5,7
	MSS	80	4	1972	1,2,3,4,7
	ETM	30 & 15	8	1999	1,2,3,4,5,7
SPOT	XS	20	3	1985	1,2,3,4,5,7
	P	10	1	1985	1,6,3,7
IRS1-C/D	Multispectral	23	3	1997	1,4,3,7
	P	6	1	1997	1,6,3,7
Ikonos	Multispectral	3	4	1999	1,2,3,4,5,7
	P	1	1	1999	1,6,3,7

1 Environmental monitoring, *2* bathymetry, *3* change detection, *4* water bottom conditions, *5* sediment patterns/turbidity, *6* coastal processes, *7* land-water interface mapping

detailed analysis. Another source for high-resolution image data for many regions of the world is the Russian KVR-1000 imagery with a 2 m spatial resolution. Table 5.1 shows the commonly used satellite data and their potential application in water resources. The satellite's sensor observes a small portion of the earth at a time. This small area is usually called a pixel, and its size is represented by the spatial resolution. The pixel size is a function of the satellite sensor and so far has ranged from 0.5 m to 1 km.

2.2 *Satellite Data Processing*

Satellite imagery is nothing more than a grid of numbers. In order to know the location of an image, the pixel rows and columns must be oriented to a known geographic coordinate system such as Universal Transverse Mercator (UTM) or Albers. The main step involved in turning raw imagery, whether satellite or aerial, into a resource from which useful products can be derived is orthorectification. In order to orthorectify imagery, a digital elevation model (DEM) and a transformation model which takes into account the various sources of image distortion generated at the time of image acquisition are required. These include, but are not limited to, sensor orientation, topographic relief, earth shape and rotation, satellite orbit and attitude variations, and systematic error associated with the sensor.

Satellite imagery is usually handled by special software; however, many of the same techniques are used in other imaging software packages. Software packages that specialize in satellite images include PCI Easi/Pace, ENVI, Erdas Imagine, ER Mapper, and Idrisi. In image processing, the computer is used to detect information about the area recorded in the images that cannot be seen by the eye. The most common procedure here is image classification. This procedure determines the land cover of pixels in a scene. The classification usually identifies land cover types such as water, forest, grassland, urbanized area, and snow. Colwell [4] and Jesen [5] give more information on satellite image processing methods.

3 Fundamentals of GIS

A GIS is best described as a system of hardware, software, data, people, organizations, and institutional arrangements for collecting, storing, analyzing, and disseminating information about areas of the earth [6]. According to Congalton and Green [7], GIS analysis can be divided into four main categories: overlaying, buffering, modeling, and network analysis. These four components of analysis represent the basic tools of GIS. Each one of these tools is simple, yet they can be combined to produce complex, spatial analysis.

The process of overlaying usually involves combining two or more different layers of a particular area to create a new layer that combines the attributes of individual layers or compares the layers to calculate summary statistics. Various types of overlay operations exist. Layers can be added, subtracted, divided, or multiplied according to the values of certain attributes. The two datasets shown in Fig. 5.2 depict the conditions of a floodplain in Kentucky during flood and post-flood periods. The (D:\BOOK\7append\glossary.html) - polygon polygon datasets shown in Fig. 5.2c and d were derived from Landsat imageries and indicate the wet and land areas in the floodplain at the different discharge conditions. If we overlay the two layers and compare the floodplain attributes (by subtracting the values of Fig. 5.2c from Fig. 5.2d), we can identify areas where the floodplain has changed in the period. After the subtraction operation, the new data layer (Fig. 5.2e) which was created shows change as a positive or negative and no change as zero value. As demonstrated in the figure, the overlay results quickly show where in the floodplain changes have occurred due to flow changes. Overlaying is frequently combined with other analysis methods to produce even more valuable results. For instance, one could combine this operation with a model that automatically calculates the water surface elevation from existing data.

Buffering is an important pre-analysis technique of identifying objects within a specified distance of a reference object. It combines spatial data query techniques and cartographic modeling. The reference object may be a point location, a line, or a polygon. A simple environmental example would be to create a buffer of groundwater pollution zone around a waste site. This buffer could be used to assess health risks to the affected population. The buffering operation can also be used to predict coastline erosion by estimating the distance of the coastline at any time from the buffer.

One of the most useful transformations in GIS has been the incorporation of modeling. This involves linking the GIS database to a computer model of some process. The GIS prepares input data for the model by combining all the relevant data for every object. This allows spatial data to be processed in large quantities using powerful, complex algorithms. There are two types of modeling: simulation and predictive. Generally simulation modeling requires a high degree of technical expertise and involves using GIS to simulate a complex phenomenon in nature. Predictive modeling is a more powerful modeling tool than the simulation modeling.

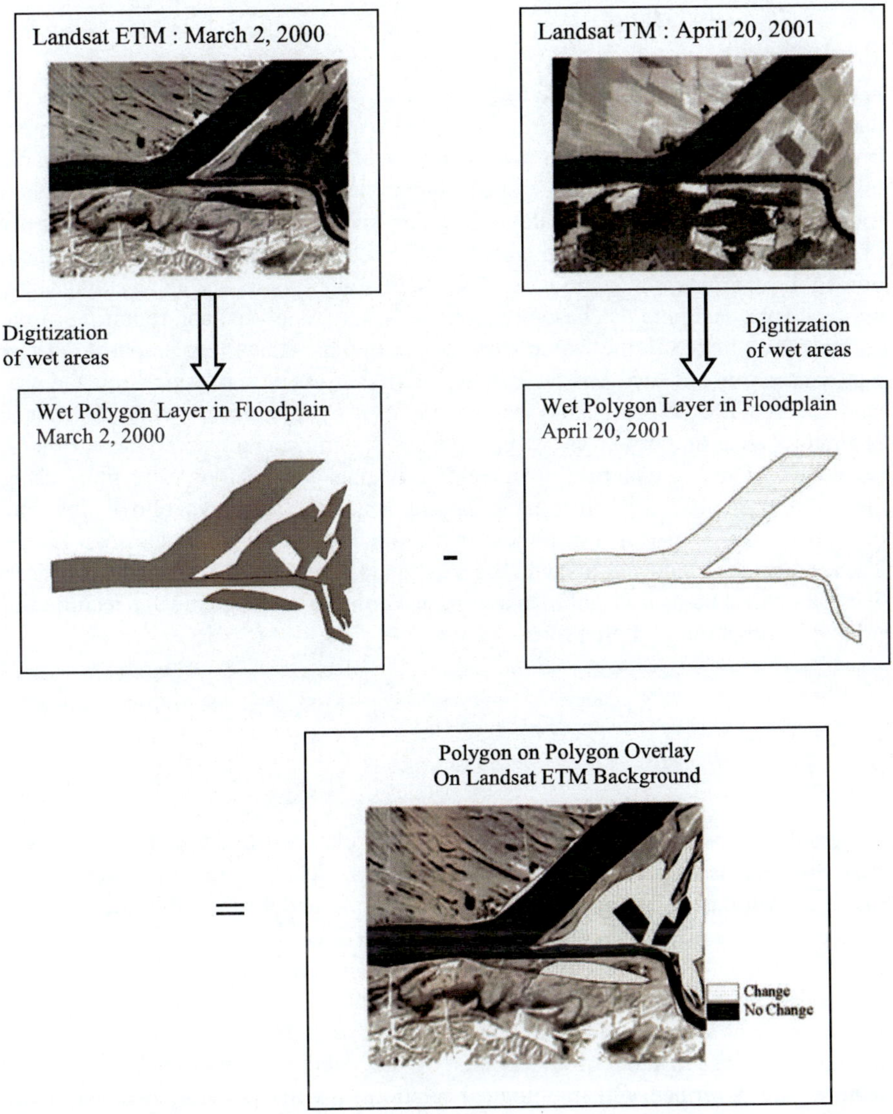

Fig. 5.2 Simple illustration of GIS overlay function

In this form of modeling, a modeler builds a statistical prediction model from available data through regression analysis. This type of modeling has been used to predict processes like flooding, groundwater contamination, and soil loss.

Network analysis allows GIS users to solve a variety of problems using geographic networks (rivers, pipelines, streets, highways, etc.).

3.1 GIS Subsystems

A GIS is composed of four basic subsystems: data input, data storage and retrieval, data manipulation and analysis, and a reporting system [8]. The data input subsystem involves the conversion of data from maps, field observations, processed satellite images, and aerial photographs into compatible digital form. Data conversion routines are available to scan or digitize analog maps for input to a GIS. The quality of input data will influence the quality of GIS products regardless of the sophistication of its hardware and software. The main functions of the data storage and retrieval subsystem are database organization, query, analysis, and reporting of the attribute data linked to the features on the maps. This subsystem prepares data by removing errors and analyzes the data to provide answers to the questions the user puts to the GIS. Typical analyses include overlaying different thematic maps, computing areas and distances, acquiring statistical information about the attributes, projection of maps, and making three-dimensional perspective view plots using elevation data. The data manipulation and analysis subsystem allows for data conversion and to perform analytical and modeling functions on the map layers. The reporting subsystem deals with the way the information is displayed to the user. This can either be as a visual display or hard copy or as magnetically recorded or printed information in digital form.

3.2 Data Models

Geographic data can be classified into two main classes: spatial data and attribute data. Spatial data are represented in two basic ways, either as a raster or a vector. In a raster or cell-based system, spatial data is represented by a geometric array of rectangular or square cells, each with an assigned value. Raster data is typically used for remote sensing data. In a vector-based system, spatial data is represented by a set of straight-line segments called vectors. The coordinates at the end of each vector segment are digitized and explicitly stored, and the connections are implied through the organization of the points in the database. Relationships between features are described with topological relationships that include area definition, adjacency, connectivity, and nestedness. A newer, non-topological data structure is the shapefile, which stores the geometry and attribute information for a geographic feature in a dataset that includes a main file, an index file, and a database table [8]. Time can also be considered a data element, since geographic information often changes over time. For instance, a river course may meander over time, or river dimensions may undergo sudden changes due to floods, and land use may change due to agricultural or industrial use.

4 Applications to Water Resources

The main applications of remote sensing and GIS in water resources are data collection, data management, analysis functions, physical and mathematical representation of hydrologic data, surface hydrology, and groundwater management. These can be summarized as mapping, monitoring, and modeling.

4.1 Mapping and Monitoring

GIS and remote sensing have been commonly used to map and monitor floods, floodplain, suspended sediment, turbidity, chlorophyll, and total phosphorus. An important step in these analyses is the correct mapping of land-water interface. The land-water interface is located in regions where both aquatic and terrestrial resource systems co-exist in space and time.

4.1.1 Mapping Land-Water Interface

The ultimate objective of flood mapping is to delineate precisely where the land-water interface is located, preferably at multiple stages of inundation. A typical procedure for mapping land-water interface using remote sensing and GIS techniques involves a number of steps shown in Fig. 5.3.

The land-water categorization begins with the identification of the spectral band which can separate land and water bodies. The distinction between water and land varies among the spectral bands, as can be seen in Fig. 5.4. The figure shows the visible and near infrared bands of Ikonos satellite imagery of an area in Michigan, USA. The visible bands (wavelength, 0.45–0.70 μm), as often the case, show variation within water bodies but do not show sharp differentiation between water and land surfaces. However, the reflected infrared bands (wavelength, 0.7–1.10 μm) can detect water areas as solid bodies with little reflective variation and are ideal for land-water categorization. The actual land-water categorization is usually performed using slicing operation or supervised and unsupervised classification. The slicing operation is based on a cutoff value which distinguishes between water and land. All pixels with values below the cutoff value are categorized as water and those with higher values as land.

Supervised classification and unsupervised classification are two important and very basic methods in the processing of remote sensing image. In supervised classification, the user indicates the characteristic spectral signatures of known surface types such as water or land. The system then assigns each pixel in the image to the surface type to which its signature is most similar. Unsupervised classification is more computer-automated. It allows the specification of parameters that the computer uses as guidelines to uncover statistical patterns in the data.

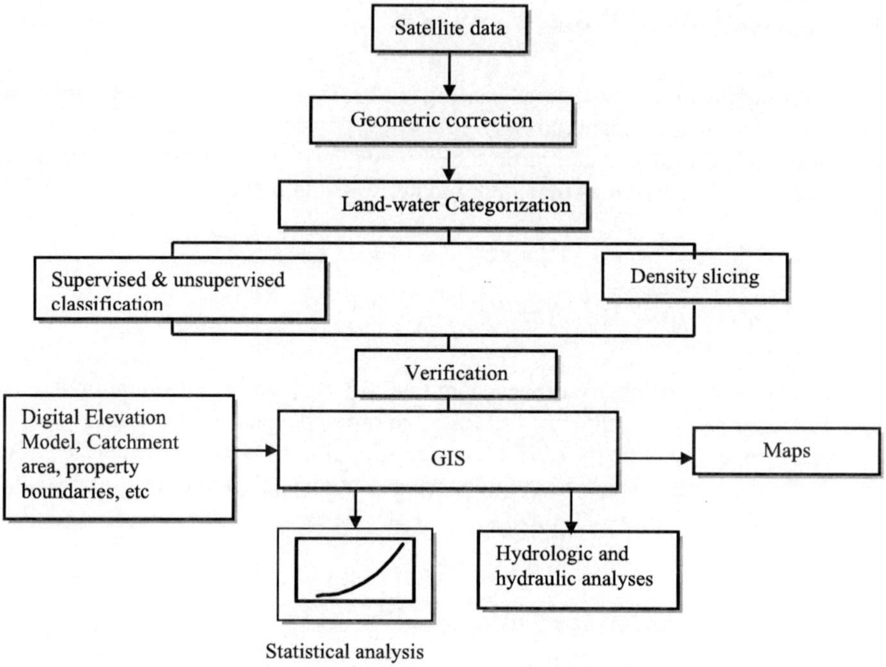

Fig. 5.3 Flow diagram illustrating remote sensing and GIS techniques for flood monitoring and floodplain mapping

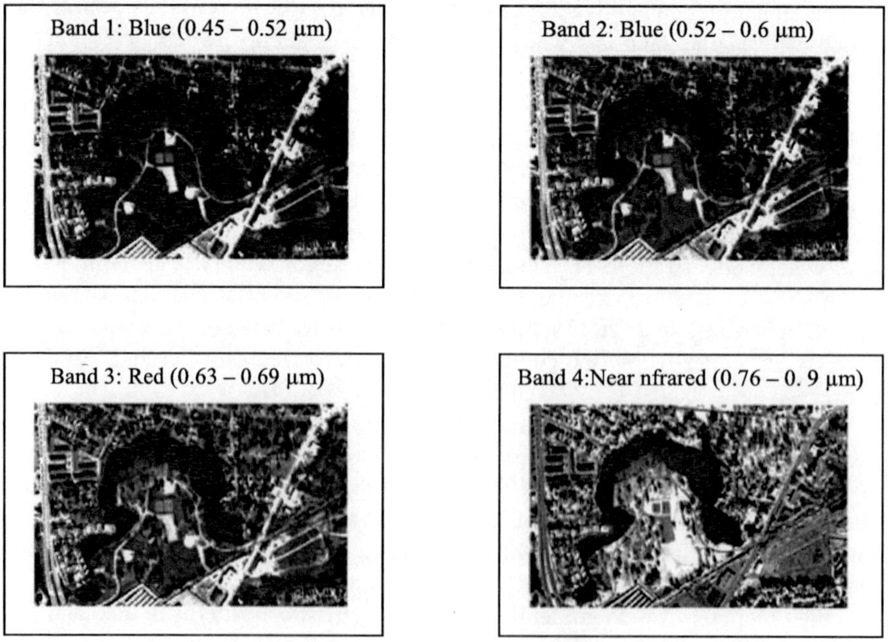

Fig. 5.4 Ikonos spectral bands

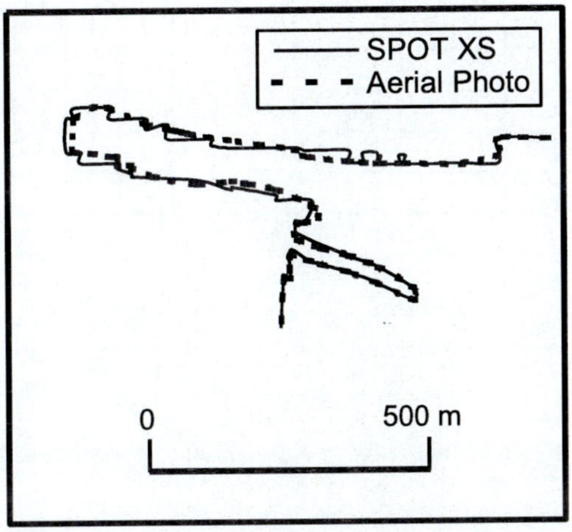

Fig. 5.5 Evaluation of land-water mapping from satellite imagery [9]

Ground truthing is an important component in determining the cutoff value to separate water from non-water areas on the images. It is also needed to check the accuracy of supervised and unsupervised classification. This is done using field observations and through the analysis of aerial photographs. The interpretations from satellite imageries are edited and digitized using GIS software into vector flood extent coverage for further analysis such as model calibration or creation of model input data.

The algorithm presented in Fig. 5.3 has been evaluated by Darkwah [9]. He used four sets of SPOT imageries of a coastal area in Miyazaki, Japan, in the evaluation. The near-infrared images were categorized into land and water cover types using cutoff values. Due to atmospheric influences cutoff values for images acquired for the same area at different times may be different. In GIS environment, he digitized and edited the satellite-derived land-water interface for comparison with existing map derived from aerial photographs. As shown in Fig. 5.5 a good agreement was found between the two results. A maximum deviation of one pixel size was found between the two datasets.

The land-water categories mapped using remote sensing and GIS tools have a number of applications in water resources. These applications include but not limited to floodplain mapping [10–12], updating flood maps [13], shoreline change analysis [9], bathymetric mapping [14], wetland mapping [15], estimation of water surface area and storage capacity [16], and water quality modeling [17, 18].

4.1.2 Floodplain Mapping

Floodplain inundation and flood hazard maps have been prepared by hydrologists, scientists, and engineers all over the world from aircraft and satellite data, mostly from the visible and infrared bands [19]. Landsat data have also been combined with digital elevation model to develop stage-area relationship of flood-prone areas

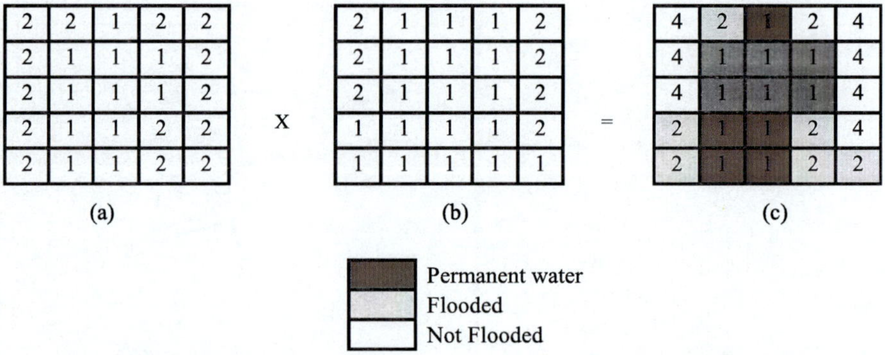

Fig. 5.6 Illustration of GIS and remote sensing technique in floodplain mapping. (**a**) Pre-flood data. (**b**) Flood data. (**c**) Change detection

[20]. Darkwah and Scoville [21] used a times series of Landsat TM and MSS to document the dynamic changes in the Kalamazoo River basin in Michigan. The GIS and remote sensing approach to floodplain mapping presents a quick and operational method for flood mapping. The technique involves using multi-temporal imageries and a basic mathematical logic to identify the extent and amount of flood. The mathematical operations usually performed include addition, subtraction, multiplication, and division. The multi-temporal change detection approach usually requires a low flow image dataset (base image data) and an image during the flood period [21]. The technique is illustrated in Fig. 5.6. Suppose Fig. 5.6a and b are land-water category maps derived from pre-flood and flood satellite images. On the figures, the numbers 1 and 2 denote water and land categories, respectively. Figure 5.6c is the output of change detection analysis performed by multiplying the pre-flood image data by the flood image data. As demonstrated in the figure, the change detection analysis resulted in three possible outcomes. A cell can have a value of 1, 2, or 4 and the values are defined as follows:

- 1 indicates that an area was classified as water on both the pre-flood and the flood image. These areas are considered to be permanent water features such as river channel, ponds, and lakes.
- 2 indicates that an area was classified as land on the pre-flood image, but classified as water on the flood image. These areas are considered to be flooded.
- 4 indicates that an area was land on both images. These areas are considered not flooded.

4.1.3 Monitoring Coastal Environment

Historic rates of shoreline change provide valuable data on erosion trends and permit limited forecasting of shoreline movement. Sequential satellite imageries and aerial photographs serve as a useful tool for compiling quantitative shoreline changes

[22, 23]. The shoreline change analysis can be automated in GIS to provide rapid, high-resolution evaluation of multiple temporal shoreline delimitation. Two approaches, raster-based and vector-based, are generally employed to model spatial and temporal shoreline change. The vector approach to analyzing historic shoreline change data contrasts with a raster approach in its sampling flexibility and temporal scalability.

The vector approach can accept any number of temporal linear representations of the shoreline and can flexibly sample those shorelines to calculate past variability and project future changes. Baselines must be constructed for all historic shorelines to provide a starting point for the transecting operation and must be digitized parallel to the general trend of the historic shorelines. There are a number of GIS extensions that can be used to generate transects, perform the analysis, and deliver results in GIS format [24]. Satellite imageries in conjunction with GIS can also be used to assess coastal storm damage. Time series of remotely sensed data have been used to investigate sediment transport processes in the coastal areas [7, 25].

4.1.4 Water Quality Monitoring

In the area of water quality, remote sensing and GIS techniques can be used to assess and map several water quality parameters including suspended sediments, chlorophyll, turbidity, total phosphorus, Secchi depth, temperature, and dissolved organic matter. These water quality parameters are important in defining total maximum daily loads (TMDLs). The remote sensing technique involves the derivation of algorithms which relate the spectral reflectance to in situ field measurements of water quality parameters. The exact relationship will vary regionally and depends on the optical properties of the water. Once an algorithm is derived, the water quality parameters for each pixel of water in the image can be calculated using only the spectral reflectance values. GIS is usually used to present the results in two-dimensional thematic maps and in dynamic sequential images. The GIS visualization helps to enhance understanding of water quality conditions.

Significant relationships have been shown between water quality parameters and reflectance values from spectral wave bands or combinations of wave bands on satellite or aircraft sensors. These algorithms are site specific, and therefore applying the same algorithms to other sites may not be appropriate due to the very different characteristics and compositions of column material in these waters. However, the same procedure used to develop these algorithms may be followed to develop algorithms for other sites. Table 5.2 lists some of the algorithms that have been derived from remote sensing data. The sources for these algorithms are also provided in the table.

Suspended sediments increase the radiance emergent from surface waters in the visible and near-infrared proportion of the electromagnetic spectrum [31]. The amount of reflected radiance tends to saturate as suspended sediment concentrations increase [32–34]. The point of saturation depends on wavelength with the shorter wavelength saturating at low concentrations. Since suspended sediments increase

Table 5.2 5.2 Water quality algorithms derived from remotely sensed data

Water quality parameter	Data source	Algorithm	Reference
Suspended sediments	Landsat TM	$SS = a_0 + a_1R_1 + a_2R_2 + a_3R_1R_2 + a_4R_1^2 + a_5R_2^2 + a_6R_1^2R_2 + a_7R_1R_2^2 + a_8R_1^2R_2^2 + \ldots$	[26]
	Airborne videography	$SS = a_aDN + b_a$	[27]
Chlorophyll-a	SPOT-XS	$Log_e\ C = a_s + b_sLog_e(X_3/X_2)$	[28]
	aircraft	$Log_e\ C = a_h + b_h(-Log_eR_2^2/(R_1*R_3)$	[29]
	Landsat TM	$Log_e\ C = a_k + b_kLog_eTM_2 + d_kLog_eTM_4$	[30]
Phosphorus	Landsat TM	$Log_e\ P = a_{ks} + b_{ks}Log_e\ TM_2 + d_{ks}Log_eTM_4$	[30]
	SPOT XS	$Log_e\ P = a_{ss} - b_{ss}Log_e\ (X_1/X_2)$	[28]
Secchi depth	SPOT XS	$Log_e\ SD = a_{sd} + b_{sd}Log_e\ (X_1/X_2)$	[28]
	Landsat TM	$Log_e\ SD = a_{kd} - b_{kd}Log_eTM_2 - d_{kd}Log_eTM_4$	[30]

a_i, b_i, and d_i, regression constants; X_1, X_2, and X_3, digital values of SPOT XS bands 1, 2, and 3; TM_2 and TM_4, digital values of Landsat TM bands 2 and 4; R_i, radiance of Landsat TM band$_i$

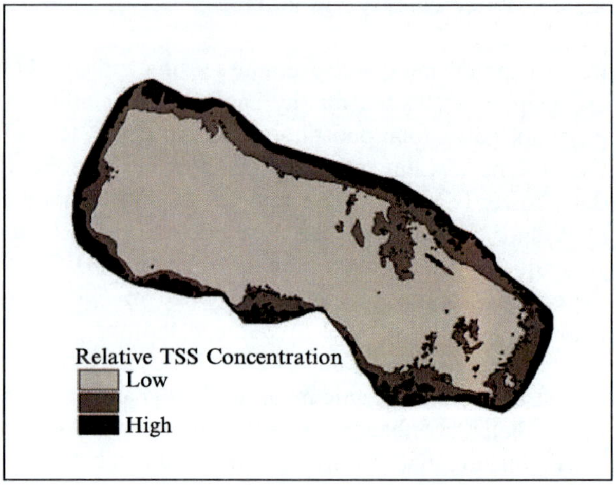

Fig. 5.7 Suspended sediment descriptive map derived from Ikonos satellite imagery

with reflectance, satellite imagery can be utilized to plan and manage in situ suspended solid sampling program. The raw satellite imagery can be automatically stenciled into different categories to represent relative suspended sediment concentrations as shown in Fig. 5.7. The suspended sediment descriptive map was derived from Ikonos satellite data for a lake in Michigan. Such analysis will help visualize the suspended sediment concentration changes within the water body and help plan where to sample. The best time period to conduct in situ measurement can be established from seasonal suspended sediment information derived from satellite imageries.

Monitoring the concentration of chlorophyll is needed to manage eutrophication in lakes [35]. While measuring chlorophyll by remote sensing technique is possible, studies have shown that the broad wavelength spectral data available on current satellites do not permit discrimination between chlorophyll and suspended sediments when suspended sediment concentrations are high due to the spectral dominance of the spectral signal from suspended sediment [36]. Recent studies have concentrated on the relationship between chlorophyll-a and the narrow band spectral details at the red edge of the visible range of the electromagnetic spectrum.

4.2 *Modeling*

Accurate modeling requires the estimation of spatial and temporal distribution of water resources parameters. Remote sensing and GIS are efficient tools used to extract and integrate the relevant spatial and non-spatial data needed for groundwater, hydrologic, and hydraulic modeling.

4.2.1 Groundwater Modeling

GIS have been increasingly used in geohydrologic investigation as they provide a variety of spatial analysis tools for groundwater modeling. GIS can be linked to groundwater models such as MODFLOW to simulate groundwater dynamics in an aquifer. Groundwater database can be imported into GIS environment to analyze hydraulic conductivity and transmissivity of the aquifer by raster overlay techniques. By using the GIS raster database, an effective and improved pre- and post-processing can be obtained.

Geological methods, involving interpretation of geologic data and field reconnaissance, represent an important first step in any groundwater investigation. Remote sensing has become a valuable tool for detecting and mapping surface features related to potential groundwater accumulation. The satellite imageries are used to map lithologies, lineament, soils, vegetation, and structure. When these maps are combined with meteorological information and local hydrologic data, groundwater potential can be assessed and prospective targets ranked according to their probable suitability as aquifers. The various thematic and interpretive surface parameters derived from remotely sensed data together with subsurface ancillary data derived from wells can be integrated and analyzed through GIS to predict potential groundwater zones. The groundwater potential zone can be calculated from thematic layers using a modified form of the DRASTIC model, which is used to assess groundwater pollution vulnerability by the Environmental Protection Agency of the United States [38]. The formula of the groundwater potential model (GP) is shown in Eq. 5.1:

$$\mathrm{GP} = R + L_{\mathrm{t}} + L_{\mathrm{d}} + L_{\mathrm{u}} + T_{\mathrm{e}} + S_{\mathrm{s}} + D_{\mathrm{d}} + S_{\mathrm{t}} \tag{5.1}$$

where R, annual rainfall; L_{t}, lithology; L_{d}, lineament density; L_{u}, land use; T_{e}, elevation; S_{s}, slope steepness; D_{d}, drainage density; and S_{t}, soil type.

4.2.2 Spatial Modeling

A lot of water resources data such as precipitation, streamflow, and water pollution measurements are observed at point locations. A raster format of these measurements is user-friendly and allows hydrologists and engineers a quick overview of the data. The most common method used to convert the point measurements to raster format is spatial interpolation. This method uses the location and magnitude of the measured data to determine estimates at the unmeasured locations. Spatial interpolation allows water resources planners and managers to analyze the spatial variability of hydrologic processes such as precipitation [39]. Such analyses are used to determine water budget at different spatial and temporal scales and validation of different hydrologic models.

Spatial interpolation routines can be performed in GIS environment to save time in exporting and importing data. The output of the interpolation can be viewed as a grid or a vector map. Some of the commonly used interpolation methods in GIS include inverse distance weighting, spline, and kriging. Inverse distance weighting is based on the assumption that the nearby values contribute more to the interpolated values than distant observations. Inverse distance weighting is by definition a smoothing technique and that the maximum and minimum values can only occur at the measured points. This technique will therefore never produce a value that is higher than the maximum value in the observed data set. The spline method can be thought of as fitting a rubber-sheeted surface through the known points using a mathematical function. Advantages of spline functions are that they can generate sufficiently accurate surfaces from only a few sampled points and they retain small features. A disadvantage is that spline functions can produce estimates that are above and below the measured minimum and maximum values. This is not always desired as maximum and minimum values are often produced where they do not occur in nature. Kriging is a stochastic technique similar to inverse distance weighted averaging in that it uses a linear combination of weights at known points to estimate the value at an unknown point. Kriging depends on spatial and statistical relationships to calculate the surface. Some advantages of this method are the incorporation of variable interdependence and the available error surface output. A disadvantage is that it requires substantially more computing and modeling time.

It has been shown that there is no single preferred method for data interpolation. Aspects of the algorithm selection criteria need to be based on the actual data, the level of accuracy required, and the time and/or computer resources available. This assessment is important because much of geographic research includes the creation of data for spatial analysis. The problem of interpolation is thus a problem of choosing a plausible model to suit the data [40]. Selecting an appropriate spatial interpolation method is key to surface analysis since different methods of interpolation can result in different surfaces and ultimately different results. The results of a study performed on August 1990 rainfall data in Michigan are presented in Fig. 5.8. The two interpolation techniques evaluated are inverse distance weighting and universal kriging. The figure illustrates that depending on the accuracy required the results could be different.

Fig. 5.8 Spatial interpolation precipitation data

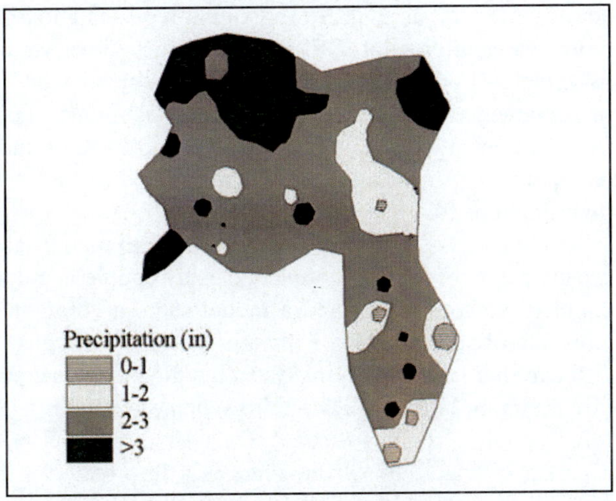

Surface derived by Inverse Distance Weighting

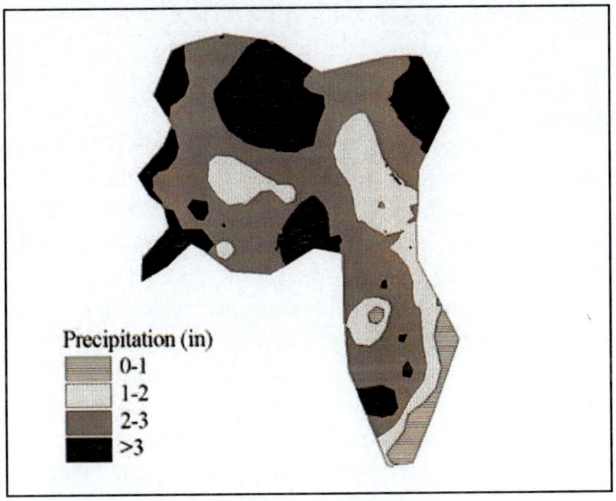

Surface derived by Universal Kriging

4.2.3 Hydrologic and Hydraulic Modeling

Distributed watershed models are commonly used to investigate water resources issues. The watershed models require watershed and subbasin geometric properties, configuration of the stream network, basin and stream slope, stream length and slope, land use, and soil type information. Traditionally, these parameters were derived from paper maps or field surveys. Recently, this information has been derived directly from digital representations of the topography [41–43] using automated methods in GIS. The digital representation of the topography is called a digital elevation model (DEM). The automated derivation of topographic watershed information from DEM is faster and less subjective and provides more reproducible measurements than traditional manual techniques applied to topographic maps [44]. The technological advances provided by GIS and the increasing availability and quality of DEM have greatly contributed to water resources and environmental investigations [42].

The methods for linking GIS to analytical models are usually divided into three broad categories: tightly coupled, closely coupled, and loosely coupled. In a tightly coupled method, a GIS and a model share a common interface and the model is embedded within the GIS software. In closely coupled method, the model and the GIS can share a common interface, but the model is created as a program in the GIS programming language. The third scheme for linking GIS to a model is known as loose coupling. In this method, the GIS and model remain separated, including a separate interface. However, there is a link between them that allows data to be transferred from one to the other. A GIS dataset can be imported to the model, analyzed, and sent back to the GIS for interpretation and mapping. This is probably the cheapest method of integration, but controlling the data flow from two different interfaces can be difficult and inefficient.

Watershed is the basic unit of hydrology and can be defined as an area contributing flow to a specified outlet. GIS is widely used to support water quantity and quality modeling. Figure 5.9 presents the automated modeling system that incorporates both hydrologic and hydraulic models. This modeling system links the results from the hydrology (discharge) to the input for the hydraulic model. After the hydraulic models are executed, model output such as flood elevations, floodway encroachments, and velocities can be extracted, displayed, and analyzed in a GIS environment.

One of the first steps in hydrologic modeling is watershed delineation to a point or a reach of interest. Using DEM or triangulated irregular networks (TIN), GIS can automatically delineate a watershed for any outlet point. Further, watersheds can be easily divided into subbasins by simply designating outlet points in any watershed. As part of the delineation, basin data such as slope, area, mean elevation, distance to centroid, and other common hydrologic parameters such as time of concentration and Soil Conservation Service (SCS) runoff curve numbers are automatically computed. The curve number is a hydrologic parameter used to describe the stormwater runoff potential for drainage area. It is a function of land use, soil type, and soil moisture. SCS composite curve numbers can automatically be computed from land use and soil data.

GIS automated watershed delineation consists of creating a depressionless DEM, flow direction, flow accumulation, watershed outlet points, and delineating watersheds. The primary method involved in determining flow paths from DEM data is the eight-point pour method developed by Puecker et al. [45]. The basic idea in Puecker's method is to assign a flow direction to each grid cell based on the neighboring cell with the lowest elevation. Based on steepest slope, each cell is assigned a flow direction in relation to one of its eight neighboring cells. The flow

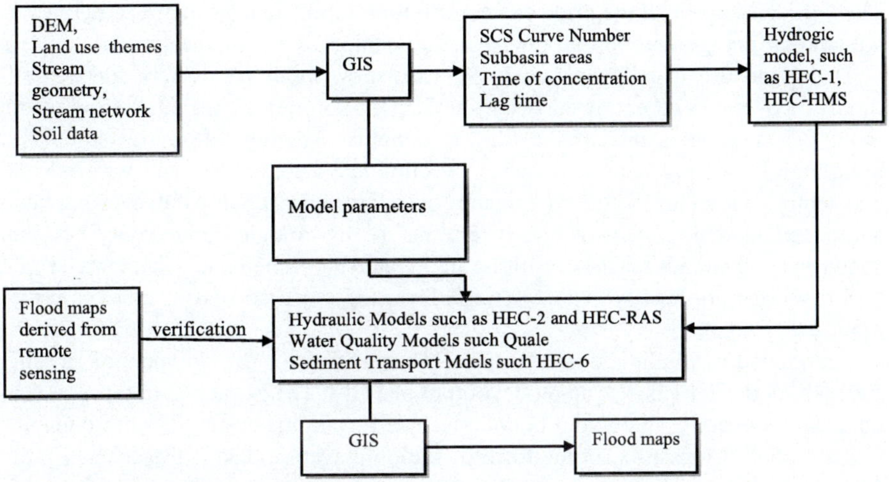

Fig. 5.9 Remote sensing and GIS application to modeling

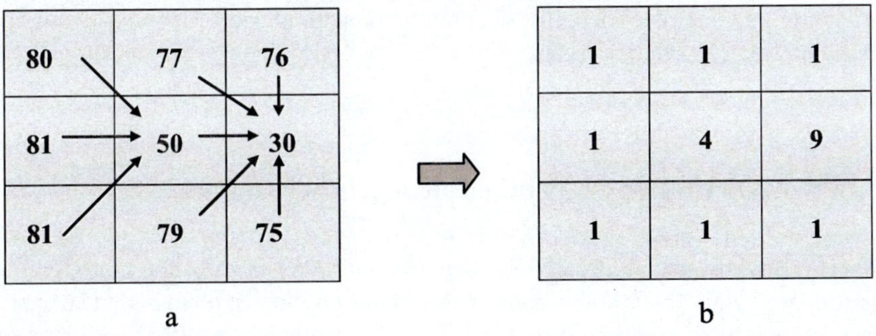

Fig. 5.10 Derivation of flow direction and flow accumulation. (**a**) Estimation of flow direction. (**b**) Estimation of flow accumulation

direction analysis is used to determine the number of upstream cells that accumulate at each individual cells. The cells with the largest flow accumulation values are the ones where the most water is flowing. Figure 5.10 illustrates the derivation of flow direction and flow accumulation grids. In Fig. 5.10a, the numbers represent elevation. The flow direction is determined from differences in elevation between neighboring cells. Water in any cell is modeled as flowing to only one of the neighboring cells, in the direction of steepest descent (lowest elevation). Using the flow direction, the number of cells draining to each cell is then totaled to produce a new data layer. In the new layer, the value of each cell represents the total number of cells draining through it as shown in Fig. 5.10b. The GIS automated techniques use the flow direction and accumulation grids to determine basin boundaries. All the cells that flow through a given outlet point define the extent of a basin draining to that point.

The GIS determines all the grid cells contributing runoff to a basin outlet, delineates a bounding polygon around all the cells, and computes a drainage area.

Using the delineated watershed characteristics, input parameters for rainfall-runoff models can be extracted from the GIS environment. This automation of the input file creation is one of the biggest benefits of using GIS-based automated hydrologic modeling techniques and can eliminate errors associated with manual data entry. In addition to the transfer of hydrologic input parameters, several automated modeling systems can determine the hydrologic connectivity between subbasins and stream reaches resulting in a watershed schematic. There are several GIS tools that are readily available to perform these pre-processing steps to create input file.

Automated hydraulic modeling typically focuses on the development of an input file for a model and displaying model output in a GIS. The extraction of cross section data and delineating floodplain boundaries can be automated in GIS environments. Cross section information in the form of station-elevation data is the primary input for most hydraulic models used for floodplain analyses. Automated hydraulic modeling tools extract cross section data from digital terrain data and format it for use with industry-standard hydraulic models such as HEC-2 or HEC-RAS. Other hydraulic parameters such as reach lengths, channel bank stations, and Manning's roughness coefficients can also be extracted from GIS data and written to an input file.

5 US Environmental Protection Agency Programs

The US Environmental Protection Agency (USEPA) is actively promoting and assisting any individuals and organizations in this country in planning and executing the geographic information system (GIS) and remote sensing (RS) activities [46–53].

There are many USEPA applications that utilize locational information from the Facility Registry Service (FRS) Geospatial Reference Tables. Any application that displays facility points from the National Shapefile Repository or the EPA Spatial Data Library System (ESDLS) is utilizing coordinates from the geospatial structures. In addition to the maps displayed within the Envirofacts and FRS Queries Web queries, these applications also make use of Facility Coordinate Information in the FRS Geospatial Reference Tables.

My Environment is a GIS application that provides a wide range of federal, state, and local information about environmental conditions and features in an area of an engineer or planner's choice.

The USEPA's GIS Data Integration document describes the GIS data integration process for collaboration. Engineers and planners are allowed to view some of the very useful files [51].

The USEPA Geospatial Dataset contains data on wastewater treatment plants, based on the agency's Facility Registry Service (FRS) [50].

The USEPA GIS technical guidance is also very useful [53]. Its geospatial policies and standards are intended to provide support of all USEPA Region 2 GIS-related projects [53]. Engineers and planners should contact their own USEPA Regional Office for any needed technical support [47, 53].

EnviroMapper provides a map interface to discover facility locations and associated environmental information for the data within the Envirofacts Warehouse.

The EPA Geospatial Data Access Project provides downloadable files of USEPA facility locations and associated information in a variety of data formats.

Glossary [37, 46–53]

Abiotic Nonliving characteristic of the environment; the physical and chemical components that relate to the state of ecological resources.

Accuracy The degree to which a calculation, a measurement, or set of measurements agree with a true value or an accepted reference value. Accuracy includes a combination of random error (precision) and systematic error (bias) components which are due to sampling and analytical operations; a data quality indicator. USEPA recommends that this term not be used and that precision and bias be used to convey the information usually associated with accuracy.

Acid deposition "A complex chemical and atmospheric phenomenon that occurs when emissions of sulfur and nitrogen compounds and other substances are transformed by chemical processes in the atmosphere, often far from the original sources, and then deposited on earth in either a wet or dry form. The wet forms, popularly called 'acid rain,' can fall as rain, snow, or fog. The dry forms are acidic gases or particulates."

Adaptive sampling strategy A sampling strategy that allows modification of sampling design and analysis to adapt to changing objectives or to changing circumstances.

Agroecosystem A dynamic association of crops, pastures, livestock, other flora and fauna, atmosphere, soils, and water. Agroecosystems are contained within larger landscapes that include uncultivated land, drainage networks, rural communities, and wildlife.

Albers map projection A standard map projection for representing the US states. This is a conic equal area projection.

Ancillary data Data collected from studies within EMAP but not used directly in the computation of an indicator. Ancillary data can help characterize parameters and assist in the interpretation of datasets; time, stage of tide, and weather conditions are examples of ancillary data.

Annual statistical summary A document that presents a brief and comprehensive report of EMAP data collected on a single EMAP resource for a specific year. Annual statistical summaries may include cumulative frequency distributions, estimates of the extent of nominal or subnominal condition, comparisons among regions, or comparisons of data over time.

Area frame A sampling frame obtained by dividing a region into well-defined, identifiable subregions that in aggregate comprise the total area of the region of interest. The subregions are sampling units defined on maps or other cartographic materials.

Area sample The units selected for measurements from an area frame.

Arid ecosystems Terrestrial systems characterized by a climate regime where the potential evapotranspiration exceeds precipitation, annual precipitation is not less than 5 cm and not more than 60 cm, and daily and seasonal temperatures range from 40 °C to 50 °C. The vegetation is dominated by woody perennials, succulents, and drought-resistant trees.

Assessment Interpretation and evaluation of EMAP results for the purpose of answering policy-relevant questions about ecological resources, including (1) determination of the fraction of the population that meets a socially defined value and (2) association among indicators of ecological condition and stressors.

Assessment endpoint Formal expressions of the actual environmental value that is to be protected. Risk Assessment Forum defines this as an explicit expression of the environmental value that is to be protected. Operationally in EMAP, an assessment endpoint is the range, proportion, or percentage of a resource that is known with statistical confidence to be in a specified condition.

Association rule A rule that unambiguously links a single resource sampling unit with a grid point. Several association rules have been identified in selecting a Tier 2 sample via the EMAP grid.

Attribute Any property, quality, or characteristic of sampling unit. The indicators and other measures used to characterize a sampling site or resource unit are representations of the attributes of that unit or site. A characteristic of a map feature (point, line, or polygon) described by numbers or text; for example, attributes of a tree, represented by a point, might include height and species.

Augmented sample A grid-based sample whose size has been increased by using a denser grid.

Auxiliary data Data derived from a source other than EMAP, that is, from an experiment or from another monitoring or sampling program, either federal or state. The sampling methods and quality assurance protocols of auxiliary data must be evaluated before the data are used. It is always important to establish the population represented by auxiliary data.

Azimuthal map projection One of a class of map projections formed by projecting coordinates from the earth's surface directly onto a plane that is either tangent at a point on the surface or that intersects the surface in a circle. When the plane intersects the surface, the projection is said to be the secant case.

Baseline grid The fixed position of the EMAP grid as established by the position of the hexagon overlaying the United States. This is distinguished from the sampling grid, which is shifted a random direction and distance from the baseline grid.

Bias In a sampling context, the difference between the conceptual weighted average value of an estimator over all possible samples and the true value of the quantity being estimated. An estimator is said to be unbiased if that difference is zero. The systematic or persistent distortion of a measurement process which deprives the result of representativeness. (i.e., the expected sample measurement is different than the sample's true value).

Bioaccumulants "Substances that increase in concentration in living organisms as they take in contaminated air, water, or food because the substances are very slowly metabolized or excreted."

Bioassay A laboratory or field test in which living organisms are used to detect the presence of or test the effect of a particular substance, factor, or condition. Results are compared to a standard preparation or control to determine the relative strength of the substance, factor, or condition.

Biodiversity The variety and variability among living organisms and the ecosystems in which they occur. Biodiversity includes the numbers of different items and their relative frequencies; these items are organized at many levels, ranging from complete ecosystems to the biochemical structures that are the molecular basis of heredity. Thus, biodiversity encompasses expressions of the relative abundances of different ecosystems, species, and genes.

Biogeographic province Geographic areas characterized by specific plant formations and associated fauna.

Biomarker Biochemical, physiological, or histological indicators of either exposure to or effects of xenobiotic chemicals at the suborganismal or organismal level.

Biomass "All of the living material in a given area; often refers to vegetation".

Biome "Entire community of living organisms in a single major ecological area."

Biotic Of or pertaining to living organisms.

Calibration Statistical transformation of a variable to correct for scale or bias or otherwise to impose the properties of another variable. Example: let two labs use different protocols of chemical analysis. Designate one protocol as the standard and generate a transformation function to convert the data from the second protocol into representation of the standard. Calibration will be required as a routine statistical procedure in many aspects of EMAP. New protocols will replace old ones. Surrogate attributes will be extensively measured, to be calibrated to represent the target attribute. Use of multiple laboratories invariably involves laboratory bias, which can be reduced by calibration. Analytic equipment requires periodic calibration, and the data generated by such a process can also benefit from statistical calibration. Double-sample methods can utilize calibration in many ways to enhance the precision of estimates based on small subsamples.

CDF Cumulative distribution function.

Change As used in EMAP, the difference in the distribution of measurements of condition indicators between two time periods.

Characterization Determination of the attributes of resource units, populations, or sampling units. A prominent use in EMAP is characterization of 40-hexes.

Classification The process of assigning a resource unit to one of a set of classes defined by values of specified attributes. For example, forest sites will be classified into the designated forest types, depending on the species composition of the forest. Systematic arrangement of objects into groups or categories according to established criteria.

Community "All of the populations occupying a given area"; "an assemblage of populations of different species within a specified location in space and time." "In ecology, a group of interacting populations in time and space. Sometimes, a particular subgrouping may be specified, such as the fish community in a lake or the soil arthropod community in a forest."

Comparability "The degree to which different methods, datasets and/or decisions agree or can be represented as similar; a data quality indicator."

Completeness "The amount of valid data obtained compared to the planned amount, and [it is] usually expressed as a percentage; a data quality indicator."

Conceptual model A "conceptual model describes a series of working hypotheses of how the stressor might affect ecological components. The conceptual model also describes the ecosystem potentially at risk, the relationship between measurement [endpoints] and assessment endpoints, and exposure scenarios."

Condition The distribution of scores describing resource attributes without respect to any societal value or desired use, that is, a state of being.

Condition indicator A characteristic of the environment that provides quantitative estimates of the state of ecological resources and is conceptually tied to a value.

Confidence coefficient The probability statement that accompanies a confidence interval and is equal to unity minus the associated type I error rate (false positive rate). A confidence coefficient of 0.10 implies that 90% of the intervals resulting from repeated sampling of a population will include the unknown (true) population parameter.

Confidence interval An interval defined by two values, called confidence limits, calculated from sample data using a procedure which ensures that the unknown true value of the quantity of interest falls between such calculated values in a specified percentage of samples. Commonly, the specified percentage is 95%; the resulting confidence interval is then called a 95% confidence interval. A one-sided confidence interval is defined by a single calculated value called an upper (or lower) confidence limit. "The numerical interval constructed around a point estimate of a population parameter, combined with a probability statement (the confidence coefficient) linking it to the population's true parameter value. If the same confidence interval construction technique and assumptions are used to calculate future intervals, they will include the unknown population parameter with the same specified probability."

Conformal map projection One of a class of map projections that preserves angular relationships between the map and the surface of the earth. At any point on the map, measurement of orientation angles is correct. A conformal map projection cannot be equal area.

Conical (or conic) map projection One of a class of map projections formed by projecting coordinates from the earth's surface onto a cone that either intersects the surface of the earth or is tangent to a circle on the surface. When the cone intersects the sphere, the projection is said to be the secant case.

Continuous A characteristic of an attribute that is conceptualized as a surface over some region. Examples are certain attributes of a resource, such as chemical stressor indicators measured in estuaries.

Crosscutting group In EMAP's Integration and Assessment section, one group of scientific and administrative personnel headed by a technical coordinator (TC) and charged with addressing specific cross-program, integrative issues in EMAP, such as Landscape Characterization,

Design and Statistics, Indicator Development, Information Management, Assessment and Reporting, Logistics, Methods, and Quality Assurance.

Cumulative distribution A means of representing the variation of some attribute by giving running totals of the resource with attribute values less than or equal to a specified series of values. For example, a cumulative areal distribution of lakes would give, for any value of area, the total area covered by lakes with individual area less than or equal to alpha. A cumulative frequency distribution for lake area would give the total number of lakes with area less than or equal to alpha. The cumulative distribution function (cdf) of some specified attribute of a population is the function F(x) that gives the proportion of the population with value of the attribute less than or equal to x, for any choice of x. For example, if the attribute was lake area in hectares, F (a) would give the proportion of lakes with area less than or equal to a ha. (In some cases, the word "cumulative" may be omitted in discussions of the cdf, and the cdf is called the distribution function.)

Cylindrical (or cylindric) map projection One of a class of map projections formed by projecting coordinates from the earth's surface onto a cylinder that either intersects the surface in two small circles or is tangent to a great circle on the surface. When the cylinder intersects the surface, the projection is said to be the secant case.

Data quality The totality of features and characteristics of data that bears on their ability to satisfy a given purpose; the sum of the degrees of excellence for factors related to data.

Data quality indicators "Quantitative statistics and qualitative descriptors that are used to interpret the degree of acceptability or utility of data to the user. The principal data quality indicators are bias, precision, accuracy, comparability, completeness, and representativeness."

Data quality objective (DQO) "Quantitative and qualitative statements of the overall level of uncertainty that a decision-maker is willing to accept in results or decisions derived from environmental data DQOs provide the statistical framework for planning and managing environmental data operations consistent with the data user's needs." A data quality objective may include goals for accuracy, precision, and limits of detection. It may also include goals for completeness, comparability, and representativeness. Data quality objectives are established before sampling is begun and may influence the level of effort required to select a sample.

Deconvolution The process of removing the influence of extraneous variation from an apparent cumulative distribution. Extraneous variation – such as random errors in measurement – has the effect of inflating observed variation relative to true population variation. The cumulative distribution that will be estimated when extraneous variation is present is the convolution of the population distribution (which is the cumulative distribution of interest) and the distribution of the extraneous variable. The convolution cumulative distribution will be flatter (have longer tails) than the population cumulative distribution.

Demonstration field program A collection of demonstration research projects to provide preliminary estimates of resource condition for one or more indicators applied to one or more resource classes over a standard federal region. A demonstration field program is likely to include a group of pilot research projects to increase the cost-effectiveness of the program. The demonstration field program does not have a data quality objective, but the individual pilot and demonstration research projects do have DQOs.

Demonstration project A field research project designed to provide preliminary estimates of a resource condition for a single indicator over a standard federal region for one or more resource classes. Separate demonstration projects are defined for each indicator of condition; in a demonstration project, quality data objectives are stated for preliminary statistical estimates.

Design-based Statistical inferences using methodology based on the sampling design. Such inferences derive their properties from the design protocols.

Design-unbiased Indicates that an estimator is unbiased under the design protocol. This property does not, however, prevent bias from entering estimates from other sources. Design-unbiased estimators of variance do not always exist under EMAP designs.

Digital line graph (DLG) Digital data produced by the US Geological Survey. These data include digital information from the USGS map base categories such as transportation, hydrography, contours, and public land survey boundaries.

Discrete resource A resource consisting of a collection of distinct units, such as lakes or stream reaches. Such a resource will be described as a finite population of such units.

Domain The areal extent of a resource; the region occupied by a resource.

Double sample A sample of a sample. Specifically in EMAP, resource attributes from remote sensing or cartographic materials can be measured on a larger sample than those attributes requiring field measurements. Attributes on the former sample can be used to guide selection of the latter sample.

Ecology "The relationship of living things to one another and their environment, or the study of such relationships."

Ecological health A metaphor used to invoke ideas about the integrity, complexity, and autonomy of an ecosystem. A definition of ecosystem health as an actual entity with specific scale has not been resolved; the terms are useful because they facilitate "a set of rules of thumb, which can be thought of analogically, for analyzing what is going wrong when environing (sic) systems undergo rapid change."

Ecological risk assessment A process that evaluates the likelihood that adverse ecological effects may occur or are occurring as a result of exposure to one or more stressors.

Ecoregion A relatively homogeneous geographic area "perceived by simultaneously analyzing a combination of causal and integrative factors including land surface form, soils, land uses, and potential natural vegetation. It defines a mapped classification of ecosystem regions of the U.S....Ecoregions are generally considered to be the regions of relative homogeneity in ecological systems or in relationships between organisms and their environments."

Ecosystem The biotic community and its abiotic environment within a specified location in space and time. "The interacting system of a biological community and its non-living environmental surroundings."

Ecosystem functions	Energy circuits, food chains, diversity patterns in time and space, nutrient cycles, development and evolution, and control within an ecosystem.
Ecotone	A habitat created by the juxtaposition of distinctly different habitats; an edge habitat; an ecological zone or boundary where two or more ecosystems meet.
Environmental Monitoring and Assessment Program (EMAP)	It was a research program run by USEPA's Office of Research and Development to develop the tools necessary to monitor and assess the status and trends of national ecological resources. EMAP developed indicators to monitor the condition of ecological resources.
Entire	Being whole, not convoluted or divided into distinct spatial parts. In EMAP, this property affects the precision of certain sample statistics.
Environment	"The sum of all external conditions affecting the life, development, and survival of an organism."
Environmental assessment	An environmental analysis prepared pursuant to the National Environmental Policy Act to determine whether a federal action should significantly affect the environment and thus require a more detailed environmental impact statement.
EPA Spatial Data Library System (ESDLS)	It is utilizing coordinates from the geospatial structures.
Epidemiologic ecology	The study of the effects of human, physical, biological, and chemical stressors on conditions of ecosystems, broadly conceived. This area of ecology uses epidemiological concepts and methods to associate ecological effects or responses with stressors through assessments, long-term monitoring data, and results of research.
Equal-area (equivalent) projection	One of a class of map projections that preserves area relationships between the map and the surface of the earth. On an equal-area projection, two areas that are equal on the surface of the earth will also be equal on the map. A projection cannot be both conformal and equal area.
Estuary	"Regions of interaction between rivers and nearshore ocean waters, where tidal action and river flow mix fresh and salt water. Such areas include bays, mouths of rivers, salt marshes, and lagoons. These brackish water ecosystems shelter and feed marine life, birds, and wildlife." In EMAP, large estuaries are defined as those estuaries greater than 260 km2 in surface area and with aspect ratios (i.e., length/average width) of less than 20. Large tidal rivers are defined as that portion of the river that is tidally influenced (i.e, detectable tide >2.5 cm), greater than 260 km^2, and with an aspect ratio of greater than 20. Small estuaries and small tidal rivers are those systems whose surface areas fell between 2.6 km^2 and 260 km^2.
Extensive resource	A resource covering a large area that is not ecologically integrated and is not a collection of natural units, for example, large marshes or rangelands. Characterization of an extensive resource is scale-dependent.
40-hex	The landscape description hexagon that is established on each of the grid points in the EMAP grid. Actual size of these hexagons is 634.5/16 = 39.7 km2.
Forest	Land with at least 10% of its surface area stocked by trees of any size or formerly having had such trees as cover and not currently built up or developed for agricultural use.

Fragmented	Being divided or convoluted into distinct parts, rather than entire. In EMAP, the spatial fragmentation of resources and the spatial/temporal fragmentation of resource attributes affect the precision of certain population statistics, so that attention must be given to this state.
Frame	A representation of a population, used to implement a sampling strategy as, for example, (1) a list frame that lists the identifying units in the population, for instance, a list of all the lakes in the United States between 10 and 2000 ha, or (2) an area frame that consists of explicit descriptions of a partition of the areal extent of an areal universe like the NASS frame.
Frame, conceptual	An explicit definition of a frame that does not involve a tangible realization.
Geographic information system (GIS)	A collection of computer hardware, software, and geographic data designed to capture, store, update, manipulate, analyze, and display geographically referenced data.
Great Lakes	In EMAP, the resource that encompasses the five Great Lakes – Superior, Michigan, Huron, Erie, and Ontario, including river mouths up to the maximum extent of lake influence; wetlands contiguous to the lakes; and the connecting channels, Lake St. Clair, and the upper portion of the St. Lawrence Seaway.
Grid	A data structure commonly used to represent map features. A cellular-based data structure composed of cells or pixels arranged in rows and columns (also called a "raster").
Grid enhancement	Increasing the grid density, a method for augmenting the sample. When the sample size is too small, as will occur for rare resources, the grid density may be increased in order to obtain a sample size adequate for population description. Grid enhancement is one method of producing an augmented sample.
Grid, hierarchical	Having nested levels and structure, the density of the EMAP grid is readily increased or reduced in a regular manner into hierarchical levels of density. Adjacent levels may differ in density by a variety of factors: 3, 4, 7, or many functions of these base factors. The grid of points at one level will be contained in the grid at a higher density.
Grid randomization	The process of randomly positioning the grid so that each (discrete) unit of area of fixed size is equally likely to contain a grid point. This process is the basis for the probability sample designation for EMAP monitoring.
Grid, triangular (EMAP)	A lattice of points in exact equilateral triangular structure on a plane. The EMAP grid points are 27.1 km apart.
Habitat	"The place where a population (e.g., human, animal, plant, microorganism) lives and its surroundings, both living and non-living."
Heuristic method	In EMAP, a process for exploring and organizing information to conceptually discover and describe or postulate relationships among indicator variables and key processes of an ecosystem or resource.
Hierarchical geometric decomposition	The mathematical process of breaking down a geometric structure into component parts. In the EMAP grid, the process by which the triangular grid is subdivided into higher density grids is hierarchical geometric decomposition. Related notions are enhancement and disaggregation; the opposite notion is composition of components into a larger whole.

Hierarchical model A construct in which component variables and systems of ecological entities are defined to have the smallest unit of spatial and temporal detail in the lowest level; each higher level reflects processes and interactions occurring at ever larger spatial and temporal scales. Behaviors of a particular level in the hierarchical model are determined both by constraints imposed by the levels above and filtered or averaged behaviors of components and subsystems of levels below.

Implementation field program The decision by EMAP to make a long-term commitment to provide scientifically defensible estimates of current status, trends, and changes in condition indicators and association of these indicators with selected stressors. The decision will include consideration of (1) the link between values and indicators proposed for implementation, (2) the scientific basis for the indictor, (3) the scientific validity of the proposed national sampling design, (4) a demonstrated ability to meet the EMAP status and trend program quality objectives, and (5) explicit plans that ensure the implementation can be conducted and produce annual statistical summaries.

Inclusion probability The probability of including a specific sampling unit within a sample.

Index Mathematical aggregation of indicators or metrics.

Index period The period of the year when measurement of an indicator yields meaningful information.

Index sample A standardized judgment sample for which explicit rules for generating the index measure are formally prescribed. An index sample is appropriate only for ecologically integrated systems. For example, in the National Lake Survey, a lake in the probability sample of lakes was field sampled at a prescribed season of the year (index period), in a prescribed standardized location in the lake (index site), and in a prescribed manner (field protocol).

Indicator In EMAP, characteristics of the environment both abiotic and biotic that can provide quantitative information on ecological resources. "In biology, an organism, species, or community whose characteristics show the presence of specific environmental conditions, good or bad."

Indicator development The process through which an indicator is identified, tested, and implemented. A candidate indicator is identified and reviewed by peers before it is selected for further evaluation as a research indicator. Existing data are analyzed, simulation studies are performed with realistic scenarios, and limited field tests are conducted to evaluate the research indicator. In the past, this research indicator was called a "probationary core indicator" or a "development indicator" as it was evaluated in regional demonstration projects. An indicator is considered a core indicator when it is selected for long-term, ecological monitoring as a result of its acceptable performance, demonstrating ability to satisfy the data quality objectives.

Integration The formation, coordination, or blending of units or components into a functioning or unified whole. In EMAP, integration refers to a coordinated approach to environmental monitoring, research, and assessment, both among EMAP resource groups and with other environmental monitoring programs. Integration in EMAP also refers to the technical processes involved in normalizing and combining data for interpretation and assessment.

Interpenetrating subsamples If a sample is partitioned into subsamples, each of which closely reflects the structure of the full design, and then the subsamples are said to mutually interpenetrate the frame. The EMAP design prescribes four interpenetrating grids, each with density one fourth of the base grid. Together, these four grids constitute the baseline grid.

Judgment sample A form of non-probability sample in which the sample is chosen according to the judgment of the sampler.

Kriging A weighted, moving-average estimation technique based on geostatistics that uses the spatial correlation of point measurements to estimate values at adjacent, unmeasured points. A sophisticated technique for filling in missing data values, kriging is named after a South African engineer, D.G. Krige, who first developed the method. The kriging routine preserves known data values, estimates missing data values, and also "estimates the variance at every missing data location. After kriging, the filled matrix contains the best possible estimate of the missing data values, in the sense that the variance has been minimized."

Lake In EMAP, a standing body of water greater than 1 hectare (about 2.5 acre) that has at least 1000 m^2 (about 0.25 acre) of open water and is at least 1 meter (about 3 feet) deep at its deepest point.

Lambert's azimuthal map projection The map projection used for laying out the EMAP grid. This is an equal-area projection.

Landscape The set of traits, patterns, and structure of a specific geographic area, including its biological composition, its physical environment, and its anthropogenic patterns. An area where interacting ecosystems are grouped and repeated in similar form.

Landscape characterization Documentation of the traits and patterns of the essential elements of the landscape, including attributes of the physical environment, biological composition, and anthropogenic patterns. In EMAP, landscape characterization emphasizes the process of describing land use or land cover, but also includes gathering data on attributes such as elevation, demographics, soils, physiographic regions, and others.

Landscape ecology The study of distribution patterns of communities and ecosystems, the ecological processes that affect those patterns, and changes in pattern and process over time.

Map projection A mathematical formula or algorithm for translating the coordinates of features on the surface of the earth to a plane for representation on a flat map.

Map resolution The accuracy with which the location and shape of map features are depicted for a given map scale.

Map scale A statement of a measure on the map and the equivalent measure on the earth, often expressed as a representative fraction of distance, such as 1:24,000.

Marginal condition The state that exists when the nominal and subnominal criteria are not contiguous.

Measurement A quantifiable attribute that is tied to an indicator.

Measurement endpoint A measurable ecological characteristic that is related to the valued characteristic chosen as the assessment endpoint. " Measurement endpoints are often expressed as the statistical or arithmetic summaries of the observations that comprise [sic] the measurement."

Meridian One of a group of abstract lines on the surface of the earth formed by the longitude and latitude coordinate system. Meridians represent lines of equal longitude and thus converge at the poles. All meridians are great circles.

Meridional zones	Areas on the surface of the earth formed by adjacent meridians.
Modeling	"Development of a mathematical or physical representation of a system or theory that accounts for all or some of its known properties. Models are often used to test the effect of changes of components on the overall performance of the system."
Monitoring	In EMAP, the periodic collection of data that is used to determine the condition of ecological resources. "Periodic or continuous surveillance or testing to determine the level of compliance with statutory requirements and/or pollutant levels in various media [air, soil, water] or in humans, plants, and animals."
My Environment	It is a geographic information system (GIS) application that provides a wide range of federal, state, and local information about environmental conditions and features in an area of your choice.
National Academy of Sciences (NAS)	The National Academy of Sciences/National Research Council (NRC) performs level 2 peer review to determine if EMAP projects have overall scientific merit and integrate both internally and with other government-sponsored monitoring programs. Two commissions of the NRC – the Commission on Geosciences, Environment, and Resources (specifically, its Water Science and Technology Board) and the Commission on Life Sciences – jointly organized the Committee to Review USEPA's Environmental Monitoring and Assessment Program in 1991. This NAS/NRC committee holds about 12 meetings and produces two or three reports every 2 years; its primary purpose is to consider the scientific and technical aspects of EMAP as designed as well as considering ways to increase EMAP's usefulness in monitoring conditions and trends in six representative types of ecosystems. The Committee also reviews the overall design objectives of the program, the indicator strategies, data collection methods, data analysis interpretation, and communication plans. Preparation for NAS/NRC reviews is coordinated by the director of OMMSQA, USEPA-ORD, who is also responsible for funding.
Nominal	Referring to the state of having desirable or acceptable ecological condition. The quantified standard established for a condition indicator to represent the desirable or acceptable condition is called a nominal assessment endpoint.
Office of Modeling, Monitoring Systems, and Quality Assurance (OMMSQA)	The office within USEPA's Office of Research and Development responsible for EMAP management within the agency.
On-frame data	Data acquired by a sampling approach that provides a probability sample.
Parallel	One of a group of abstract lines on the surface of the earth formed by the latitude and longitude coordinate system: parallels represent lines of equal latitude. Only the parallel at the equator is a great circle; other parallels are small circles.
Parameter	"Any quantity such as a mean or a standard deviation characterizing a population. Commonly misused for 'variable,' 'characteristic,' or 'property'."
Pattern	In EMAP, the location, distribution, and composition of structural landscape components within a particular geographic area or in a spatial context.
Peer review	In EMAP, peer review means written, critical response provided by scientists and other technically qualified participants in the process. EMAP documents are subject to formal peer review procedures at

laboratory and program levels. In EMAP, Level 1 peer reviews are performed by USEPA's Science Advisory Board, level 2 by the NAS National Research Council, level 3 by specialist panel peer reviews, and level 4 by internal USEPA respondents.

Pilot field program A collection of pilot projects whose objectives are to provide the scientific information required prior to a decision to implement a monitoring design for a resource. Preliminary estimates of resource condition are not planned and will not be made from a pilot field program, which does not have a data quality objective; the individual pilot projects and demonstration projects do have DQOs.

Pilot project A research project that requires fieldwork to meet a stated research project quality objective, and the objective is not to provide preliminary estimates of resource condition. Each research activity on an indicator should be described as an individual pilot project usually during a single index period. Pilot projects are used to evaluate indicators, sampling strategy, methods, and logistics.

Population "A group of interbreeding organisms occupying a particular space; the number of humans or other living creatures in a designated area. In statistics and sampling design, the total universe addressed in a sampling effort; an assemblage of units of a particular resource, or any subset of extensive resources, about which inferences are desired or made. RAF defines population to be "an aggregate of individuals of a species within a specified location in space and time."

Population estimation Classic survey estimation of population parameters. Such estimates will not reflect spatial configuration except through identification of the population, or of subpopulations, which may be defined by spatial attributes.

Population units The entities that make up a target population. The units can be defined in many ways, depending on the survey objectives and the type of measurement to be made. Typically, definitions of environmental units include (1) an explicit statement of the characteristics each population unit must possess in order to be considered a member of the target population and a (2) specification of location in space and time.

Precision The degree to which replicate measurements of the same attribute agree or are exact. "The degree to which a set of observations or measurements of the same property, usually obtained under similar conditions, conform to themselves; a data quality indicator."

Probability sample A sample chosen in such a manner that the probabilities of including the selected units in the sample are known and all population units have a positive probability of selection. This implies that the target population is represented by the sample and that the target population is explicitly defined.

Projection A mathematical model that transforms the locations of features on the earth's surface to locations on a two-dimensional surface.

Quality assessment "The evaluation of environmental data to determine if they meet the quality criteria required for a specific application."

Quality assurance (QA) "An integrated system of activities involving planning, quality control, quality assessment, reporting and quality improvement to ensure that a product or service meets defined standards of quality with a stated level of confidence." In EMAP, quality assurance consists of multiple steps taken to ensure that all data quality objectives are achieved.

Quality control (QC) "The overall system of technical activities whose purpose is to measure and control the quality of a product or service so that it meets the needs of users. The aim is to provide quality that is satisfactory, adequate, dependable, and economical." In EMAP, quality control consists of specific steps taken during the data collection process to ensure that equipment and procedures are operating as intended and that they will allow data quality objectives to be achieved.

QA/QC Quality Assurance/Quality Control. "A system of procedures, checks, audits, and corrective actions to ensure that all USEPA research design and performance, environmental monitoring and sampling, and other technical and reporting activities are of the highest achievable quality."

Quantile The value of an attribute indexing a specified proportion of a population distribution or distribution function. Quartiles (25th, 50th, and 75th percentiles), the median (50th percentile), and other percentiles are special cases of quantiles.

Recovery The partial or full return of a population or community to a condition that existed before the introduction of the stressor.

Randomization The process of imposing an element of chance on the selection of a sample. Randomization is a step in the design protocol and may take many forms; it is the basis for determining the design-based properties of the resulting probability sample.

Reference condition The set of attributes of ecological resources that assist in identifying the location of a portion of the resource population along a condition continuum from the worst possible condition to the best possible condition given the prevailing topography, soil, geology, potential vegetation, and general land use of the region. Reference condition typically refers to the best resource condition, but it is used more broadly in EMAP.

Reference site One of a population of benchmark or control sampling locations that, taken collectively, represent an ecoregion or other large biogeographic area; the sites, as a whole, represent the best ecological conditions that can be reasonably attained, given the prevailing topography, soil, geology, potential vegetation, and general land use of the region or clearly subnominal condition

Region Any explicitly defined geographic area. In the EMAP objectives, region refers to the ten standard federal regions.

Relation The concept of function, correlation, or association between or among attributes, which may be qualitative as well as quantitative.

Representativeness "The degree to which data accurately and precisely represent the frequency distribution of a specific variable in the population; a data quality indicator."

Research project A single, focused activity with a single primary research objective. The project includes the entire research process of initial planning, conduct of research (including fieldwork if necessary), and evaluation (analysis and reporting). A research project involving fieldwork may require several years to complete. Many EMAP research projects will not require fieldwork: such projects include development of conceptual models for an indicator, development of a proposed national sampling design strategy, analysis of existing data for indicator development, and others.

Resource In EMAP, an ecological entity that is identified as a target of sampling and is a group of general, broad ecosystem types or ecological entities sharing certain basic characteristics. Seven such categories currently are identified within EMAP: estuaries, great lakes, inland surface waters, wetlands, forests, arid ecosystems, and agroecosystems. These categories define the organizational structure of monitoring groups in EMAP and are the resources addressed by EMAP assessments. A resource can be characterized as belonging to one of two types, discrete and extensive, that pose different problems of sampling and representation.

Resource class A subdivision of a resource; examples include small lakes, oak-hickory forests, emergent estuarine wetlands, field cropland, small estuaries, and sagebrush dominated desert scrub.

Resource domain The areal extent of a resource; the region occupied by a resource.

Resource group In EMAP's Resource Monitoring and Research section, a group of scientific and administrative personnel, headed by a technical director (TD), responsible for research, monitoring, and assessments for a given EMAP resource. There are seven such groups in EMAP: estuaries, great lakes, inland surface waters, wetlands, forests, arid ecosystems, and agroecosystems. Landscape ecology is also considered a resource group in EMAP.

Resource unit A unit of a discrete resource, for example, a lake. A population of such a resource will be an explicit set of resource units.

Risk "A measure of the probability that damage to life, health, property, and/or the environment will occur as a result of a given hazard." In statistics, "the expected loss due to the use of a given decision procedure."

Risk assessment "Qualitative and quantitative evaluation of the risk posed to human health and/or the environment by the actual or potential presence and/or use of specific pollutants."

Risk characterization Determination of the nature of a given risk and quantifying of the potential for adverse change to the environment from that risk. "A phase of ecological risk assessment that integrates the results of the exposure and ecological effects analyses to evaluate the likelihood of adverse ecological effects associated with exposure to a stressor. The ecological significance of the adverse effects is discussed, including consideration of the types and magnitudes of the effects, their spatial and temporal patterns, and the likelihood of recovery."

Risk communication "The exchange of information about environmental risks among risk assessors, risk managers, the general public, news media, special interest groups, and others."

Risk management "The process of evaluating and selecting alternative regulatory and non-regulatory responses to risk. The selection process necessarily requires the consideration of scientific, legal, economic, and behavioral factors.

Sample A subset of the units from a frame. A sample may also be a subset of resource units from a population or a set of sampling units.

Sampling strategy A sampling design, together with a plan of analysis and estimation. The design consists of a frame, either explicit or implicit, together with a protocol for selection of sampling units.

Sampling unit An entity that is subject to selection and characterization under a sampling design. A sample consists of a set of sampling units or sites that will be characterized. Sampling units are defined by the frame; they may correspond to resource units, or they may be artificial units constructed for the sole purpose of the sampling design.

Science Advisory Board (SAB) A peer review panel internal to USEPA. The Ecological Effects Committee of the SAB conducts reviews of EMAP's overall program and the conceptual framework for integrating EMAP with ORD's Ecological Risk Assessment program. Preparation for SAB reviews is coordinated by the director of the Office of Modeling, Monitoring Systems, and Quality Assurance (OMMSQA); the assistant administrator for ORD is responsible for funding. SAB review is considered level 1 peer review.

Spatial model A set of rules and procedures for conducting spatial analysis to derive new information that can be analyzed to aid in problem solving and planning.

Spatial statistics Statistical methodology and theory that accounts for spatial aspects of a spatially distributed dataset. Conventional population estimation does not normally account for spatial attributes, except perhaps for spatial identity of subpopulations.

Status The distribution of scores for condition indicators with relation to the reference condition associated with specific social values or desired uses for a specific time period.

Stratum (strata) A sampling structure that restricts sample randomization/selection to a subset of the frame. Inclusion probabilities may or may not differ among strata.

Stressor "Any physical, chemical, or biological entity that can induce an adverse response."

Stressor indicator A characteristic of the environment that is suspected to elicit a change in the state of an ecological resource, and they include both natural and human-induced stressors. Selected stressor indicators will be monitored in EMAP only when a relationship between specific condition and stressor indicators is known or if a testable hypothesis can be formulated.

Subnominal Having undesirable or unacceptable ecological condition. The quantified standard established for a condition indicator to represent unacceptable or undesirable ecological condition is called the subnominal assessment endpoint.

Subpopulation Any subset of a population, usually having a specific attribute that distinguishes its members from the rest of the population, for example, lakes from a specified population that are above 1000 m in elevation. Subpopulations are important entities in the EMAP plan. Any defined subpopulation is subject to characterization via estimation of subpopulation attributes and comparison to other subpopulations. It is this focus that imposes the greatest restrictions on the EMAP design and establishes the primary directions of the EMAP analyses.

Surface fitting A statistical procedure of estimating the parameters of a surface model or of approximating an implied surface by distribution free methods from a spatially distributed sample. A two-dimensional generalization of regression.

Surface waters The inland surface waters consisting of all the nation's lakes (other than the Great Lakes), rivers, and streams. Lakes are distinguished from wetlands by depth and by size. Streams (and rivers) will be identified from stream traces on maps and confirmed in field visits.

	Streams are operationally defined as any first- or higher-order stream that is represented as a blue line on a USGS 1:100,000 topographic map. "All water naturally open to the atmosphere (rivers, lakes, reservoirs, ponds, streams, impoundments, seas, estuaries, etc.) and all springs, wells, or other collectors directly influenced by surface water."
Systematic sample	A sampling design that utilizes regular spacing between the sample points, in one sense or another. The EMAP design selects samples via the triangular grid. Spatial arrangement of the selected resource units is not always strictly systematic, but the systematic grid is an important aspect of the design.
Target population	A specific resource set that is the object or target of investigation.
Technical coordinator (TC)	The individual responsible for directing the activities of an individual crosscutting group.
Technical director (TD)	The individual responsible for directing the activities of an individual resource group.
Terminology services (TS)	The centralized approach for maintenance and display of glossaries at USEPA.
Tessellation	A pavement or tiling of a space by a mosaic pattern. The EMAP design prescribes a regular tessellation on the United States, consisting of hexagons of 634.5 km^2.
Tier 1 resource	A specific designation given an EMAP resource class that will be treated as a stratum at Tier 2.
Tier 1/Tier 2	The EMAP design is a multistage, or multiphase, design. These two terms describe the multistage structure of EMAP's monitoring activities. Tier 1 refers to studies that reflect the structure of the sampling grid, and these studies characterize the extend of the resource and its geographical distribution. Tier 2 refers to a double sample from a Tier 1 sample; these studies quantify the ecological condition on a national or regional basis. (See related: Tier 3/Tier 4.)
Tier 3/Tier 4	The general investigation of status and nature of environmental issues involves other levels of investigation below those of monitoring: in EMAP, these activities are designated as Tier 3 or Tier 4. Tier 3 studies are specialized, intensive studies of a finite duration to help diagnose or determine the extent of a degraded condition. Tier 4 studies are basic research studies that support EMAP, and they complement Tier 1, Tier 2, and Tier 3 studies.
Total quality management (TQM)	A system that is implemented in every aspect of an organization with the focus of providing quality, that is, highly valued products. The system provides a framework for planning, documentation, communication, etc. and strongly emphasizes a client-oriented perspective. "The process whereby an entire organization, led by senior management, commits to focusing on quality as a first priority in every activity. TQM implementation creates a culture in which everyone in the organization shares the responsibility for continuously improving the quality of products and services in order to satisfy the customer."
Trends	The changes in the distribution of scores for condition indicators over multiple time periods.
Trophic levels	A functional classification of taxa within a community that is based on feeding relationships (e.g., aquatic and terrestrial green plants comprise (sic) the first trophic level and herbivores comprise (sic) the second.)

Universal Transverse Mercator (UTM) projection The map projection used in many standard map series. This projection uses the Mercator projection formula on meridional zones; it is a cylindrical conformal projection.

Universe The total entity of interest in a sampling program, often together with some structural features. The EMAP universe is the entire United States, together with adjoining waters.

Value A characteristic of the environment that is desired. In the past, the term "environmental value" was defined to mean characteristic of the environment that contributes to the quality of life provided to an area's inhabitants; for example, the ability of an area to provide desired functions such as food, clean water and air, aesthetic experience, recreation, and desired animal and plant species. Biodiversity, sustainability, and aesthetics are examples of environmental values. A quantity's magnitude.

Variance A measure of the variability or precision of a set of observations.

Vector A coordinate-based data structure commonly used to represent map features. (Term added 1993.)

Watershed "The terrestrial area of the landscape contributing to flow at a given stream location. The land area that drains into a stream."

Weights In a probability sample, the sample weights are the inverses of the inclusion probabilities; these are always known for a probability sample.

Wetlands Lands transitional between terrestrial and aquatic systems where the water table is usually at or near the surface or where shallow water covers the land and where at least one of the following attributes holds: (1) at least periodically, the land supports aquatic plants predominantly; (2) undrained hydric soils are the predominant substrate; and (3) at some time during the growing season, the substrate is saturated with water or covered by shallow water. An area that is saturated by surface or groundwater with vegetation adapted for life under those soil conditions, as swamps, bogs, fens, marshes, and estuaries.

Xenobiotic "A chemical or other stressor that does not occur naturally in the environment. Xenobiotics occur as a result of anthropogenic activities such as the application of pesticides and the discharge of industrial chemicals to air, land, or water."

References

1. Burrough PA, McDonnell RA (1998) Principles of geographical information systems. Oxford University Press, New York
2. Lillesand TM, Kiefer RW (1979) Remote sensing and image interpretation. Wiley, New York
3. Barrett EC, Curtis LF (1982) Introduction to environmental remote sensing. Chapman & Hall, London
4. Colwell RN (ed) (1983) Manual of remote sensing, vol I and II. American Society of Photogrammetry, Falls Church
5. Jensen JR (1986) Introductory digital image processing: a remote sensing perspective. Prentice Hall, New Jersey, pp 138–157
6. Dueker KJ, Kjerne D (1989) Multi-purpose cadastre: terms and definition. ASPRS and ACSM, Falls Church

7. Congalton RG, Green K (1995) The ABCs of GIS: an introduction to geographic information systems. In: Lyon JG, Mccarthy J (eds) Wetland and environmental applications of GIS. CRC Press, Inc., Boca Raton, p 16
8. DeMers MN (2000) Fundamentals of geographic information systems, 2nd edn. John Wiley & Sons, Inc., New York, p 498
9. Darkwah SO (1997) Analyses of Beach Changes Around A River Mouth Using Satellite Data and Numerical Model, Dissertation presented to the united graduate school, Kagoshima University
10. Blasco F, Bellan MF, Chaudhury MU (1992) Estimating the extent of floods in Bangladesh using SPOT data. Remote Sens d 39(3):167–178
11. Philipson WR, Hafker WR (1981) Manual versus digital Landsat analysis for delineating river flooding. Photogramm Eng 47(9):1351–1356
12. Wang Y, Colby J, Mulcahy K (2001) An efficient method for mapping flood extent in a coastal floodplain by Landsat TM and DEM data. Int J Remote Sens:20
13. Jone JL, Haluska TL, Williamson AK, Erwin ML. Updating flood maps efficiently: Building on existing hydraulic information and modern elevation data with a GIS, U.S. Geological Survey Open-File Report 98–200
14. Darkwah SO, Deguchi C, Sugio S, Kunitake M (1998) A simple method to estimate nearshore zone water depth. J Remote Sens Soc Japan 18(2):13–27
15. Jensen JR, Huang X, Mackey HE (1997) Remote sensing of successional changes in wetland vegetation as monitored during a four-year drawdown of a former cooling lake. Appl Geogr Stud 1(1):31–44
16. Bryant ES, Pangburn T, Bolus RL, Pedrick GA, Peacock G, Tracy BG, Evelyn JB (1999) Painted rock reservoir: water surface area and storage capacity estimate derived from Landsat data classification US Army Corps of Engineers Special Report 99–6
17. Feng J (1995) Suspended sediment concentrations analysis using TM, spot images and GIS technique: a case study of the Chao Phraya estuary, Masters thesis presented to Asian Institute of Technology, Thailand
18. Silberbauer MJ (1997) The application of geographic information systems to water quality monitoring. Remote sensing and geographic information systems for design and operation of water resources systems. Proceedings of Rabat symposium S3, April 1997, IAHS Publication, No. 242, 189–195.
19. Deutsch M, Ruggles FH (1974) Optical data processing and projected applications of the ERTS-1 imagery covering the 1973 Mississippi River Valley floods. Water Res Bull 10(5):1023–1039
20. Struve H, Judson FE (1979) Development of stage area tables for the Yazoo backwater area using Landsat data. Proceedings of the U.S. Army Corps of Engineers Remote Sensing Symposium, Reston, Virginia, 193–199
21. Darkwah SO, Scoville MS (2001) Floodplain mapping using remote sensing, GIS and hydrological/hydraulic models. Proceeding of AWRA/UCOWR on Decision Support Systems for Water Resources Management, Snowbird, Utah, 315–320
22. Sonu CJ (1965) Study of shore processes with the aid of aerial photogrammetry. Photogram Eng 30(6):404–406
23. El Ashry MT, Wanless HR (1967) Shoreline features and changes. Photogram Eng 33 (2):184–189
24. DanDanforth WW, Theiler ER (1992) Digital Shoreline Analysis System Users Guide, Reston, Virginia, U.S. Geological Survey Open-File Report Number 92-355, 42
25. Darkwah SO, Kunitake M, Sugio S, Deguchi C, Kondo F, Imayama K (1998) Estimation of longshore sediment transport rate using deep water wave data and sediment grain size, Transaction of the Japanese Society of Irrigation, Drainage and Reclamation Engineering, No 193, 159–167
26. Abdullah K, Din ZB, Mahamod Y, Rainis R, MatJafri MZ (2000) Remote sensing of total suspended solids in Penang coastal waters, Malaysia. Proceeding Asian Conference on Remote Sensing, 312

27. Liedtke JRA, Luternauer J (1995) Practical remote sensing of suspended sediment concentration. Photogramm Eng Remote Sens 61:167–175
28. Yang MD, Merry CJ, Sykes RM (1996) Adaptive short-term water quality forecasts using remote sensing and GIS. AWRA Symposium on GIS and Water Resources, Ft. Lauderdale, FL
29. Harding LW, Itsweire EC, Esaias WE (1995) Algorithm development for recovering chlorophyll concentrations in Chesapeake Bay using aircraft remote sensing. Photogramm Eng Remote Sens 61:177–185
30. Cheng K-S, Lei T-C (2001) Reservoir trophic state evaluation using Landsat TM images. J Am Water Res Assoc:1321–1334
31. Ritchie JC, Schiebe FR (2000) Water quality. In: Schulz GA, Engman ET (eds) Remote sensing in hydrology and water management. Springer, Berlin, pp 287–303
32. Ritchie JC, Schiebe FR, McHenry JR (1976) Remote sensing of suspended solids in surface water. Photogramm Eng Remote Sens 42:1539–1545
33. Ritchie JC, Cooper CM, Schiebe FR (1990) The relationship of MSS and TM digital data with suspended sediments, chlorophyll, and temperature in Moon Lake, Mississippi. Remote Sens Environ 33:137–148
34. s PJ, Novo EMM (1988) The relationship between suspended sediment concentration and remotely sensed spectral radiance: a review. J Coast Res 4:351–368
35. U.S. Environmental Protection Agency (2000) Nutrient Criteria Technical Guidance Manual, Lakes and Reservoirs, EPA-822-F-00-002
36. Ritchie JC, Schiebe FR, Cooper CM, Harrington JA Jr (1994) Chlorophyll measurements in the presence of suspended sediment using broad band spectral sensors aboard satellites. J Fresh Water Ecology 9:197–206
37. Gitelson A, Mayo M, Yacobi YZ, Paroarov A, Berman T (1994) The use of high spectral resolution radiometer data for detection of low chlorophyll concentrations in Lake Kinneret. J Plankton Reasearch 16:–993, 1002
38. Aller L, Bennett T, Lehr JH, Petty RJ (1985) DRASTIC: a standard system for evaluating groundwater pollution potential using hydrogeologic settings, EPA/600/2-85/018. R.S. Kerr Environmental Research Laboratory, U.S. Environmental Protection Agency, Ada
39. Tabios GQ, Salas JD (1985) A comparative analysis of techniques for spatial interpolation of precipitation. Water Res Bull (3)
40. Burrough PA (1986) Principles of geographical information systems for land resources assessment. Clarendon Press, Oxford, UK
41. Jenson SK, Domingue JO (1988) Extracting topographic structure from digital elevation data for geographic information system analysis. Photogramm Eng Remote Sens 54(11):1593–1600
42. Moore ID, Grayson RB, Ladson AR (1991) Digital terrain modeling: a review of hydrological, geomorphological and biological applications. Hydrol Process 5:3–30
43. Martz LW, Garbrecht J (1992) Numerical definition of drainage network and subcatchment areas from digital elevation models. Comput Geosci 18(6):747–761
44. Tribe A (1992) Automated recognition of valley lines and drainage networks from grid elevation models: a review and a new method. In J Hydrolology 139:263–293
45. Puecker TK, Douglas DH (1975) Detection of surface-specific points by local parallel processing of discrete terrain elevation data. Comput Gr Image Process 4:375–387
46. USEPA (1988) Geographic information system (GIS) guidelines document. US Environmental Protection Agency, Washington, DC, p 20460
47. USEPA (2020) Environmental acronyms, abbreviations and glossary and terms. US Environmental Protection Agency, Washington, DC, p 20460
48. USEPA (2012) Terminology services glossary management guide. US Environmental Protection Agency, Washington, DC, p 20460
49. USEPA Region 2 (2020) GIS deliverables guidance. US Environmental Protection Agency, Washington, DC, p 20460
50. USEPA (2018) EPA geospatial data and geospatial resources at EPA. US Environmental Protection Agency, Washington, DC, p 20460

51. USEPA (2018) GIS data integration. US Environmental Protection Agency, Facility Registry Service (FRS), Washington, DC, p 20460
52. UESPA (2020) EnviroAtlas interactive map. US Environmental Protection Agency, Washington, DC, p 20460
53. USEPA (2020) Geospatial policies and standards. US Environmental Protection Agency, Washington, DC, p 20460

Chapter 6
Investigation and Management of Water Losses from Wet Infrastructure

Nazih K. Shammas, Lawrence K. Wang, Mohamed A. Khadam, and Yousef Al-Feraiheedi

Contents

1 Introduction 238
2 Management and Control of Leakage 240
3 The Study City 243
4 Leak Detection and Quantification 244
4.1 Water Distribution Network 244
4.2 Sanitary Drainage Sewers 247
4.3 The Storm Drainage System 247
5 Discussion 249
5.1 Leakage from the Water Supply System 249
5.2 Leakage from the Sanitary Drainage System 251
5.3 Leakage from the Storm Drainage System 252
6 Conclusions and Recommendations 252
Acronyms 253
References 254

Abstract During the past few decades, the city of Ar-Riyadh, the capital of Saudi Arabia, has seen a significant increase in urbanization, from a gross residential area of 8500 ha in 1980 to 62,712 ha in 1990 and up to 155,500 ha in 2015 (33, 242 and 600 mi^2, respectively). This increase in urbanization in one and three decades was accompanied by a corresponding increase in population from 850,000 to 1,622,000 and up to 6,500,000. This has led to a great increase in water consumption that has reached 620 L/cap/day (164 gpcd). Most of this water is imported into the Ar-Riyadh environment, but the capability of that environment to dispose of the excess water or wastewater by means of infiltration or evaporation is limited.

N. K. Shammas · L. K. Wang (✉)
Lenox Institute of Water Technology, Newtonville, NY, USA

M. A. Khadam
Civil Engineering Department, University of Khartoum, Khartoum, Sudan

Y. Al-Feraiheedi
Ar-Riyadh Development Authority (ADA), Ar-Riyadh, Saudi Arabia

L. K. Wang, M. -H. S. Wang, Y. -T. Hung, N. K. Shammas (eds.), *Integrated Natural Resources Management*, Handbook of Environmental Engineering 20,
https://doi.org/10.1007/978-3-030-55172-8_6

Consequently the city has begun to experience a rise in the groundwater level that is causing a variety of problems for its citizens.

This chapter outlines the methodologies and discusses the results from a field study of water losses from the municipal wet infrastructures in six selected areas of Ar-Riyadh that included the potable water, the sanitary sewer, and the storm drainage networks. More emphasis was put on the field identification of leakage from the potable water network that approached 16% of the water fed into the areas. Thus Ar-Riyadh compares well with other cities around the world where leakage has been reported to be in the range between 5% and 50%. Losses from the sanitary and storm drainage systems were 7% and 4.5%, respectively. Besides the impacts on buildings and infrastructure services, the cost of potable water lost by leakage on its own, estimated at a minimum value of US$50 million per year, is a considerable economic loss.

Keywords Water losses · Infrastructure · Methodology · Potable water · Wastewater · Drainage water · Water table · Ar-Riyadh · Percent water loss · Field study and problems

1 Introduction

Information from many cities around the world shows that water systems are never completely watertight. In pressurized potable water networks, losses may vary between 5% and 50% (see Table 6.1), depending on the local conditions of the system. The factors that may contribute to leaks in pipes include the following [1–3]:

1. Elevated pressures in the water network
2. Pipe material, quality of fittings, and workmanship
3. Soil characteristics and movement
4. Corrosion and deterioration of pipes
5. Age of pipes
6. Heavy traffic
7. Excavations and damage by others
8. Leakage through abandoned connections

Safe drinking water is a necessity for life. Every day billions of gallons of this precious commodity are delivered to millions of people across the world. Thousands of independent water utilities around the world are dedicated to producing, treating, and delivering safe water to the public. Significant resources are required to install, operate, and maintain the infrastructure of a public water system (PWS). PWSs are facing more obstacles and challenges today than they have in the past with more resource and funding constraints especially so in developing countries. Key issues contributing to the poor operation and maintenance in developing countries have been identified as follows [4]:

1. Lack of funds and inadequate data on operation and maintenance

Table 6.1 Water losses in various cities and countries [1–3]

Country/city	Water loss (%)
Canada	
Toronto	18
Germany	
Munich	12
Hagen	8
Hamburg	2
Hong Kong	30
India	
Mumbai	33
Kolkata	25
Delhi	17.7
Chennai	36
9 cities	20–35
Korea	
Daegu City	36.6
The Philippines	
Metro Manila	51
Sri Lanka	
Greater Colombo	30
Sweden	
Range	1–40
Stockholm	19
Ostergotlands	6
Kalmar	5
Malmohus	12
Goteborg	23
Gotland	31
Thailand	
Bangkok	49
Hat Yai/Songkhla	50
United Kingdom	
Range	10–50
Average	24
United States	
Massachusetts	
46 communities	2–50
Boston	43 and 33
Cambridge	18.5
Springfield, IL	25
Westchester, NY	16.3
Miami, FL	26.6
Philadelphia, PA	31.2
Cleveland, OH	28.7
Los Angeles	3.9

2. Inappropriate system design and poor workmanship
3. Overlapping responsibilities
4. Inadequate training of personnel
5. Inadequate emphasis on preventive maintenance
6. Lack of operation manuals
7. Lack of real-time field information

The infrastructure of many of the drinking water systems in many cities was built decades ago and is currently in need of attention. PWSs are not only expected to produce safe drinking water at a low cost but must also address current growing concerns such as limited water availability, increasing water demands, climate change, increasing regulatory requirements, and limited resources and funding. *The reasons to find and repair leaks are* [5]:

1. Leaks get bigger with age.
2. Repairing leaks reduces growing water losses.
3. Repairing leaks with regularly scheduled maintenance reduces overtime costs of unscheduled repairs.
4. Leak repairs provide more treated, pressurized water to sell to customers.
5. Leak detection and repair can reduce power costs to deliver water and reduce chemical costs to treat water.
6. Leaks have been known to cause damage to nearby roads, other infrastructure, and sometimes buildings. Some water utilities conduct frequent leak detection and repair programs near unstable geologic areas to reduce their legal liability against expensive lawsuits.
7. Leak detection and repair improves public relations. The public appreciates seeing that its water systems are being maintained.
8. The utility gains credibility by putting its own house in order before asking the customers to conserve water.

2 Management and Control of Leakage

A PWS must balance use of its resources to address the financial and personnel demands of economic restrictions, water availability, population and climate changes, regulatory requirements, operational costs, and public and environmental stewardship. A water loss control program can help identify and reduce actual water losses along with apparent losses resulting from metering, billing, or accounting errors. Water loss control programs can potentially defer, reduce, or eliminate the need for a facility to expend resources on costly repairs, upgrades, or expansions. A water loss control program will also protect public health through reduction in potential entry points of disease-causing pathogens.

Maintaining system infrastructure to deliver clean and safe drinking water to customers is often a significant challenge for the operators of public water systems (PWSs). Much of the estimated 880,000 miles of drinking water infrastructure in the

United States has been in service for decades and can be a significant source of water loss [2]. In addition to physical loss of water from the distribution system, water can be "lost" through unauthorized consumption (theft), administrative errors, data handling errors, and metering inaccuracies or failure. Water is a commodity that is produced by a PWS; therefore, lost or unaccounted-for water can be equated to lost or unaccounted-for revenue. A water loss control program can help to locate and reduce these water losses and thus maintain or increase revenue [6–13].

A water loss control program is an iterative process that must be flexible and customizable to the specific needs of a PWS. There are three major components of an effective water loss control program that must be repeated on a periodic basis to continually evaluate and improve the performance of a PWS. These three components are [2]:

1. Water audit
2. Intervention
3. Evaluation

Conducting a water audit is a critical first step in developing a water loss control program. A water audit quantifies the amount of water that is being lost. Most states have regulatory policies that set acceptable losses from PWS distribution systems at a maximum of between 10% and 15% of the water produced by the PWS. This percentage of unaccounted water provides estimated losses and does not adequately quantify how or why this water is categorized as "unaccounted-for." Lack of standardized terminology has historically added to difficulties in comparing water losses from different PWSs. The International Water Association (IWA) and the American Water Works Association (AWWA) have developed standard methods and terminology to perform water audits and to assist water utilities in tracking their distribution system losses [2]. The AWWA/IWA water audit methodology is based on the water balance table, which specifies different types of water consumption and losses. Through the water audit, options will become apparent regarding how to proceed with further identifying where losses are occurring or where efforts to control or eliminate the losses should be concentrated.

The intervention process begins to address the findings of the water audit and can include a variety of actions such as gathering of further information, implementing metering programs, adding or changing metering, and detecting and repairing leaks. The selected intervention option should provide the highest potential benefit value for the resources available that will help to alleviate a flaw or deficiency in the distribution system.

The evaluation portion of the program consists of assessing the success of the audit and intervention actions. The evaluation of an intervention action provides detailed quantification of the implemented methodology. For example, the pipe replacement resulted in a reduction of water losses of so many gallons per customer per year of performance indicators. Performance indicators numerically evaluate different aspects of the distribution system and need to be consistent, repeatable, and presented in meaningful standardized units. A performance indicator (or collection of several) can be used to establish a benchmark. A benchmark allows a PWS to evaluate its performance over a period of time by repeating the performance indicating tests and comparing them with previous results. Performance indicators and benchmarks also allow comparisons between public water providers.

Accurate metering is crucial in a water loss control program. Metering establishes production and customer use volumes as well as provides historic demand and consumption data that is useful not only for auditing but for planning future needs. There is no single type of meter that will accurately measure flow for all applications, but there are a variety of meters that have been developed using different operating principles, designed to perform within required tolerances under different circumstances. The cost of meters typically ranges from a few hundred dollars to thousands of dollars per meter depending on size, complexity, and operating conditions. A PWS must select the meters they use carefully according to intended use, flow rates, and the environment where it will be installed. How the meters will be read is also a decision that a PWS has to decide when considering metering programs. Meters can be read manually but most PWSs are moving toward a variety of different Automatic Meter Reading (AMR) systems that reduce reading errors and allow labor to be reduced or reallocated.

While it is possible to spot losses through billing data discrepancies or abrupt changes in amounts of water that have been historically used, it is typically necessary to physically pinpoint the leak in the field. The location of a leak is not always obvious unless it is large. An array of techniques and equipment are available to assess leakage from distribution lines within a geographic area or pinpoint a leak within a suspected segment of pipe. Flow monitoring of a District Meter Area (DMA) can be used to determine leakage within an area that can be isolated and may encompass 1500 to 2000 service connections. These techniques monitor flow to specific areas and compare water flowing into the area with known or estimated night usage to determine losses in the DMA or along a branch water line.

There are several different types of leak detection equipment that use different operating principles. Acoustic equipment detects a leak through noise made by water as it leaks from the pipe. Electromagnetic field detection is used on pre-stressed concrete pipe and locates defective reinforcing steel in the pipe. Thermal detection devices look for the temperature differences in the surrounding ground caused by saturation due to the leaked water. Chemical detection relies on locating substances added to the treated water such as chlorine or fluoride that do not occur naturally. A tracer gas may also be introduced into dewatered lines. If there is a leak, a special instrument can detect it at the surface. The different styles of leak detection equipment require varying levels of skill and experience to operate with accuracy. Capital costs for typical leak detection equipment range from less than hundred to several thousand dollars depending on its complexity.

Once a leak is located it should be repaired. Some repair techniques include wrapping or using repair clamps. Replacement can be done by installing new pipe in an excavated trench or by use of a trenchless method, such as sliplining or pipe bursting, where a new pipe of the same size or larger is pulled through the existing pipe with special equipment. Micro tunneling and hydraulic jacking are other trenchless techniques where pipe is either pushed or pulled underground without the necessity of large amounts of excavation.

Operations and maintenance (O&M) procedures and standards should also be a part of any water loss control program. Along with ensuring proper design and installation of new distribution components, maintenance and operation measures such as system flushing, valve exercising, meter assessment testing and replacement programs, system modeling, and pressure management all contribute to improved efficiency, reduction in water losses, and often cost savings.

Developing a complete water loss control program requires careful consideration of the water loss reduction goals a PWS wishes to achieve. The program should be customized for the unique features of the PWS and be flexible enough to update as the PWS conducts future water audits [2, 14, 15]. In gravity drainage systems such as sanitary and storm sewers, smaller amounts of losses are expected.

3 The Study City

Quantification and assessment of impacts of losses from municipal water services in Ar-Riyadh (see Fig. 6.1) were verified in a research project. For an estimated population of 6.5 million inhabitants in 2015, the amount of water supplied to the city has reached 4 million m^3/day. More than two-thirds of this amount is desalinated seawater transported from the Arabian Gulf, 400 km away, and the remaining portion is extracted from local deep well fields within a radius of about 100 km from the city. The urban limit of the city covers an area of 155,500 ha that is fully covered by the water supply system. Sanitary sewers serve about 54% of the population and 27% of the urban area [16]. Two main storm water channels follow the natural courses of two wadis (valleys) and discharge southward into Wadi Hanifah.

Water losses from municipal wet infrastructures percolate into the subsurface environment of the city that is incapable of admitting such an excess amount of water. With the exception of minor alluvial deposits, the geological formations of the city are characterized by thin soil layers, which overlie impervious limestone. As a result of increasing water losses, the city has experienced a rise in groundwater level at a rate of 0.6 in. per year, rising from more than 20 in. in depth to 0.5 in. below the ground surface in urban areas at present [17]. This water rise is causing a variety of problems to Ar-Riyadh's citizens. A brief description of the fieldwork and a discussion of the results of leakage from the wet systems are outlined in the following sections.

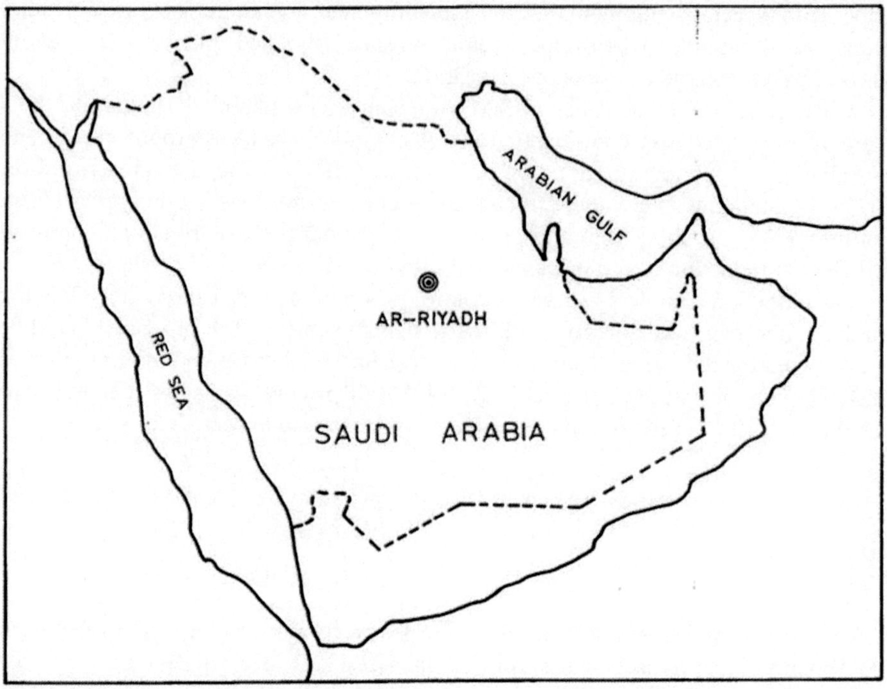

Fig. 6.1 Map showing location of Ar-Riyadh City

4 Leak Detection and Quantification

To obtain factual data on water losses from municipal water infrastructures, detailed field measurements and inspections were performed in six study areas, each of 25 ha (see Fig. 6.2). Selection of these areas was made in such a way that they would represent the diverse conditions in the city.

Leakage from the following wet utilities was assessed:

- The water distribution network
- The sanitary drainage sewers
- The storm water drainage system

4.1 Water Distribution Network

The water distribution network of Ar-Riyadh is relatively new. The oldest parts of the network date only from around 1973. The large feeder mains (with diameters of 600–1200 mm) are cylindrical concrete pipes; mains (300–500 mm) are either

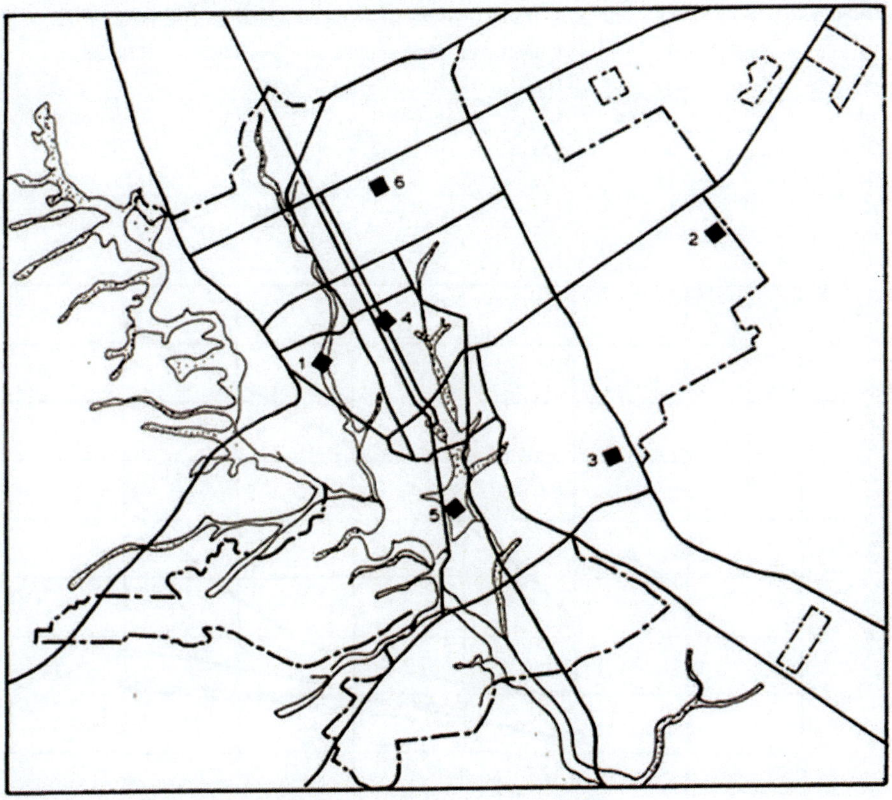

Fig. 6.2 Location of study areas in Ar-Riyadh

asbestos cement or ductile iron; distribution lines (100 and 150 mm) are polyvinyl chloride (PVC); and house connections (19–32 mm) are either PVC or polyethylene (PE).

The leak detection study was carried out in five localized areas (pipe diameters 150 mm and smaller) using MicroCorr leak detection equipment [18]. All identified leaking pipe stretches were then subjected to leak quantification by pressure testing. After isolation of the stretch by closing off the appropriate valves, the required amount of water was pumped to maintain testing pressures of 2, 4, and 6 bars for durations of 30, 15, and 15 minutes, respectively. The quantity of water needed for each test represents the amount of leak from the stretch at the specified pressure and time duration. Leakage rates from each stretch were calculated as shown in Table 6.2.

Average values of leakage, from all stretches, at the three values of pressure were computed and plotted for each study area. Five graphs – representing leak variation with pressure – were produced for the study areas as shown in Fig. 6.3.

In order to determine the actual operating conditions, the daily pressure variations in the water network within the study areas were monitored. The average working

Table 6.2 Leakage from a pipe stretch (length = 123.3 m and diameter = 100 mm)

Pressure applied	Time	Quantity of water applied	Leakage	Leakage
(*P*)	(*t*)	(*v*)	(*v*/*t*)	
Bar	minute	L	L/minute	L/km/mm/day
2	30	167	5.57	651
4	15	202	13.47	1577
6	15	221	14.73	1725

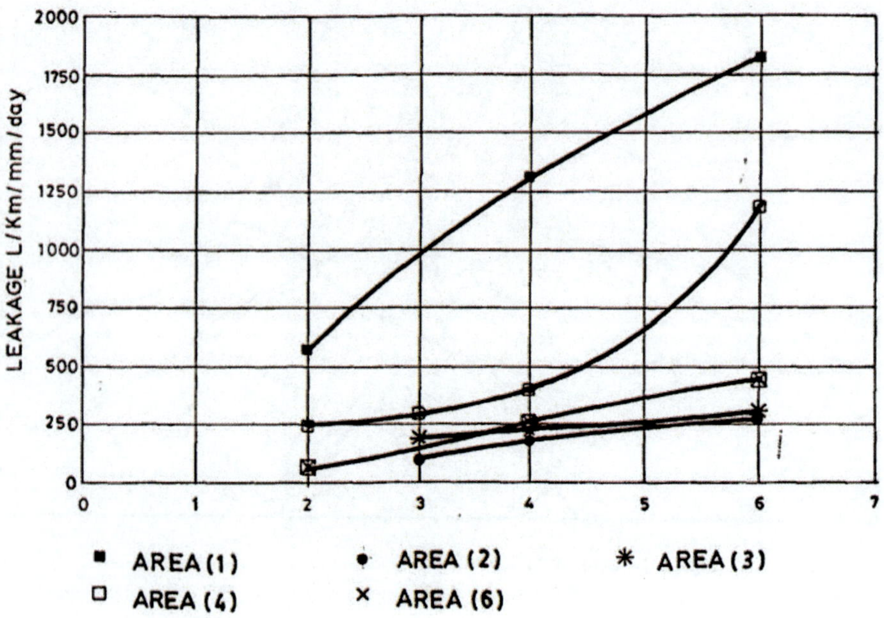

Fig. 6.3 Variation of leakage rate with pressure

pressures in the five areas were 2.5, 3.2, 3.9, 1.8, and 2.8 bars, and the corresponding values of leakage rates obtained from Fig. 6.3 were 800, 122, 230, 240, and 284 L/km/mm/day (see Table 6.3).

The percentage of water lost by leakage was computed by the following relationship:

$$\%\text{Leakage} = \frac{100q}{(q + Q)}$$

where q = total amount of leakage from the network in m^3/day. It was computed from the following parameters: average leakage rate, number of leaking stretches, average length of stretches and the pipe size.

Q = amount of water consumed in the area in m^3/day.

Table 6.3 Summary of quantification of leakage from domestic water networks in five areas

Area	Average operating pressure Bars	Leakage rate L/km/mm/day	Leakage q m^3/day	Flow in residents in the area Q m^3/day	Leakage %
1	2.5	800	156	488	24
2	3.2	122	81	490	14
3	3.9	230	78	442	15
4	1.8	240	48	447	10
5	2.8	284	88	368	19
Overall weighted average					16

A demographic survey was conducted and the average water consumption per household was measured by calibrated flow meters. Results of the survey are shown in Table 6.4.

4.2 Sanitary Drainage Sewers

Three out of the six study areas in the city are served by sanitary drainage sewers (see Table 6.4). Field tests were conducted in these areas to determine wastewater leaking from, and water infiltrating into, the sewers. The testing was performed on pipe sections from manhole to manhole (40–80 m in length) using pneumatic balls to plug the sections [33].

For assessment of leakage, water was supplied to the section through a bypass ball that was connected to a barrel to maintain a constant head of about 1.5 m of water above the crown of pipe. The quantity of water supplied to maintain a constant head for a period of 1 hour was considered to be the leakage from the section.

For quantification of infiltration, the sewer section was plugged from the upstream side, while the downstream side was left open in order to collect and measure the infiltrating water. The amount of water collected over a period of 1 hour would represent the infiltration into the pipe. The results are summarized in Table 6.5. All tested sewers are less than 15 years old and have been constructed of PVC pipes 200 and 250 mm in diameter at slopes ranging between 0.5% and 1%.

4.3 The Storm Drainage System

In two of the areas (Table 6.4) where storm drainage sewers were available (about 25% of Ar-Riyadh is sewered), tests for leakage were conducted on selected stretches. The testing was performed by means of an ISCO Model 2300 Flow Meter [19] that determines the rates of flow by measuring the water levels at

Table 6.4 Demographic survey of study areas

						All values are in %								
						Monthly income (SR)[a]				Education of household head				
Area No.	Location	Sanitary sewerage	Strom sewerage	Number of sampled house	Size of plot m^2	Persons per house	Less than 4000	4000–5999	6000–9999	More than 10000	Elem. School	2ndary School	Univ. Graduate	Water use L/cap/day
1	Um Al-Hamam	No	No	31	455	11.5	22.6	9.7	54.8	12.9	32.3	51.6	16.1	215
2	Al-Naseem	No	No	32	506	12.0	41.9	22.6	32.8	3.2	45.2	54.8	0.0	242
3	Al-Faisaliyah	No	No	34	308	14.6	79.0	18.0	3.0	0.0	84.4	12.5	3.1	158
4	Al-sulaymaniyah	Yes	Yes	31	694	9.0	12.9	12.9	38.7	35.5	16.7	30.0	53.3	468
5	Al-Dirah	Yes	Yes	28	155	9.1	88.5	3.8	7.7	0.0	80.8	19.2	0.0	297
6	Al-Maseef	Yes	No	33	616	10.5	9.0	27.3	45.5	18.2	45.4	36.4	18.2	324

[a]USD = SR 3.75

Table 6.5 Leakage from sanitary drainage sewers

	Length of sewers in network	Leakage out	Infiltration	Net leakage	Flow in network	Leakage
	L			q	Q	
Area	m	m^3/day	m^3/day	m^3/day	m^3/day	%
6	2275	153	7	146	2844	5
4	6242	1106	5	1101	18240	6
5	1440	1019	8	1011	10416	10
Weighted average						7

Table 6.6 Summary of leakage from storm drainage sewers

	Water depth in sewers	Leakage rate	Flow in storm network	Leakage	
		q	Q		
Area	mm	L/km/mm/day	m^3/day	m^3/day	%
4	120	2468	5616	304	5.4
5	100	3096	12210	495	4.1
Weighted average					4.5

consecutive sections of the storm sewers. A total of nine stretches (40–105 m in length) of asbestos cement storm sewers having 400 and 600 mm diameters were tested in each area. Water was supplied via a constant head device to the stretch's upstream manhole. The flow rate was calculated using Manning equation. Table 6.6 summarizes the weighted average values of leakage from storm sewers in the two study areas.

5 Discussion

5.1 Leakage from the Water Supply System

1. The leakage rate was calculated on the basis of the average operating pressure in each area. It is apparent from leakage-pressure curves that the variation of leakage rate with pressure does not follow a specific exponential relationship. This deviation of leakage flow from the orifice formula conforms to the findings of other studies [20–24].

   The nonuniformity in the leakage-pressure relationship appears to result from several factors, the most important ones being:

   - Pressure variations during the day cause a change in the width of cracks, particularly in plastic PE and PVC pipes; hence the orifice sizes also vary with pressure.
   - Thermal fatigue due to extreme changes in temperature causes defective PVC and PE pipes to have various shapes of cracks and holes.

- Poor construction workmanship causes the joints between pipes to separate, resulting in an opening that will widen and pass increasing flows with high pressures. It was also observed that the backfill materials were not of uniform sizes. Sometimes large pieces of stone were found in the fill materials causing splits, particularly in plastic pipes.

2. Leakage rates in the five areas ranged between 10% and 24%, with an overall weighted average of 16%, for pipe sizes between 19 and 150 mm. According to Riyadh Region Water and Sanitary Drainage Authority (RRWSDA), the leakage in water mains greater than 150 mm is negligible. The reason for this is that failures in large supply mains are easily detected and consequently they are promptly repaired. Although water losses in Ar-Riyadh are not excessive in comparison to other cities in the world (see Table 6.1), yet a leakage control program is essential because of the scarcity and high cost of water in Saudi Arabia.
3. Desalinated water is imported to Ar-Riyadh from the Arabian Gulf at a relatively high cost. The unit cost of desalinated water is in the range of SR 3–6 (USD 0.8–1.6) per cubic meter [25]. For a total water supply of about 6.5 million m^3/day in 2015, the annual cost of lost water at a leakage rate of 16% is around USD 300 million as a minimum figure. This is a considerable economic loss, notwithstanding problems created by the rising groundwater table in the city. As part of the preventative solutions adopted to control the rising groundwater table, a 2-year crash program for leak detection and repair was implemented to reduce leakage rate to an acceptable level (see Tables 6.7 and 6.8).

The program was sponsored by Ar-Riyadh Development Authority (ADA) and executed by RRWSDA. After 2 years of fieldwork initiation and an expenditure of USD 2 million, the first phase of the program has been completed. The water network has been surveyed (using sound detectors and pressure tests) and all major leaks have been repaired. The success of the program has been assessed by two studies sponsored by ADA [27] and King Abdulaziz City for Science and Technology (KACST) [28]. ADA has reported that the leakage in Area 2 has decreased from 14% to 13% and in Area 6 from 19% to 12%. Two other areas that were reported by the KACST study indicated very low leakage rates of 1% and 7% after the program was completed. The encouraging results led ADA to sponsor Phase 2 of the leakage program, whereby a second round of testing will soon be

Table 6.7 Classification of leakage levels [26]

Leakage level (%)	Network condition
5% or Less	Very good performance
5–10%	Acceptable performance
10–15%	Reasonable performance
15–25%	Average to low performance
More than 25%	Unsatisfactory

Table 6.8 Unaccounted for water standard for selected US states [2]

State	Agency	Standard
Arizona	Department of Water Resources	10% (large) 15% (small)
California	Urban Water Conservation Council	10%
Florida	Southwest Florida Water Management District	12%or less
Florida	St. Johns River Water Management District	10%
Georgia	Environmental Protection Division	Less than 10%
Indiana	Department of Environ mental Management	10–20%
Kansas	Kansas Water Office	15%
Kentucky	Department of Energy, Water and Sewer Branch	15%
Louisiana	Department of Environmental Quality	15%
Massachusetts	Department of Environmental Protection	15%
Minnesota	Department of Natural Resources	10%
Missouri	Department of Natural Resources	10%
North Carolina	Division of Water Resources	15%
Ohio	Public Utility Commission and Environmental Protection Agency	15%
Oregon	Water Resources Division	10–15%
Pennsylvania	Public Utility Commission	20%
Pennsylvania	Bureau of Water and Wastewater Management	10–15%
Rhode Island	Water Resources Board	10–15%
South Carolina	Public Service Commission	7.5%
South Carolina	Department of Health and Environmental Control	10%
Texas	Water Development Board	10–15%
Texas	Natural Resources Conservation Commission	20%
Washington	Department of Health	20% (10% proposed)
West Virginia	Public Service Commission	15%
Wisconsin	Public Service Commission	15% (large) 25% (small)
Delaware River Basin Commission	Delaware River Basin Commission	15%

initiated to recheck and repair remaining, as well as newly developed, leaks in the water network.

5.2 *Leakage from the Sanitary Drainage System*

1. Testing and flow calculations (using the Manning equation) for the sanitary drainage sewers were performed when the pipes were flowing full. Therefore, the weighted average percentage of leakage of 7% based on "flowing full

conditions" could reasonably be assumed to represent the leakage value at other normal flow conditions in the sanitary drainage system.

2. Routine inspection of sanitary sewers is essential. Leakage from sanitary sewers has serious health impacts because there is a possibility of cross-connection with the potable water system [29]. The best method of inspection is to pass a small television camera through the sewer, which is also necessary to make sure that the sewer is well maintained to avoid blockage and leakage.

5.3 *Leakage from the Storm Drainage System*

1. The leakage rates in two districts, both of which being served by storm drainage sewers, were 5.4% and 4.1% with a weighted average of 4.5%.
2. The rainfall intensity (i) in Ar-Riyadh for a 50-year return period and duration of 1 hour is about 28 mm/hour. Thus, the quantity of storm water drained from an urbanized catchment area (A) of 25 ha, based on the rational formula ($Q = 2.76$ CiA) [30], would be = (2.76) (0.36) (28 mm/hour) (25 ha) = 690 L/s = 2500 m^3 for a duration of 1 hour. Consequently, in 2015, the amount of leakage from the storm sewers could reach 4.5% of 2500 = 113 m^3/storm in a typical 25 ha area and up to 700,000 m^3/day in the sewered parts of the city.
3. Storm drainage sewers in Ar-Riyadh are now frequently being used for discharging groundwater collected by subsurface drains and by dewatering projects. Consequently, it is expected that considerable leakage from these storm sewers could occur in the lower reaches where the water depth in the sewer is high.
4. For storm drainage sewers, which carry major flows, it is essential that routine inspection be carried out. As for sanitary sewers, inspection can be made by passing a small television camera through the sewer to locate leaks, incrustations, and failures. Repairs should be made promptly.

6 Conclusions and Recommendations

1. Leakage from municipal utilities varies greatly from area to area even in the same city. Leakage from the potable water network in five locations of Ar-Riyadh varied between 10% and 24%.
2. Leakage increased exponentially with rising pressure to an even greater extent than would be predicted by the orifice equation.
3. The impacts of water losses from networks are very significant from multiple aspects:

   - Direct economic loss (estimated at around a minimum of USD 300 million per year in Ar-Riyadh)
   - Loss of an important resource especially important where water is scarce, as in Saudi Arabia [31, 32]

- Indirect economic loss, environmental problems, and damage to structures, public utilities, and roads due to the rise in the groundwater levels (this is particularly critical in areas where groundwater rise occurs after planning and as a result of rapid urbanization)
- Possible contamination of drinking water and its associated health hazards

4. The establishment of a leakage control program is not only essential but it proved to be practical and cost-effective. It should consist at least of the following steps:
   - Pressure control (lowering of pressure during the night will effectively decrease the leakage rates).
   - Survey of the water network for leaks using sound detectors and/or pressure testing
   - Repair of leaks.
   - Installation of "district meters" at appropriate locations in the city in order to detect abnormal increase in water consumption and obtain early warning of possible water losses.
   - Surveying for leak detection should be an ongoing process, repeated at 2–5 years interval, in order to locate and repair newly developed leaks.
5. Routine inspection of sanitary and storm sewers is essential and can be carried out by passing a small video camera through the sewers.

Acknowledgments This chapter is a contribution of the Environmental Management and Protection Program, Rising Groundwater Study, Ar-Riyadh Development Authority (ADA), Saudi Arabia. This updated version is being published by permission from the journal *Water International*, 16, 4, 254–261, 1991, and published online 22 Jan 2009.

Acronyms

ADA	Ar-Riyadh Development Authority
AWWA	The American Water Works Association
DMA	District Meter Area
IWA	The International Water Association
KACST	King Abdulaziz City for Science and Technology
PE	Polyethylene
PVC	Polyvinyl Chloride
PWS	Public Water System
O&M	Operations and maintenance
RRWSDA	Riyadh Region Water and Sanitary Drainage Authority

References

1. Al-Dhowalia KH, Shammas NK, Quraishi AA, Al-Muttair FF (1989) Assessment of leakage in the Riyadh Water Distribution Network, First Progress Report. King Abdulaziz City for Science and Technology, Ar-Riyadh
2. U.S. Environmental Protection Agency (2010, November) Control and mitigation of drinking water losses in distribution systems. U.S. EPA Office of Water (4606M), EPA 816-R-10-019
3. Naik KS, Glickfeld M (2015) Water distribution system efficiency: an essential or neglected part of the water conservation strategy for Los Angeles County water retailers? UCLA Water Resources Group, Institute of the Environment and Sustainability, University of California Los Angeles, Los Angeles
4. Dighade RR, Kadu MS, Pande AM (2014) Challenges in water loss management of water distribution systems in developing countries. Int J Innov Res Sci Eng Technol 3:6
5. California Department of Water Resources, Leak detection, http://www.water.ca.gov/wateruseefficiency/leak. Accessed February 2018
6. Arscott A (2007) International Water Association, Water Loss Conference 2007. http://www.waterloss2007.com/pdf_vortraege/Mittwoch/C9-3.pdf. Accessed January 2018
7. American Water Works Association (AWWA) (2009) Water audits and loss control programs. In: Manual of water supply practice M-36, 3rd edn. AWWA, Denver
8. Brothers K (2003) A practical approach to water loss reduction. In: IWA water loss task force. IWA Publishing, London. http://www.iwapublishing.com/pdf/WaterLoss-Jun.pdf. Accessed February 2018
9. Georgia Environmental Protection Division (EPD) (2007) Water loss control program: EPD guidance document, Georgia Environmental Protection Division, Watershed Protection Branch, http://www1.gadnr.org/cws/Documents/Water_Loss_Program.pdf. Accessed January 2018
10. Kalinkov V, Vladov G, Radovanov V (2011) Studies and procedures of water loss reduction in the water supply system of the town of Vidin. Water Util J E.W. Publica 1:11–17
11. Alkasseh JMA, Adlan MN, Abustan I, Abdul Azziz H, Hanif ABM (2013) Applying minimum night flow to estimate water loss using statistical modeling: a case study in Kinta Valley, Malaysia. Water Resour Manag 27(5):1439–1455
12. Trojan F, Morais DC (2015) Maintenance management decision model for reduction of losses in water distribution networks. Water Resour Manag 29(10):3459–3479
13. Mutikanga H, Sharma S, Vairavamoorthy K (2013) Methods and tools for managing losses in water distribution systems. J Water Resour Plan Manag 139:2
14. Farley M (2003, September) Understanding and managing losses in water distribution networks. Water & Wastewater International. http://www.pennnet.com/display_article/189881/20/ARTCL/none/none/1/Understanding-andmanaging-losses-in-water-distribution-networks/. Accessed January 2018
15. Heneghan & Associates (2004, September) Water loss – What is it? Civil talk. http://www.haengr.com/Newsletters/newsletter92004.htm. Accessed February 2018
16. Ar-Riyadh Development Authority (ADA) (1991) Appraisal of the sanitary drainage system in Ar-Riyadh City, Wastewater Management Plan and Construction Program, ADA Report, Ar-Riyadh, Saudi Arabia
17. Ar-Riyadh Development Authority (ADA) (1990) Study of rising groundwater levels in Ar-Riyadh. Report prepared for the High Commission for Development of Ar-Riyadh, Saudi Arabia
18. Palmer Environment Services Limited (1987) Microcorr Super Model MC 200, 6 Knightwood Court, Gapton Hall Industrial Estate, Great Yarmouth, Norfolk NR31 ONG, England
19. Instruction manual, Model 2300 Flow Meter ISCO, Inc. (1986) 531 Westgate Boulevard, Lincoln, Nebraska
20. Jeffcoate P, Saravanapavan A (1987) The reduction and control of unaccounted-for water: World Bank Technical Paper No. 72. World Bank, Washington, D.C.

21. Ferrante M, Meniconi S, Brunone B (2014) Local and global leak laws, the relationship between pressure and leakage for a single leak and for a district with leaks. Water Resour Manag 28 (11):3761–3782
22. Dai PD, Li P (2016) Optimal pressure regulation in water distribution systems based on an extended model for pressure reducing valves. Water Resour Manag 30(3):1239–1254
23. Tabesh M, Asadiyani AH, Burrows A (2009) Integrated model to evaluate losses in water distribution systems. Water Resour Manag 23:477–492
24. Fantozzi M, Lambert A (2007, September) Including the effects of pressure management in calculations of short run economic leakage levels. In: Paper presented at the IWA International Specialized Conference, Water Loss 2007, Bucharest. http://www.leakssuite.com/Documents/Papers/003_Fantozzi.pdf. Accessed January 2018
25. Ukayli MA, Husain T (1988) Comparative evaluation of surface water availability, wastewater reuse and desalination in Saudi Arabia. Water Int 13(4):218–225
26. Al-Dhowalia KH, Shammas NK (1991) Leak detection and quantification of losses in a water network. Int J Water Resour Dev 7(1):30–38
27. Ar-Riyadh Development Authority (ADA) (1991) Leakage control program for Ar-Riyadh Water Distribution Network. ADA Internal Report. ADA, Ar-Riyadh
28. Al-Dhowalia KH, Shammas NK, Quraishi AA, Sendil U (1991) Assessment of leakage in the riyadh water distribution network, Fourth Progress Report. King Abdulaziz City For science and Technology, Ar-Riyadh
29. Besner MC, Gauthier V, Servais P, Camper A (2002) Explaining the occurrence of coliforms in distribution systems. J AWWA 94(8):95–109
30. Shammas NK, Wang LK (2011) Water supply and wastewater removal, 3rd edn. Wiley, New York, USA, p 388
31. Bourg J (2008) Water conservation, Whole building design guide (WBDG). National Institute of Building Sciences. http://www.wbdg.org/resources/water_conservation.php. Accessed January 2018
32. Dickinson MA (2008) A vision for advanced water conservation. Alliance for Water Efficiency
33. http://www.texaswatermatters.org/pdfs/Mary_Ann_part1.pdf. Accessed February 2018

Chapter 7
Lake Restoration and Acidic Water Control

Lawrence K. Wang, Mu-Hao Sung Wang, Nazih K. Shammas, R. Derrick I. Kittler, and Donald B. Aulenbach

Contents

1 Importance of Lakes 258
2 Characteristics of Lakes 259
3 Importance of Biological Activity 261
4 Considerations in Remediation 264
5 Treatment to Prevent Nutrient Discharges 265
6 Recovery of Eutrophic Lakes 269
6.1 Aeration 269
6.2 Weed Harvesting 270
6.3 Dredging 271
6.4 Sediment Fixation 271
7 Hypolimnetic Phosphorus Removal by DAF 272
8 Sources, Chemistry, and Control of Acid Rain 278
8.1 Effects of Acid Rain 279
8.2 History and Regulations 283
8.3 Causes of Acid Rain 286
8.4 Reducing Acid Rain 289
8.5 Acid Rain Permit Program 291
8.6 Acid Rain Monitoring 292
8.7 Acid Rain Allowances 292
8.8 Acid Rain Measurement 293
8.9 Trends 295
8.10 Acid Rain and Other Pollution Problems 297
8.11 Future Efforts 298
9 Case Histories 299

L. K. Wang (✉) · M.-H. S. Wang · R. D. I. Kittler
Lenox Institute of Water Technology, Newtonville, NY, USA

N. K. Shammas
Lenox Institute of Water Technology, Newtonville, NY, USA

Krofta Engineering Corporation, Lenox, MA, USA

D. B. Aulenbach
Rensselaer Polytechnic Institute, Troy, NY, USA

Lenox Institute of Water Technology, Newtonville, NY, USA

L. K. Wang, M. -H. S. Wang, Y. -T. Hung, N. K. Shammas (eds.), *Integrated Natural Resources Management*, Handbook of Environmental Engineering 20,
https://doi.org/10.1007/978-3-030-55172-8_7

9.1 Lake Brazos, Waco, TX 299
9.2 Water Treatment from Lake Roine, Tampere, Finland 300
9.3 Restoration of Lake Apopka, FL, USA 301
9.4 Water Treatment from Lake DeForest in Clarkstown, NY, USA 304
10 Summary 304
Glossary 306
Appendix A 307
Appendix B 308
Appendix C 309
Appendix D 310
Appendix E 311
Appendix F 312
Appendix G 315
References 317

Abstract The steps leading to the demise of a lake are discussed. One of the primary causes of the death of lake is excessive biological growth, called eutrophication. Biological growth is limited primarily by the availability of the nutrients necessary for growth. It has been shown that phosphorus is most frequently the limiting nutrient to control biological growth in a lake, but nitrogen is also commonly limiting. Phosphorus may be permanently removed from a lake by various processes, whereas nitrogen is difficult to remove permanently due to the fact that certain blue-green algae can fix atmospheric nitrogen as a nitrogen source. Thus, emphasis has been placed on removal of phosphorus. There are various methods for treatment of wastewaters to remove the nutrients before being discharged to a body of water. Once in a lake, phosphorus removal is most frequently achieved by producing an insoluble aluminum salt of the phosphorus, but iron salts are effective under aerobic conditions. Calcium salts are effective in removing phosphorus, but they generally adversely increase the pH of the lake. Precipitated aluminum phosphate salts may be allowed to settle to the bottom of the lake, or they may be removed from the water column. A study showed that removing the phosphate-rich hypolimnetic waters from a summer-stratified temperate climate lake, precipitating the phosphorus as either aluminum or iron salts, separating the precipitate by DAF, and returning the phosphate-reduced water to the lake were very effective in controlling the phosphorus nutrient content in Devils Lake, WI, USA.

Acid rain is formed when sulfur dioxide and nitrogen oxides reach the air and are transformed into sulfate or nitrate particles. When combined with water vapor, they are converted into sulfuric or nitric acids. Acid rain can adversely affect aquatic life at all levels of the food chain that can be harmed by acid rain. Destruction begins at the lowest level of the food chain, when the tiny microorganisms that are food for minnows and other small organisms die. As food sources dwindle, more and larger fish die. Acid in the water may also interfere with oxygen circulation, harm fish gills, and cause heart problems in fish. The chemistry and control of acid rain are also discussed. A case history involving the use of lime or sodium aluminate for neutralization of acid rain contaminated reservoir water is also presented.

Keywords Dedication · Donald B. Aulenbach · Nutrients · Productivity · Biological activity · Stratification · Eutrophication · Remediation · Phosphorus precipitation ·

Acid rain control · Algae separation · Dissolved air flotation · DAF · Acid rain · Neutralization · Sulfur dioxides · Nitrogen oxides · Causes · Monitoring · Regulations · Environmental effects · Reduction and control · AquaDAF

Nomenclature

ANC	Acid neutralization capacity
ARP	Acid Rain Program
CEM	Continuous emissions monitor
COD	Chemical oxygen demand
CSAPR	Cross-State Air Pollution Rule
DAF	Dissolved air flotation
HAPs	Hazardous air pollutants
NAAQS	National Ambient Air Quality Standards
NADP	National Atmospheric Deposition Program (NADP)
NAPAP	National Acidic Precipitation Assessment Program
USEPA	US Environmental Protection Agency
USGS	US Geological Survey
WDNR	Wisconsin Department of Natural Resources

1 Importance of Lakes

All lakes and reservoirs have a finite life. That life may be measured in geological time or in human lifetimes. The general pattern of aging of a lake is for the lake to fill in with either allochthonous materials (carried into the lake from inlet streams or direct runoff) or autochthonous materials (generated by biological growth within the lake). As a lake ages, the water becomes more shallow. The decreased volume of water concentrates the same nutrient input. This encourages more biological growth, which further fills in the lake with dead biomass. When the depth decreases to about 2 m, rooted aquatic plants proliferate due to their access to direct sunlight. This further increases the filling in of the lake. Most frequently (but not always) when a lake reaches this point, it becomes a wetland or a bog. At this point emergent plants and eventually trees appear. These tend to take up the moisture, drying out the system. The wet organic material may progress to peat, a useful source of fuel. In geologic time, with the aid of pressure, this progressed to coal, a very valuable source of energy.

However useful peat and coal may be as a source of energy and raw materials, lakes are considered more important for their water. All life depends upon water and its unusual characteristics. In addition to water for drinking, water is essential for irrigating crops. Irrigation is the largest consumer of water on Earth today. As the Earth's population grows, there will be a greater demand for food, much of it needing irrigation. Other industries require water, including process water and cooling water. Thus water is essential for life as well as for the living of modern-day life.

In addition, water in its place, such as a lake, is important to our livelihood. Besides its use in transportation, many recreational activities, such as swimming and boating, depend upon lakes and streams. Further, lakes have an aesthetic quality. Many poems and stories have been written about lakes. The beauty and tranquility of lakes adds to our consolation. Storms on a lake inspire awe. Thus it may be seen that lakes are essential to our way of life.

In this chapter, reservoirs are considered in the same manner as lakes. By definition, reservoirs are artificial lakes, generally constructed to serve a specific purpose. That purpose may include drinking water supply, flood control, low flow augmentation, travel enhancement, storage for periods of low precipitation, recreation, and any combination of these. The life of a reservoir mimics that of a lake, although the factors that influence the life of a reservoir may be somewhat different, or in different magnitude, from those that impact the life of a lake. In both cases these factors are so variable that to predict the life of a lake, each lake must be studied as its own entity. No two lakes or reservoirs are exactly the same, nor have the same needs.

It may be concluded that the preservation of lakes and the extension of their lives is important to the continuation of human life on Earth.

2 Characteristics of Lakes

Even though each lake has its own characteristics, we can make generalities on the factors that influence the life of a lake or reservoir. By understanding these characteristics, we can devise means of slowing the aging process and in some cases even reversing that process. By studying ancient lakes and terminal lakes, we can describe the factors that have either preserved the lake or hastened its demise.

There are numerous factors that control the life of a lake. Not in any preferred order, the morphology of a lake is an initial factor. Deep lakes with steep sides seem to have greater longevity. Large shallow areas tend to encourage rooted plant growth, which leads to the more rapid filling in of the lake. Steep sides may even limit human habitation as experienced in Crater Lake, Oregon, formed in the caldera of a former volcano. The smaller the ratio of the watershed to the lake surface area, the longer its life; again Crater Lake is a prime example. Larger lakes such as the Great Lakes of North America have a long life. The underlying geological formations in the lake may contribute essential nutrients that may allow biological growth. A forested watershed will lessen the amount of nutrients being carried into the lake. Conversely, farmed areas contribute large amounts of nutrients from fertilizers. Human development may contribute significantly to the demise of a lake. Whereas everyone enjoys the beauty and the recreation attractions of a lake, more inhabitants result in more direct surface runoff to the lake and more domestic wastes containing nutrients to ultimately reach the lake. A significant impact is lakeside homeowners who pride themselves with their green lawns, right down to the water's edge, kept green with fertilizers, which readily reach the lake. Again, no two lakes are identical, and the combination of factors affecting a lake's life is infinite.

Deep lakes in temperate climates exhibit an interesting circulation pattern. Under ordinary conditions there is a period of stratification during ice cover in the winter and another period of stratification during the summer. There are also two periods during which the water is completely mixed from top to bottom by the impact of wind at the surface. This occurs when the water temperature is uniform and usually occurs just before ice formation (fall turnover) and just after ice-out (spring turnover). Such a lake is called dimictic. This pattern is the result of the temperature-density relationship and the anomalous condition of water's being most dense at 4 ° C. Thus in winter the bottom temperature is 4° while the ice on the surface is at zero. During the summer the surface is warmed by the sun while the bottom may remain at or near 4 ° C. During the summer thermal stratification usually occurs as a result of a combination of solar heating of the surface, the impact of wind, and the temperature-viscosity relationship of the water. This summer stratification prevents surface reaerated water from being carried to greater depths, a factor that also contributes to the long-term demise of the lake.

The temperature succession in a lake may be shown starting with ice cover in winter. The surface ice is at or below 0 ° C, while the bottom is at 4 ° C. There is no mixing of this water, because the ice cover prevents any wind effects. Biological activity is also at a minimum. As spring comes, the sun melts the ice and then begins to warm the surface of the lake. As all of the water approaches 4 ° C, even a gentle wind will mix this isothermal water from top to bottom, called the spring overturn. As the sun warms the surface of the lake, the warmer water will tend to float on the surface due to its lower density. If this heating occurs during a period of strong wind, there may still be complete mixing and the entire lake will be heated to the temperature of the surface. However, if warming occurs during a period of light or no wind, a point is reached at which the wind does not have sufficient energy to mix the upper warmer water with the cooler lower layer of water with greater density and viscosity. This forms a period of summer stratification where there is circulation near the surface, but none below a certain depth. Frequently in large deep temperate lakes, the level of stratification occurs at about 10 m depth. Further, the shape and orientation of the lake to the wind have an influence on the depth of the upper mixed zone. During the summer in a typical temperate lake, there is a warm upper layer that is equally mixed by the wind, then a zone in which there is a rapid decrease in temperature with depth, and finally a layer of relatively cold uncirculating water near the bottom. Thus the lake is divided into three layers in which the upper layer is called the epilimnion, the middle layer the metalimnion or the thermocline, and the bottom layer the hypolimnion, as shown in Fig. 7.1.

The summer stratification may last up to 5 months, during which time there is little to no mixing in the hypolimnion and no opportunity for oxygen from surface aeration to reach this area. So long as only a little decomposable organic matter is present at the onset of stratification, the available oxygen present may not be entirely consumed. The colder bottom water temperature also contributes to a slower biological activity, thereby conserving the oxygen supply. This condition is conducive to supporting a cold-water fish habitat. However, if large amounts of decomposable organic matter settle into the hypolimnion, the limited amount of oxygen available

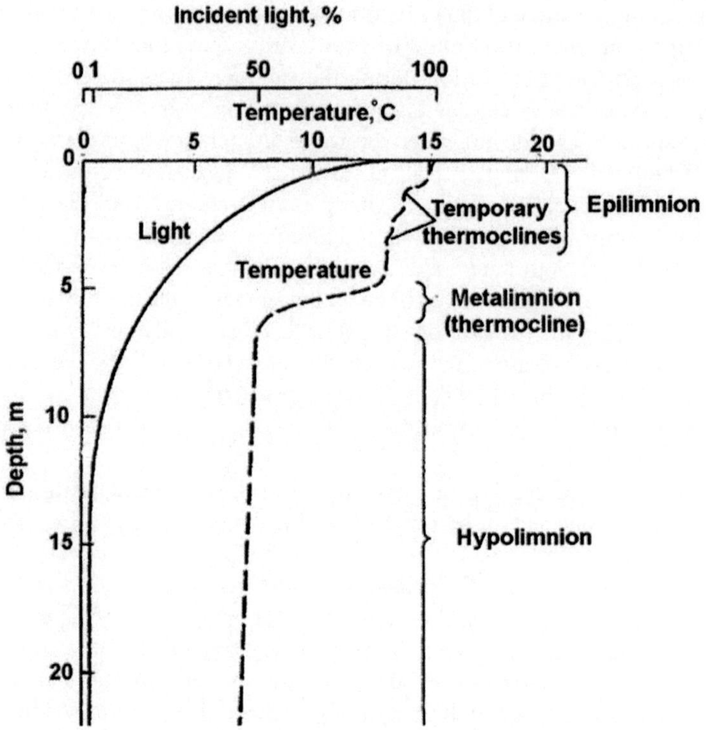

Fig. 7.1 Temperature and light profiles in a temperate climate lake during summer stratification

may be consumed and the hypolimnion will become anaerobic. Not only will this interfere with fish life, it also results in the release of certain nutrients, specifically phosphorus, that are insoluble in aerobic conditions, but soluble under anaerobic conditions. The presence of more nutrients may increase oxygen-consuming biological activity that will further create anaerobic conditions.

As fall approaches, the surface of the lake is cooled and the cooler water circulates to a depth of equal temperature and/or density. This tends to lower the thermocline until the lake becomes uniform in temperature. Now even a light wind can circulate the water from top to bottom and the period of fall overturn occurs. During this time complete oxygen saturation of the water usually occurs and aerobic reactions persist.

As the air temperature reaches 4 ° C and becomes colder, the surface of the lake will approach 0 ° C, but the denser 4 ° C water will remain on the bottom. When ice covers the surface, the period of winter stagnation begins. The duration of this depends upon latitude, altitude, weather conditions, and numerous specific lake conditions. Lakes with significant warm underground springs have been found to have less ice cover and, in some instances, have holes in the ice above the location of the spring. Very deep lakes such as Lake Baikal, Crater Lake, and Lake Tahoe contain so much heat energy in the water that they do not freeze. Figure 7.2 summarizes the circulation/depth patterns during the seasons in a deep temperate climate lake.

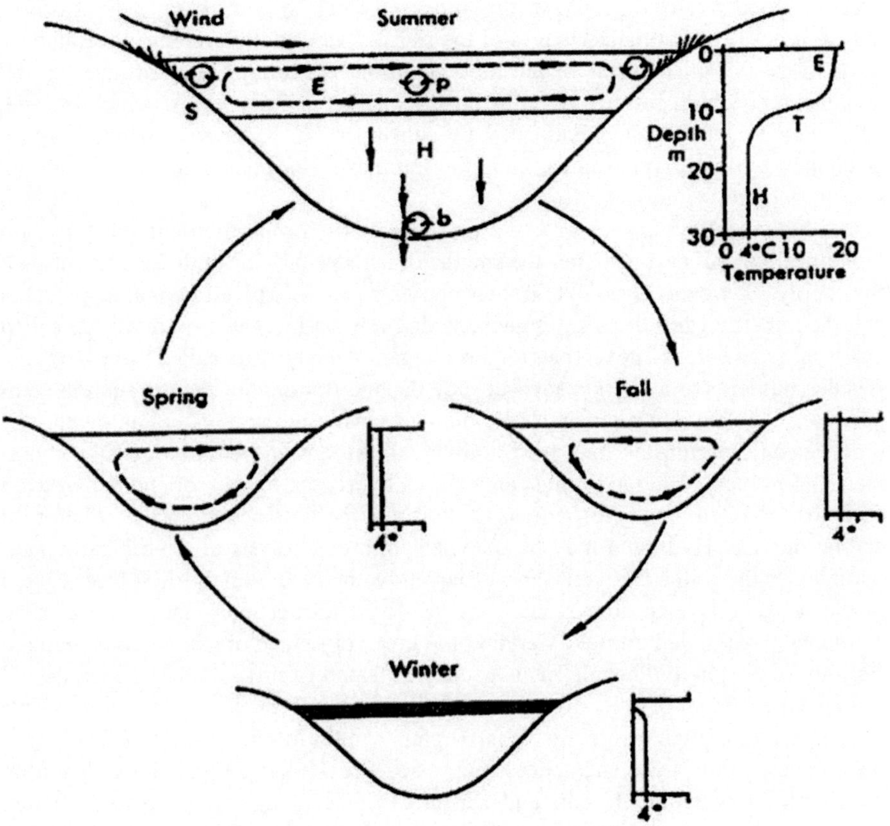

Fig. 7.2 Seasonal circulation patterns in a deep temperate climate lake

3 Importance of Biological Activity

It may be noted in this discussion that the interrelationship between nutrients and biological activity represents a continuing thread in the study of the life of a lake. Thus an understanding of the relationship between biological activity in a lake and its aging process is essential [1].

A lake contains many biological communities. Within the water column are numerous organisms of microscopic size. The floating microscopic organisms are called plankton, which may be subdivided into two groups: the phytoplankton or plant life, which includes algae, fungi, and pollens that fall into the lake, and the zooplankton or animal forms. The plankton may also be broken down into the nekton, or free swimming organisms, and the benthon, which exist on the bottom.

A prime concern is the algae, the microscopic green plants floating in the water column. These organisms represent the base of the food chain in that they can convert simple inorganic matter into organic matter with the aid of sunlight in the process called photosynthesis. In this process of cell growth, oxygen is also produced. It has been estimated that ¾ of the Earth's supply of oxygen is generated by algae in the ocean. In terms of the food chain, the algae may be consumed by the zooplankton, which in turn are consumed by larger animal forms, which may be consumed by small fish, which may be consumed by larger fish, which may be consumed by larger vertebrates, including humans. The microscopic algae are the start of this food chain.

All biological systems require the presence of the proper nutrients to grow and reproduce. For larger organisms, the smaller organisms provide both the nutrients and the energy. However, algae obtain their nutrients from dissolved inorganic materials and their energy from the sun. Organisms that rely on inorganic nutrients are called autotrophic, whereas those that rely on organic matter are called heterotrophic. Besides nutrients and energy, growth may depend upon other factors such as temperature, light, etc. Nutrients in a lake may vary with location, including depth, and time. Specific organisms may have individual nutrient and environmental requirements. However, common to most are carbon, hydrogen, oxygen or another electron acceptor, nitrogen, and phosphorus. Carbon may be obtained from the solution of carbon dioxide. Hydrogen may be obtained from electrolysis or from bicarbonates dissolved in the water. Oxygen is most frequently obtained from the dissolved oxygen in the water. Nitrogen is secured from dissolved nitrogenous materials including ammonia, nitrites, and nitrates. Certain blue-green algae can obtain gaseous nitrogen from the atmosphere. Phosphorus is usually obtained from geological materials and from the breakdown of other organic materials. A general rule for the ratio of nutrients to support the growth of organisms is 60 parts carbon to 15 parts nitrogen to 1 part phosphorus. Some trace substances may also be essential. One of these is sulfur, which may be present in the soil, and is available in decaying organic matter. Another is silicon, which is required to form the shell case, called the frustule, of diatoms.

Every species of organism has a specific requirement for nutrients. Other factors being satisfactory, organisms will continue to grow until one of the essential nutrients has been completely utilized. Then growth may be retarded or completely stopped. Conversely, providing the limited nutrient will encourage additional growth. Frequently limiting nutrients such as nitrogen and phosphorus are contained in wastes, including human wastes. Conventional wastewater treatment does not remove nitrogen and phosphorus. Thus additional treatment to remove these nutrients is frequently required before discharge into a lake.

Productivity in a lake is commonly expressed as the amount of fishable fish in a lake. Since the number of fish is directly related to the fish's food and the food ultimately is a function of algae in the food chain, which in turn is a function of the available nutrients, we can use measurement of the nutrients to estimate the potential productivity of a lake. Whether or not productivity is desirable is up to individual taste. A lake that is low in productivity will be clear and have a low fish population. A lake that is high in fish population tends to be turbid and frequently accompanied by extensive shoreline weed growths. Moreover, the fish population will vary in each case with game fish such as trout and salmon predominant in less productive lakes and pan fish such as bass, pickerel, and catfish predominant in highly productive lakes.

The term oligotrophic has been used to describe lakes low in nutrients and consequently low in productivity. Lakes high in productivity are termed eutrophic. As a general rule, lakes proceed from oligotrophic to eutrophic as the lake ages. Some researchers add the word mesotrophic to designate lakes on the verge of becoming eutrophic. These terms are not intended to imply that all eutrophic lakes are undesirable or that all oligotrophic lakes are desirable. The desirability of a specific level of productivity is a function of the specific use of the lake. Probably what is desirable is a mixture of lakes of the different types. The long-range problem is that as lakes age, the nutrients accumulate within the lake. New nutrients are brought into the lake from allochthonous inputs. Siltation may decrease the volume of water within the lake, thus concentrating the nutrients. Anthropogenic inputs such as wastewaters and fertilizers add significantly to the nutrient level. Deforestation results in more rapid runoff, which carries both silt and nutrients into the lake. All these combine to increase eutrophication in a lake.

4 Considerations in Remediation

In order to prolong the life of a lake, actions must be taken to reduce the rate of eutrophication. Very little can be done to overcome the natural process of eutrophication. However, much can be done to overcome the anthropogenic impacts. It is easy to say just stop any human activities that contribute to the eutrophication, but that is difficult to achieve. The best that can be done is to determine what activity will provide the best return for the effort and/or expenditure.

Sakamoto [2] showed a direct correlation between the phosphorus concentration in a lake at the time of spring turnover and the amount of productivity as measured by the amount of chlorophyll-α present during the summer (Fig. 7.3). Correspondingly, the greater the chlorophyll-α content, which indicates the presence of algae, the greater the turbidity of the water, and, therefore, the lower the clarity of the water as measured by the Secchi disk depth. Whereas there was good coordination between the phosphorus content and the chlorophyll-α, there was poor correlation between chlorophyll-α and the clarity of the water. Substances other than chlorophyll-α can impact the turbidity of the water. These include the presence of zooplankton that feed on the phytoplankton and particulate matter, such as fine clay or silt that is carried into the lake in the runoff.

Numerous models have been derived to correlate certain specific parameters with the trophic state of a lake. Two stand out as being quite reliable and simple. Both relate total phosphorus loading to the trophic state of the lake as a function of the body of water. In the original work by Vollenweider [3], he showed a correlation between the total phosphorus loading and the mean depth of the lake. Many lakes were studied and there was a good correlation between these two parameters. Later Vollenweider and Dillon [4] improved the model by comparing phosphorus loadings with the mean depth and the retention time of the lake (Fig. 7.4). The correlation was poor with lakes that were not phosphorus limited.

Fig. 7.3 Total phosphorus concentrations at spring turnover vs. average summer chlorophyll-α concentrations

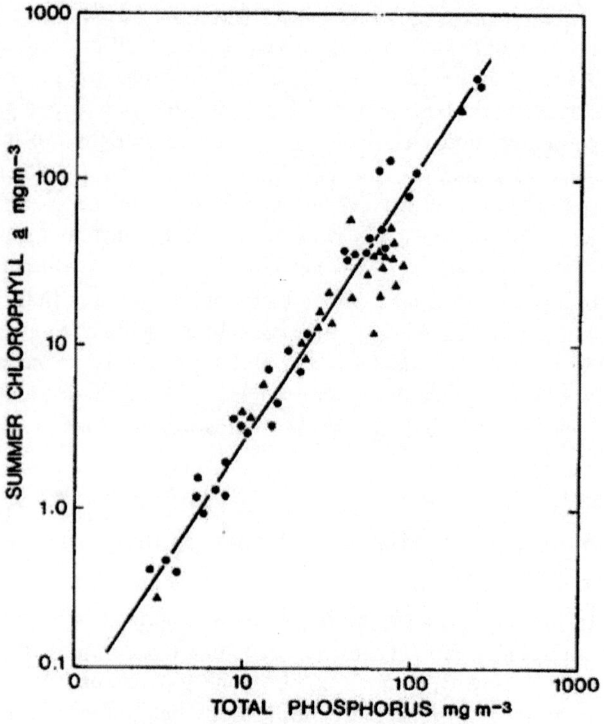

Notes: Dots represent data from Sakamoto, 1966
Triangles are for other lakes in the literature
Line is the regression line for Sakamoto points
Correlation coefficient (r) is 0.97

5 Treatment to Prevent Nutrient Discharges

It is apparent that the most effective measure to control eutrophication would be to control the nutrient inputs. However, this is not always possible nor practical. It is nearly impossible to lower the total carbon inputs to a lake, because there is always some dissolved carbon dioxide present from the atmosphere that can become available as a carbon source. It is not desirable to limit the oxygen, as that would encourage anaerobic decomposition with its odors and other undesirable conditions. Nitrogenous materials can be removed from a wastewater treatment plant effluent, but certain blue-green algae can utilize nitrogen from the atmosphere. Phosphorus can also be removed from wastewater effluents. Unless there is a large phosphate deposit in the watershed or the lake bed, this can result in a permanent removal of the phosphate so long as the lake maintains aerobic conditions. Thus phosphorus removal has received much attention in the effort to limit primary productivity. Furthermore, in his study of lakes around the world, Vollenweider [3] observed that the nutrient most frequently limiting productivity in lakes was phosphorus.

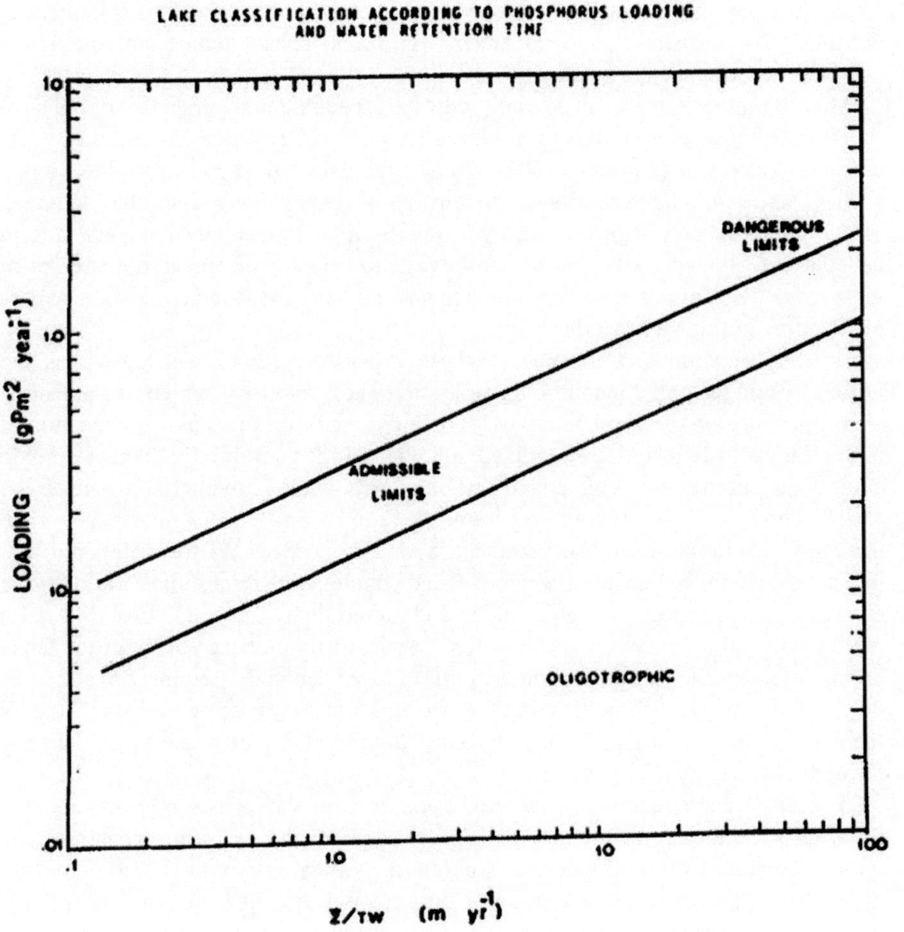

Fig. 7.4 Trophic state of a lake based on its mean depth and hydraulic residence time

Since phosphorus is most frequently the limiting nutrient in a lake, more efforts have been directed toward finding means of reducing phosphorus inputs to a lake. Means that have been applied include diversion of all stormwater runoff from the lake, installation of stormwater infiltration basins, removal of phosphorus from treatment plant effluents, the use of land application of wastewaters, and passing treatment plant effluents through wetlands before they enter the lake.

Another reason phosphorus has been chosen as the nutrient to be removed is the ease of precipitating phosphorus with iron, aluminum, or calcium salts, with the subsequent removal of the solids. Phelps [5] showed that limiting the phosphorus concentration in a lake at the time of spring turnover to less than 10 μg/L would limit excess productivity in most lakes.

Removal of phosphorus from wastewaters in treatment processes is important in limiting phosphorus discharges to streams and lakes. These include both biological and chemical treatment systems.

Most biological treatment systems rely on a peculiar trait of many organisms, specifically those present in typical biological wastewater treatment systems, especially activated sludge systems. When these organisms are starved for phosphorus, such as under anoxic conditions, and then subjected to normal aerobic activated sludge aeration, they take up more phosphorus than immediately needed, a term called luxury uptake. Thus treatment involves alternate anoxic and aerobic treatment in separate tanks or alternate conditions in a single tank, with removal of the excess phosphorus in the waste sludge.

Wilson [6] summarized some of these processes, sometimes known as the Ludzack-Ettinger and Johannesburg or Bardenpho processes, which are patented. Variations include the number and order of anoxic and aerobic tanks, the location of both return activated sludge and mixed liquor suspended solids to help create anoxic conditions, and the use of an added carbon source, such as methanol, to create the anoxic conditions. If effluent requirements require phosphorus levels less than 0.3 mg/L, additional chemical treatment is usually needed. Wilson compared biological and chemical phosphorus removal and concluded that multiple aeration tanks consume energy; return activated sludge and mixed liquor suspended solids require more energy; the cost of a carbon source (methanol) may be great; multiple tanks require more space; and for low phosphorus effluent demands, chemical treatment is needed anyway. He also pointed out that the additional volume of sludge created by the addition of chemicals is small compared to the volume of waste sludge already created.

In order to achieve total phosphorus levels in wastewater discharges of less than 0.1 mg/L, chemical precipitation is very useful. Phosphorus forms insoluble salts with aluminum, iron, and calcium. Aluminum is most commonly used. The iron phosphate sediment must be kept aerobic to prevent the release of the phosphorus when less soluble iron sulfide is created. Calcium is usually applied as lime, which has a high pH. This may be detrimental under certain circumstances. Availability and cost of the chemicals has a large role in the choice of chemical. Eberhardt [7] has published a report on calculating the optimum aluminum dose.

Tabor [8] evaluated two patented treatment systems for phosphorus removal. The Actiflo process consists of coagulant addition with rapid mix, polymer and sand addition, slow mix for particle agglomeration and floc formation, plate settlers for solids/liquid separation, separation of the sand from the solids in a hydrocyclone, and return of the sand to the system. The DensaDeg process consists of coagulant with rapid mix, polymer and thickened return activated sludge addition, a plug flow zone for particle agglomeration and floc formation, tube settlers for solids/liquid separation, and thickening of solids for recycle and disposal. Both systems are capable of removing total phosphorus to less than 0.2 mg/L.

Patoczka [9] described upgrading an existing conventional activated sludge treatment plant utilizing a backwashable sand filter to achieve an effluent total phosphorus content of less than 0.1 mg/L. Chemical addition was shown to be effective. Both alum and iron salts were studied, and the optimum dosages and pH for each were determined for the particular waste. The effects of chemical addition at the primary settling tank, the aeration tank, and the final clarifier were studied. The most effective location for adding the chemicals and the most effective chemical for phosphate removal were the addition of alum at the final clarifier, but some chemical savings could be achieved by addition to the aeration unit due to the return of some of the chemical in the return activated sludge. Alum addition increased the sludge generation in the range of 0.5 to 0.7 lb of dry sludge per lb of alum used. Chemical addition aided sludge settling in the final clarifier and also increased BOD and TSS removal.

The Federal Highway Authority has issued a report for the best management practices for stormwater management [10]. A simple method is an alum injection system that adds alum directly to a stormwater channel at a flow-controlled rate. The precipitated chemicals are merely discharged to the receiving stream or lake where thy settle to the bottom (under appropriate flow conditions). The added solids in lake sediment are considered insignificant. Total phosphorus in Lake Ella, Florida, was reduced by 89%, and total nitrogen by 78% [11]. Pitt [12] described a multichamber treatment train that consists of a series of treatment units that mimic a conventional wastewater treatment plant. In the first tank mild aeration separates the heavy solids from the lighter ones. In the bottom of the second tank, most of the solids are settled out by an inclined tray settler, and above this a DAF system lifts floatables and oil to the surface. The final tank uses a sand/peat filter for final treatment. Total phosphorus removal was determined to be 88%. Allard et al. [13] patented the StormTreat System for treating stormwater. It consists of a circular holding tank 1.2 m deep with discharge to the subsurface of a surrounding wetland. Overall the system removed total phosphorus by 89%. Claytor and Schueler [14] have described a constructed vegetated rock filter for biological treatment of stormwater, with application to the subsurface of the filter. This achieved 82% removal of total phosphorus.

Farming is a major source of nutrient discharges to streams and ultimately to lakes. Runoff from fertilized fields carries the excess fertilizer off the field. This can be controlled by establishing an unfertilized buffer zone between an active field and the waterbody. Also the trend toward large feedlots has exacerbated runoff problems. A large combined animal and plant farm in the United Kingdom has installed an environmentally sound water and wastewater system [15]. The collected liquid wastes are treated in a DAF system before entering a reedbed treatment system. The effluent flows into a lake whose overflow passes into a willow plantation. Water from the lake is used for irrigation and pig wallowing. Seepage under the lake is pumped out a sufficient distance away to allow for reuse. The lake also serves as a fish and wildlife habitat.

In studies at the Lake George Village, NY, sewage treatment plant using trickling filters and alum addition before the secondary clarifiers, with the final effluent being dosed onto deep natural sand beds, Aulenbach [16] found that total phosphorus was reduced to less than 1 mg/L within 7 m of vertical transport through the sand. In another study of phosphate removal in the soil, Aulenbach et al. [17] traced a septic tank effluent in shallow soil and found removal to less than 1 mg/L within 35 ft of horizontal transport.

6 Recovery of Eutrophic Lakes

The best way to prolong the life of a lake is to control the nutrient inputs to the lake before it progresses through the mesotrophic state to the eutrophic state. This is sometimes difficult or even impossible. If upon study of a lake recovery is considered possible, numerous methods are available [18–20].

6.1 Aeration

Several variations of aeration are available to prevent the hypolimnion from becoming anaerobic. This will tie up the phosphorus in an insoluble form and keep the surface of the bottom deposits aerobic to prevent resolubilization of the phosphorus. Aeration is generally more applicable to small lakes. The pressure to pump air to the bottom of a deep lake requires special equipment.

When air is used, the system is designed to create a circulation within the lake so that anaerobic hypolimnetic water is brought to the surface where natural reaeration occurs. Whereas some reaeration results from the addition of the air, the surface aeration is responsible for most of the reaeration. More than one air system may need to be placed in a lake depending upon the shape of the lake. A disadvantage of the complete circulation system is that the thermocline is destroyed and the lake becomes isothermal from top to bottom at a mean temperature. Air systems must be turned on before the hypolimnion becomes anaerobic. These systems are relatively inexpensive.

A modification of the plain aeration system is a hypolimnetic aeration system. This consists of two concentric vertical tubes normally placed entirely in the hypolimnion. The top of the larger tube is sealed. Water from near the bottom of the lake enters the smaller inner tube where an aerator both lifts the water and aerates it at the same time. At the top of the inner tube, the water overflows into the larger outer tube and is carried back downward. The aerated discharge from the larger tube is generally above the intake to minimize short circuiting back to the inlet tube. Since the entire device is placed in the cold hypolimnion, there is little impact on the temperature in the hypolimnion. Judicious placement of the intake and the discharge minimizes the impact on the lake bottom, and the system maintains the normal thermal stratification of the lake.

Oxygen has also been used instead of air. In this case, the oxygen provides the source of the reaeration. This usually requires on-site generation of the oxygen.

6.2 Weed Harvesting

A common situation in eutrophic lakes is to have a shallow (<2 m deep) shoreline filled with both submerged and emergent growths. These are considered unsightly, interfere with boating, make swimming undesirable, and make fishing nearly impossible. At the same time they provide a breeding ground for fish. Weed harvesting has been used under the guise of reducing the nutrient inputs to a lake. However, it has been estimated that they represent only in the order of 1% of the phosphorus content of the lake. They are usually harvested by a special boat that may not be able to reach the shallowest portion or certain bays in a lake. Here weeds may be removed by rake or hand pulling while wading in the shallow water. Also, the weeds harvested must be removed from the shore, or the nutrients will return to the lake as the weeds decompose. Harvesting the extensive Eurasian watermilfoil in Lake Wingra, WI, resulted in the reduction of only a small fraction of the lake's metabolically active nutrient pool [21]. This is a relatively expensive treatment for the amount of nutrient reduction accomplished. It does remove the unsightly and undesirable weeds.

Related to weed harvesting is the use of herbicides to kill the weeds. This must be applied before the weeds reach full growth and may have to be repeated during the growing season. Any dead weeds should be removed. The use of herbicides may have other undesirable environmental impacts, and they are not recommended if the water is used for drinking.

6.3 Dredging

The principle of dredging is to remove the organic sediments on the bottom of the lake that add to the nutrient supply when the hypolimnion becomes anaerobic [22, 23]. This is an expensive technology and is impractical for deep lakes. It also destroys the natural bottom of the lake. It is somewhat practical in artificial lakes or reservoirs where the water level can be drawn down (usually during the winter), and surface equipment such as bulldozers can be used for the dredging. Any dredged material must be handled in an environmentally safe way. If any hazardous contaminants are shown to be present, this could be costly. Starting in mid-August of 2012, the US Environmental Protection Agency (US EPA) has targeted up to US$57 million in Great Lakes Restoration Initiative funds for two projects in the Sheboygan River, focusing on dredging contaminated sediment from the Great Lakes' river area. [42]. Dissolved air flotation (DAF) is one of the best processes for treatment of the dredged materials [43–45]. Appendixes A, B, C, and D document the pollutant contents of the dredge materials from Ashtabula and Fairport of Ohio, USA. The pollutants in the dredged materials contain high concentrations of total phosphorus, nitrogen, oil and grease, chemical oxygen demand (COD), toxic heavy metals, and toxic volatile solids. Appendix E is a US EPA control technology summary for dissolved air flotation [46]. It appears that DAF can adequately treat the dredged materials for removal of nutrients, heavy metals, and other conventional and volatile pollutants.

6.4 *Sediment Fixation*

Eutrophic lakes are synonymous with significant organic bottom deposits. When these become anaerobic, they release their nutrients, specifically phosphorus. As the lake overturns, these nutrients are distributed throughout the lake, enabling more biological growth, which ultimately dies and settles to the bottom. Instead of trying to remove these sediments, chemicals may be added to more permanently precipitate the phosphorus. Aluminum salts have been found to be most effective since the aluminum phosphate remains insoluble so long as the surface of the sediments, in contact with aerobic water, remains aerobic [24, 25]. Iron salts are effective in precipitating phosphates, but in the deep anaerobic sediments, the iron combines with reduced sulfur to form ferrous sulfide that is more insoluble than the iron phosphate, thus releasing the phosphate back into solution. Calcium salts are also capable of forming precipitates of calcium phosphate; however, their high alkalinity may undesirably raise the pH of the water. This may be desirable in acid lakes. Thus, aluminum salts have been found to be most effective in tying up the phosphate permanently in the sediments. As more organic material settles to the bottom, reapplication may be necessary in future years. This becomes extremely expensive for large lakes.

One difficulty in binding the sediment phosphate is establishing adequate contact. The alum must be spread fairly uniformly over the bottom to be effective. This is usually achieved by the use of boats crisscrossing the lake. A novel system was set up in a sewage oxidation pond in California [26]. A mechanical mixer was installed in the middle of the pond, providing both mixing and aeration. Alum was applied at the mixer, which was solar powered. This eliminated a long power cord. The alum combined with both the sediment phosphorus and the soluble or suspended phosphorus in the pond, settling to the bottom. Excessive biological growth was eliminated, and the upper liquid layer met the phosphorus discharge limits to the receiving water.

7 Hypolimnetic Phosphorus Removal by DAF

A different approach is to remove the excess phosphorus from the anaerobic hypolimnion. Here the phosphorus level may be high enough to be removed by conventional precipitation by aluminum, iron, or calcium salts. A flocculation/filtration system located on the shore could accomplish this. Successful use of such a program at three lakes in Germany has been reported [27]. Further, a DAF system could be installed at the lakeshore without the cost and obstruction of a conventional sedimentation basin.

A study conducted [28] using water from eutrophic Laurel Lake in Massachusetts, adding 40 mg/L ferric chloride and subjecting it to DAF with sand filtration, showed removal of 96% to 98% of the phosphate (Tables 7.1 and 7.2.) with no iron residual. This was used to set up a pilot study for the removal of hypolimnetic phosphorus in Devils Lake, Wisconsin.

Devils Lake is surrounded by ancient bluffs in the east, west, and south [28]. The preglacial Wisconsin River flowed through a gap between these bluffs in the south range of the Baraboo Hills. Devils Lake was formed at the end of the last ice age by terminal moraines deposited at the north and the southeast ends of the gap, diverting the Wisconsin River to the east around the Baraboo Hills.

Figure 7.5 shows the depth profile of the lake [29]. Its surface area is 149 ha and its maximum depth is 14.3 m. Its mean depth averages about 9.3 m. The east and west shorelines between the bluffs are steep, while the lake's littoral zones are mostly at the north and south ends of the lake. The watershed area is relatively small, 6.86 km^2, and the ratio of watershed to lake surface area is only 4.6. Most of the watershed is forested [30]. There is only one small inlet that drains through a small wetland and no outlet. The lake water level is maintained by fluctuations in ground water level and the balance of precipitation and evapotranspiration [28].

In 1991 the Wisconsin Department of Natural Resources (WDNR) began evaluating whether hypolimnetic withdrawal and phosphorus removal would reduce sediment phosphorus concentrations with concomitant lower sediment phosphorus release during anaerobic hypolimnion periods. WDNR measured iron-bound phosphorus concentrations in profundal sediments around the lake both before and after

Table 7.1 Results of Laurel Lake bench-scale DAF studies – summer

Parameter	Units	Raw influent	DAF/filtration	Removal efficiency
Temperature	°C	16		
pH	unit	7.3	7.1	
Turbidity	NTU	7.9	0.7	91%
Color	PCU	48	<1	99%
Iron, Fe	*mg/L*			
PO_4^{3-}	*mg/L*	0.72	0.03	96%

Table 7.2. Results of Laurel Lake bench-scale DAF studies – late fall

Parameter	Units	Raw influent	DAF/filtration	Removal efficiency
Temperature	°C	8		
pH	unit	7.0	7.0	
Turbidity	NTU	17	0.3	99%
Color	PCU	132	3	98%
Fe Residual	*mg/L*	0.61	0.47	*
PO_4^{3-}	*mg/L*	1.16	0.02	98%

*Note: Iron residual after DAF is 23% below *influent* concentration

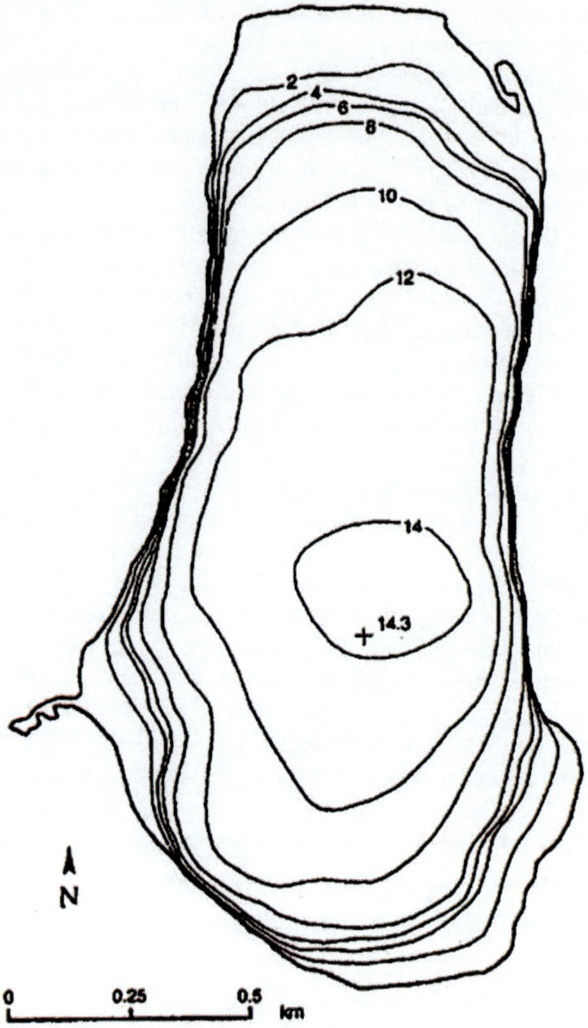

Fig. 7.5 Depth profile of Devils Lake, WI

hypolimnetic anoxia occurred in order to estimate the amount of phosphorus released to the overlying water during each season. Similar long-term laboratory column studies were conducted to support those results. The US Geological Survey (USGS) also studied lake level and water budgets to model the impact of removal of water from the hypolimnion [28]. Although a temperature-depth profile of the lake was not available, data from the phosphorus concentrations in Table 7.3. indicate that the thermocline was located at about 13 m depth on September 20, 1996. This indicates that the hypolimnion existed in only approximately 1.3 m of the bottom of the lake. It is likely that some lake cooling had occurred before September 20 and that during the warmer summer period, the thermocline was higher.

Table 7.3. Profile of Devils Lake – September 20, 1966

Depth (m)	pH (unit)	TP (mg/L)	Fe (mg/L)
0 (surface)	7.9	0.007	–
4	–	0.007	–
10.3	–	0.013	–
11.3	–	0.018	0.09
12.3	–	0.021	0.13
12.8	–	0.077	0.84
13.3	–	0.216	2.6
13.8	6.9	0.408	3.9
14.1 (0.5m off bottom)	–	0.432	4.1

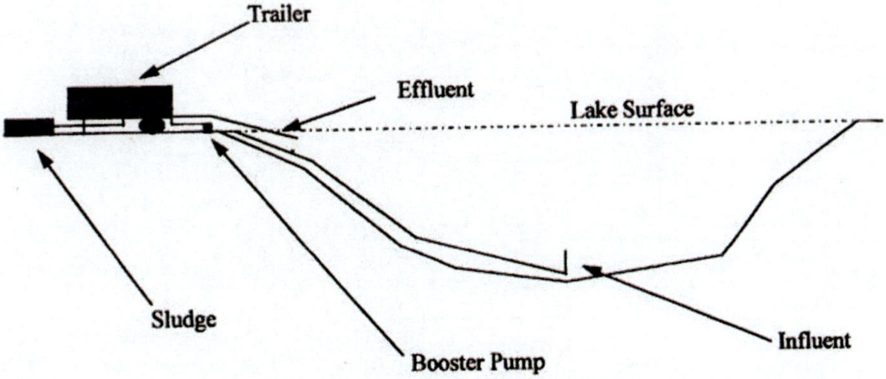

Fig. 7.6 Pilot plant setup for removal of phosphate from the hypolimnion of Devils Lake by DAF

A pilot DAF system with sand filtration was set up on the shore of Devils Lake (Fig. 7.6) and operated from September 25, 1996 through October 3, 1996 [28]. A 150 hp pump brought the hypolimnetic water to the treatment plant by means of an approximately 0.5 mile pipe that terminated approximately 14.5 m deep in the lake. The water intake had a vertical intake covered with a screen mesh to keep out bottom debris. Treated water was returned to the surface of the lake.

The coagulants used were alum, aluminum chlorohydrate (A/C), and ferric chloride, with Percol added as a coagulant aid to all tests. Each coagulant was studied individually. Dosages were varied to provide a range of results that would indicate an optimum dose. Alum dosages varied between 13.2 and 49.5 mg/L, ferric chlorides varied between 5 and 50 mg/L, and A/C varied between 6.6 and 23 mg/L. The Percol dosages varied between zero and 0.7 mg/L. Flows through the pilot plant were varied between 35 and 60 gpm.

The results of the 9-day operation of the pilot plant are shown in Table 7.4. Figures 7.7, 7.8, and 7.9 depict the results for the use of ferric chloride, A/C, and alum, respectively. It may be seen that effective phosphorus removal required a minimum of 40 mg/L of ferric chloride. Doses as low as 7 mg/L A/C resulted in

Table 7.4. Results of DAF pilot studies of phosphate removal from Devils Lake

Chemical addition mg/L	Date mo/d/yr	Influent TP mg/L	Effluent TP mg/L	Flow rate gpm	Percent removal %
49.5 Alum	092596	0.496	<0.007	40	99
47.6 Alum	092596	–	<0.007	40	99
23.0 A/C	092596	–	<0.007	40	99
20.8 Ferric	092696	0.495	0.036	35	93
13.2 Ferric	092696	–	0.045	35	93
19.8 Ferric	092696	–	0.058	30	88
15.8 Ferric	092696	–	0.204	50	59
5.0 Ferric	092796	0.436	0.130	35	70
9.4 Ferric	092796	–	0.222	35	49
10.2 Ferric	092796:	–	0.326	52	25
32 Ferric	092796	0.411	0.096	45	77
29 Ferric	092796	–	0.082	50	80
26.4 Alum	092896	0.492	0.007	45	99
17.6 Alum	092896	–	0.016	45	97
13.2 Alum	092896	–	0.013	45	97
14.4 Alum	093096	0.575	0.087	55	83
22.8 Alum	093096	–	0.019	55	96
20.4 Alum	093096	–	0.022	65	96
6.6 A/C	100196	0.588	<0.007	60	98
8.8 A/C	100196	–	0.008	60	98
7.5 A/C	100196	–	0.011	55	98
35 Ferric	100396	–	0.028	55	95
50 Ferric	100396	–	<0.007	55	99
40 Ferric	100396	–	<0.007	55	99
40 Ferric	100396	0.516	<0.007	55	99
38.1 Ferric	100396	–	<0.007	45	99

Note: alum as aluminum sulfate, ferric as ferric chloride, and A/C as aluminum chlorohydrate

effective phosphorus removal. An alum dose of 25 mg/L or more is needed to achieve effective phosphorus removal. There did not seem to be any correlation of flow rate with treatment efficiency at the flow rates studied. Considering that flocculation is slower in the cold hypolimnion waters, this represents satisfactory operation for phosphorus removal.

Based upon WDNR Table 7.1, the depth of the thermocline on September 20 was estimated to be at 13 m. Thus, at this time the volume of water in the hypolimnion was relatively small. However, the results of the phosphorus content of the inlet to the treatment system showed that hypolimnetic water was consistently used during this study. From the contour map of the lake (Fig. 7.5), the volume of the lake at its normal level would be 13,641 million m^3 (481,660 million ft^3 or 3,602,817 MG). The volume below 13 m depth was only 83,040 MG. Nevertheless, at an average pumping rate of the treatment system of 50 gpm, it would take 1,153 days to deplete

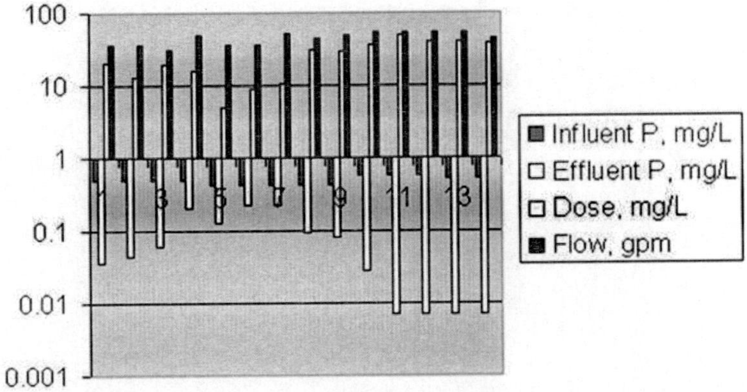

Fig. 7.7 Results of DAF pilot plant study for removal of phosphorus from the hypolimnion of Devils Lake using ferric chloride

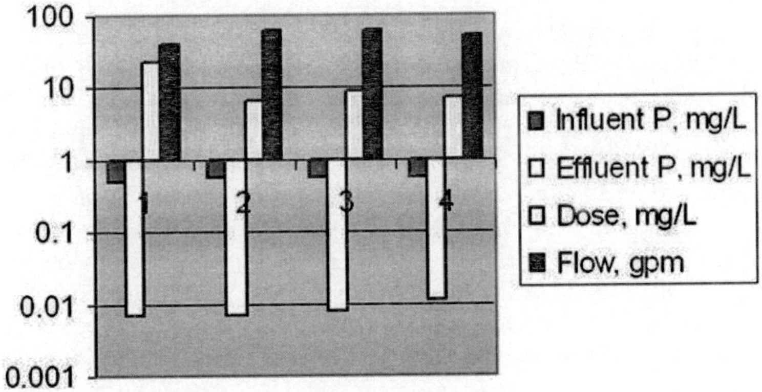

Fig. 7.8 Results of DAF pilot plant study for removal of phosphorus from the hypolimnion of Devils Lake using aluminum chlorohydrate (A/C)

the volume in the hypolimnion. Thus, it was considered that the water removed by the pilot study had minimal impact on the available water in the hypolimnion.

An estimate was made of the relative costs of the coagulants studied. Based on the 2015 US cost and the concentration needed, the following comparison was made:

Coagulant	Cost, cents per 1,000 gal (or 3,785 L)
Aluminum sulfate	0.98
Aluminum chlorohydrate	3.63
Ferric chloride	20.75

Appendix G is a US Army Corps of Engineers Civil Works Construction Yearly Average Cost Index for utilities, which has been used for the above cost estimation. An advantage of using A/C is that it does not result in any aluminum residual.

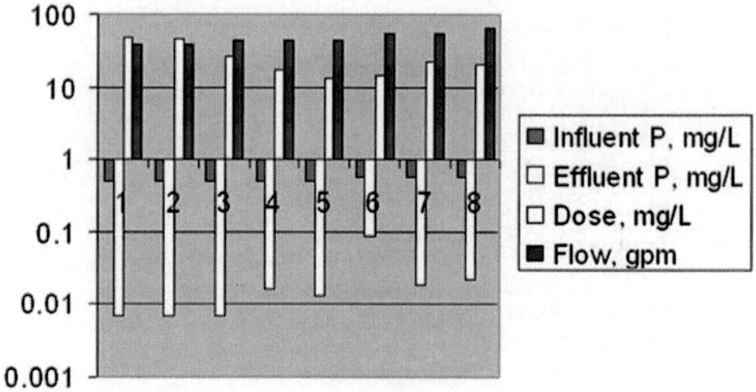

Fig. 7.9 Results of DAF pilot plant study for removal of phosphorus from the hypolimnion of Devils Lake using alum

Aluminum is toxic to some fish. Ferric chloride is not recommended due to its high cost and its potential to leave a residual color.

In order to apply the technique of phosphate removal from a hypolimnion, the first step would be to determine the volume of the hypolimnion. DAF/filtration systems of the type used in this study are available up to 13,000 gpm (49,205 L/min). Knowing the existing phosphorus concentration and the treated effluent concentration a calculation can be made of how much volume of water would have to be treated to bring the phosphorus concentration down to an acceptable level. This may require several years of operation. However, if the lower nutrient level will reduce the biological growth to a level where the hypolimnion may remain aerobic, there will be less release of phosphorus from the benthic deposits. A further consideration is that DAF involves aerating the water. If the effluent is discharged to the hypolimnion, it may provide sufficient additional oxygen to maintain aerobic conditions. This should enter into the calculation and influence the final decision to utilize DAF/filtration (DAFF) to control lake eutrophication. The US Environmental Protection Agency (US EPA) has summarized the performance data of DAF alone (Appendix E) and supplemental filtration (Appendix F).

8 Sources, Chemistry, and Control of Acid Rain

Acid rain is a serious environmental problem that affects large parts of the United States and Canada. Acid rain is particularly damaging to lakes, streams, and forests and the plants and animals that live in these ecosystems. Acid rain is rain consisting of water droplets that are unusually acidic because of atmospheric pollution [47]; it is rain with a higher concentration of positively charged atomic particles (ions) than normal rain. Acid rain and its frozen equivalents, acid snow and acid sleet, are part of a larger problem called acid deposition. Acid deposition also includes direct deposition, in which acidic fog or cloud is in direct contact with the ground, and dry deposition, in which ions become attached to dust particles and fall to the ground. "Normal" or unpolluted rain has an acidic pH, but usually no lower than 5.7, because carbon dioxide and water in the air react together to form carbonic acid, a weak acid according to the following reaction:

$$H_2O(\text{liquid}) + CO_2(\text{gas}) \rightleftharpoons H_2CO_3(\text{aqueous})$$

Carbonic acid then can ionize in water forming low concentrations of hydronium and carbonate ions:

$$H_2O(\text{liquid}) + H_2CO_3(\text{aqueous}) \rightleftharpoons HCO_3^-(\text{aqueous}) + H_3O^+(\text{aqueous})$$

However, unpolluted rain can also contain other chemicals, which affect its pH (acidity level). A common example is nitric acid produced by electric discharge in the atmosphere such as lightning [48]. Acid deposition as an environmental issue (discussed later in the chapter) would include additional acids to H_2CO_3.

Acid rain is one type of atmospheric deposition. Atmospheric deposition includes any precipitation, airborne particles, or gases deposited from the atmosphere to the Earth's surface. Other forms of atmospheric deposition may also be by wet or dry methods. Much of the material in atmospheric deposition may be a nuisance but does not harm the environment. Some air pollutants, such as those in acid rain, can cause environmental problems (Fig. 7.10). It was not until the late 1960s that scientists began widely observing and studying the acid rain phenomenon [49]. Over many decades, the combined input of contaminants to sensitive environments can lead to widespread environmental problems. Smaller particles with a diameter of 10 μ (.004 in.) or less are too light to be deposited and so remain in the atmosphere where they can cause health problems. They pose a different problem and are regulated as particulates, or PM.

Acid rain occurs when sulfur dioxide and nitrogen oxides [50] are emitted into the atmosphere, undergo chemical transformations, and are absorbed by water droplets in clouds. The droplets then fall to the Earth as rain, snow, or sleet (see Fig. 7.11). This can increase the acidity of the soil and affect the chemical balance of lakes and streams. Decades of enhanced acid input has increased the environmental stress on high elevation forests and aquatic organisms in sensitive ecosystems. In extreme cases, it has altered entire biological communities and eliminated some fish species from certain lakes and streams. In many other cases, the changes have been subtler, leading to a reduction in the diversity of organisms in an ecosystem. This is particularly true in the northeastern United States, where the rain tends to be most acidic and often the soil has less capacity to neutralize the acidity. Acid rain also can damage certain building materials and historical monuments. Some scientists have suggested links to human health, but none have been proven. Public awareness of acid rain in the United States increased in the 1970s after *The New York Times* published reports from the Hubbard Brook Experimental Forest in New Hampshire of the myriad deleterious environmental effects shown to result from it [51]. Industrial acid rain is also a substantial problem in China and Russia [52].

Fig. 7.10 Atmospheric pollution (US EPA)

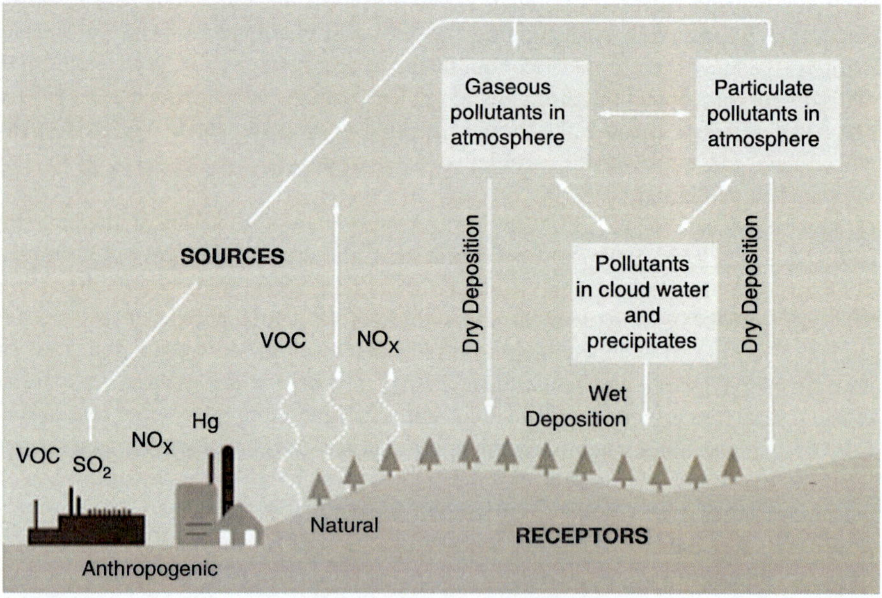

Fig. 7.11 Processes involved in acid deposition, http://www.epa.gov/acidrain/images/origins.gif (US EPA)

Acidity is measured on the per-hydrogen or pH scale. This is a measure of the concentration of positively charged ions in a given sample. It ranges from 14 (alkaline or negatively charged ions) to 0 (acidic or positive ions). Pure water has a pH of 7 (neutral). Most rainwater is slightly acidic (pH about 6). A change in the pH scale of one unit reflects a tenfold (10X) change in the concentration of acidity. Generally, rain with a pH value of less than about 5.3 is considered acid rain. Most of the rainwater, which falls in the Eastern United States, has a pH between 4.0 and 5.0. This is generally lower (more acidic) than the national average. The use of tall smokestacks installed to reduce local pollution has contributed to the spread of acid rain by releasing gases into regional atmospheric circulation, with deposition occurring at a considerable distance downwind of the emissions [53].

8.1 Effects of Acid Rain

The impacts of acid rain and deposition are varied and often interrelated, creating complex and far-reaching consequences to aquatic and terrestrial ecosystems, visibility, and public health:

1. Acid precipitation can increase the acidity of lakes and streams by either passing through soils or falling directly on water bodies. Changes in the acidity of lakes and streams can impact the survival of fish and amphibian populations by impairing the ability of certain fish and water plants to reproduce, grow, and ultimately survive.
2. Terrestrial ecosystems can also be altered by increasing acidity of precipitation and heavy metal deposition. Acids strip forest soils of essential nutrients needed to sustain plant life. This process threatens the reproduction and survival of trees and other forest vegetation.
3. Acid deposition of acidic particles is known to contribute to the corrosion of metals and to the deterioration of stonework on buildings, statues, and other structures of cultural significance, resulting in depreciation of the objects' value to society. Acid deposition can also damage paint on buildings and cars.
4. Additionally, the same gases that cause acid deposition are responsible for the formations of small particles in the air that greatly reduce visibility and can adversely affect human health. Sulfate aerosol particles and, to a lesser extent, nitrate particles in the atmosphere produced from SO_2 and NO_x emissions account for more than 50% of the visibility reduction in the Eastern United States and heavily influence concentrations of small particles or PM. These particles are small enough in size to be inhaled deeply into lung tissue, aggravating the reparatory and cardiopulmonary systems, especially in sensitive populations (people with asthma, emphysema, or other respiratory illnesses).

The most obvious environmental effect of acid rain has been the loss of fish in acid-sensitive lakes and streams. Many species of fish are not able to survive in acidic water. Acid rain affects lakes and streams in two ways: chronic and episodic

[47]. Chronic or long-term acidification results from years of acidic rainfall. It reduces the alkalinity (buffering capacity) and increases the acidity of the water. Chronic acidification may reduce the levels of nutrients and minerals such as calcium and magnesium, which, over time, may weaken the fish and other plants and animals in an aquatic ecosystem:

$$2\,H^{+}(\text{aqueous}) + Mg^{2+}(\text{clay}) \rightleftharpoons 2\,H^{+}(\text{clay}) + Mg^{2+}(\text{aqueous})$$

$$2\,H^{+}(\text{aqueous}) + Ca^{2+}(\text{clay}) \rightleftharpoons 2\,H^{+}(\text{clay}) + Ca^{2+}(\text{aqueous})$$

Most of the effects on forests are subtle. Acid deposition may influence forest vegetation and soils. Acid rain weakens the trees' natural defenses, making them more vulnerable to diseases. Acid rain has been cited as a contributing factor to the decline of the spruce-fir forests throughout the Eastern United States. Acid rain may remove soil nutrients such as calcium and magnesium from soils in high elevation forests and cause damage to needles of red spruce. Acid rain may also help weaken natural defenses of some trees, making them more vulnerable to some diseases and pests.

Episodic acidification is a sudden jump in the acidity of the water. This can result from a heavy rainstorm. It also happens in the spring, because the sulfates and nitrates will concentrate in the lowest layers of a snowpack. In the spring, when that snow melts, it will be more acidic than normal. Episodic acidification can cause sudden shifts in water chemistry. This may lead to high concentrations of substances such as aluminum, which may be toxic to fish.

Acid rain deposits nitrates that can lead to increases in nitrogen in forests. Nitrogen is an important plant nutrient, but some forest systems may not be able to use all they receive, leading to nitrogen saturation. In the Eastern United States, there is evidence of nitrogen saturation in some forests. Nitrates can remove additional calcium and magnesium from the soils. Continued nitrogen deposition may alter other aspects of the nutrient balance in sensitive forest ecosystems and alter the chemistry of nearby lakes and streams.

Excess nitrogen may cause eutrophication (over nourishment) in areas where rivers enter the ocean. This may lead to unwanted growth of algae and other nuisance plants. As much as 40% of the total nitrogen entering coastal bays on the Atlantic and Gulf Coasts may come from atmospheric deposition. Table 7.5 shows estimates of the percentage of nitrogen deposition, which comes from the atmosphere.

Acid rain can react with aluminum in the soil. Trees cannot absorb naturally occurring aluminum, but acid rain may convert it to aluminum sulfate or aluminum nitrate. These can be absorbed by the trees and may adversely affect them. The effects of acid rain, combined with other environmental stressors, leave trees and plants less able to withstand cold temperatures, insects, and disease [54]. The pollutants may also inhibit trees' ability to reproduce. Some soils are better able to neutralize acids than others. In areas where the soil's "buffering capacity" is low, the harmful effects of acid rain are much greater.

Table 7.5 Nitrogen input from the atmosphere to New England bays and estuaries [1]

Bay	Atmospheric contribution as % of total nitrogen input
Casco Bay (ME)	About 40%
Massachusetts Bay (MA	5–27%
Waquoit Bay (MA)	29%
Narragansett Bay (RI)	4–12%
Long Island Sound (CT)	About 20%

Acid rain has not been shown to be harmful to human health, but some of the particles, which can be formed from sulfate and nitrate ions, can affect respiration. They can be transported long distances by winds and inhaled deep into people's lungs. Fine particles can also penetrate indoors. Many scientific studies have identified a relationship between elevated levels of fine particles and increased illness and premature death from heart and lung disorders, such as asthma and bronchitis.

Acid deposition has also caused deterioration of buildings and monuments. Many of these are built of stone that contains calcium carbonate. Marble is one such material. The acid rain can turn the calcium carbonate to calcium sulfate (gypsum). The calcium sulfate can crumble and be washed away:

$$CaCO_3(\text{solid}) + H_2SO_4(\text{aqueous}) \rightleftharpoons CaSO_4(\text{solid}) + CO_2(\text{gas}) + H_2O\ (\text{liquid})$$

Acid rain also increases the corrosion rate of metals, in particular iron, steel, copper, and bronze. Figure 7.12 shows how Harvard University wraps some of the bronze and marble statues on its campus with waterproof covers every winter, in order to protect them from erosion caused by acid rain and acid snow.

8.2 *History and Regulations*

Acid rain was first observed in the mid-nineteenth century, when some people noticed that forests located downwind of large industrial areas showed signs of deterioration. The term "acid rain" was coined in 1872 by Robert Angus Smith, an English scientist [47]. Smith observed that acidic precipitation could damage plants and materials.

Acid rain was not considered a serious environmental problem until the 1970s. During that decade, scientists observed the increase in acidity of some lakes and streams. At the same time, research into long-range transport of atmospheric pollutants, such as sulfur dioxide, indicated a possible link to distant sources of pollution. Many power plants use coal with a relatively high concentration of sulfur as fuel. Scientists realized that sulfur dioxide emitted from many of these plants could be transported to the Northeast. When we began to see acid rain as a regional, rather than a local, problem, the federal government had to become involved.

Fig. 7.12 Harvard University wraps some of the bronze and marble statues on its campus, with waterproof covers every winter, in order to protect them from erosion caused by acid rain and acid snow (Wikipedia) https://en.wikipedia.org/wiki/Acid_rain#/media/File:Bixi_stele_(wrapped),_Harvard_University,_Cambridge,_MA_-_IMG_4607.JPG

In 1980, the US Congress passed an Acid Deposition Act. From the start, policy advocates from all sides attempted to influence NAPAP (National Acid Precipitation Assessment Program) activities to support their particular policy advocacy efforts or to disparage those of their opponents [55]. This Act established a 10-year research program under the direction of the NAPAP program. NAPAP looked at the entire problem. It enlarged a network of monitoring sites to determine how acidic the precipitation actually was and to determine long-term trends and established a network for dry deposition. It looked at the effects of acid rain and funded research on the effects of acid precipitation on freshwater and terrestrial ecosystems, historical buildings, monuments, and building materials. It also funded extensive studies on atmospheric processes and potential control programs. Significant impacts of NAPAP were lessons learned in the assessment process and in environmental research management to a relatively large group of scientists, program managers, and the public [56].

In 1991, NAPAP provided its first assessment of acid rain in the United States. It reported that 5% of New England Lakes were acidic, with sulfates being the most common problem. They noted that 2% of the lakes could no longer support brook trout and 6% of the lakes were unsuitable for the survival of many species of minnow. Subsequent reports to Congress have documented chemical changes in soil and freshwater ecosystems, nitrogen saturation, decreases in amounts of nutrients in soil, episodic acidification, regional haze, and damage to historical monuments.

Meanwhile, in 1990, the US Congress passed a series of amendments to the (CAA) Clean Air Act. One was the inclusion of section 112(m), Atmospheric Deposition to Great Lakes and Coastal Waters (ADGLCW). The biennial report required by this section of the CAA amendments is to cover the following [57]:

1. The contribution of atmospheric deposition to pollution loadings in the Great Waters
2. The environmental and public health effects of any pollution attributable to atmospheric deposition to these waterbodies
3. The sources of any pollution attributable to atmospheric deposition to these waterbodies
4. Whether pollution loadings in these waterbodies cause or contribute to exceedances of drinking water or water quality standards or, with respect to the Great Lakes, exceedances of the specific objectives of the Great Lakes Water Quality Agreement
5. Descriptions of any revisions of the requirements, standards, and limitations of relevant CAA and federal laws to ensure protection of human health and the environment

The *First* and *Second Great Waters Reports to Congress* on atmospheric deposition to the Great Waters were published in May 1994 (US EPA 1994) and June 1997 (US EPA 1997). The first two reports presented the programmatic background and covered the scientific issues that are addressed by the Great Waters program. The *Third Great Waters Report to Congress* provides an update to the information presented in previous reports and specifically highlights progress made since the *Second Report to Congress*, including changes in pollutant emissions, deposition, and effects, as well as recent advancements in the scientific understanding of relevant issues. In addition, the report discusses recent activities and accomplishments of the many different initiatives that help protect the Great Waters from pollutants deposited from the atmosphere.

The amendments also established research, reporting, and potential regulatory requirements related to atmospheric deposition of HAPs (hazardous air pollutants) to the "Great Waters. Title IV of these amendments established a program designed to control emissions of sulfur dioxide and nitrogen oxides. Title IV called for a total reduction of about 10 million tons of SO_2 emissions from power plants. It was implemented in two phases. Phase I began in 1995 and limited sulfur dioxide emissions from 110 of the largest power plants to a combined total of 8.7 million tons of sulfur dioxide One power plant in New England (Merrimack) was in Phase I. Four other plants (Newington, Mount Tom, Brayton Point, and Salem Harbor) were added under other provisions of the program. Phase II began in 2000 and affects most of the power plants in the country.

Emissions of nitrogen oxide and nitrogen dioxide, generally called NO_x, have been reduced by a variety of programs required under the Clean Air Act. NO_x is emitted by anything burning fuel, such as power plants, large factories, automobiles, trucks, and construction equipment.

In New England, between 1990 and 2000, we have seen a 25% decrease in NO_x emissions from all sources (from approximately 897,000 tons to 668,000 tons). Between 2000 and 2006, NO_x emissions from acid rain-affected power plants in New England have further decreased by more than 31,000 tons. During that same period, SO_2 emissions from those power plants have decreased by 54% (from approximately 211,000 tons to 96, 500 tons).

During the 1990s, research has continued and gradually developed a better understanding of acid rain and its effects on the environment. A closer look at soil chemistry showed how acid rain has changed the balance of calcium, aluminum, and other elements. Since acid rain makes waters acidic, it causes them to absorb the aluminum that makes its way from soil into lakes and streams. Sulfur dioxide pollution mostly from coal-fired power plants was causing acid rain and snow, killing aquatic life and forests. A debate ensued: Regulation would direct all plant owners to cut pollution by a set amount, but this method, critics argued, would be costly and ignore the needs of local plant operators. The solution was devised to cap-and-trade approach, written into the 1990 Clean Air Act. It required cutting overall sulfur emissions in half, but let each company decide how to make the cuts. Power plants that lowered their pollution more than required could sell those extra allowances to other plants. A new commodities market was born. Sulfur emissions went down faster than predicted and at one fourth of the projected cost. Since its launch, cap-and-trade for acid rain has been regarded widely as highly effective at solving the problem in a flexible, innovative way [58]. Since this first historic success, efforts were expanded to help create new market mechanisms that account for the impact to the environment. This solution has served as the inspiration behind one of the most powerful tools we have to fight climate change: carbon markets [58].

The success of the Acid Rain Program has led to consideration of other programs based on setting an emissions cap. The NO_x budget program, which began in 1999, places a limit on NO_x emissions from power plants and some other sources during the warmer months of the year. Its purpose is to control ground level ozone, but it will have some effect on acid rain also. Massachusetts, New Hampshire, and Connecticut have designed their own programs to further limit emissions of NO_x and SO_2. Connecticut's rule contributed to a 68% decrease in SO_2 emissions from large sources from 2001 to 2002 [47].

On March 10, 2005, US EPA issued the Clean Air Interstate Rule (CAIR). This rule provides states with a solution to the problem of power plant pollution that drifts from one state to another. CAIR permanently capped emissions of SO_2 and NO_x in the Eastern United States. US EPA's CAIR addressed regional interstate transport of soot (fine particulate matter) and smog (ozone), which are associated with thousands of premature deaths and illnesses each year. CAIR required 28 eastern states to make reductions in sulfur dioxide (SO_2) and nitrogen oxide (NO_X) emissions that contribute to unhealthy levels of fine particle and ozone pollution in downwind states. Once it was fully implemented, CAIR reduced SO_2 emissions in 28 eastern states and the District of Columbia by over 70% and NO_x emissions by over 60% from 2003 levels [59]. CAIR was replaced by the Cross-State Air Pollution Rule (CSAPR), as of January 1, 2015.

On July 6, 2011, the US Environmental Protection Agency finalized the rule that protects the health of millions of Americans by helping states reduce air pollution and attain clean air standards. This rule, known as the Cross-State Air Pollution Rule (CSAPR), requires states to significantly improve air quality by reducing power plant emissions that contribute to ozone and/or fine particle pollution in other states [60]. In a separate, but related, regulatory action, US EPA finalized a supplemental rulemaking on December 15, 2011 to require five states – Iowa, Michigan, Missouri, Oklahoma, and Wisconsin – to make summertime NO_X reductions under the CSAPR ozone season control program. CSAPR requires a total of 28 states to reduce annual SO_2 emissions, annual NO_X emissions, and/or ozone season NO_X emissions to assist in attaining the 1997 ozone and fine particle and 2006 fine particle National Ambient Air Quality Standards (NAAQS).

8.3 *Causes of Acid Rain*

Two elements, sulfur and nitrogen, are primarily responsible for the harmful effects of acid rain. Sulfur is found as a trace element in coal and oil. When these are burned in power plants (see Fig. 7.13) and industrial boilers, the sulfur combines with oxygen to form sulfur dioxide (SO_2). Because SO_2 does not react with most chemicals found in the atmosphere, it can travel long distances. Eventually, if it comes in contact with ozone or hydrogen peroxide, it can be converted to sulfur trioxide. Sulfur trioxide can dissolve in water, forming a dilute solution of sulfuric acid. In the gas phase sulfur dioxide is oxidized by reaction with the hydroxyl radical via an intermolecular reaction:

$$SO_2 + OH\bullet \rightarrow HOSO_2\bullet$$

which is followed by:

$$HOSO_2\bullet + O_2 \rightarrow HO_2\bullet + SO_3$$

In the presence of water, sulfur trioxide (SO_3) is converted rapidly to sulfuric acid:

$$SO_3(\text{gas}) + H_2O\ (\text{liquid}) \rightarrow H_2SO_4(\text{aqueous})$$

Nitrogen makes up about 78% of the atmosphere. When heated to the temperatures found in steam boilers and internal combustion engines, it can combine with oxygen from the atmosphere to form nitrogen oxide and nitrogen dioxide (NO_x). NO_x is the sum of nitrogen oxide and nitrogen dioxide in a given parcel of air. These can dissolve in water, forming weak solutions of nitric and nitrous acids. Nitrogen dioxide reacts with OH to form nitric acid:

Fig. 7.13 The coal-fired Gavin Power Plant in Cheshire, Ohio https://upload.wikimedia.org/wikipedia/commons/7/75/Gavin_Plant.JPG (Wikimedia) Clouds of sulfuric acid coming from the vertical column stacks. The emissions from the Cooling Towers are just water vapor

$$NO_2 + OH\bullet \rightarrow HNO_3$$

NO_x and SO_2 can come from natural or human-made (anthropogenic) sources. Volcanoes and sea spray are typical natural sources of SO_2. Lightning is the most common natural source of NO_x. Contributions from natural sources are generally small compared to those from anthropogenic sources.

US EPA classifies the sources of anthropogenic emissions of pollutants into three groups: point (or stationary) sources, area sources, and mobile sources. Point sources include factories, power plants, and any other large "smokestack" facilities. Area sources consist of smaller facilities, which occur in greater numbers. These include residential heating equipment, small industry, and other categories in which it is impractical to analyze each individual emission source. Mobile sources include anything that can move. They can be divided into on-road sources (including cars, trucks, buses, motorcycles, etc.) and non-road (tractors, snowmobiles, boats, airplanes, lawnmowers, etc.).

Point sources emit the largest amount of SO_2. Of these, coal-fired power plants are the highest emitters. The Brayton Point Station in southeastern Massachusetts is the largest point source for SO_2 in New England. In 2006, 16 units at eight facilities emitted a total of 82,129 tons of sulfur dioxide.

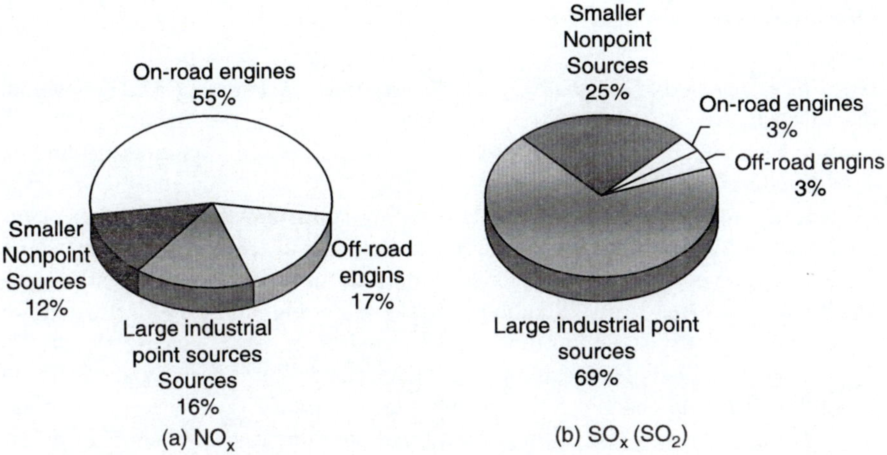

Fig. 7.14 Emissions of NO_x and SO_x (SO_2) in New England in 2002 [15]

Figure 7.14 charts show how much each group contributed to emissions of NO_x and SO_x (SO_2) in New England in 2002 [47, 61]: Once SO_2 and NO_x have been released into the air, they can be transported by the wind. The prevailing winds above most of the United States flow from west to east. Storm systems and other meteorological events may alter this flow. The final effects of these pollutants may occur as much as 1000 miles from where they were released. Eventually, these elements will dissolve in water droplets and be converted into nitrate and sulfate ions. In this form, they may return to Earth through acid deposition. Acid deposition occurs when these ions are deposited to the ground. It may be in the form of wet deposition, either indirect (acid rain, acid snow) or direct (acid fog), or dry deposition.

Prior to the mid-1990s, most scientists felt that the most common method of acidic deposition was by rain and other forms of wet deposition. This is the most visible and best understood means of deposition. It is easy to measure, and its effects are most obvious. Acid snow tends to carry less acidity per unit of water, since the chemistry is slower in cold weather, but since snow accumulates over the course of a season, when it melts, it releases a surge of acidity. At higher altitudes, direct deposition can occur when clouds descend to the surface. This can cause a severe problem because it may last for hours [47].

Dry deposition occurs when sulfate or nitrate ions do not dissolve in water, but rather fall to the surface as small particles or go directly from gaseous form in the atmosphere to soil or water. Unlike wet deposition, dry deposition is not easily measured. Very little falls at one time or at one location, but since dust is constantly settling to the Earth's surface and the atmosphere is constantly in contact with the Earth, it can potentially have a large impact. As we have learned more about it, we realize it is an important part of acid deposition, and as the amount of acidity in rain decreases, dry deposition has become a more prominent route for deposition. Dry deposition now accounts 20–60% of the total deposition.

8.4 Reducing Acid Rain

Reducing emissions of SO_2 and NO_x is necessary if we are to reduce acid deposition. The first attempts at reducing SO_2 took place in 1936 at the Battersea Plant in London, England. In recent years, we have made considerable progress in finding ways to reduce emissions of both SO_2 and NO_x.

There are two principal methods for reducing sulfur emissions at power plants and other facilities that burn coal or oil: fuel switching and scrubbing. Fuel switching means replacing coal or oil, which contains more sulfur, by fuels such as natural gas, which has little or no sulfur. Scrubbing means removing sulfur by electrostatic or chemical (wet or dry scrubbing) means. Electrostatic involves placing electrically charged plates, called electrostatic precipitators, inside the industry's stack. These attract the positively charged sulfur particles to the surface. The surface is periodically cleaned, removing the sulfur before it gets into the air. Wet scrubbing means injecting water or a chemical solution into the exhaust gases. Dry scrubbing involves a chemical such as lime, which reacts with the gases without the use of water. The sulfur will react with the water or chemical and fall out. All types of scrubbing do pose a problem; we must find an environmentally acceptable way of disposing of the sulfur after we have removed it.

There are several methods of reducing NO_x emissions. Some are mechanical: changing the ratio of air to fuel or changing the temperature of the combustion. The cooler the flame is, the less NO_x the furnace gives off. Others are chemical: injecting chemicals such as ammonia, which will react with the NO_x and convert it back into nitrogen and oxygen.

US EPA's Acid Rain Program (ARP) has given the utility industry a reason to reduce SO_2 and NO_x emissions. This program was established by Title IV of the 1990 Clean Air Act Amendments. It set a cap on the amount of SO_2 power plants can emit. The program also addressed NO_x emissions, but only set maximum emission rates based on the type of boiler [62].

The Acid Rain Program focused on power plants, the largest single source of SO_2 emissions and a major source of NO_x emissions. The plants affected by the program submitted permit applications explaining how they planned to comply with the program. US EPA issues permits to each facility. The program also requires the use of continuous emissions monitors (CEMs), which measure their emissions and transmit the information directly to US EPA.

A unique element of the program is its use of emissions trading as a compliance option. Although the national cap on emissions limits the total SO_2 released into the air, companies may decide the most cost-effective method. Each plant is assigned a number of "allowances," based on their average annual SO_2 emissions during the period from 1985 to 1987. These could be bought, traded, sold, or held. At the end of the year, each facility has to surrender one allowance for each ton of SO_2 it emitted. US EPA set up an allowance tracking system and also set up auctions and direct sales to enable plants, which were not assigned allowances to obtain those they needed to operate. In 1990, the 263 units designated as part of the Phase I program emitted 10.0 million tons of SO_2. In 1995, the first year in which the units were required to comply with Phase I of the program, they reduced their emissions to 5.3 million tons. This is a 47% reduction in emissions over 5 years [62].

Phase II of the program began in 2000. It affects more power plants and gives all of the plants fewer allowances. Merrimack Station in New Hampshire had 31,343 allowances per year in Phase I. They have 13,530 allowances in Phase II. This will lead to further reductions in SO_2 emissions.

The SO_2 cap and allowance trading system proved successful and is now being used or considered for use to regulate several other pollutants, including NO_x and carbon dioxide.

The Acid Rain Program (ARP) also monitors and limits emission of NO_x. In New England, the acid rain NO_x rules have not had a noticeable effect, because other programs have imposed more stringent limits.

The Acid Rain Program also promotes the use of renewable energy and energy conservation. Some allowances were set aside to award to companies, which encouraged residential, commercial, and industrial conservation or used certain forms of renewable energy. Six New England companies were awarded 4,186 allowances for energy conservation and for using biomass and landfill gas as a source of energy. Allowances are also available for solar, wind, and geothermal energy.

New England states have also been active in the fight against acid rain. Prior to 1990, Massachusetts and New Hampshire had passed laws limiting the emissions from power plants. In 2001, Massachusetts adopted regulations that will further limit emissions from large power plants by as much as 75% of SO_2 and 50% of NO_x. Connecticut has adopted regulations that will further limit emissions by as much as 50% of the SO_2 and 30% of the NO_x currently being emitted.

The United States has been working with Canada to reduce transboundary effects of acid deposition. A bilateral Air Quality Agreement was signed in March 1991. The third biennial report, published in 1996, focused on the progress made by the United States and Canada in achieving emissions reduction goals. Their 2006 Annual Progress Report reported that both countries have made progress in reducing emissions, which lead to acid rain. The Integrated Atmospheric Deposition Network (IADN) was established in 1990 to collect data that can be useful in assessing the relative importance of atmospheric deposition. The IADN program established a database for atmospheric deposition data in both countries.

Individuals can also help prevent acid rain by conserving energy. The less electricity people use in their homes, the fewer chemicals power plants will emit. Vehicles are also major fossil fuel users, so drivers can reduce emissions by using public transportation, carpooling, hybrid and electric cars, biking, or simply walking wherever possible [2].

8.5 *Acid Rain Permit Program*

US EPA's Acid Rain Program was developed in response to Title IV of the 1990 Clean Air Act Amendments. The program requires most power plants and other facilities that choose to participate in the program to obtain a permit [63, 64].

An Acid Rain Permit tells us how the power plant plans to comply with the Acid Rain Program. It may be issued separately from other permits required by US EPA, but eventually becomes part of the plant's Title V Permit [65]. The owners of the plant send an application to the appropriate agency. The agency, after checking to see if the application is complete, reviews the application and issues the permit.

During Phase I of the Acid Rain Program (1995–2000), US EPA Regional Offices issued permits to those power plants, which participated. In New England, Merrimack Station in Bow, NH, was the only station required to participate. Four other power plants (Newington Station, Portsmouth, NH; Mount Tom, Chicopee, MA; Salem Harbor, Salem, MA; Brayton Point, Somerset, MA) also participated in Phase I under other provisions of the program. US EPA New England issued Phase I permits to each of these plants. The Phase I permits expired on January 1, 2000.

During Phase II (2000–present), the permits are issued by the states. Each New England State has a program, which issues permits. In Vermont, the Acid Rain Permit is generally issued as part of the Title V Permit. Other states generally issue the Acid Rain Permit first and later incorporate it into the Title V Permit. Acid Rain Permits are usually issued for a period of 5 years [66]. Each of these facilities has filed an Acid Rain Permit application with the appropriate state, and the state has issued a permit.

8.6 Acid Rain Monitoring

Continuous emissions monitors (CEMs) tell us how much of a pollutant a power plant (or other affected facility) emits. Under the Acid Rain Program, each affected unit must monitor emissions of sulfur dioxide and nitrogen oxides. Most also measure carbon dioxide. CEMs are the most common way to monitor emissions, but in some cases, utilities may use an approved alternate method. For example, if a unit burns oil and they know the sulfur content of the oil and the amount they used, they can compute the amount of sulfur dioxide they will emit.

Each plant in the Acid Rain Program must submit a monitoring plan to US EPA and revise it as their monitoring system changes. This document tells which monitors will be used, where they will be located, and how the data will be gathered and sent to US EPA. If a plant's monitoring system changes, they must revise their monitoring plan.

There are several ways to monitor emissions from a plant. Nitrogen oxides are measured by taking a sample of the gas emitted thought the plant's smokestack and analyzing it. In some cases, sulfur dioxide is measured this way, but it can also be computed sulfur emissions from the fuel's sulfur content. When plants use natural gas, which contains only traces of sulfur, it may be estimated by assuming the gas contains a certain low amount of sulfur. Sulfur emissions from such plants are so low that an actual measurement of sulfur dioxide is impractical.

Power plants must submit data whenever they are operating. If their monitor is not operating, they must report a "default" value, which is generally the maximum amount of the pollutant they can emit. US EPA gathers the data and tabulates the emissions from each plant. The latest, 2013, ARP, and CAIR emission reductions are as follows [67]:

1. CAIR and ARP Annual SO_2 Emissions: 3.2 million tons (69% below 2005)
2. CAIR and ARP Annual NO_x Emissions: 1.7 million tons (53% below 2005)
3. CAIR Ozone Season NO_x Emissions: 470,000 tons (41% 2005)
4. Ambient particulate sulfate concentrations have decreased since the ARP was implemented, with average concentrations decreasing by 60 to 65% in observed regions from 1989–1991 to 2011–2013
5. The Northeast and Mid-Atlantic have showed the greatest improvement with an overall 64% reduction in wet sulfate deposition from 1989–1991 to 2011–2013

8.7 *Acid Rain Allowances*

The Acid Rain Program uses a market-based approach to control sulfur dioxide emissions from power plants. Under the current system, known as a "cap and trade" system, it limits, or caps, sulfur dioxide emissions by issuing a fixed number of allowances and allowing only one ton of SO_2 to be emitted per allowance. The total number of allowances is distributed by US EPA each year. Each power plant that was operating during the years from 1985 to 1987 is assigned a number of allowances based on their emissions of sulfur dioxide during that period. Each power plant must hold one allowance for each ton of SO_2 they emit. At the end of the year, these facilities must surrender a number of allowances equal to the number of tons of SO_2 they emitted during the year. If they emit more SO_2 than they have allowances, they must buy additional allowances. Conversely, if they can further reduce their emissions, they can sell their excess allowances. Some allowances were withheld by US EPA and are made available through an annual auction. By limiting the number of allowances, US EPA can limit the amount of SO_2 the power plants release.

The Acid Rain Program was implemented in two phases. During Phase I (1995–2000), 11 units at five New England Stations were allocated a total of 167,779 allowances. During Phase II (starting in 2000), 27 facilities (about 80 units) in New England received a total of 273,150 allowances. The number per unit was reduced during Phase II, so the total amount of SO_2 was also reduced. The five units in Phase I received only 91,136 allowances per year in Phase II [68].

New units, which were not operating in 1987, are not assigned allowances. They must obtain the necessary allowances from other units or buy them.

In 1995, US EPA issued about 8.7 million allowances to Phase I units. If all were used, this would have resulted in a 2.2-million-ton reduction from the 1980 level. Actual emissions in 1995 were 5.3 million tons. That means that in 1995, SO_2 was reduced by 5.6 million tons. Of the 3.4 million unused allowances in that year, about 36,000 were sold or donated to private individuals or environmental groups and "retired." This effectively lowered the cap for SO_2 emissions for that year. The other allowances were held either by brokerage firms or by the utilities themselves.

Holding allowances does not automatically allow a plant to emit SO_2. They are still subject to other state and federal permit conditions, which may further limit their emissions. Also, they may not cause exceedances of US EPA's National Ambient Air Quality Standards (NAAQS) or Prevention of Significant Deterioration (PSD) limits.

8.8 Acid Rain Measurement

In the 1970s, we realized the importance of monitoring the acidity of rainfall. The National Atmospheric Deposition Program (NADP) was formed in 1977 to investigate and measure acid deposition. There were 22 stations nationwide in 1978. Currently, the National Trends Network, which now conducts measurements for NADP, has nearly 220 stations. Each station (Fig. 7.15) collects a sample once a week. Samples are collected in a container. This container is covered until rain begins. The presence of water automatically opens the container, and it remains open until the rain stops. These samples are analyzed and sent to a central location. The data go into a database, which now stores data from over 20 years, and give us a feel for how acidity varies with location and also the trend over this period. This picture above shows a typical acid rain monitor. The container on the left represents an attempt to measure dry deposition. When precipitation begins, a sensor causes the cover to move from the right container to the left. Rainwater will then collect in the right container [69].

Fig. 7.15 Acid rain monitor, Glacier National Park, MT, USA [23]

Dry deposition is difficult to measure directly. There have been attempts to use a method similar to that for wet deposition, but it usually takes too long to cover the collector when rain starts. It takes only a few drops of rain to contaminate a dry deposition sample. The Clean Air Status and Trends Network (CASTNet) calculates dry deposition using atmospheric conditions, meteorological data, and information on land use, vegetation, and surface conditions.

Other programs have looked at changes in the acidity of lakes, streams, and soil. Monitored data for some lakes are available from as early as 1980. Facilities such as Hubbard Brook in New Hampshire have been measuring the acidity of streams and lakes on their 7800-acre property. They also have looked at soil chemistry and effects on the entire ecosystem.

In addition to specific efforts to measure acid rain, EPA maintains monitoring networks for monitoring SO_2 and NO_2 in the air for other programs. Although these programs are not intended to consider acid rain, they do give us an idea of the amount of these chemicals in the air at certain locations.

8.9 *Trends*

Figures 7.16 and 7.17 show how SO_2 and NO_x emissions from power plants in New England in the Acid Rain Program have decreased since 1990 to 2011. Data for 1995 and later years come from continuous emissions monitor data as reported by US EPA's Clean Air Markets (CAM) Division's Acid Rain Progress Reports 2013 or Emissions Data and Compliance Reports [70].

Data for 1990–1994 are estimates based on fuel usage and US EPA's conversion factors. It is provided by the Clean Air Markets Division.

The graphs presented above show how emissions (i.e., what we put into the air) have changed over time. We next look at some observations of how deposition (i.e., what comes out of the air) has changed over time.

As emissions of sulfur dioxide have decreased, we have noticed some improvement in the atmosphere in recent years. A recent report prepared for EPA by Paul Miller of the Northeast States for Coordinated Air Use Management (NESCAUM) shows a decrease of about 25% in average sulfate deposition in Maine between 1980 and 1999. Similar decreases in sulfate deposition have been observed through much of the Northeast.

Data for 1993 through 1998 show no marked systematic trend of pollutants in precipitation for the New England sites. Sulfate deposition in 1995 showed a marked decline at every monitoring site in New England, except the Acadia NP site in Maine. The average 1995 sulfate deposition in New England was the lowest ever recorded during the 20 years of sulfate deposition monitoring. The 1996 data indicate that sulfate deposition increased in eight of the ten sites. Nonetheless, sulfate levels in 1996 were about 16% below historical averages (1979–1995). The 1996 data show similar increases occur for nitrate at most sites, and these increases appear to be further enhanced in 1997. In 1998, however, nitrate deposition decreased at

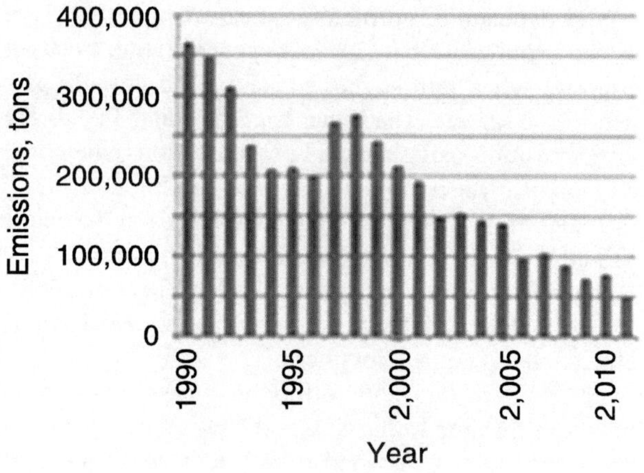

Fig. 7.16 SO_2 emissions from all New England facilities in US EPA Acid Rain Program, 2013 [24]

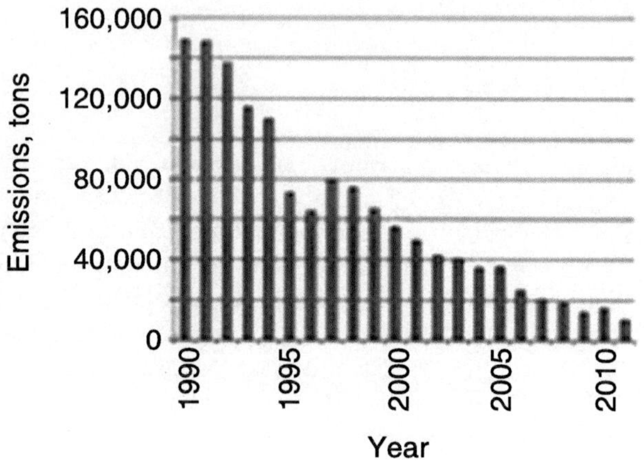

Fig. 7.17 NO_x emissions from all New England facilities in US EPA's Acid Rain Program, 2013 [24]

each of the ten trend sites. Nitrate deposition during 1998 remained virtually the same as the historical average (1979–1998). In the years since 1998, the we have seen a trend toward less sulfate in precipitation in New England. Until about 2000, trends in nitrate remained fairly constant, but since 2000, even the nitrate deposition in New England has begun to decrease.

The year-to-year variation that is recorded at the NADP sites can be attributed to differences in precipitation and prevailing wind patterns, as well as emission sources. This is characteristic of precipitation data and should be viewed as normal. Trends in pollutant deposition within New England, whether due to long-term changes in pollutants emitted to the atmosphere or trends in climate, will be evident only after many years (10–20+ years).

Lakes and streams have been slower to respond, but we have begun to see improvement. A report in *Nature* Magazine in 1998 showed some improvement in lakes in Maine, but not elsewhere in New England. In 2003, US EPA compiled an assessment of the surface chemistry of lakes and streams in the Northern and Eastern United States. This assessment showed that the concentration of sulfate in New England lakes is decreasing. About 30% of the lakes in New England that were acidic are no longer considered acidic. In New England, the "acid neutralization capacity" (ANC), the ability of the environment to neutralize acidic precipitation, has not significantly improved, but this was observed in the Adirondacks and in other areas. The soil in New England is slower to react than that in the Adirondacks, so it will take longer for the ANC in New England to improve.

The full recovery of New England's aquatic ecosystems will take more time. In some cases, it will be decades before we see ecosystems fully restored to their pre-industrial condition.

8.10 Acid Rain and Other Pollution Problems

In addition to the direct effects of acid rain, both the emissions and the attempts to control them affect other forms of environmental pollution.

Sulfur and nitrate emissions can also lead to the formation of small particles. These particles are too small to fall out of the atmosphere. They can get deep into your lungs, and some may even get into your bloodstream. Exposure to such particles can affect both the lungs and the heart. Larger particles are of less concern, although they can irritate the eyes, nose, and throat [71].

Also, small particles scatter light, which can reduce visibility. This detracts from our appreciation of the many magnificent scenic vistas in New England and the rest of the world. This is called Regional Haze problem [72]. US EPA and other agencies have been monitoring visibility in national parks and wilderness areas since 1988. In 1999, the US Environmental Protection Agency announced a major effort to improve air quality in national parks and wilderness areas. The Regional Haze Rule calls for state and federal agencies to work together to improve visibility in 156 national parks and wilderness areas such as the Grand Canyon, Yosemite, the Great Smokies, and Shenandoah [72].

The rule requires the states, in coordination with the Environmental Protection Agency, the National Park Service, US Fish and Wildlife Service, the US Forest Service, and other interested parties, to develop and implement air quality protection plans to reduce the pollution that causes visibility impairment.

Nitrogen oxides are also important in the formation of ground-level ozone. In the presence of sunlight, they can react with other chemicals to increase the concentration of ozone to a point where it can lead to health effects. Ground-level or "bad" ozone is not emitted directly into the air, but is created by chemical reactions between oxides of nitrogen (NOx) and volatile organic compounds (VOC) in the presence of sunlight. Emissions from industrial facilities and electric utilities, motor vehicle exhaust, gasoline vapors, and chemical solvents are some of the major sources of NOx and VOC. Breathing ozone can trigger a variety of health problems, particularly for children, the elderly, and people of all ages who have lung diseases such as asthma. Ground-level ozone can also have harmful effects on sensitive vegetation and ecosystems [73].

Coal-fired power plants also emit mercury, another trace element found in coal. Once in the atmosphere, the mercury can be deposited on the Earth. It finds its way into lakes and streams, where it is absorbed by microscopic plants and animals. These can be consumed by fish. The fish will store the mercury. When humans eat the fish, they can get a potentially harmful dose of mercury.

Burning of fossil fuels can also release other toxic chemicals such as cadmium, benzene, and formaldehyde. Usually these are released in very small amounts, but over time, and when combined with emissions from other sources, they can accumulate.

The same actions, which will help reduce acid rain, will also affect the climate. In addition to nitrates, power plants emit large amounts of carbon dioxide. Gases, which hold heat better than dry air, are called "greenhouse gases." Carbon dioxide and most oxides of nitrogen are both greenhouse gases, and so increasing the concentration of them makes the atmosphere hold more heat. This can lead to an increase in the global average temperature. Many scientists feel they already have evidence that this is happening and feel that increased concentrations of greenhouse gases will have serious consequences. By reducing our use of electrical energy, increasing the efficiency of power plants, and using sources of energy, which do not require combustion (wind, solar, etc), we can reduce acid rain and emissions of greenhouse gases [74].

8.11 Future Efforts

While we have made progress in reducing the acidity of rainfall and are beginning to see positive effects in our environment, it is not yet possible to declare victory. We still must continue to reduce our emissions of atmospheric pollutants. Clean fuels, renewable energy, and increased efficiency are important ways to reduce our dependence on coal and oil as a source of energy [75].

Fuels such as natural gas and wood chips emit fewer acid rain causing pollutants per unit of electricity. By converting some of our existing plants to run on these fuels and building new facilities that can replace some of our older, less efficient plants, we can generate electricity with fewer emissions.

Renewable energy sources such as wind, hydroelectric, and solar are contributing to cleaner air in New England and slowing the region's increase in fossil fuel consumption. Some states are calling for increasing amounts of electricity from renewable resources. Wind power, in particular, has become a more important source of electricity due to the fact that its costs are now similar to that of traditional fossil fuel resources.

Conservation programs such as US EPA's Energy Star, Green Buildings, and the New England Community Energy Challenge will help us reduce the demand for electricity. Energy Star [76] identifies products that are more energy efficient. Green Buildings encourages the planning and construction of more energy efficient buildings. Recently, in New England, US EPA launched the New England Community Energy Challenge, a municipal energy efficiency program.

Cleaner fuels, renewable sources of energy, and conservation programs will help us conserve energy. The result will be a cleaner environment, and this will benefit all of the inhabitants of the Earth.

9 Case Histories

9.1 Lake Brazos, Waco, TX

On the Brazos River, a few hundred yards downstream from the La Salle Avenue Bridge lies the Low Water Dam or, more recently, the Lake Brazos Dam. About 40 years ago city leaders decided that a dam below Waco would significantly widen the river and stabilize the river level. These improvements would enhance the natural beauty of the Brazos River through Waco. Additionally, the dam would create an impoundment from which up to 5,600 acre-ft of water can be withdrawn annually for municipal purposes [31].

Since the original dam was constructed, there have been many new developments. McLennan Community College has built the Bosque River Stage and Amphitheater. Baylor University has developed all along the eastern edge near the river including a new law school, natural history museum, and a soon to be finished science building. These improvements along with a world-class athletic complex have been built immediately adjacent to the Brazos River. In addition, Baylor operates a marina on the Brazos River to facilitate sailing and canoeing opportunities. Other colleges from Colorado and Kansas routinely travel to Waco for early spring rowing practices.

The City of Waco has also made many improvements including miles of riverwalk and various parkway improvements that include landscaping, lighting, and a boat dock in Cameron Park. Recent improvements include the Texas Sports Hall of Fame and upgrades to the Texas Ranger Museum. There have also been many private developments such as the Brazos Queen (dining boat), the Spirit of the Rivers paddleboat, Lake Brazos Steakhouse, Dock's Restaurant, and kayaking. Even with the enormous amount of improvements, the unmet potential remains tremendous.

With so much focus and activity along the Brazos River, there is a tremendous need to ensure a reliable, constant-level town lake. The existing dam was completed in 1970. The structure, which originally consisted of two drum gates, has quite a history for poor or non-performance. The original design, despite several modifications, did not function as intended for a reliable town lake. In 1985, hydraulic cylinders were attached to the underside of the massive gates (117 ft each) to provide a positive control mechanism for maintaining the level of Lake Brazos. This modification, while expensive to maintain, has restored quite a bit of reliability. Through decades of modifications and expensive maintenance, it was time to replace the structure with a new dam, which will offer reliability while requiring less maintenance. The new dam will be a concrete labyrinth weir. The new dam is scheduled for completion in fall 2007.

The City of Waco remains committed to providing a safe and plentiful supply of water for its citizens. Work has begun on 80 million USD in improvements to the water treatment system. The first phase has been completed with the completion of construction on the 42″ water distribution line and upgrades at the Mount Carmel Water Treatment Plant. This will improve the treatment capacity at that plant from 45 to 66 MGD. Similar improvements will take place at the Riverside Plant, and when the entire project is completed, the City of Waco will be able to treat a maximum of 130 MGD of water.

The City of Waco is also making the transition to a new dissolved air flotation (DAF) treatment process. This will address issues with the taste and odor problems caused by by-products of algae in the North Bosque River Watershed. The second phase of the Water Quality and Quantity project is designed to dramatically improve the taste of the finished water. In order to complete this task, the City of Waco is constructing a new clarification facility, featuring DAF for the removal of algae and other suspended particles from the raw water supply. DAF is particularly effective on waters with significant amounts of lightweight particles such as algae. Algae are the primary source of taste and odor-causing compounds in Texas waters.

DAF works by attaching air bubbles to particles suspended in the raw water and floating them to the surface of a tank for removal. The process includes flocculation to bind particles suspended in the raw water into larger flocs that can more easily be removed, a saturator that entrains air into a side stream for injection into the process, an air nozzle header that releases extremely fine air bubbles that attach to the flocculated particles, a skimmer which removes the suspended particles after they float to the surface of the basin, and effluent laterals which collect the clarified water off the bottom of the basin.

The odor-causing compounds MIB and geosmin are found inside the algae cells and get released to the water when the algae is killed or damaged. By removing the algae at the lake site (see Fig. 7.18) before it has a chance to be killed or damaged during piping to the treatment plant or the water treatment process itself, water quality will be significantly improved. The reduced levels of MIB and geosmin in the raw water once the algae is removed will allow any residual taste and odor compounds to be more easily removed in later stages of the treatment process. After DAF clarification, the water will be treated with chlorine dioxide for primary disinfection and then sent to the two treatment plants where it will be filtered through sand and granular activated carbon for removal of residual taste and odor-causing compounds and any remaining suspended materials.

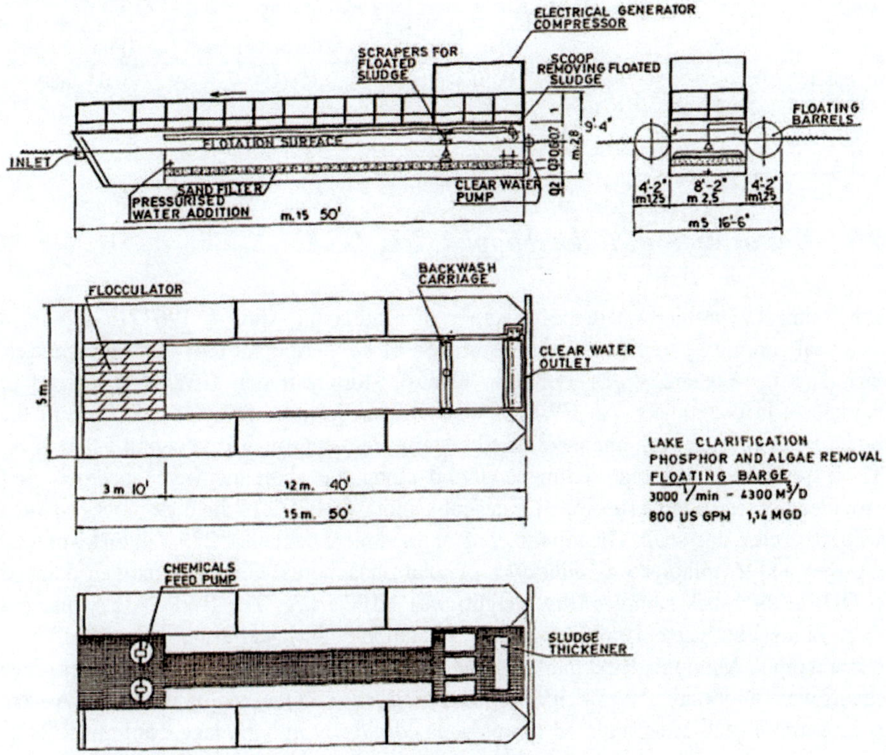

Fig. 7.18 Floating barge for algae removal from lake water

9.2 *Water Treatment from Lake Roine, Tampere, Finland*

In the 1960s, the City of Tampere, Finland, began drawing water from Lake Roine when the quality of water from its original lake source deteriorated. A horizontal sedimentation basin was constructed in the early 1970s, and a laminar flotation process was implemented in 1989. Activated carbon replaced sand filtration in 1996, but the City was still not satisfied with its drinking water quality [33].

In 1997, a pilot dissolved air flotation (AquaDAF) system was retrofitted in one of the conventional sedimentation basin flocculators. The DAF pilot demonstrated much higher flotation rise rates than those previously achieved with the laminar DAF units. In 2000, retrofitting of all laminar DAF units was completed. Today, the plant uses only the AquaDAF for clarification. The system was retrofitted in the original basins and sludge channels. The new structure is constructed entirely of wood.

The AquaDAF™ system utilizes hydraulic flocculation underneath the flotation area. Two very small unpacked saturators operate at a design recycle rate of 10%. The only submerged moving part is the effluent weir used for desludging. Water quality and system performance is shown in Table 7.6.

Table 7.6 System performance (Lake Roine water temperature range = 0.1°–17 ° C)

Parameter	Raw influent	Clarified effluent	Filter effluent
Turbidity, NTU	0.40–0.60	0.15–0.90	0.05–0.20
Dissolved organic carbon (DOC), mg/L	4.5–5.9	<2.0	
pH	6.2–7.5	5.0–7.0	

9.3 *Restoration of Lake Apopka, FL, USA*

Environmental problems led the governor of Florida on April 4, 1967, to appoint a technical committee to evaluate the restoration of Lake Apopka [34]. Sixteen agencies, including the Federal Water Pollution Control Administration (FWPCA), agreed to participate in the project. An FWPCA study begun in 1968 revealed that 90% of the bottom was covered with unconsolidated bottom sediment (muck) averaging 1.5 m thick. These sediments and peat sediments found along the shoreline were anaerobic and provided limited suitable substrate for desirable biota. Only 5% of the bottom was covered with sand, clay, and shell. The top meter of lake sediment contained 225 million kg of total nitrogen and 2 million to 4 million kg of total phosphorus. Chemical oxygen demand (COD) in the muck samples (dry weight) was 1,100 mg/g. The FWPCA also made a crude nutrient budget and emphasized that restoration of the lake must include reduction of nutrient input. Although direct rainfall on the lake and high nutrient input from citrus grove runoff were important, the principal controls on inputs emphasized by the FWPCA were point sources such as agricultural runoff pumped directly into the lake from muck farms and municipal and industrial wastes. In addition to control of external nutrient sources, several solutions for improving lake water quality are listed below. These include [35]:

1. Dredging to remove nutrient-rich unconsolidated bottom sediments to increase lake depth and reduce internal nutrient recycling.
2. Using lake drawdown to expose and subsequently consolidate large areas of lake bottom by oxidation and compaction.
3. Adding an inert sealing material to stabilize bottom sediments.
4. Engaging in hydroponic farming to remove dissolved nutrients.
5. Harvesting to remove algae by flotation, filtration, precipitation (not within the lake), or centrifugation (recovered algae could be used as a feed supplement).
6. Harvesting fish to remove nutrients "on a large scale." Harvested fish could be used as a protein supplement.

The governor of Florida assigned complete responsibility for a 1970 restoration of Lake Apopka to the Florida Air and Water Pollution Control Commission. This agency decided to proceed with the lake drawdown approach by allowing gravity drainage to lower the lake level 60 cm beginning in December 1970. The effect of this lowering was to be evaluated, and the lake would then be drained further by pumping to 25% of its original area. This final drawdown would occur in the spring of 1971. It was anticipated that two beneficial effects would result from the drawdown:

1. Nutrient recycling would be reduced or eliminated from dried, compacted sediments.
2. Suitable substrate for rooted aquatic vegetation would also be a result.

This plan to lower the lake about 7 ft below normal water level was not implemented, however, because of the projected cost (20 million USD) and because of concern about environmental and economic impacts [36]. For example, the loss of lake volume would minimize the freeze protection citrus growers received from the large heat capacity of the lake.

In the 1970s, additional studies were conducted on water quality problems and on restoration of Lake Apopka [37]. Studies of techniques that might be used to restore the lake have continued. Biomanipulation of algal standing crops with gizzard shad may actually increase standing crops of undesirable algae. A multimillion-dollar feasibility study on growing and harvesting water hyacinths to remove nutrients from the lake was launched [38]. The field test of this project in Lake Apopka was abandoned when the enclosure that was to have been used for the experiment was destroyed by water movements in the lake.

The St. Johns Water Management District began a feasibility study on using marsh restoration to improve water quality in the lake [39]. The water management district purchased muck farmland that will be flooded to restore the wetland by using the wetland as a filter to remove nutrients. The hydrology of the wetland will be manipulated so that highly nutrient-enriched water will flow from the lake into the wetland and nutrient-depleted water from the wetland will be directed back to the lake. If successful, this project will result in both a restored wetland and a restored lake.

There seem to be two divergent views about Lake Apopka. One group contends that the lake can be restored. This viewpoint is supported by the need to reduce nutrient inputs to prevent accelerated eutrophication. Schneider and Little [35] commented that the history of Lake Apopka "is not atypical" because other lakes in Florida and reservoirs all over the south were being subjected to similar attacks. They stated that the lake could be restored, but only with great expense and difficult decisions (e.g., the extent to which a 10 million USD plus marginal muck farming operation could expend money for nutrient removal). "The technical capabilities to prevent accelerated eutrophication are and have been available for some time. The planning and foresight needed to prevent the early demise of our lakes, however, has come into being only lately. Today, we must consider the full ecological impact of all our resource development activities if we are to eliminate the Lake Apopka syndrome from our aquatic environment," they emphasized.

At the other extreme is the viewpoint that restoration should not be attempted because it will meet with failure or it is too expensive. This viewpoint can be supported to a certain extent with results of studies on Lake Tohopekaliga (Lake Toho), Florida. A number of restoration measures have been instituted on Lake Toho since 1971, with little evidence of improvement in water quality [39]. In this lake, nutrient inputs have been reduced by sewage treatment and by stormwater detention and filtration. In addition, drawdown has been used as a restoration measure. What is not known is whether water quality would have been degraded even more if remedial measures had not been instituted. Dierberg et al. [40] point out that evaluation of restoration practices in Florida lakes has been hampered by the lack of long-term data and the consequent limitation on the use of robust statistical approaches in evaluating effectiveness.

9.4 Water Treatment from Lake DeForest in Clarkstown, NY, USA

United Water New York draws about 80% of its water supply from wells throughout Rockland County. The remaining 20% is supplied from a surface source, Lake DeForest in Clarkstown, NY [41].

The process of Lake Water treatment begins by pumping the water from the Lake DeForest Reservoir into the treatment plant. During the pumping process chemical is added to oxidize inorganic material, and the water passes through screens which remove large objects prior to entering the pumps. The water then proceeds to the new DAF (dissolved air flotation) system. A coagulant (aluminum sulfate) is added as the water enters the DAF system. This allows smaller particles to form larger flocs, getting the water ready to enter the final step of the DAF process. In this step, millions of microbubbles are added to the water floating the flocs to the top. The float (residuals) is removed and the clean water proceeds to the filters. It is during this step that chlorine is added. Chlorine destroys bacteria and viruses in the water. The water passes through the filters (layers of coal, sand, and gravel) to remove the smallest remaining particles. Next, the water receives another small dose of chlorine to be sure that the water remains pure and safe. Finally, corrosion control chemicals are added. This step helps prevent corrosion of the water pipes and plumbing. It also reduces the chance of lead dissolving in the water from plumbing.

10 Summary

A major factor in the death of a lake is eutrophication, which is the result of increased biological growth within the lake, or acidification, which is the result of increased release of sulfur dioxide and nitrogen oxides from industrial plants, or acid mine drainage waters from mines. Biological life may be controlled by limiting the nutrients or pH in the lake.

Reduction or elimination of external pollutant sources certainly is important for lake restoration and water quality improvement. Many eutrophic and hypereutrophic lakes, however, exhibit a very slow improvement in water quality even several years after major external pollutant sources were eliminated.

Lakes with high hydraulic retention times and/or having received excessive pollutant (note: mainly nutrient, acid, heavy metals, spills, etc.) loadings can experience significant internal loading of nutrients and pollutants from the sediments during anoxic periods. In such cases effective treatment of lake water and/or lake sediments may be warranted.

Among the lake water improvement technologies, physical and chemical processes are feasible for pollutant inactivation, precipitation, and biotic harvesting.

Lake water aeration is a physical process. There are many lake water aeration systems available. They increase the oxygen content of the lake water through mechanical mixing/agitations, air injection, or pure oxygen injection. These aeration systems either aerate lake waters at all depths causing thermal destratification or they preserve the thermal gradient and aerate the bottom lake waters only applying hypolimnetic aeration. For lake restoration, thermal destratification is beneficial for most warm water fisheries, and hypolimnetic aeration can create or greatly expand the cold water fishery potential of a lake.

Phosphorus is generally the nutrient that can be significantly removed. Phosphate removal is usually the primary objective of lake restoration. Using pollutant (mainly nutrient) inactivation or precipitation for treatment of standing bodies of lake water has been practiced since 1983 when 2 mg/L of ferric chloride was applied to a few selected reservoirs and lakes in Berkshire County, MA, USA. The first lake-wide application of aluminum sulfate for nutrient inactivation occurred at Langston, Sweden, in 1968. Since 1970 most of the larger lake water treatments using aluminum sulfate have occurred in the United States, such as Horseshoe Lake, Wisconsin, Dollar and Twin Lakes in Ohio, and Liberty Lake and Medical Lake in Washington, USA. In 1971 to 1974, the highly eutrophic Cline Pond near Corvallis, Oregon, USA, was treated by sodium aluminate and zirconium tetrachloride. In all the aforementioned cases, chemicals were dosed to lake waters by barge distribution and/or manifold injection. Most of the lakes treated by chemicals have shown reduced phosphorus content and less nuisance algal growth as well as higher hypolimnetic dissolved oxygen [45]. There are many ways to control the nutrients entering a lake, to control the existing nutrients within a lake, and to reduce the existing nutrients within a lake. Each lake must be studied individually to determine the best method to control eutrophication.

Biotic separation and harvesting is an important technique of nutrient removal that can lead to a reversal of eutrophic conditions in lakes. Algae and other aquatic plants such as water hyacinths, *Typha latifolia*, etc., can produce more biomass containing phosphorus and nitrogen. Harvesting of these aquatic plant species thus is an efficient method for nutrient removal from lakes. However, even the greatest potential harvest will not remove enough nutrients to offset moderate to heavy nutrient loading to lakes. Biotic harvesting as a lake restoration technique may work only where phosphorus loading has already been reduced to less than 1 g/ m^2/year [45]. A wide variety of mechanical harvesters has been designed for aquatic weed harvesting. The most efficient and cost-effective technique for water-algae separation and algae harvesting appears to be dissolved air flotation [43–46].

Whereas algae may be undesirable in lake water treatment, algae are also being harvested in some places for their nutrient value. The use of dissolved air flotation for harvesting cultivated algae for animal feed has been shown to be an efficient means of separating algae from their watery growth medium. The algae have been used as a livestock feed since they have a high nutrient and protein content and are reported to be highly acceptable to livestock. Another important benefit of the algae harvesting from lake water is reduction of chlorine dosage as an algicide, in turn, reduction of trihalomethanes, taste, and odors in drinking water when the lake water is used as the freshwater supply [45].

Acid rain and acid mine drainage pollute lake waters everywhere and have become a serious international problem. In order to develop alternative remedial methods for recovery of acid lakes believed due to acid rain and/or acid mine drainage, many engineering and management studies have been conducted on various chemical manipulation methods for lake water neutralization [45]. The materials evaluated for lake water neutralization include calcium hydroxide, sodium carbonate, agriculture limestone, fly ash, water treatment softening sludge, cement plant bypass dust, di-calcium silicate, calcium oxide, and magnesium oxide. It has been known [45] that the use of lime products is an effective way to neutralize acidified lake waters if properly administrated.

If acid lakes are contaminated by acid mine drainage, not only the water's high acidity shall be neutralized, but also its heavy metal concentrations (copper, zinc, cadmium, lead, iron, manganese, etc.) shall be reduced to safe or desirable levels. Chemical precipitation is a common process for inactivation of soluble metal ions. The precipitated insoluble metals in floc forms can be further separated by sedimentation, dissolved air flotation, and/or filtration.

The dissolved air flotation (DAF) investigated in this research is of an innovative process [83–87]. The treated effluent containing an almost saturated concentration of dissolved oxygen can be discharged to the lake bottom for hypolimnetic aeration. The use of aluminum sulfate, sodium aluminate, and lime in a flotation system can accomplish phosphate removal, biotic separation, and harvesting. Either sodium aluminate or lime can be used in the flotation system for neutralization of acidity caused by acid rain and acid mine drainage. A DAF plant built on a boat is one method that should be considered for phosphate removal, acid neutralization, or algae harvesting in a lake. Lake restoration may be effective in extending the service life of a lake. Other lake restoration technologies, such as chemical feeding, mixing, acid neutralization, chemical precipitation, coagulation, flocculation, filtration, nitrogen removal processes, dredging, algae harvesting, etc. can all be found from the literature [32, 42, 43, 77–83, 88]. AquaDAF was investigated for lake restoration. The AquaDAF process is introduced in detail in the literature [87]. Although AquaDAF has been demonstrated successfully for lake water treatment, all other manufacturers' DAF equipment, such as Supracell, Sandfloat, KAMET, Clari-DAF, etc., may also be used for successful lake restoration under the conditions that the types and dosages of chemicals must be correct, and the operators must know what they are doing.

Glossary [83–87]

Acid rain	Precipitation having a pH lower than the pH range commonly found in natural waters, caused by absorption from the atmosphere of sulfur dioxide (SO_2) gas and/or other acid-causing substances, such as nitrogen oxides NOx, which then forms sulfuric acid and nitric acid, respectively, in solution. It is also called acid precipitation.
AquaDAF	A rectangular dissolved air flotation clarifier (DAF) manufactured by and commercially available from SUEZ Water Technologies and Solutions, 8007 Discovery Drive, Richmond, VA 23229, USA
Clari-DAF	A rectangular dissolved air flotation clarifier (DAF), manufactured by and commercially available from Xylem Water & Wastewater, 227 S. Division St, Zelienople, PA 16063, USA
Dissolved air flotation (DAF)	One of dissolved gas flotation (DGF) processes when air is used for generation of gas bubbles. See dissolved gas flotation (DGF).
Dissolved gas flotation (DGF)	It is a process involving pressurization of gas at 25–95 psig for dissolving gas into water and subsequent release of pressure (to one atm) under laminar flow hydraulic conditions for generating extremely fine gas bubbles (20–80 microns) which become attached to the impurities to be removed and rise to the water surface together. The impurities or pollutants to be removed are on the water surface are called float or scum which scooped off by sludge collection means. The clarified water is discharged from the flotation clarifier's bottom. The gas flow rate is about one percent of influent liquid flow rate. The attachment of gas bubbles to the impurities can be a result of physical entrapment, electrochemical attraction, surface adsorption, and/or gas stripping. The specific gravity of the bubble-impurity agglomerate is less than one, resulting in buoyancy or non-selective flotation (i.e., Save-All).
KAMET	Krofta Advanced Municipal Effluent Treatment, which is a circular combined Supracell-Sandfloat (DAF-D AFF) package plant manufactured by and commercially available from Krofta-related companies worldwide.
Lake	It is an inland body of water, fresh or salt, of considerable size (usually more than 50 acres (more than 200,000 square meters)) and occupying a basin or hollow on the Earth's surface.
Lake restoration	Clean the lake water or soil in order to restore it to original natural conditions.
LIWT	Lenox Institute of Water Technology, which is a nonprofit educational organization with a goal of scientific development and world peace.
Sandfloat	A circular combined dissolved air flotation and sand filtration package plant (DAFF) manufactured by and commercially available from Krofta Engineering Corporation-related companies worldwide.
Supracell	A circular dissolved air flotation clarifier (DAF) manufactured by and commercially available from Krofta-related companies worldwide.

Appendix A

Pollutant contents as percentage of total solids in dredged materials

	Ashtabula samples				Fairport samples			
Parameters	1A	2A	3A	4A	1F	2F	3F	4F
Chemical oxygen demand	4.334	2.483	1.987	4.037	0.237	16.567	4.567	4.010
Oil and grease	0.210	0.094	0.065	0.265	TRACE	0.275	0.468	0.520
Kjeldahl nitrogen	0.138	0.043	0.048	0.152	0.016	0.065	0.167	0.020
Total phosphorus	0.004	0.001	0.008	0.030	0.007	0.009	0.006	0.003
Mercury	0.00004	0.00001	0.00002	0.00007	0.00001	0.00003	0.00004	0.00006
Iron	2.104	1.750	1.299	2.100	1.105	1.000	1.599	1.790
Chromium	0.008	0.010	0.007	0.014	0.002	0.002	<0.018	0.020
Cadmium	<0.0001	<0.0001	0.0001	<0.0001	0.001	0.0001	<0.0001	<0.0008
Arsenic	<0.0001	<0.00008	<0.0002	<0.0005	<0.00009	<0.0003	<0.0002	<0.0001
Zinc	0.018	0.015	0.013	0.017	0.006	0.006	<0.005	0.008
Lead	0.010	0.010	0.010	0.002	0.002	0.035	0.005	0.009
Copper	0.003	0.001	0.002	0.031	0.0009	<0.0009	0.002	0.003
Nickel	0.001	0.001	0.001	0.001	0.0009	0.009	0.001	0.001
Manganese	0.055	0.045	0.040	0.060	0.035	0.030	0.040	0.040
Volatile solids	4.905	4.150	18.100	—	2.400	—	—	8.499
Dissolved solids	0.093	0.082	0.116	0.693	0.053	0.139	0.188	0.589

Appendix B

Settling-test data, Sample 2A from Ashtabula Harbor, Ohio

	Before settling	Remaining after 1 h of settling	Remaining after 4 h of settling	Remaining after 18 h of settling	Remaining after 40 h of settling
Chemical oxygen demand, mg/L	14,900	110	108.8	81.5	77.6
Total solids, mg/L	600,000	800	—	780	872
Dissolved solids, mg/L	490	490	490	490	490

(continued)

Settling-test data, Sample 2A from Ashtabula Harbor, Ohio					
	Before settling	Remaining after 1 h of settling	Remaining after 4 h of settling	Remaining after 18 h of settling	Remaining after 40 h of settling
Suspended solids, mg/L	599,510	480	—	290	—
Turbidity, JTU[a]	87,500	500	—	150	39
Total phosphorus, mg/L	8.0	0.03	0.09	0.05	0.10
Surfactants, mg/L	—	0.06	—	—	—
Sediment, ml/L	—	600	—	—	—
Kjeldahl nitrogen, mg/L	259	—	29.0	25.1	25.6
Oil and grease, mg/L	565	<1.00	—	nil < 1	nil < 1
Iron, mg/L	10,500	41.00	—	—	—
Chromium, mg/L	60	0.50	0.045	0.020	0.020
Cadmium, mg/L	<0.6	0.04	0.055	0.020	0.022
Zinc, mg/L	90	1.1	0.170	0.095	0.075
Manganese, mg/L	270	3.10	—	0.021	0.0187
Copper, mg/L	8.0	0.12	0.055	0.040	0.020
Nickel, mg/L	<6.0	—	—	<0.015	—
Chlorides, mg/L	—	25.0	—	—	—

[a]*JTU* Jackson Turbidity Unit

Appendix C

Settling-test data, Sample 4F from Fairport Harbor, Ohio (analyzed by Calspan Corporation, Buffalo, NY)					
	Before settling	Remaining after 1 h of settling	Remaining after 4 h of settling	Remaining after 18 h of settling	Remaining after 40 h of settling
Chemical oxygen demand, mg/L	14,540	202	133	97	101
Total solids, mg/L	363,000	4,240	—	2,712	2,460
	2,140	2,140	2,140	2,140	2,140

(continued)

Settling-test data, Sample 4F from Fairport Harbor, Ohio (analyzed by Calspan Corporation, Buffalo, NY)

	Before settling	Remaining after 1 h of settling	Remaining after 4 h of settling	Remaining after 18 h of settling	Remaining after 40 h of settling
Dissolved solids, mg/L					
Suspended Solids, mg/L	360,860	2,130	—	672	320
Turbidity, JTU[a]	155,000	780	—	195	52
Total phosphorus, mg/L	10	0.280	—	0.58	0.16
Surfactants, mg/L	—	0.06	—	—	—
Sediment, ml/L	—	700	—	—	—
Kjeldahl nitrogen, mg/L	696	56.5	—	38.2	—
Oil and grease, mg/L	1,865	4	—	<10	<10
Iron, mg/L	6,500	22.5	—	—	—
Chromium, mg/L	73	0.20	—	0.040	0.035
Cadmium, mg/L	<0.4	0.03	0.052	0.051	0.048
Zinc, mg/L	29	0.15	0.099	0.025	0.025
Manganese, mg/L	145	2.20	—	0.021	0.019
Copper, mg/L	11	0.07	—	0.060	0.070
Nickel, mg/L	<3.6	—	—	<0.015	—
Chlorides, mg/L	—	284	—	—	—

[a]*JTU* Jackson Turbidity Unit

Appendix D

Dredging supernatant water quality versus US drinking water standards

Parameters	Water quality after 18 h of settling (mg/L)		Proposed federal drinking water standards (mg/L)
	Sample 2A Ashtabula	Sample 4F Fairport	
Oil and grease	nil < 1	<10	absent

(continued)

Parameters	Water quality after 18 h of settling (mg/L)		Proposed federal drinking water standards (mg/L)
	Sample 2A Ashtabula	Sample 4F Fairport	
Chromium	0.020	0.040	0.5
Cadmium	0.020	0.051	0.01
Zinc	0.095	0.025	5.0
Copper	0.040	0.060	1.0

Appendix E

Control technology summary for flotation (Source: US Environmental Protection Agency, September 1981)

Pollutant	Effluent Concentration		Removal efficiency %	
	Range	Median	Range	Median
Classical pollutants, mg/L:				
BOD (5-day)	140–1000	250	4–87	68
COD	18–3200	1200	8–96	66
TSS	18–740	82	6–98	88
Total phosphorus	<0.05–12	0.66	50–>99	98
Total phenols	> 0.001–23	0.66	3–>94	12
Oil and grease	16–220	84	57–97	79
Toxic pollutants, μg/L:				
Antimony	ND-2300	20	4–95*	76
Arsenic	ND-18	<10	8–>99	45
Xylene	ND-1000	200	95–>99	97
Cadmium	BDL-< 72	3	0–>99	98*
Chromium	2–620	200	20–99	52
Copper	5–960	180	9–98	75
Cyanide	<10–2300	54	0–<62	10
Lead	ND-1000	70	9–>99	98
Mercury	BDL-2	BDL	33–88	75
Nickel	ND-270	41	29–>99	73
Selenium	BDL-8.5	2		NM
Silver	BDL-66	19		45
Thallium	BDL-50	14		NM
Zinc	ND-53000	200	12–>99	89
Bis(2-ethylhexyl) phthalate	30–1100	100	10–98	72
Butyl benzyl phthalate	ND-42	ND	97–>99	>99
Carbon tetrachloride	BDL-210	36		75
Chloroform	ND-24	9	20–>99	58
Dichlorobromomethane		ND		>99

(continued)

Control technology summary for flotation (Source: US Environmental Protection Agency, September 1981)

	Effluent Concentration		Removal efficiency %	
Pollutant	Range	Median	Range	Median
2,4-Dichlorophenol		6		NM
Di-n-butyl Phthalate	ND-300	20	0–>99	97
Diethyl phthalate		ND		>99
Di-n-octyl phthalate	ND-33	11	61–>99	78
N-nitrosodiphenylamine		620		66
N-nitroso-di-n-propylamine		84		NM
2-Chlorophenol		2		NM
2,4-Dimethylphenol	ND-28	14		>99
Pentachlorophenol	5–30	13		19
Phenol	9–2400	71	0–80	57
2,4,6-Trichlorophenol		3		NM
Benzene	5–200	120		NM
Chlorobenzene		57		NM
Dichlorobenzene	18–260	140		76
Ethylbenzene	ND-970	44	3–>99	65
Toluene	ND-2100	580	10–>99	39
Fluoranthene	0.5-<10	5.2		NM
Fluorene		14		NM
Naphthalene	ND-840	96	33–>99	77
Pyrene	0.3–18	9.2		0
Anthracene/phenanthrene	0.2–600	10	45–>98	81
2-Chloronaphthalene		17		0

Blanks indicate data not available; BDL, below detection line; ND, not detected; NM, not meaningful; *, approximate value

Appendix F

Control technology summary for filtration

	Data points		Effluent concentration		Removal efficiency, %	
Pollutant	Pilot scale	Full scale	Range	Median	Range	Median
Classical pollutants, mg/L:						
BOD(5)	3	6	5.3–23,000	23	8–48	22
COD	11	9	29–260,000	150	0–59	24
TSS	19	15	<1–7,300	17	6–>99	76
TOC	8	6	17–25,000	43	6–49	14
Total phosphorus	10	4	<0.07–14	1.1	7–86	52

(continued)

Control technology summary for filtration

Pollutant	Data points		Effluent concentration		Removal efficiency, %	
	Pilot scale	Full scale	Range	Median	Range	Median
Total phenols	15	9	0.001–64	0.02	0–60	23
Oil and grease	7	10	<0.5–9,900	12	0–>98	39
Fluoride		3	2.2–11	9.0	0–63	32
Aluminum	10	4	0.04–47	0.5	15–99	34
Manganese	11	5	0.007–3.2	0.12	0–88	15
Vanadium	10	17	0.001–0.52	0.02	0–67	4
Barium	10	3	<0.001–11	0.01	0–>99	24
Iron	11	4	0.06–100	0.34	0–92	32
Tin	8	3	0.02–0.48	<0.02	0–56	3
Titanium	9	4	>0.001–15	0.004	0–99	23
TS		1		12,000		45
TDS		1		4,900		36
Sulfides	4	1	<0.005–<0.2	0.01	0–94	50
Palladium		1				NM
Calcium	9	4	4.5–800	7.6	0–98	13
Magnesium	10	3	1–37	3.7	0–62	6
Sodium	9	3	54–400	180	0–6	0
Molybdenum	10	2	0.01–0.2	0.01	0–0	0
Cobalt	10	3	0.006–0.11	0.007	10–70	20
Boron	10	1	0.006–2	0.27	0–33	0
Silicon	10		1.3–15	3.4	0–48	3
Ammonia	5	2	0.22–23	2.6	0–99	21
Nitrate	5	2	0.12–6.5	3.4	0–52	10
Strontium	9	1	0.02–0.26	0.05	0–29	0
Tellurium		1		<0.004		NM
Ammonia nitrogen	1			0.58		17
Nitrate nitrogen	1			0.25		11
Yttrium	1			<0.003		<50
Platinum	1			<0.005		NM
Gold	1			<0.002		NM
TVS	1			7,500		38
Toxic pollutants, μg/L:						
Antimony	13	13	BDL – 700	<25	0–92	>44
Arsenic	14	8	BDL – 100	7	0–>99	55
Asbestos, total (fibers/L)	1			4.7E8		90
Asbestos, chrysotile (fibers/L)	1			3.3E8		>99
Beryllium	17	1	<0.02–<10	1.1	0–7	71
Cadmium	15	14	ND – 97	<2	0–>99	67

(continued)

Control technology summary for filtration						
	Data points		Effluent concentration		Removal efficiency, %	
Pollutant	Pilot scale	Full scale	Range	Median	Range	Median
Chromium	17	15	<4–320	34	0–99	43
Copper	17	18	<4–4,500	32	0–>99	42
Cyanide	11	11	3–260	11	0–>99	54
Lead	19	18	BDL – 2,100	50	3–>99	47
Mercury	8	5	0.1–2,900	0.5	0–89	57
Nickel	17	13	BDL – 700	36	0–>99	19
Selenium	10	6	BDL – 100	9	0–10	0
Silver	15	12	BDL – <100	<5	0–91*	17
Thallium	11	5	0.1–<50	<15	NM	NM
Zinc	20	18	16–18,000	160	0–>99	55
Hexavalent chromium	2		20–20	20	0	0
Bis(2-ethyl hexyl) phthalate	9	12	BDL – 16,000	29	21–98	77
Butyl benzyl phthalate	3	4	ND – 10	2.5	52– >99	77
Di-n-butyl phthalate	7	12	ND – 1,300	6	0–96	11
Diethyl phthalate	3	4	ND – 11,000	1	50– >99	61
Dimethyl phthalate	2	1	ND – BDL	BDL	99*– >99	99*
Di-n-octyl phthalate		3	ND – 4	2	50– >99	64
Xylene		1		12,000		75
N-nitrosodi phenylamine	1	1	ND – 0.4	0.2	>99	>99
N-nitroso-di-n-propylamine	1	1	ND – BDL	0.05	>99	>99
2-Chiorophenol		2	2–17	10	0	0
2, 4-Dichlorophenol	3	2	ND – 2	BDL	67– >99	83
2, 4-Dimethylphenol	2	1	BDL – 29	0.9	NM	NM
Pentachlorophenol	2	4	ND – 12	10	>99	>99
Phenol	8	4	ND – 34,000	5	22– >99	32
2, 4, 6-Trichiorophenol		1		69		80
p-chloro-m-cresol	3		BDL – 1.1	0.6	NM	NM
Benzene	4	4	ND – 200	6.9	29– >99	41
Chlorobenzene	2	1	0.1–470	4.8	98	98
1, 2-Dichlorobenzene	4		BDL – 5.8	3.5	44–97*	55
Ethylbenzene	9	2	ND – 2	BDL	0–>99	89*

(continued)

Control technology summary for filtration

	Data points		Effluent concentration		Removal efficiency, %	
Pollutant	Pilot scale	Full scale	Range	Median	Range	Median
Nitrobenzene		1		ND		>99
Toluene	10	10	ND – 200	1.2	0–>99	67
1,2,4-Trichlorobenzene	1	2	ND – 94	84	3.7	3.7
Acenaphthene	1	2	ND – 10	0.6	73–>99	86
Acenaphthylene		1		500		NM
Anthracene	6	4	ND – <32,000	0.5	0–99*	59
Benz(a)anthracene		1		7,300		NM
Benzo(a)pyrene	2		0.2–0.8	0.5	NM	NM
Benzo(k) fluoranthene	1			0. 1		NM
Fluoranthene	3	2	0.05–93	0.2	20–50	29
Fluorene	2	1	BDL – 10,000	0.05	NM	NM
Naphthalene		5	BDL – 160	1.5	86	86
Phenanthrene	2	4	ND – 3, 200	5.3	67	67
Pyrene	3	2	0.09–32,000	0.3	0–10	0
Anthracene/phenanthrene		3	0.1–3.5	2	50–65	58
Aroclor 1016		1		480		16
Aroclor 1221		1		650		20
Aroclor 1232		1		480		16
Aroclor 1424		1		650		20
Aroclor 1248		1		480		16
Aroclor 1254		1		650		20
Aroclor 1260		1		480		16
Carbon tetrachloride		3	ND-55	15	89–>99	93

Appendix G

US Army Corps of Engineers Civil Works Construction Yearly Average Cost Index for Utilities*

Year	Index	Year	Index
1967	100	1995	439.72
1969	112.17	1996	445.58
1970	119.75	1997	454.99
1971	131.73	1998	459.40

(continued)

Year	Index	Year	Index
1972	141.94	1999	460.16
1973	149.36	2000	468.05
1974	170.45	2001	472.18
1975	190.49	2002	486.16
1976	202.61	2003	497.40
1977	215.84	2004	563.78
1978	235.78	2005	605.47
1979	257.20	2006	645.52
1980	277.60	2007	681.88
1981	302.25	2008	741.36
1982	320.13	2009	699.70
1983	330.82	2010	720.80
1984	341.06	2011	758.79
1985	346.12	2012	769.30
1986	347.33	2013	776.44
1987	353.35	2014	791.59
1988	369.45	2015	786.32
1989	383.14	2016	782.46
1990	386.75	2017	807.94
1991	392.23	2018	827.52
1992	399.07	2019	844.07
1993	410.63	2020	860.96
1994	424.91		

*US ACE, Yearly Average Cost Index for Utilities. In: *Civil Works Construction Cost Index System Manual*, 1110-2-1304, US Army Corps of Engineers, Washington, DC. Amendment #1 30 September 2017. DF file is available on the Internet at http://www.publications.usace.army.mil/Portals/76/Publications/EngineerManuals/EM_1110-2-1304.pdf?ver=2017-11-17-073237-627 (2018).

Acknowledgment and Dedication This book chapter is one of many book chapters and technical papers published in memory of our co-author, Dr. Donald B. Aulenbach, who worked closely with Professors Lawrence K. Wang, Mu-Hao Sung Wang, Nazih K. Shammas, William A. Selke, Milos Krofta, and Daniel Guss from 1981 to 2019 for developing an environmental humanitarian engineering program at the Lenox Institute of Water Technology (LIWT), teaching/researching there as an adjunct professor for almost two decades, and publishing many research papers and textbooks for academic contribution to the humanity. The senior author, Dr. Lawrence K. Wang, was also a colleague with Drs. Donald B. Aulenbach, Lenore S. Clesceri, and Nicholas L. Clesceri at Rensselaer Polytechnic Institute (RPI), Troy, NY, USA, in 1973–1978, where/when the happy RPI faculty team conducted so many lake investigations together. Dr. Aulenbach's lifetime accomplishments at both LIWT and RPI are duly acknowledged. The following is a Warm Tribute from RPI faculty representatives:

A Warm Tribute to Our Dear Friend and Colleague, Professor Emeritus Donald Bruce Aulenbach (Aulie) from Professor Emerita Lenore S. Clesceri and Professor Emeritus Nicholas L. Clesceri of Rensselaer Polytechnic Institute, Troy, NY, USA

When we returned from our post-docs in Switzerland in 1965, to begin teaching at Rensselaer, Don was among the first friends we developed. His help and friendship will always be highly valued. His good humor is legendary.

He was very involved in the research at Lake George and the development of the Lake George Water Research Center (LGWRC) with Nick.

Reflecting on the early days, Don and Nick made quite a few trips to Gull Bay in Putnam County on the eastern shore of Lake George, specifically, to Smith Bay. It seemed that each time we ventured forth, we encountered snow, which we could have anticipated at that time of the year. But, Nick and Don never faltered! Lots of Fun...

As things started to take shape, President Richard Folsom and Provost Clayton Dohrenwend encouraged us to see what might occur. Nick and Don had many meetings with them, keeping them "in the loop." Then, Professor Steve Wiberley (and later, Provost) was tasked by Provost Dohrenwend to work with us on developing the LGWRC. As fate would have it, Steve viewed the Gull Bay area for suitability for a fledgling LGWRC with Nick and Don...In a SNOW STORM!

Through President Folsom, we made connections with RPI Board of Trustee Member, Mr. Harold Strang, retired GE VP, who lived in summers at Smith Bay. In addition, Mr. Strang owned an adjacent former camp for youngsters. Mr. Strang offered his camp as a location to initiate LGWRC research. As was standard, and expected, at that stage, Nick and Don were the principle salesmen.

BUT, this historical sojourn would take many more pages to describe; however, a few things alert the brain cells, namely, "around-the-clock" lake water sampling, "through-the-ice" water sampling, and stream sampling, to name just a few things that Don did so well.

Bringing things to the present, this ultimately led to RPI's Darrin Fresh Water Institute, located in Bolton Landing, NY, on the western shore of Lake George.

Don's research was far-ranging, from Radiological Health, to Public Health to his forward-thinking work with his students on the Village of Lake George Wastewater Treatment Facility. He literally probed the subsurface as he studied the sand infiltration beds, which functioned as tertiary treatment for this system. This research placed Don in a small cadre of fellow researchers from around the world who were seeking understanding of sand infiltration capability for tertiary treatment. He collaborated with them and presented at specialty conferences.

Lastly, Don's contributions in the Department's educational endeavors were manifold. He was a key figure in the Department gaining the first-in-country ABET accredited degrees in Environmental Engineering. Nick and Don team-taught a number of courses, which was always constructive for the Department's reputation. Don also taught a Limnology course at the Lake George facility, which Lenore inherited.

On a personal note from Nick, in addition to the brief notes above, Don was a Father, Family man, Educator, Patriot and God-Fearing human being. He faithfully served his Church in Clifton Park, NY, for decades. How full a life is that!!

I consider him a BROTHER.

Aulie, Rest In Peace, dear friend.

References

1. Sawyer CN (1966) Basic concepts of eutrophication. J Water Pollution Control Fed 38(5):737
2. Sakamoto M (1966) Primary production by phytoplankton community in some Japanese lakes and its dependence on lake depth. *Arch Hydrobiol* 62:1

3. Vollenweider RA (1968) The scientific basis of lake and stream eutrophication, with particular reference to phosphorus and nitrogen as factors in eutrophication, Technical Report to OECD, Paris, DAS/CSI/68.27, 1
4. Vollenweider RA, Dillon PJ (1974) The application of the phosphorus loading concept to eutrophication research. Canada Centre for Inland Waters, NRCC 13690, Burlington
5. Phelps EB (1944) *Stream sanitation*. Wiley, New York
6. Wilson TE (2006) Upgrades to BNR: optional approaches, Earth Tech, Alexandria, VA, Presented at 78th NY Water Environment Association Conference
7. Eberhardt TE (1997) Aluminum dose calculation. Sweetwater Technology, Aitkin
8. Tabor CW (2006) Comparison of high rate clarifiers in tertiary phosphate removal applications, Stantec Consulting, Inc., Richmond, VA, Presented at 78th NY Water Environment Association Conference, USA
9. Patoczka J (2006) Reducing phosphorus to 0.1 mg/L with existing facilities at the East Windsor Plant", Hatch Mott MacDonald, Millburn, NJ, Presented at 78th NY Water Environment Association Conference, NY, USA
10. FHA (2006) Stormwater best management practices in an ultra-urban setting: selection and monitoring. Federal Highway Administration, USA
11. Harper HH (1990) Long-term performance evaluation of the alum stormwater treatment system at Lake Ella, FL, Final Report for FDER Project WM339
12. Pitt RF (1996) The control of toxicants at critical source areas. Proceedings of effects of watershed development and management on aquatic ecosystems symposium, Engineering Foundation
13. Allard LA, Graham W, Platz W, Carr R, Wheeler J (1996) The storm treat system used as a storm water best management practice. Proceedings of Watershed '96
14. Claytor RA, Schueler TR (1996) Design of stormwater filtering systems. The Center for Watershed Protection, Silver Spring, USA
15. Sheepdrove Organic Farm – Sustainable Water Management Project, Winner of National Farmers Union, UK (2006)
16. Aulenbach DB (1978) Chemical interactions in a rapid infiltration system accomplishing tertiary treatment of wastewaters. In: Rubin AJ (ed) Chemistry of wastewater technology. Ann Arbor Science Publishers, MI, USA, pp 227–244
17. Aulenbach DB, Clesceri NL, LaDuc M, Peachy B, Sliwinski R (1982) Monitoring a septic tank Leaching field. In: Schadler MH (ed) The Lake George ecosystem, Lake George Association, New York, vol II, pp 142–149
18. US EPA (1980) Our Nation's Lakes. US Environmental Protection Agency, Washington, DC, USA. EPA 440/5-80-009
19. NYS Museum (1986) Understanding New York's Lakes. New York State Museum, Education Leaflet 26, Albany, NY, USA
20. Barko JW, Adams MS, Clesceri NL (1986) Environmental factors and their consideration in the management of submersed aquatic vegetation: a review. J Aquat Plant Manage 24:1
21. Carpenter SR, Adams MS (1977) The macrophyte tissue nutrient pool of a hardwater eutrophic lake: implications for macrophyte harvesting. Aquat Bot 3:239
22. EOA (1980) Restoration of Lake Temescal, EOA-625/2-80-026
23. George CJ, Tobiessen PL, Snow D, Jewell TK (1982) The monitoring of the restorational dredging of Collins Lake, Scotia, New York, USA. EPA-600/S3-81-017
24. US EPA (1980) Restoration of Medical Lake. US Environmental Protection Agency, Washington, DC, USA. EPA-625/2-80-025
25. US EPA (1980) Lake restoration in Cobbossee Watershed. US Environmental Protection Agency, Washington, DC, USA. EPA-625/2-80-027
26. Myers J, Christensen M, Bleth J, Tormaschy W (2005) Evaluation of phosphorus reduction in wastewater ponds by injecting aluminum sulfate (alum) into a high flow solar powered circulator. Water Environ Technol 17(12):99

27. Clasen J, Bernhardt JH (1987) Chemical methods of p-elimination in the tributaries of reservoirs and lakes. *Schwez Z Hydrol* 49:249
28. Kittler RDI (1997) The use of dissolved air flotation to reduce the internal phosphorus supplies in Eutrophic Devils Lake, WI, Masters Thesis, Lenox Institute of Water Technology, Lenox, MA, USA
29. WDNR (2018) Great Lakes restoration initiative. Wisconsin Department of Natural Resources, Madson. http://dnr.wi.gov/topic/greatlakes/restore.html
30. Lillie RA, Manson JW (1986) Historical changes in water quality and biota of Devils lake, Sauk County, 1866–1985. Trans Wis Acad Sci Arts Lett 74:81
31. City of Waco (2018) Lake Brazos Dam... A more reliable water level. City of Waco Water Utility Services, Waco, USA. http://www.wacowater.com/news.html
32. Krofta M, Wang LK (2000) Flotation engineering, 1st edn. Lenox Institute of Water Technology, Lenox, MA, USA. Technical Manual No. Lenox1-06-2000/368
33. Ondeo-Degremont, Inc. (2018) Water treatment case study. Rusko Water Treatment Works, Tampere, Finland, Richmond, www.onedo-degremont-usa.com
34. CGER (2018) Restoration of aquatic ecosystems: science, technology, and public policy. Commission on Geosciences, Environment and Resources (CGER), http://www.nap.edu/books/0309045347/html/395.html
35. Schneider RF, Little JA (1973) Rise and fall of Lake Apopka: A case study in reservoir mismanagement. In: Ackerman C (ed) Man-made lakes: their problems and environmental effects. American Geophysical Union, Washington, DC, pp 690–694
36. Lowe EF, Adams J, Bateman J, Fell C, Graetz D, Hulbert J, Ingram W, Johnson W, Muller P, Nall L, Reddy R, Snyder L, Williams V (1985) Proposed scope of work for the Lake Apopka Pilot Project, Report, p 55
37. Brezonik PL, Pollman CD, Crisman TC, Allenson JN, Fox JL (1978) Limnological and water quality studies on Lake Apopka and Downstream Lakes in The Upper Oklawaha River Basin, historical trends and current status, Report to Florida Department of Environmental Regulation, Env-07-78-01. Department of Environmental Engineering Science, University of Florida, Gainesville, FL, USA, p 283
38. Amasek, Inc. (1985) A practical concept for the restoration of Lake Apopka, Report by Amasek, Inc., Cocoa, p 38
39. Lowe EF, Stites DL, Battoe LE (1989) Potential role of marsh creation in restoration of hypereutrophic lakes. In: Hammer DA (ed) Constructed wetlands for wastewater treatment: municipal, industrial and agricultural. Lewis Publishers, Chelsea, USA, pp 710–717
40. Dierberg FE, Williams VP, Schneider WH (1988) Evaluating water quality effects of lake management in Florida. Lake Reservoir Manage 4:101–111
41. United Water (2018) Surface water treatment processes. Lake DeForest in Clarkstown, Located at Web Site: https://www.unitedwater.com/LocalNavigation.aspx
42. US EPA (2018) Great Lakes Sheboygan River area of concern. US Environmental Protection Agency, Washington, DC, USA. www.epa.gov
43. Wang LK, Leonard RP (1976) Dredging pollution and environmental conservation in the United States. Environ Conserv 3(2):123–130. Switzerland
44. Wang LK (1982) Compact nitrogen removal options for Lake Como. Effluent Water Treat J 22 (11):435–437
45. Krofta M, Wang LK (1984) Algae separation, phosphate removal and acid neutralization for lake water purification. Lenox Institute of Technology (formerly Lenox Institute of Water Technology), Lenox, MA, USA
46. Krofta M, Wang LK (1984) Tertiary treatment of secondary effluent by dissolved air flotation and filtration. Civ Eng Pract Des Eng 3:253–272
47. United States Environmental Protection Agency (U.S. EPA) (2017) Acid rain in New England, http://www3.epa.gov/region1/eco/acidrain
48. Likens GE, Keene WC, Miller JM, Galloway JN (*1987*) Chemistry of precipitation from a remote, Terrestrial Site in Australia. J Geophys Res *92:13299*
49. Likens GE, Bormann FH, Johnson NM (1972) Acid rain. *Environment* 14(2):33–40

50. Weathers KC, Likens GE (2006) Acid rain. In: Rom WN, Markowitz S (eds) Environmental and occupational medicine, 4th edn. Lippincott-Raven, Philadelphia, pp 1549–1561
51. Likens GE, Bormann FH (1974) Acid rain: a serious regional environmental problem. *Science* 184(4142):1176–1179
52. Chandru, CHINA: industrialization pollutes its countryside with acid rain, *Southasiaanalysis.org,* 2006-09-09. Accessed Feb 2016
53. Likens GE, Wright RF, Galloway JN, Butler TJ (1979) Acid rain. *Sci Am* 241(4):43–51
54. National Geographic, Acid rain*:* effects felt through the food chain. Web: http://environment.nationalgeographic.com/environment/global-warming/acid-rain-overview/. Accessed Feb 2017
55. Lackey RT (1997) Science, policy, and acid rain: lessons learned. Renew Resour J 15(1):9–13
56. Winstanley D, Lackey RT, Warnick WL, Malanchuk J (1998) Acid rain: science and policy making. *Environ Sci Pol* 1:51
57. United States Environmental Protection Agency (U.S. EPA) (2000) Deposition of air pollutants to the great waters. 3rd Report to congress 2000, according to the Clean Air Act Amendments of 1990, 42 U.S. Code 7651
58. Environmental Defense Fund (EDF) (2017) Acid rain pollution solved using economics. Web: https://www.edf.org/approach/markets/acid-rain?utm_source=ggad&utm_medium=cpc&utm_campaign=gr-AcidRain&gclid=CNbkmOTN-8oCFY4lgQodFxgO4w
59. United States Environmental Protection Agency (U.S. EPA) (2005) Clean Air Interstate Rule (CAIR), http://archive.epa.gov/airmarkets/programs/cair/web/html/index.html. Accessed Feb 2017
60. United States Environmental Protection Agency (U.S. EPA) (2015) Cross-State Air Pollution Rule (CSAPR), http://www3.epa.gov/crossstaterule/. Accessed Feb 2017
61. United States Environmental Protection Agency (U.S. EPA) (2006) Acid rain and related programs. Progress report to congress, EPA-430-R-07-011. Office of Air and Radiation, Clean Air Markets Division, Washington, DC
62. United States Environmental Protection Agency (U.S. EPA) (2017) Reducing acid rain. EPA. http://www3.epa.gov/region1/eco/acidrain/reducing.html. Accessed Feb 2017
63. United States Environmental Protection Agency (U.S. EPA) Acid rain permit program, http://www3.epa.gov/region1/eco/acidrain/permitsprog.html, Accessed February 2017
64. United States Environmental Protection Agency (U.S. EPA) Clean air market, http://www.epa.gov/airmarkets. Accessed Feb 2017
65. United States Environmental Protection Agency (U.S. EPA) Title V operating permits, http://www3.epa.gov/region1/eco/permits/title5/index.html. Accessed Feb 2017
66. United States Environmental Protection Agency (U.S. EPA) Power plants in New England in the Acid Rain Program – 2011, http://www3.epa.gov/region1/eco/acidrain/powerplants.html. Accessed Feb 2017
67. United States Environmental Protection Agency (U.S. EPA), 2013 Program progress – clean air interstate rule. Acid Rain Program, and Former NO_x Budget Trading Program, http://www3.epa.gov/airmarkets/progress/reports/index.html. Accessed Feb 2017
68. United States Environmental Protection Agency (U.S. EPA) Acid rain allowances, http://www3.epa.gov/region1/eco/acidrain/allowances.html. Accessed Feb 2017
69. United States Environmental Protection Agency (U.S. EPA) (2017) Acid rain measurement, http://www3.epa.gov/region1/eco/acidrain/measurement.html
70. United States Environmental Protection Agency (U.S. EPA) Trends, http://www3.epa.gov/region1/eco/acidrain/trends.html. Accessed Feb 2017
71. United States Environmental Protection Agency (U.S. EPA) Acid rain and other pollution problems, http://www3.epa.gov/region1/eco/acidrain/pollproblems.html. Accessed Feb 2017
72. United States Environmental Protection Agency (U.S. EPA) EPA's Regional Haze Program, http://www3.epa.gov/visibility/program.html. Accessed Feb 2017
73. United States Environmental Protection Agency (U.S. EPA) Ground-level ozone, http://www3.epa.gov/ozonepollution/. Accessed Feb 2017

74. United States Environmental Protection Agency (U.S. EPA) Energy and global climate change in New England, http://www3.epa.gov/region1/eco/energy/index.html. Accessed Feb 2017
75. United States Environmental Protection Agency (U.S. EPA) Acid rain in New England, http://www3.epa.gov/region1/eco/acidrain/future.html. Accessed Feb 2017
76. United States Environmental Protection Agency (U.S. EPA) Energy star: the simple choice for energy efficiency, http://www.energystar.gov/?_ga=1.86346840.176297172.1454363977. Accessed Feb 2017
77. Wang LK, Shammas NK, Selke WA, Aulenbach DA (eds) (2010) Flotation technology. Humana Press, Totowa, NJ, USA, p 680
78. Shammas NK, Wang LK (2016) Water engineering: hydraulics, distribution and treatment. Wiley, Hoboken, NJ, USA, p 806
79. US EPA (2018) Acid rain program. US Environmental Protection Agency, Washington, DC, USA. www.epa.gov
80. US EPA (2018) Consumer factsheet on nitrates/nitrites. US Environmental Protection Agency, Washington, DC, USA. www.epa.gov
81. CFD (2018) Phosphorus fact sheet. City of Fond deLac, WI, USA. https://www.fdl.wi.gov/cofuploads/FACT_SHEET.pdf
82. US EPA (1976) Process design manual for phosphorus removal. US Environmental Protection Agency, Washington, DC, USA. EPA-625/1-76-001a
83. Wang MHS, Wang LK (2015) Environmental water engineering glossary. In: Yang CT, Wang LK (eds) Advances in water resources engineering. Springer Science + Business Media, USA, pp 471–556
84. Wang LK, Wang MHS (2021) A new wave of flotation technology advancement for wastewater treatment. In: Wang LK, Wang MHS, Shammas NK (eds) Environmental flotation engineering. Springer, Nature Switzerland
85. Wang LK (2021) Humanitarian engineering education of the Lenox Institute of Water Technology and its new potable water flotation processes. In: Wang LK, Wang MHS, Shammas NK (eds) Environmental flotation engineering. Springer, Nature Switzerland
86. Wang LK, Wang MHS, Shammas NK, Holtorff MS (2021) Independent physicochemical wastewater treatment system consisting of primary flotation clarification, secondary flotation clarification and tertiary treatment. In: Wang LK, Wang MHS, Shammas NK (eds) Environmental flotation engineering. Springer, Nature Switzerland
87. Wong JM, Hess RJ, Wang LK (2021) Operation and performance of the AquaDAF process system for water purification. In: Wang LK, Wang MHS, Shammas NK (eds) Environmental flotation engineering. Springer, Nature Switzerland
88. Wong JM, Farmerie JE, Wang LK (2021) Operation and performance of the Clari-DAF process system for water purification. In: Wang LK, Wang MHS, Shammas NK (eds) Environmental flotation engineering. Springer, Nature Switzerland

Chapter 8
Biohydrogen Production Through Mixed Culture Dark Anaerobic Fermentation of Industrial Waste

Abdollah Hajizadeh, Noori M. Cata Saady, Sohrab Zendehboudi, Rajinikanth Rajagopal, and Yung-Tse Hung

Contents

1 Introduction 325
1.1 Quantities of the Organic Waste Suitable for H_2 Production 327
1.2 General Description of Anaerobic Digestion 328
1.3 Mixed Culture Hydrogen Dark Fermentation 332
2 Metabolic Pathways of Dark Hydrogen Fermentation 333
3 Industrial Wastewater Characterization 335
3.1 Food Waste 335
3.2 The Organic Fraction of Municipal Solid Waste (OFMSW) 338
3.3 Food Processing Wastewater 338
3.4 Biodiesel Industrial Waste 344
3.5 Dairy Wastewater and Cheese Whey 344
3.6 Manure from Livestock Industry 346
4 Inocula for Hydrogen Production from Waste 347
4.1 Sources of Inocula 347
4.2 Culture Pretreatment 350
4.3 Organic Loading Rate (OLR) 352
4.4 Comparative Studies on the Effects of Pretreatments 352

A. Hajizadeh · S. Zendehboudi
Department of Process Engineering, The Memorial University of Newfoundland, St. John's, NL, Canada
e-mail: ahajizadeh@mun.ca; szendehboudi@mun.ca

N. M. C. Saady (✉)
Department of Civil Engineering, The Memorial University of Newfoundland, St. John's, NL, Canada
e-mail: nsaady@mun.ca

R. Rajagopal
Agriculture and Agri-Food Canada, Science and Technology Branch, Sherbrooke Research and Development Centre, Quebec, QC, Canada
e-mail: rajinikanth.rajagopal@canada.ca

Y.-T. Hung
Department of Civil and Environmental Engineering, Cleveland State University, Cleveland, OH, USA

L. K. Wang, M. -H. S. Wang, Y. -T. Hung, N. K. Shammas (eds.), *Integrated Natural Resources Management*, Handbook of Environmental Engineering 20,
https://doi.org/10.1007/978-3-030-55172-8_8

5 Bioreactors Types and Configuration for H_2 Production 354
6 Cost of Dark Fermentative H_2 Production from Industrial Waste 355
7 Future Research on Biohydrogen Production from Industrial Waste Dark Fermentation 356
8 Conclusions 356
Glossary 357
References 357

Abstract Industrial organic waste from food processing, livestock production, brewery, bakery, and other related industries is a renewable substrate for anaerobic digestion to produce methane (CH_4) or with some process manipulation and control to produce hydrogen (H_2). Type of waste, its strength, presence of any toxic compounds, and other specific characteristics affect the operating conditions such as organic loading rate, hydraulic retention time, substrate pretreatment, as well as the yield and the rate of H_2 production from industrial waste. Therefore, they need to be optimized for each waste. Research is required on the modeling, cost analysis, economic evaluation, comparative studies about the effect of bioreactor design, as well as on combining several industrial wastes to prepare a well-balanced substrate for H_2-producing mixed culture dark fermentation.

Keywords Hydrogen · Industrial waste · Mixed cultures · Dark fermentation

Nomenclature

AnSBBR	Anaerobic sequencing batch biofilm reactor
AFBR	Anaerobic fluidized bed reactor
ASBR	Anaerobic sequence batch reactor
BESA	2-Bromoethanesulfonic acid sodium salt
BOD	Biochemical oxygen demand
CH_4	Methane
CO_2	Carbon dioxide
COD	Chemical oxygen demand
CSTR	Continuous stirred tank reactor
EGSBR	Expanded granular sludge bed reactor
GHG	Greenhouse gas
H_2	Hydrogen
HCl	Hydrochloric acid
HRT	Hydraulic retention time
LBR	Leaching bed reactor
LCFAs	Long-chain fatty acids
MCF	Mixed culture anaerobic fermentation
NAD	Nicotinamide adenine dinucleotide
NaOH	Sodium hydroxide
OFMSW	Organic fraction of municipal solid waste
OLR	Organic loading rate

OMW	Olive mill waste
P_{H_2}	Hydrogen partial pressure
POME	Palm oil mill effluent
SBR	Sequencing batch reactor
UASB	Upflow anaerobic sludge blanket
VFAs	Volatile fatty acids
VS	Volatile solids
VSS	Volatile suspended solids

1 Introduction

Recovering energy and nutrients from industrial waste enhances the sustainability of the treatment methods used to mitigate the environmental impact of such waste. Converting organic industrial wastes to hydrogen (H_2) contributes to pollution control and economic gain (resources recovery and offset part of waste management cost by revenue generated from biofuel production).

Biological treatment methods use biocatalyst to stabilize organic waste. These methods are reliable, cost-effective, and environmental-friendly. Anaerobic digestion is the most important biological process which could treat waste and generate bioenergy. Energy could be extracted from the anaerobic digestion of domestic, industrial, and agricultural waste as methane (CH_4) or with slight operation control as hydrogen (H_2). H_2 is more energetic than CH_4 (based on the energy content per unit mass), and it does not produce CO_2 when combusted. The latter characteristic of H_2 is advantageous and allows for more efficient CO_2 capture and sequestration at the fermentation site; therefore, H_2 is preferred over CH_4 as the end product of anaerobic digestion. The energy yield of anaerobic digestion is increased when H_2 is recovered instead of CH_4. To be sustainable, all the components of the H_2 production process have to be sustainable. Therefore, the feedstock, biocatalyst (inoculum), and any other input material should be obtained from renewable and sustainable sources. Organic industrial wastes from processes such as food processing, brewery, winery, slaughterhouses, bakery, etc. are ideal substrates for H_2 production. Anaerobic mixed culture can easily be obtained from many natural environments such as agricultural soil, animal dung, and/or man-made environments such as compost and anaerobic digesters in municipal and industrial wastewater treatment plants. The use of anaerobic mixed culture to digest organic industrial waste closes the cycle and makes the H_2 production process more sustainable. This chapter provides the current state of the art of H_2 production through anaerobic mixed culture dark fermentation of industrial waste.

Hydrolytic and fermentative bacteria can convert waste rich in carbohydrates to H_2 and short-chain volatile fatty acids. Therefore, utilizing carbohydrate-containing waste for biohydrogen production via mixed-culture dark fermentation has been investigated frequently. The wastes from food and agro-product industries such as

Table 8.1 Characteristics of some typical food industrial wastewater used for $BioH_2$ production

Industry wastewater	BOD (g/L)	COD (g/L)	References
Food processing[a]	0.6–4.0	1.0–8.0	[1]
Palm oil mill effluent	NR	50	[2]
Sugar beet processing	NR	6.6	[3]
Dairy	0.65–6.3	0.4–15.2	[4]
Dairy processing effluents	NR	0.2–23	[5]
Whey wastewater	35–45	65–80	[6]
Raw cheese whey	NR	68.8	[6]
Ice cream factory	NR	9.5	[7]
Palm oil mill effluent	10–44	15–100	[6]
Cheese processing plant	63.3	NR	[6]
Rice winery wastewater	18.7	35.4	[8]
Sugar factory wastewater	NR	6–30	[9]
Cassava starch wastewater	NR	8.3–10	[10]
Vinasse	17–50	20–60	[11]
Sugar mill press mud	NR	86.9	[12]
Corn milling	3.0	4.9	[13]
Potato chips	5.0	6.0	[14]
Baker's yeast industry	NR	6.1	[13]

[a]Contains flour, soybean, tomato, pepper, and salt

distilleries, sugar factories, dairies, and pulp and paper mills are rich in organic pollutants, mainly carbohydrates. Therefore, they can be treated biologically through the process of anaerobic digestion to combine waste stabilization with energy generation. The main characteristics of typical agro-food industrial wastewater (biochemical and chemical oxygen demand (BOD and COD, respectively)) are given in Table 8.1. Generally, any waste which was converted to CH_4 was implicitly converted to H_2 because it is an intermediate and precursor to CH_4 formation. The composition and strength of agro-food waste in terms of organic content vary from low strength such as wash water from sugar mill or dairy effluents to high strength such as dairy, cheese, and winery wastewaters.

The main characteristics of suitable feedstock for H_2 dark fermentation include:

1. High concentration of biodegradable organics
2. High fraction of readily fermentable organics
3. Does not require pretreatment
4. Absence or low levels of inhibitory compounds to H_2 producing microorganisms
5. Can be co-digested with other substrates

Agricultural and agri-food industrial wastes have a high content of carbohydrates because they contain starch, cellulose, and hemicellulose. The crystallinity of ligno-cellulosic residues decreases their biodegradability. However, the processing during food manufacturing usually produces waste streams that contain soluble and fine particulate residues, which are readily biodegradable. Pretreatments such as chemical, physical (mechanical), and enzymatic hydrolysis render the waste treatment energy drain and costly. Compounds such as metal monomers, metal oxides, nanoparticles, and synergistic factors have been investigated for their potential to increase the activities of microorganisms and increase hydrogen production [15]. Of course, such processes can be used when the prime objective is biofuel production rather than waste treatment only.

A variety of wastes or their combinations have been investigated for their potential to produce sustainable H_2 during dark anaerobic fermentation in batch and continuous studies. Many industrial wastewaters have been investigated for their potential to produce hydrogen by mixed culture dark fermentation. These wastes include bean curd manufacturing waste [16], jackfruit peel [17], noodle making waste [18], olive pulp [19], olive pulp wastewater [20–24], pulp and paper industry wastewater, rice bran, wheat bran [25], rice slurry [26], rice winery wastewater [8], starch-manufacturing waste [27], starch-rich wastewater [28], sugar factory wastewater [29], sweet potato starch residue [27], and tofu wastewater [30].

1.1 Quantities of the Organic Waste Suitable for H_2 Production

Industrial processes handling organic plant material generate large quantities of waste containing fractions of the raw and processed material. For example, the percent of the annually wasted fraction out of the total processed material is around 30–50% for fruits and vegetables processing industry; 30% for cereals; 40–50% for root crops; 20% for oilseeds, meat, and dairy; and 30% for fish [31]. Globally, around 1.3 billion tonnes of the food produced in the world for human consumption is wasted every year [31]. In the European countries, the wastes generated (tonnes) by some food processing industries are sugar beet pulp (4,840,000), brewers' spent grains (340,0000), fruits' pulp (apple and peach) (310,000), and onion peeling waste (500). Figure 8.1 shows the percentage of non-utilized raw material in carbohydrate-rich food processing wastes which could be used as substrates for H_2 production.

The food industry supply chain generates a large amount of organic waste which is biodegradable and could be fed to bioreactors to sustain H_2 production. In the UK, for example, the manufacturing element of the food industry supply chain generates about 2.5 million tonnes of food waste annually [33]; this is almost 30% of the amount of domestic food waste generated by the UK households [33]. The global whey production is 150–200 million tonnes, and it is increasing at a rate of 2% per year [34].

Production of palm and olive oil generates waste suitable for H_2 production. Around 2.5 tonnes of palm oil mill effluent (POME) and 1.3 tonnes of empty bunch are generated from the production of a tonne of palm oil [35]. POME is produced in large amounts worldwide. For example, Malaysia generated around 44×10^6 tonnes

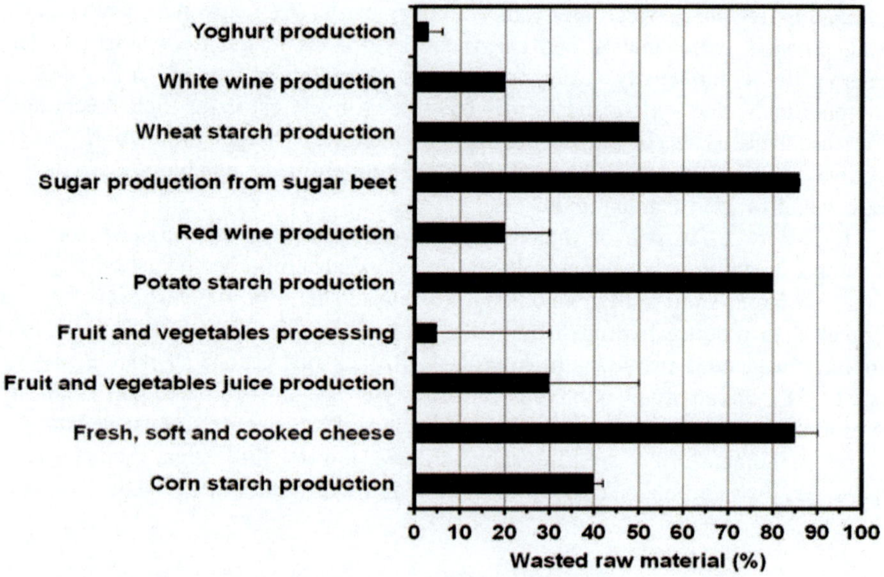

Fig. 8.1 Comparative fractions of non-utilized raw material in food processing wastes which could be used as substrates for H_2 production. (Adapted from [32])

of POME in 2008 [36]. In 2016, the total production of oil palm fruits was about 300 million tonnes [37]. Olive mill waste (OMW) is produced annually in large quantities; around 5.4×10^6 m^3 of OMW are produced annually worldwide [38]. OMW has COD and BOD of 40–220 and 23–100 g/L [39, 40], respectively, with around 2–8% sugar.

1.2 General Description of Anaerobic Digestion

Anaerobic digestion is a well-established bioprocess for treating organic waste and producing methane. In this process, a series of biochemical reactions degrade and convert reduced (rich in electrons) organic materials to CH_4 through the metabolic interactions of various anaerobic microorganisms (Fig. 8.2). Several groups of trophically distinct microorganisms work together to degrade the organic material and produce CH_4, carbon dioxide (CO_2) and water in four successive stages [42]. Anaerobic digestion depends on a balanced symbiotic relationship between metabolically distinct microbial groups [43]. The major groups of microorganisms are as follows [44]:

1. Hydrolytic and fermentative (acidogenic) bacteria
2. Hydrogen-producing acetogenic bacteria
3. Methanogenic Archaea (both acetoclastic and hydrogenotrophic)

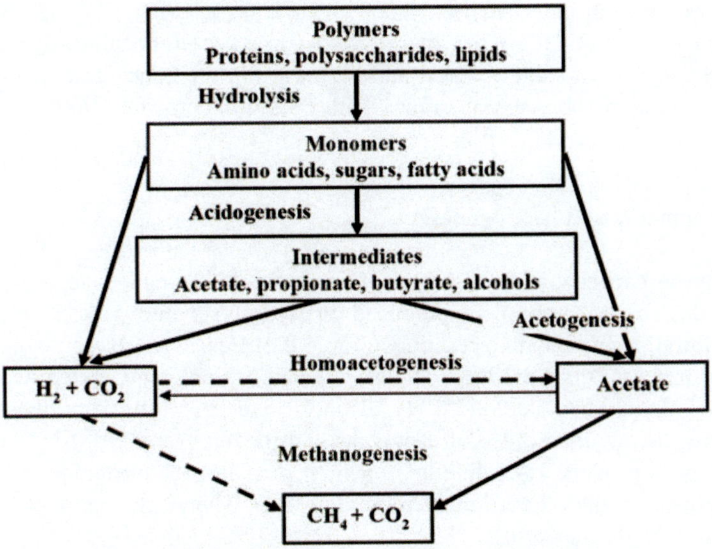

Fig. 8.2 Stages of the anaerobic digestion process. (Adapted from [41])

4. Homoacetogenic bacteria

Anaerobic fermentation includes four different stages: hydrolysis, acidogenesis, acetogenesis, and methanogenesis (Fig. 8.2) [41]. A physiologically unique microbial population performs each stage. The four stages occur synergistically in a successive fashion such that the product of one reaction becomes the substrate for the next reaction.

1.2.1 Hydrolysis or Liquefaction

Hydrolysis is an extracellular process that converts large complex organic polymers into small soluble monomers that can be transported into the microbial cells and subsequently metabolized. The hydrolytic microorganisms catalyze the hydrolysis of polymers by secreting extracellular enzymes. These enzymes include cellulase, cellobiase, xylanase, amylase, protease, and lipase [45].

The rate of hydrolysis depends on pH, temperature, composition and particle size, type and concentration of the substrate, biomass concentration, enzyme production and adsorption, and the concentration of intermediate products [45]. Hydrolysis produces dissolved products that are taken up and fermented by the fermentative/acidogenic bacteria. The hydrolysis of carbohydrates such as glucose, sucrose, and starch proceeds quite fast, whereas hydrolysis of lignocellulosic material is rate-limiting in anaerobic digestion [46]. Physicochemical hydrolysis of lignocellulosic substrates suffers from the formation of inhibitory by-products such as furan derivatives (furfural and 5-hydroxy methyl furfural (5-HMF)) [47, 48]. The relative content of 5-HMF in the hydrolysates of some lignocellulosic substrates was corn cob > corn stalk > wheat and barley while the concentration of furfural in media was corn cob > wheat straw > barley straw > corn stalk [48].

1.2.2 Fermentation/Acidogenesis

This stage is carried out by a group of microorganisms called acidogens. They convert the simple monomers produced in hydrolysis into volatile fatty acids (VFAs) through fermentative reactions (Table 8.2: Eqs. 1 to 9). Acidogenesis is a complex process that produces VFAs (acetate, propionate, butyrate, lactate, caproate, valerate, heptoate, etc.), H_2, alcohols, and CO_2 [49]. Acidogens have a higher growth rate than those of acetogens and methanogens [50, 51]. Acidogens and acetogens produce H_2, while hydrogenotrophic and acetotrophic methanogens consume it to produce CH_4 in anaerobic digestion. Therefore, acidogens are very important to H_2 fermentation.

Table 8.2 Some hydrogen producing and consuming anaerobic bioreactions during dark fermentation

Acidogenic reactions	$\Delta G^{o\prime}$ (kJ)	Eq. No.
$C_6H_{12}O_6 + 2H_2O \rightarrow 2CH_3COOH + 4H_2 + 2CO_2$	−206.0	1
$C_6H_{12}O_6 \rightarrow CH_3CH_2CH_2\ COOH + 2CO_2 + 2H_2$	−254.0	2
$3C_6H_{12}O_6 \rightarrow 4CH_3CH_2COOH + 2CH_3COOH + 2CO_2 + 2H_2O$	−694.7	3
$C_6H_{12}O_6 + 2H_2 \rightarrow 2CH_3CH_2COOH + 2H_2O$	−279.4	4
$C_6H_{12}O_6 + 2H_2 \rightarrow 2CH_3COOH + 2H_2O$	−198.1	5
$C_6H_{12}O_6 \rightarrow 2CH_3CHOHCOOH + H^+$	−225.4	6
$C_6H_{12}O_6 \rightarrow 2CH_3CH_2OH + 2CO_2$	−164.8	7
$3CH_3CHOHCOOH \rightarrow 2CH_3CH_2COOH + CH_3COOH + HCO_3^- + H^+$	−165.0	8
$2CH_3CHOHCOOH + 2H_2O \rightarrow CH_3CH_2CH_2COOH + 2HCO_3^- + H^+ + 2H_2$	−56.3	9
Acetogenic reactions		
Propionate: $CH_3CH_2COOH + 2H_2O \rightarrow CH_3COOH + 3H_2 + CO_2$	+76.2	10
Butyrate: $CH_3CH_2CH_2COOH + 2H_2O \rightarrow 2CH_3COOH + 2H_2$	+48.4	11
Lactate: $CH_3CHOHCOOH + 2H_2O \rightarrow CH_3COOH + HCO_3^- + 2H_2$	−4.2	12
Ethanol: $CH_3CH_2OH + H_2O \rightarrow CH_3COOH + 2H_2$	+9.6	13
Homoacetogenesis: $4H_2 + 2CO_2 \rightarrow CH_3COOH + 2H_2O$	−104.0	14
Methanogenic reactions		
Hydrogen: $4H_2 + CO_2 \rightarrow CH_4 + 2H_2O$	−135.0	15
Acetate: $CH_3COOH \rightarrow CH_4 + CO_2$	−31.0	16
Sulfidogenic reactions		
$SO_4^{2-} + 4H_2 + H^+ \rightarrow HS^- + 4H_2O$	−152.2	17

1.2.3 Acetogenesis

Acetogenesis is the conversion of VFAs such as propionic, butyric, and valeric acids as well as alcohols to acetate, H_2, and CO_2 (Table 8.2: Eqs. 10 to 13). Acetogenesis is carried out by a specific group of bacteria called proton-reducing acetogenic bacteria. These bacteria produce the substrates (acetate and H_2) utilized by methanogens [49]. Acetogens suffer from product inhibition under elevated H_2 levels [52]. Generally, acetogenesis from propionic and butyric acids is desired because it releases H_2 and contributes to increasing its yield.

1.2.4 Homoacetogenesis

Homoacetogens are fast-growing anaerobic bacteria that grow on H_2 and CO_2 and/or on the fermentation of organic compounds to produce acetate [53]. They oxidize or synthesize acetate depending on the external H_2 concentration [54]. At high P_{H_2} (>500 Pa), acetogenesis is favored (from H_2 and CO_2), and at low P_{H_2} (<40 Pa), acetate oxidation occurs [55]. Homoacetogenesis (Table 8.2, Eq. 14) is undesirable during H_2 fermentation because it decreases the H_2 yield.

1.2.5 Methanogenesis

Methanogenesis is the final stage of anaerobic digestion. Methanogens produce CH_4 either by cleavage of acetic acid molecules to CO_2 and CH_4 or by reduction of CO_2 with H_2 (Table 8.2; Eqs. 15 and 16). They grow directly on acetate and H_2/CO_2 and other one-carbon compounds such as formate, methanol, and methylamine [56]. Acetoclastic methanogens can utilize only acetate, while hydrogenotrophic methanogens use H_2/CO_2. Therefore, they live in syntrophic association with acetogens because the latter produce the substrate for the former. Hydrogenotrophic methanogenesis regulates the anaerobic process by maintaining low P_{H_2}, affects the syntrophic acetogens, and influences the whole fermentation process [56]. In anaerobic digestion, up to 70% of CH_4 is produced from acetate, while 30% is derived from H_2/CO_2 [57]. Methanogens are sensitive to oxygen concentration and to changes in operational parameters and environmental conditions such as pH, temperature, hydraulic loading rate, organic loading rate (OLR), and feed composition [58].

Hydrogenotrophic methanogens are more resistant to environmental changes and grow faster than acetoclastic methanogens [49]. Inhibition of methanogens by low pH or toxic chemicals results in the accumulation of H_2 and VFAs; such conditions increase concentrations of reduced products [54].

1.3 Mixed Culture Hydrogen Dark Fermentation

Usually, the by-products (H_2, VFAs, and alcohols) during mixed culture dark fermentation are ultimately converted to CH_4. In balanced mixed culture fermentation, H_2 is produced through the activity of the hydrogenase of acidogenic bacteria as an intermediate by-product with acetate and butyrate production (Table 8.2; Eqs. 1 and 2) [59]. If H_2-consuming microorganisms are inhibited, a maximum of 4 or 2 mol H_2/mol glucose could be produced when acetate or butyrate (Table 8.2; Eqs. 1 and 2, respectively) is the only final product, respectively. Generally, acidogenic bacteria prefer to produce acetate because it regenerates the reducing equivalents and allows them to synthesize ATP [60]; however, under elevated P_{H_2} (>60 Pa), acidogens produce butyrate to avoid the accumulation of inhibitory reducing equivalents [61].

Anaerobic mixed culture contains different species of strict anaerobes and facultative anaerobic bacteria. These bacteria produce H_2 during their metabolism when they dispose the excess electrons in the form of molecular H_2 through the activity of hydrogenase [62]. Strict anaerobes such as *Clostridia*, methylotrophs, and rumen bacteria produce H_2 as a by-product with their main product which is the VFA [63]. The most effective H_2-producing microorganisms in dark fermentation are the species of *Clostridia* [62, 64] such as *C. pasteurianum*, *C. butyricum*, and *C. beijerinckii* [65–67]. *Clostridium* species (*Clostridiaceae*) constituted 64.6% of the clones detected in H_2-producing mesophilic mixed cultures [68]. Because *Clostridia* and *Bacillus* species are spore formers, then heat treatment can be used to enrich them, while methanogens are eliminated because they are not spore formers. Other anaerobes such as *Actinomyces* sp. and *Porphyromonas* sp. have also been detected in addition to *Clostridia* in H_2-producing cultures [69]. Species of facultative anaerobes such as *Enterobacteriaceae* [70], *Citrobacter* [71], *Lactobacillaceae, Bacillaceae* [72], and *Klebsiella* [68] have been detected in H_2-producing anaerobic mixed cultures. Generally, facultative bacteria produce less quantities of H_2 than strict anaerobes.

1.3.1 Pure Versus Mixed Culture for H_2 Production

Examples of pure culture strains that were used in H_2 production include *Klebsiella* sp. TR17 [73], *Clostridium pasteurianum* [74], *Thermotoga* [75], *Escherichia* [76], *Rhodopseudomonas palustris* [77], and *Enterobacter aerogenes* EB-06 [78, 79]. However, economic H_2 production requires a cheap and sustainable process. The pure culture cannot be used to treat "waste" and generate H_2 because the substrate (i.e., the waste) itself is a major source of microbial contamination. Microbial contamination cannot be eliminated, and the strain purity cannot be maintained. Continuous heating of the substrate (waste) to bring its temperature to the range usually used during sterilization and to maintain it for a specific time is an energy drain that disqualifies the process as an economically profitable investment.

In addition, the culture in the reactor needs to be sterilized frequently (every 1–3 months as has been concluded by studies reported by Duangmanee et al. [80], Chen et al. [81], and Goud and Mohan [82]); otherwise, methanogens establish themselves again in the reactor.

Pure cultures are suitable for a pure substrate such as glucose, xylose, etc. to produce H_2. However, there is still no proof that such a process would be technically feasible at large scale, i.e., heating digesters of the same volumes as found in wastewater treatment plants to 100 or 120 °C to sterilize the culture. Moreover, such a process will be economically unprofitable. The full-scale application of heat treatment is technically complicated and questionable, as has been discussed by Hawkes et al. [83] and Valdez-Vazquez and Poggi-Varaldo [84].

Pure culture (where sterilization is used to maintain the culture's integrity for usually a single strain) is used in the production of special commodities such as flavors, antibiotics and other drugs, etc. The demand for special commodities is not as high as that for fuel such as H_2, and at the same time, the price of these products is high compared to the targeted prices of biofuels. Therefore, the use of sterilization in the case of special commodity production is still economically profitable. The reactor volume used in the production of the special commodities is not as large as the targeted reactors for H_2 production, the substrate is not waste, and the substrate flow rate is not as high as that targeted for H_2 production.

Pure cultures are expensive and require sterilized conditions (substrate, reactor, nutrients, etc.), while mixed cultures use non-sterile substrates such as waste streams. Mixed cultures can be obtained from wastewater treatment facilities, live-stock manure, agriculture soil, swamps and marshland, compost, and other natural sources. Hydrogen yields from mixed cultures utilizing carbohydrate, palm oil, and rice production facilities have been shown to reach maximum values of about 2.52 mol/mol glucose [85]. For the reasons mentioned earlier, pure or single culture studies are out of the scope of this chapter which focuses on waste streams to make the anaerobic treatment of industrial waste sustainable.

2 Metabolic Pathways of Dark Hydrogen Fermentation

The interaction of the various microorganisms in mixed culture anaerobic fermentation (MCF) results in a metabolic network that incorporates the bioreactions through which the substrate is converted to intermediate metabolites and finally into CH_4. Figure 8.3 shows a typical and generally accepted metabolic network for glucose fermentation in MCF for biohydrogen production [86]. During the fermentation of glucose, pyruvate is the central branching metabolite and the redox couple (nicotinamide adenine dinucleotide (NAD)/NADH) serves as the electron carrier.

The oxidation of excess NADH (reaction 2) produces H_2. Briefly, H_2 is produced in two catabolic steps: (1) cleavage (decarboxylation) of pyruvate to acetyl-CoA, CO_2 and H_2 (reaction 2) [86, 87], and hydrogenase [87] and (2) cleavage of formate (reaction 7), which is catalyzed by formate hydrogenlyase [86].

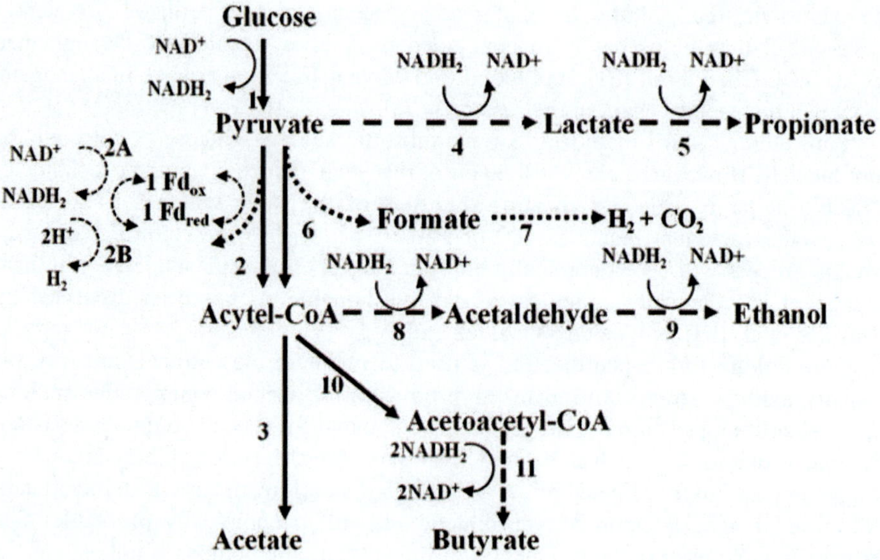

Fig. 8.3 Metabolic network for dark anaerobic fermentation of glucose in mixed culture. The solid dot arrow reactions (2, 6, and 7) produce H_2, while dash arrow reactions (4, 5, 8, 9, and 11) consume potential H_2. (Adapted from [86])

The competition between NAD^+ and H^+ for Fd_{red} (reaction 2) controls the H_2 yield [86]. Coupling Fd_{red} and NAD^+ generates $NADH_2$ which is required for biomass synthesis and lactate, propionate, acetaldehyde, ethanol, and butyrate formation. The production of lactate and propionate (reactions 4 and 5 in Fig. 8.3 and reactions 5 and 4 in Table 8.2) decreases the H_2 yield because these reactions consume $NADH_2$ (potential H_2).

Theoretically, 12 mol of H_2 can be produced by the complete oxidation of glucose to H_2 and CO_2. However, 4 mol of H_2/mol glucose is the maximum amount that is biologically possible if acetate is the only by-product with H_2 (Table 8.2; Eq. 1). The yield decreases to 2 mol H_2/mol glucose when butyrate is the by-product (Table 8.2; Eq. 2) instead of acetate. Hence, the H_2 yield depends on the fermentation pathway and end products [64]. Practically, a mixture of acetate and butyrate is produced, and the H_2 yield is between 2 and 2.5 mol H_2/mol glucose. However, the production of propionate and reduced end products, such as alcohol and lactic acid, results in low H_2 yields [64].

The type of microorganisms, concentration and the oxidation state of the substrate, and environmental conditions (pH and P_{H_2}) dictate the dominant pathway [85, 88]. Microorganisms use different pathways to degrade the same or different substrates depending on the environmental conditions. The is an important factor which affects the pathway and can be used to control H_2 production during dark fermentation.

3 Industrial Wastewater Characterization

Converting waste to biohydrogen generates income and offsets some waste disposal-associated costs. However, a question that needs to be answered is whether a waste stream is suitable for biohydrogen production. Generally, waste streams rich in carbohydrates are ideal substrates for H_2 because they are readily degradable, renewable, and sustainable and can be obtained at low cost. However, the suitability of a feedstock for CH_4 and H_2 production by anaerobic digestion depends on several factors. These factors include the feedstock's physical-chemical properties, organic content (mainly carbohydrate), cost, availability, the extent of biodegradability, absence or presence of inhibitors and toxic compounds, requirements of pretreatment, and sufficient concentration which makes the process energetically favorable [69, 89]. Wastes rich in sugar monomers or polymers such as glucose, xylose, sucrose, lactose, starch, cellulose, and hemicellulose could serve as substrates for H_2 production because they are easily biodegradable. Many waste streams such as olive and palm mill effluent contain ethanol, acetate, butyrate, and propionate; these compounds can be consumed by microorganisms in H_2-producing bioreactions. Therefore, these waste streams were frequently reported as suitable for biological H_2 production [90].

When assessing the suitability of industrial waste to anaerobic digestion, it must be noticed that industrial waste may contain toxic compounds. These toxic compounds include oxidizing agents (disinfectant), long-chain fatty acids (LCFAs), antibiotics, ammonia, detergents, pharmaceuticals, metal ions, and aromatic compounds from plants (lignin, tannins, phenolic compound, and pigments). The decision of whether waste containing these compounds is suitable for CH_4 or H_2 production depends on the extent to which these toxic compounds are inhibitory to acidogens or methanogens. For example, LCFAs are inhibitory to methanogens but not to acidogens. Therefore, they could be used as a co-substrate to inhibit methanogens in mixed culture dark fermentation. Recently, it was reported that moderate concentrations of ammonia increased the hydrogen yield from alkaline-pretreated sludge [10]. Upon increasing the initial NH_4^+-N concentration from 36 to 266 mg/L, the maximum hydrogen yield from alkaline (pH 9.5)-pretreated sludge increased from 7.3 to 15.6 mL per gram of volatile suspended solids (VSS) under the standard condition [10].

3.1 Food Waste

Food waste obtained from food catering facilities has been investigated for its potential to produce H_2 by mixed culture dark fermentation. An advantageous characteristic of such waste is that it has undergone chopping, soaking, and cooking which decreased their particle size, lowered the degree of crystallinity, and increased the soluble fraction. Careful examination of Table 8.3 reveals that wide variation exists in the yield of H_2 reported for food waste. Indeed, food waste is a heterogeneous substrate with a complex chemical composition which comprises various organic and inorganic compounds. The absence of unified characterization criteria for food waste makes it impossible to compare results from different studies or even to reproduce the results of the same study. The quality of COD or the VS is quite variable and depends on whether the solids are carbohydrate, protein, or fat. Even for carbohydrates, the quality of the organics depends on whether they are soluble or particulate, hydrolyzed, or in lignocellulosic structural form. For example, the COD value of a gram of volatile solids of carbohydrate, protein, or fat is 1.06, 1.42, and 2.9 g/g VS, respectively, assuming that the theoretical molecular formulae for the three types of organics are $C_6H_{10}O_5$, $C_5H_7O_2$, and $C_{57}H_{104}O_6$, respectively. Notice that H_2 yield from carbohydrate, protein, and fat-rich waste was 50, 2.5, and 3.3 mL/g VS, respectively [113]. The H_2 production rate of protein-rich material is slower than that obtained from carbohydrate-rich substrates [69]; thus, protein requires more time to biodegrade compared to carbohydrate-rich food waste. For example, the carbohydrate components of bean curd manufacturing waste, rice bran, and wheat bran waste degraded rapidly, whereas soluble protein degradation proceeded at very slow rate [25]. Similarly, H_2 yield from carbohydrate-rich high solid organic (rice and potato-based) waste was 20 times larger than those of fat and protein-rich waste (meat and chicken skin, egg and lean meat) [113].

A challenge in converting food waste to H_2 is that its carbohydrates – and protein – content vary in types and concentrations. The degradation of carbohydrates requires different environmental and operating conditions than that required for protein degradation during dark fermentative H_2 production; thus, the relative composition of the waste must be considered when designing the operational parameters for H_2 dark fermentation. For the fermentation of food waste (Table 8.3), the H_2 yield obtained at thermophilic conditions (overall average of 119 mL/g VS) is higher than that obtained at mesophilic conditions (overall average of 98 mL/g VS). The composition and the type of organics in the food waste affect the HRT required to produce H_2 in continuous reactors. Generally, the yields of 0.9–1.8 mol H_2/mol hexose that were obtained in batch and continuous reactor operated at mesophilic and thermophilic conditions are promising. The optimal conditions for biohydrogen production from food waste via dark fermentation using mixed culture were reported by Wongthanate and Chinnacotpong as follows: initial pH 8.0, initial food to microorganisms (F/M) ratio 4.0, initial iron concentration 100 mg $FeSO_4$/L, and thermophilic condition (55 ± 1 °C) [114]. They reported a maximum hydrogen yield of 46.19 mL H_2/g COD_{add} with 66% COD removal efficiency.

Table 8.3 Hydrogen potential of food waste

Substrate	Type of operation	Temperature (°C)	H_2 yield	H_2 production rate	References
Food waste (3% VS)	Batch	Mesophilic	0.1 mol/mol hexose		[91]
Food waste	ASBR	Not reported	80.9 mL/g VS	0.11 L/L/h	[92]
Food waste	CSTR	Not reported	25.8 mL/g VS	0.04 L/L/h	[93]
Food waste	Batch	Mesophilic	53.4–78.7 N mL/g VS		[94]
Food waste	Batch	Mesophilic	14.6–104.8 mL/g VS fed		[95, 96]
Food waste + sludge	Batch	Mesophilic	11.57–102.6 mL/g VS fed		[95, 96]
Food waste + sewage sludge	Batch	Mesophilic	162 mL/g VS fed		[97]
Food waste	Batch	Mesophilic	39 mL/g VS fed		[98]
Food waste	Batch	Mesophilic	37–101 mL/g COD		[99]
Food waste	Batch	Mesophilic	153.5 mL/g VS fed		[100]
Food waste	Batch	Mesophilic	16.5 mL/g VS fed		[101]
Food waste	Continuous	Mesophilic	34.7–310 mL/g VS fed 390 mL/g COD		[102, 103]
Food waste and sewage sludge	Batch	Mesophilic	123 mL/g COD carbohydrate 60 mL/g VS fed	2.67 L/g VSS/d	[104]
Liquid food waste	Continuous	Mesophilic	1.82 mol/mol glucose		[105]
Restaurant food waste	Batch	Mesophilic	2.82 mL/g COD		[106]
Food waste	CSTR	Mesophilic	64.7 mL/g VS fed		[107]
Food waste	Batch	Hyperthermophilic	82.9–133.5 mL/g VS fed		[108]
Food waste (6 g VS/L)	Batch	Thermophilic	1.8 mol/mol hexose	0.29 L/g VSS/d	[91]
Food waste (3% VS)	Batch	Thermophilic	0.9 mol/mol hexose		[91]
Food waste	Continuous	Thermophilic	125.0 mL/g VS fed	1.00 L/L/h	[109]
Food waste	Batch	Thermophilic	137.2 mL/g VS fed		[101]
Vegetable food waste	Batch	Thermophilic	0.48 mmol/g COD fed	-	[110]
Vegetable food waste	Semicontinuous	Thermophilic	1.7 mmol/g COD fed 72 mL/g VS fed		[111]
Food waste	Batch	Thermophilic	57 mL/g VS fed		[98]
Food waste	Semicontinuous	Thermophilic	205 mL/g VS fed		[112]

3.2 *The Organic Fraction of Municipal Solid Waste (OFMSW)*

The organic fraction of municipal solid waste (OFMSW) is a sustainable carbohydrate-rich substrate which can be fed to anaerobic mixed culture during dark fermentation to produce H_2. De Gioannis et al. [115] listed more than 80 studies published on H_2 production from OFMSW. OFMSW increasingly attracts the interest because it represents up to 70% of the total municipal solid waste produced, whereas the biodegradable fraction (containing 50–70% water) forms about 53% of the waste composition [116]. A generic theoretical molecular formula ($C_6H_{10}O_4$) has been proposed by Themelis and Kim [117] for OFMSW. The H_2 potential of such waste is quite high compared to that from glucose based on the molecular formulae. Preparing the OFMSW as a substrate for H_2 production requires initial separation, sorting, and some pretreatment which means an additional cost.

The potential of the organic fraction of municipal solid wastes (OFMSW) to produce H_2 was investigated in several studies (Table 8.4). Inconsistency and wide variation exist in the reported yield and production rate of H_2 from the OFMSW for the same or similar substrate, reactor design, and operation conditions because of the heterogeneous nature of the OFMSW and lack of its comprehensive characterization during these studies (Table 8.4). For example, most studies reported the H_2 yield per gram VS fed, gram COD fed, or gram substrate fed, but they did not analyze for soluble or particulate carbohydrate, protein, fat, and fiber contents. Overall, the general trend of the results reported (Table 8.4) indicates the potential of the OFMSW for H_2 production through dark fermentation. Generally, the H_2 yield ranged between 43 and 114 mL H_2/g VS fed for mesophilic and thermophilic operations, respectively [105, 119]. Hydrogen yield of 150 mL/g OFMSW and a yield equivalent to 1.35 mol/mol glucose have also been reported (Table 8.4). At mesophilic operation and a high F/M ratio (0.4 g OFMSW/g biomass), an H_2-specific production rate of 43 mL/g VSS/h and 125 mL/g TVS/h has been reported [67]. Co-substrates increased the H_2 yield and the rate of production both in mesophilic and thermophilic conditions (Table 8.4).

3.3 *Food Processing Wastewater*

Industrial wastes from sugar, dairy, olive mill, cheese whey, potato processing, and brewery plants are carbohydrate-rich and have good potential to produce H_2 (Table 8.5). For example, noodle, potato, and rice processing food industries produce waste containing around 0.44 ± 0.02 g carbon / kg of dry matter [160]. However, they may require pretreatment to remove undesirable components and to supplement any deficit nutrients. Generally, waste that contains lignocellulosic residues needs intensive pretreatment such as acid, alkaline, or thermal hydrolysis

Table 8.4 Hydrogen production from the organic fraction of municipal solid waste

Substrate	Type of operation	Temperature (°C)	H_2 yield	H_2 production rate	References
Kitchen waste	Continuous	Not reported	44.8 mL/g COD fed	1.34 L/L/h	[118]
Household solid waste	Continuous	Mesophilic	43.0 mL/g VS fed	1.61 L/L/h	[119]
OFMSW	Continuous	Mesophilic	99–127 mL/g VS removed		[120]
OFMSW	Batch	Mesophilic	132 mL/g TVS 150 mL/g OFMSW	0.4 L/g VSS/d	[67]
OFMSW	Batch	Mesophilic	49.1 mL/g substrate		[121]
OFMSW + Slaughterhouse waste	Semicontinuous	Mesophilic	52.5–71.3 mL/ g VS removed		[122]
OFMSW	CSTR	Thermophilic	1.35 mol/mol glucose	2.2 L/L/d	[123]
OFMSW	Semicontinuous	Thermophilic	114 mL/g VS added		[124]
OFMSW + paper waste	Continuous	Thermophilic		2.4–5.4 L/ L/d	[125]

or their combinations. For example, Zhang et al. [146] reported that acidification of corn stalk waste enhanced its bioconversion to H_2 during dark fermentation using inoculum derived from cow dung compost. Alkaline (0.2% NaOH boiling for 30 min) and acid (0.2% HCl boiling for 30 min) treatments of corn stalk increased the H_2 yield (mL/g VS) in dark fermentation from 4 (control) to 57 (alkaline) and 150 (acid) [146]. Cellulose is abundant in waste from industries such as pulp, paper, and food processing. However, the presence of increased cellulose content decreases the H_2 yields; thus, a substrate pretreatment may be required to hydrolyze the cellulose prior to dark fermentation.

Food processing industries generate waste containing high COD and readily biodegradable organics. However, it varies in nature and composition; therefore, the specific characteristics of the waste affect the operational condition (OLR, HRT, substrate pretreatment) as well as the yield and the rate of H_2 production. Yasin et al. [161] reviewed the H_2 production potential of food and food processing waste and pointed out that the lipid, protein, and carbohydrate components in food waste are convertible to H_2 with a maximum yield around 0.67 L H_2/g substrate of food waste.

As an example of how the characteristics of the waste affect the operation parameters of H_2 production by dark fermentation process, the optimum HRT for H_2 production from cheese processing wastewater (3 h) [162] is shorter than that for olive oil mill effluent (14.5 h) [24]. The production of H_2 from carbohydrate-rich

Table 8.5 Hydrogen potential of food processing waste

Substrate	Type of operation	Temperature (°C)	H_2 yield	H_2 production rate	References
Food processing and domestic wastewater	Batch	Room	100 mL/g COD		[126]
Sugar factory wastewater	CSTR	Thermophilic	2.6 mol/mol hexose	4.4 L/L/h	[29]
Press mud from sugar industry	UASB	Not reported	4.99 L/kg press mud fed 320 mL/g COD fed	187 mL/g VS removed/d	[127]
Starch wastewater	Batch	Thermophilic	92 mL/g starch		[28]
Cassava starch wastewater	UASB	Mesophilic	0.2–0.3 mol H_2/mol Carb		[15]
Potato processing wastewater	Batch	Mesophilic	2.8 L/L wastewater 6.0 mmol/g COD 0.1 L/g COD		[104] [126]
Apple processing wastewater	Batch	Mesophilic	0.1 L/g COD		[126]
Apple pomace waste	Batch	Not reported	2.3 mol/mol glucose		[128]
Apple pomace	Batch	Mesophilic	101 mL/g TS	8.08 mL/g TS/h	[129]
Alcohol wastewater and fermentation residue	ASBR	Thermophilic	128 mL/g COD removed	2.88 L/L/d	[130]
Noodle manufacturing wastewater	Continuous	Mesophilic	200 mL/g hexose		[16]
Starch and rice noodle wastewater			0.08 L/L substrate		[106]
Rice winery wastewater	UASB	Thermophilic	291 mL/g COD 1.37–2.14 mol/mol hexose	0.16 L/L/h 9.3 L/g VSS/d	[8]
Winery wastewater			0.003 L/L substrate		[106]
Rice slurry	Batch	Mesophilic	346 mL/g carbohydrate		[26]
Filtered leachate of waste biosolids	Batch	Mesophilic	184 mL/g COD		[131]
Sugar beet wastewater	Continuous	Mesophilic	231 mL/g hexose		[132]
	CSTR	Mesophilic		0.34 L/L/h	[133]

(continued)

Table 8.5 (continued)

Substrate	Type of operation	Temperature (°C)	H_2 yield	H_2 production rate	References
Coffee drink manufacturing wastewater			0.32 mol/mol hexose		
Coffee drink manufacturing wastewater	UASBR	Mesophilic	1.78 mol/mol hexose	2.76 L/L/h	[134, 135]
Diluted sugarcane vinasse	AFBR	Thermophilic	2.86 mmol/g COD fed		[136]
Tequila vinasse	ASBR	Mesophilic	200 mL/g COD	1.21 L/L/d	[137]
Starch wastewater	EGSBR	Mesophilic		0.07 L/L/h	[138]
Raw starch waste	Batch	Mesophilic	2.02 mL/g COD		[106]
Brewery waste	Continuous	Mesophilic	37.9 mL/g VS fed	2.68 L/L/h	[139, 140]
Beer lees waste	Batch	Mesophilic	68.6 mL/g TVS 54.4 mL/g dry substrate		[140]
Beer lees waste	Batch	Mesophilic	53.0 ml/g-dry beer lees		[141]
Mushroom cultivation waste	Batch	Thermophilic	0.73 mmol/g TVS	2.02 mmol/g VSS/d	[142]
Bean crude manufacturing wastewater	Continuous	Mesophilic	24.0 mL/g VS fed	1.20 L/L/h	[143]
Fresh leachate	EGSBR	Mesophilic		0.09 L/L/h	[144]
Corn stalk	Batch		144.3 mL/g corn stalk	3.6 mL/g corn stalk/h	[145]
Corn stalk waste	Batch	Mesophilic	150 mL/g TVS		[146]
Coffee pulp	Batch	Mesophilic	4.18 L/kg substrate		[147]
Sweet sorghum syrup	Batch		6.9 L/L		[148]
Sweet sorghum syrup	ASBR		1.6 mol/mol hexose	0.13 L/L/h	[149]
Sweet sorghum	CSTR	Mesophilic	10.4 L/kg substrate	2.5 L/d	[150]
Cassava wastewater	Batch	Ambient	9.4 mL/g COD	3.5 mL/g VSS/h	[151]
Cassava wastewater	AnSBBR	Mesophilic	1.1 mol/kg carbohydrate	0.7 L/L/d	[152]
Cassava stillage	Batch	Mesophilic	50.4 mL/g substrate		[153]

(continued)

Table 8.5 (continued)

Substrate	Type of operation	Temperature (°C)	H_2 yield	H_2 production rate	References
Cassava stillage + sludge	Batch	Thermophilic	74 mL/g VS added		[154]
Stillage (ethanol production wastewater)	CSTR	Thermophilic	43 mL/g VS fed	1.0 L/L/d	[155]
Sweet lime peeling extracts	CSTR	Mesophilic	21 mL/g VS		[156]
Snack wastewater Coconut milk wastewater Juice wastewater	Batch	Mesophilic	530 mL/L 520 mL/L 310 mL/L		[106]
Mushroom cultivation waste	Batch	Thermophilic	0.73 mmol H_2/g TVS		[142]
Mushroom waste	Batch	Thermophilic	51 mL/g substrate		[157]
Soybean oil extraction residue	Batch	Thermophilic	27 mL/g substrate		[157]
Pineapple waste extract			1.83 mol/mol glucose	77.31 mL/g VSS/h	[158]
Tofu-processing waste	CSTR	Thermophilic	2.3 mol/mol glucose	12.0 L/L/d	[159]
Fructose wastewater	Batch	Thermophilic	170 mL H_2/g COD	0.03 L/g VSS/h	[27]

AnSBBR anaerobic sequencing batch biofilm reactor, *ASBR* anaerobic sequencing batch reactor, *AFBR* anaerobic fluidized bed reactor, *CSTR* continuous stirred tank reactor, *EGSBR* expanded granular sludge bed reactor, *UASB* upflow anaerobic sludge blanket reactor

food processing waste requires relatively short HRT. A maximum H_2 production rate of 2.55 L H_2/d at HRT of 6 h and a maximum yield of 10.4 L H_2/kg sweet sorghum have been obtained at the HRT of 12 h [150]. Tofu-processing wastewater produced 2.3 mol H_2/mol glucose using heat- and acid-treated (110 °C, 30 min, 0.5% HCl) thermophilic mixed anaerobic cultures in a continuous reactor at 4 h HRT and pH 5.5 [159].

Processing of potato to produce starch generates waste high in sugar and is suitable for H_2 production. Production of a tonne of potato flour generates waste containing around 6.6 m^3 of potato juice and 0.73 tonne of potato pulp [163]. The COD contents of potato processing effluents are relatively high: raw wastewater from potato starch plant (13.9 g/L), potato chip wastewater (7.0 g/L), potato steam peeling (27.4 g/L), potato blanching (39.6 g/L), and potato and corn crisp (4.5 g/L) [164].

Table 8.5 presents the reported H_2 yield and production rate from typical food processing industry wastes such as rice winery, coffee drink manufacturing, noodle, potato processing, sugar, molasses, apple pomace, vinasse manufacturing, olive mill wastewater, olive pulp, and cheese whey. The variation in the chemical composition of the waste and its degree of complexity increased the variability and conflict among different studies conducted on the same waste. Promising yields (mol/mol hexose) have been reported from waste such as sugar factor (2.6), apple pomace (2.3), rice winery (2.1), and coffee drink (1.8).

3.3.1 Molasses

Sugar production from beet generates waste which contains molasses, beet pulp, and cutoffs (beet top and beet leaves). Production of a tonne of beet sugar generates 0.33 tonne of beet pulp, 0.53 tonne of grass cutoffs [165], and liquid waste (molasses) from sugar extraction processes. Molasses is a syrup which contains around 45% sugar [166] with COD and BOD content of about 55 and 27.5 g/L, respectively [167]. Around 55% of the total solids in molasses are organic [167]. Table 8.6 gives some of the reported yields of H_2 produced in mesophilic and thermophilic dark fermentation of molasses in batch and continuous reactors.

Molasses wastewater has been used for continuous production of H_2 in a mesophilic (35 °C) CSTR enhanced by granular activated carbon; at an HRT of 6 h, a 4–8 kg COD molasses wastewater/m^3/d produced H_2 at a rate of 5.9 L/d [176]. Similarly, a yield of 87.0 mL H_2/g COD and an H_2 production rate of 42.7 mL

Table 8.6 Hydrogen yield from molasses wastewater

Substrate	Type of operation	Temperature (°C)	H_2 yield	H_2 production rate	References
Molasses wastewater	EGSBR	Mesophilic	1.95 mol/mol hexose 142 mL/g COD	0.71 L/L/h	[168]
Molasses wastewater	Batch	Mesophilic	4.71 mmol/g COD removed		[169]
Molasses	Continuous	Mesophilic		0.2 L/L/h	[170]
Molasses	Continuous	Mesophilic	5.57 L/L reactor/d		[171]
Molasses wastewater	Batch	Thermophilic	190 mL/g COD	0.04 L/g VSS/h	[172]
Molasses	Continuous	Mesophilic	25 mol/mol sucrose		[173]
Molasses	CSTR	Mesophilic	2.1 mol/mol hexose		[174]
Molasses wastewater	CSTR	Mesophilic		2.41 L/L/d	[175]

H_2/g VSS/h from molasses-processing wastewater have been reported [172]. A yield of 166.8 mL H_2/g COD and H_2 production rate of 26.7 mL H_2/g VSS/h have been obtained from fructose-processing wastewater in a CSTR operated at 55 °C [172].

3.3.2 Palm Oil Mill Effluent (POME)

Palm oil mill effluent (POME) is a good substrate for H_2 production [177] and has COD and BOD of 15–100 and 10–44 g/L, respectively [6]. H_2 production rates and yields, from POME-extracted olive pulp, were better at thermophilic conditions compared to mesophilic conditions [178]. Olive mill waste (OMW) has COD and BOD of 40–220 and 23–100 g/L [39, 40], respectively, with around 2–8% sugar, but it contains a high fraction of oil and grease plus some polyphenols (1–2.4%) [39]; these compounds are inhibitory to methanogens at high concentrations. Hamdi [40] reported 5–80 g/L of polyphenols and 23 g/L of fats in olive mill wastewater (Table 8.7).

3.4 Biodiesel Industrial Waste

Glycerol is a by-product or waste of the rapidly growing biodiesel industry due to the interest in green biofuels. It is a nontoxic simple sugar alcohol compound soluble in water. Thus it is an excellent substrate for H_2 production by mixed dark fermentation. The reported H_2 yield from glycerol ranged between 0.3 and 0.4 mol/mol glycerol or 38 and 72 mL H_2/g COD glycerol for mesophilic conditions (Table 8.8). Since the theoretical H_2 yield has been estimated at 1.0 mol H_2/mol glycerol based on the reaction of glycerol conversion to H_2, CO_2, and ethanol [188, 189], then there is a need to optimize the fermentation conditions to achieve the theoretical yield perhaps by selecting a proper co-substrate or enriching specific microorganisms.

3.5 Dairy Wastewater and Cheese Whey

Wastewater from dairy industry processing such as cheese production contains readily fermentable sugars such as lactose. For example, cheese whey contains about 15% lactose [190]. Because the waste is mainly composed of a single substrate, then adding a co-substrate such as livestock manure can supplement the required nutrients for the growth of microorganisms and increase the H_2 yield, as reported in Table 8.9 [197]. Cheese whey wastewater has been used for the H_2 production using CSTR [195] and produced a yield between 0.8 and 0.9 mol H_2/mol

Table 8.7 Hydrogen yield from palm and olive mill wastewater

Substrate	Type of operation	Temperature (°C)	H_2 yield	References
Olive oil mill wastewater	Continuous	Mesophilic	196 mL/g hexose	[24]
Olive pulp	Continuous	Mesophilic	0.19 mol/kg TS	[178]
POME	ASBR	Thermophilic	6.5 L/L POME	[179]
POME	Batch	Thermophilic	243 mL/g sugar 4.82 L/L POME	[177]
POME	Batch	Thermophilic	6.33 L/L POME	[180]
POME	Batch	Thermophilic	4.7 L/L POME	[181]
POME	Batch	Thermophilic	2.3 L/L POME	[181]
POME	Continuous	Not reported	0.42 L/g $COD_{removed}$	[17]
POME + oil palm industry solid waste				
POME	Batch	Thermophilic	53.1 mL/g VS	[182]
POME + palm decanter cake	Batch	Thermophilic	50.9 mL/g VS	[182]
POME + palm press fiber	Batch	Thermophilic	60.9 mL/g VS	[182]
POME + oil palm frond	Batch	Thermophilic	30.6 mL/g VS	[182]
POME + oil palm trunk	Batch	Thermophilic	21.9 mL/g VS	[182]
POME + empty fruit bunch	Batch	Thermophilic	28.2 mL/g VS	[182]

Table 8.8 Hydrogen yield from glycerol waste in mesophilic batch dark anaerobic fermentation

Substrate	H_2 yield	H_2 production rate	References
Waste glycerol	0.41 mol/mol glycerol	0.35–0.72 L/L/h 0.98 mmol/L/h	[183]
Glycerol	38.1 mL/g COD glycerol		[184]
Waste glycerol	0.3 mol/mol glycerol	0.76 mmol/L/h	[185]
Waste glycerol + sludge	0.3 mol/mol glycerol	1.37 mmol/L/h	[185]
Glycerol from biodiesel fuel production	0.28 mol/mol glycerol 72.0 mL/g COD		[186]
Glycerol + sanitary sewage	0.35–3.01 mol/mol glycerol		[187]

hexose in continuous mesophilic operation (Table 8.9). Conflicting conclusions have been reported about the optimum pH for H_2 production from cheese whey. For mesophilic operation, pH of 6 and 7 were reported optimum for maximizing the H_2 yield and production rate, respectively [193]. In contrast, Yang et al. [192] reported a pH of 4–5 is the best in terms of both H_2 yield and rate.

Table 8.9 Hydrogen potential of dairy industry waste

Substrate	Type of operation	Temperature (°C)	H_2 yield	H_2 production rate	References
Dairy wastewater	Continuous	Mesophilic		1.59 mmol/L/d	[191]
Cheese whey Cheese processing wastewater	Batch Batch	Mesophilic Batch	10 mmol/g COD 2.4 mmol/g COD		[192]
Cheese whey	Batch	Mesophilic	5.9 mol/mol hexose		[193]
Cheese whey	Continuous	Mesophilic	0.9 mol/mol hexose	0.10 L/H_2/L/h	[194]
Cheese whey	CSTR	Mesophilic	290 mL/g VS		[195]
Cheese whey	UASB	Mesophilic	6.1 mL/g COD		
Whey permeate	Batch	Room	4.13 mol/mol lactose	86.31 mmol H_2/L/d	[196]
Cheese processing wastewater	Batch	Mesophilic	3.5 mol/mol lactose consumed		[162]
Cheese whey wastewater	CSTR	Mesophilic	0.83 mol/mol hexose	0.12 L/H_2/L/h	[195]
Cheese whey and dairy manure (45:55) mixture	Batch	Mesophilic	63.1 mmol/L substrate		[197]
Cheese whey and dairy manure (45:55) mixture	CSTR	Mesophilic	25.8 mmol/L substrate		[197]
Dairy manure	CSTR	Mesophilic	0.26 mmol/L substrate 10 mL/g COD		[197]
Dairy wastewater	Batch	Mesophilic	113.2 mmol/g COD 12.8 mmol/g carbohydrate		[198]

3.6 *Manure from Livestock Industry*

The current practice of livestock manure land application is environmentally and economically problematic because it introduces chemicals, antimicrobial, pathogens, and excessive nutrients into the soil and surface and groundwater as well as it contributes to greenhouse gas (GHG) emission. Although the land application of manure fertilizes the soil, it is not the most efficient way to achieve that. With proper treatment before manure land application, valuable resources could be recovered as nutrients and energy while stabilizing any organics and chemicals, eliminating pathogens, reducing GHG emissions, and eliminating the risk of soil and water pollution. Digestate from methane or hydrogen-producing anaerobic digestion is concentrated with nutrients and is an excellent fertilizer which contains nitrogen, phosphorus, and other nutrients but less carbon. Livestock manure is a good substrate for H_2 production because it is cheap, carbohydrate-rich (contains cellulose and hemicellulose), and biodegradable.

Manure can be utilized as a substrate, a co-substrate, a source of nutrient, and a buffer agent for H_2-producing anaerobic mixed cultures. Table 8.10 gives the reported H_2 yield and production rate from livestock and dairy waste. A yield of 59.5 mL H_2 /g VS fed has been obtained from 40 g VS/L of 70:30 cow manure/waste milk mixture in thermophilic (55 °C) batch reactors [204]. The H_2 yield from waste milk has been increased by 60 times when cow manure has been used as a co-substrate because it prevented acidogenic inhibition due to the high rate of hydrolysis [204]. Fermenting sucrose with dairy manure at ambient temperature yielded 2.9–5.3 mol H_2/mol sucrose [205]. The cattle manure provided stability to the fermentation by maintaining the pH level because it has an excellent buffering capacity [123, 205] and provided the mixed anaerobic culture with the necessary nutrients [205]. Swine manure supplemented with glucose yielded 1.48 L H_2/L substrate in mesophilic (37 °C) ASBR [208]. A stable co-fermentation of fruit-vegetable waste and swine manure produced 126 mL H_2/g VS at an HRT of 2 days [123]. No difference in H_2 yield from dairy manure could be observed when it was pretreated with acid (0.2% HCl boiling for 30 min), base (0.2% NaOH boiling for 30 min), or infrared radiation (2 h) to hydrolyze the lignocellulosic fiber [202].

4 Inocula for Hydrogen Production from Waste

4.1 Sources of Inocula

For sustainable waste treatment, the inoculum has to be obtained from sustainable open sources. Sludge from municipal and industrial wastewater treatment processes (both anaerobic digestion and activated sludge processes), compost, cow dung, and agriculture soil are sustainable natural open sources for anaerobic mixed culture.

4.1.1 Sewage Sludge

Anaerobic sludge from primary and secondary digesters in municipal wastewater treatment plants is a readily accessible and cheap source of inoculum which could be used after a proper pretreatment to produce H_2 in dark fermentation. Therefore, it has been widely used in H_2 dark fermentation studies from industrial waste [67, 91, 93,

Table 8.10 Yield of hydrogen from dark anaerobic fermentation of livestock manure and dairy industry waste

Substrate	Reactor	H_2 yield	H_2 production rate	References
Cattle manure	Batch (52 °C)	65 mL/g VS fed	Not reported	[199]
Cow feces and urine	Batch (75 °C)	18 mL/g VS	Not reported	[200]
Cow feces and urine	Batch (60 °C)	29 mL/g VS 22.1 mL/g substrate	Not reported	[200]
Cow feces and urine	Batch (37 °C)	0.7 mL/g VS	Not reported	[200]
Feedlot cattle manure	Batch (52 °C)	65 mL/g substrate		[201]
Dairy manure	Batch (36 °C)	18 mL/g VS 27 mL/g substrate	Not reported	[202]
Dairy manure	Batch (37 °C)	10 mL/g COD 0.26 mmol/L substrate	Not reported	[197]
Dairy wastewater	Continuous	0.122 mmol/g COD	Not reported	[203]
Cattle wastewater	Batch (45 °C)	53 mL/g VS fed	Not reported	[199]
Cow manure + waste milk	Batch (55 °C)	60 mL/g VS fed	Not reported	[204]
Dairy cattle manure + Sucrose	Batch (25 °C)	2.9–5.3 mol/mol sucrose 16.3 mL/g substrate	Not reported	[205]
Swine liquid manure	Semicontinuous (35 °C)	210 mL/g VS fed	Not reported	[206]
Pig slurry	CSTR (70 °C)	3.7 mL/g VS	0.1–0.15 L/L/d	[207]
Liquid swine manure and glucose	ASBR (37 °C)	1480 mL/L substrate	8.9 L/L/d	[208]
Swine manure + glucose	ASBR (37 °C)	1.5 mol/mol glucose	2.25 L/L/d	[209]
Swine manure and fruit-vegetable	CSTR (55 °C)	130 mL/g VS	3.2 L/L/d	[123]
Poultry slaughterhouse sludge	Batch (37 °C)	8.83 mL/g TS	Not reported	[203]

104, 109, 118, 122]. A pretreatment is required to eliminate methanogens because they consume hydrogen and produce methane.

The potential of mixed cultures from anaerobic and aerobic treatment processes to produce H_2 during dark fermentation was evaluated extensively. The proportions of the various trophic groups in different mixed cultures vary widely based on the type of substrate and the environmental conditions prevailing in the source of the culture [210]. Moreover, substrate degrading activity varies widely in cultures from different sources under comparable environmental conditions. Compared to inocula such as cow dung, compost, chicken manure compost, and river sludge, sewage sludge at optimum conditions (pH 5.5 and 45 °C) produced a higher yield (319 mL H_2/g COD) [199].

4.1.2 Industrial Wastewater Sludge

O-Thong et al. [177] used anaerobic sludge from palm oil mill effluent (POME) treatment plant as a source of inocula for the conversion of POME to H_2. They reported a maximum of 4820 mL H_2/L POME (243 mL H_2/g sugar) with a total sugar consumption and chemical oxygen demand (COD) removal efficiency of 98.7% and 46%, respectively. Anaerobic sludge treating palm oil mill effluent (POME) has been used as inoculum to convert POME to bioH_2 [177] and yielded 4.8 LH_2/L POME (H_2 yield of 243 mL H_2/g sugar) at initial pH 5.5 and 60 °C. Inocula derived from industrial waste such as soybean meal [25, 67, 211] and cassava treating sludge [212] have been used in several studies.

4.1.3 Compost

Compost has been used to generate inoculum for H_2 production in dark fermentation of several wastes such as food waste [213], carbohydrate, protein, and fat-rich organic waste [113]. Ueno et al. [29] used compost to ferment sugary wastewater and reported a yield of 2.52 mol H_2/mol hexose in thermophilic operation.

4.1.4 Animal Dung

Cow manure has frequently been used, after a pretreatment, to serve as inoculum for H_2 production during dark fermentation. Cow and cattle manure contains relatively less methanogens compared to anaerobic sludge from municipal and industrial wastewater treatment plants [214]. Therefore, cow manure requires less intensive pretreatment (shorter time of heat shock treatment, smaller dosages of methanogen inhibitors, and shorter duration of acid/base treatment). Song et al. [145] used microwave pretreated cow dung as inoculum to ferment corn stalk and produce hydrogen. Radjaram and Saravanane [215] investigated the start-up of a UASB reactor to ferment municipal sewage sludge to produce H_2, using cow dung as an inoculum. Fangkum and Reungsang [216] reported a yield of 2.56 mol H_2/mol sugar from glucose/xylose using heat-treated elephant dung as inoculum for dark anaerobic fermentation. Xing et al. [202] used dairy manure as inoculum (infrared-pretreated for 2 h) and substrate (pretreated with boiling in 0.2% HCl for 30 min) for dark fermentation and reported a yield 31.5 mL H_2/g VS. Vijayaraghavan et al. [17] reported that cow dung successfully digested palm oil mill effluent and produced 2.03 mol H_2/mol hexose.

4.1.5 Soil

Soil, particularly agriculture soil, has been used to prepare anaerobic mixed culture inocula for hydrogen production studies from food processing and domestic wastewater [126] and molasses and potato waste [217].

4.2 Culture Pretreatment

To be economically profitable and sustainable, dark fermentation of industrial waste to recover H_2 should yield a positive net energy gain. Typically, anaerobic mixed culture produces CH_4 rather than H_2 as the end product. H_2 is an intermediate metabolite which is produced by acidogenic and acetogenic bacteria and is later consumed by hydrogenotrophic methanogens, homoacetogens, and sulfate-reducing bacteria. Therefore, H_2-consuming microorganisms have to be eliminated or at least inhibited during dark fermentation so that H_2 accumulates and is recovered before they consume it while producing CH_4, acetate, or other by-products. Several pretreatment methods can achieve the elimination of H_2-consuming microorganisms. These methods include heat shock [96, 97], acid-base treatment, irradiation, toxic methanogens-inhibiting chemicals (e.g., 2-bromoethanesulfonic acid (BESA), chloroform, long-chain fatty acids (LCFAs), etc.) [153, 191, 210, 218–220], ultrasound sonication [221], acid/base [203, 222], and BSAE [203]. Generally, there is a consensus in the scientific literature on the importance of culture pretreatment but not on which method is the optimum relative to the others.

4.2.1 Heat Treatment

Heat shock treatment is the most widely used pretreatment, particularly for cultures used to ferment food waste [67, 93–96, 101, 104, 111, 119, 159, 178]. Heat treatment selects spore-forming microorganisms such as *Clostridia* and eliminates non-spore-forming microorganisms such as methanogens [220]. Unfortunately, heat pretreatment is not a well-controlled pretreatment; it does not exclusively select H_2-producing or kill H_2-consuming microorganisms [223]. Many H_2-consuming bacteria form spores and can survive heat pretreatment, whereas H_2-producing bacteria which cannot form spores such as *Enterobacter* sp. and *Citrobacter* spp. [223] are killed. Heat pretreatment does not eradicate methanogens; they regrow when the conditions are favorable [224]. Repeated heat treatments were necessary because the H_2 production declined after 1 month of continuous operation [80]. It must be noted that heat treatment decreases microbial diversity, particularly when the heat treatment temperature is increased from 65 to 90 °C [225]. The duration of the optimum heat treatment increases with the percent of the total solids [84]. Reungsang and Sreela-or [158] found that heat treatment of anaerobic sludge produced a good inoculum for pineapple waste extract fermentation.

Although heat treatment application in lab studies is simple, its practical application at the industrial scale is technically complicated. Moreover, heating a digester with volume of several thousand cubic meter to the temperatures usually used to induce spore-forming and maintain it for a specific duration is an energy drain which decreases the net energy yield of the waste treatment. Detailed and comprehensive energy analysis for various wastes is required to establish the net energy yield for various reactor designs at different operation conditions during H_2 production via dark fermentation. The absence of a standard protocol for heat treatment application to mixed cultures and lack of data on energy consumption and energy analysis during studies involving heat treatment limit the applicability of the results to large-scale applications.

4.2.2 Acid/Base Treatment

Treating the culture with acid (pH 3) or alkaline (pH 9 to 11.5) enriches H_2-producing bacteria in anaerobic mixed cultures [226–228]. Acid/base treatment of cultures in industrial-scale bioreactors by maintaining pH 3 or pH 9 to 11.5 would consume large quantities of acid and alkaline and requires equipment for feeding, mixing, pH control, and then neutralizing or adjusting the pH to the desired initial pH (usually 5.0–5.5). The acid and base consumption adds cost and lowers the economic profit of the process. For example, 48.3% of the overall cost of H_2 production during dark-photo fermentation was the cost of the acid required to adjust the pH [229]. Therefore, using an acidic or alkaline waste to pretreat the culture would offset the cost of acid/base and enhance the sustainability of waste treatment.

4.2.3 Methanogen Inhibitors

Inhibitors used to inhibit methanogens can be classified into two types: natural and synthetic. Long-chain fatty acids (LCFAs) at a concentration greater than some threshold levels inhibit methanogens. LCFAs can be obtained from food processing wastewater, such as from the vegetable oil industry. Obviously, such waste could be used as a co-substrate with carbohydrate-rich waste to produce H_2. Pendyala et al. [230] reported H_2 yield of 17 mL/g COD fed from fermenting a food and paper cardboard waste blend using linoleic acid (C18:2)-treated (2 g/L) anaerobic mixed culture. The efficacy of LCFAs in dark fermentation hydrogen production studies for pure substrates has been reported in several studies [53, 210, 218, 219]. More studies are required to assess the effectiveness of LCFAs with different industrial wastes. Replacing BESA and chloroform with LCFAs increases the sustainability of H_2 production from industrial wastewater treatment. Linoleic acid increased the H_2 yield to the same level in three different mixed cultures from different sources compared to their controls [210].

Inhibitors such as 2-bromoethanesulfonic acid sodium salt (BESA) and chloroform can inhibit methanogens and block CH_4 formation. They have been used during batch and continuous H_2 production from industrial waste. Only LCFAs could be obtained from renewable and sustainable sources (from food processing wastewater), while BESA and chloroform are petroleum-based derivatives. Thus, they represent a sustainable and renewable input. Mohan et al. [220] reported that treating the inoculum with 0.2 g BESA/L for 24 h increased the H_2 yield from dairy wastewater by a factor of 2.6 compared to heat shock (100 °C; 1 h).

4.3 *Organic Loading Rate (OLR)*

The organic loading rate (OLR) is an important operation parameter which has to be optimized to achieve the maximum yield per unit mass of inoculum. The OLR of waste fed to dark fermentative H_2 production process has continued to be an objective of the research. The OLR means the food/microorganism ratio which affects the microbial growth; therefore, it needs to be within the optimum range to support optimal microbial growth because H_2 production is a growth-associated reaction. Han et al. [231] reported that increasing the OLR from 8 to 24 kg molasses/ m^3/d increased the H_2 production rate and achieved a maximum value of 10.74 mmol/L/h in a CSTR. Similarly, the H_2 yield increased with the increase of cow manure and waste milk OLR in thermophilic (55 °C) batch reactors [204].

The OLR also affects the biogas composition in an upflow anaerobic sludge blanket (UASB); the higher the OLR, the greater the biogas H_2 content [232]. On the contrary, Bhaskar et al. [233] found that increasing the OLR of chemical wastewater decreased the specific H_2 yield and the chemical oxygen demand (COD) removal efficiency in biofilm-configured sequencing batch reactors (SBR). The substrate concentration affects the H_2 production; increasing the substrate (hydrolysate) concentration from 20% to 25% increased the lag phase of H_2 production in batch reactors, from 11 to 38 h, and at a level of 30% inhibited the H_2 fermentation [234]. The variation in the findings regarding the effects of the OLR on the rate, yield, and the gas composition can be due to the type and quality of the substrate fed despite that two substrates can be fed at the same OLR (based on the COD) and their metabolism rate can be different. In conclusion, the OLR needs to be optimized on a case-by-case basis.

4.4 *Comparative Studies on the Effects of Pretreatments*

Several studies investigated the effect of some culture pretreatments on the preparation of inoculum for the conversion of corn stover hydrolysate, dairy wastewater, palm oil mill effluent, and biodiesel production waste (glycerol) (Table 8.11). However, the results reported were conflicting and inconclusive. The ranking of the effect of several pretreatment on the H_2 yields (mmol H_2/g sugar utilized) from corn stover hydrolysate was as follows: heat (5.03) > base (4.5) > acid

Table 8.11 Results of comparative studies on culture pretreatments

Type of culture	Substrate	Maximum H_2 yield (mol/mol substrate)	Reported treatment trend	References
Municipal aeration tank culture	Corn stover hydrolysate	5.0	Heat > acid > base > ultrasonic > ultraviolet	[235]
Anaerobic granulated sludge	Palm oil mill effluent	19.5[1]	Heat > base > chloroform = acid > freezing-thawing > control	[236]
Anaerobic digested sludge	Glycerol	0.05[2]	Wet heat > dry heat > freezing-thawing > control > acid > base	[237]
Anaerobic digested mixed culture	Dairy wastewater	0.032[1]	BESA > acid + BESA > acid + heat > acid + heat + BESA, heat + BESA > acid	[220]

Notes
(1) The units of H_2 yield = mmol H_2/g COD. (2) The units of H_2 yield = mol H_2/g glycerol. (3) BESA = 2-bromoethanesulfonic acid

(3.2) > ultrasonic (2.9) > ultraviolet (2.9) > control (2.7) [235]. The dry heat was found the best pretreatment for preparing inoculum for palm oil mill effluent and corn stover hydrolysate. However, wet heat was found better for dairy processing wastewater, while inhibiting the methanogens by BESA was the best pretreatment for biodiesel production waste (glycerol).

For H_2 production from corn stalk waste, acid has been reported as the best culture pretreatment to prepare H_2-producing inoculum from cow dung compost compared to base (alkaline) and infrared radiation [238]. For dairy wastewater dark fermentation using anaerobic sludge, BESA has been shown efficient for H_2 evolution and substrate removal [220]. However, conflicting conclusions have been reported regarding which pretreatment is the best compared to acid, heat, and their combination [203]. Mu et al. [239] reported that compared to thermal, microwave, thermal-alkaline, and microwave-alkaline pretreatments, alkaline treatment was the best to prepare H_2-producing inoculum from anaerobic mixed culture to convert sugar beet pulp.

5 Bioreactors Types and Configuration for H_2 Production

Almost all types of reactor designs have been used to enhance the H_2 yield and production rate from industrial waste (Tables 8.1, 8.2, 8.3, 8.4, 8.5, and 8.6). These types of reactors included batch, anaerobic fluidized bed (AFBR), continuously stirred tank reactor (CSTR), anaerobic fluidized bed bioreactor (AFBR), expanded granular sludge bed reactors (EGSBR), leaching bed reactor (LBR), and anaerobic sequence batch reactor (ASBR). Generally, for commercial application of industrial waste treatment and H_2 production, the reactor should provide robust, reliable, and stable performance in terms of H_2 production rate and yield and minimum temporal fluctuations in operational parameters over the long-term operation [85]. Optimization of reactor design should aim to increase the H_2 production rate through improving the biomass immobilization by granulation, membranes, or carriers and increase the H_2 yield by effluent recycling to allow complete substrate degradation. Although operating at short HRT to match the fast growth rate of the H_2-producing acidogenic biomass would achieve the highest H_2 production rate, complex substrates such as food processing wastewater may require a longer HRT, particularly when they contain proteins and fats.

6 Cost of Dark Fermentative H_2 Production from Industrial Waste

Environmental and economic parameters measure the sustainability of industrial wastewater treatment. In this respect, the sustainability of producing H_2 as a clean and renewable biofuel from industrial waste depends on the cost of the infrastructures, input materials (substrate, inoculum, nutrients, and chemicals), and the operation. Industrial waste from food processing facilities is a renewable source of low-cost organics that can be converted to H_2 at low production costs. There are some management and operation opportunities which could help reduce the cost of H_2 production from industrial waste. For example, co-fermenting different wastes lowers the production cost. In this case, the major substrate provides the carbon source, the second substrate provides the nutrients required for the growth of microorganisms, and the third substrate provides the required buffering capacity and/or adjusts the pH. Such integration of co-substrate is economically crucial because, for example, chemicals required to adjust the pH during dark fermentative H_2 production formed 48.3% of the overall cost [229].

Limited information is available in the accessible literature on the economic aspects of dark fermentative H_2 from industrial waste. A few studies [229, 240–242] discussed the economic aspects of dark fermentative H_2 production briefly. The H_2 production cost (USD/m^3) through dark fermentation of waste has been estimated theoretically for beverage wastewater (7100), agriculture waste (7900), solid biowaste (1.52), sugarcane distillery effluent (35.4), and kitchen waste (0.1–0.12) [243–245]. Although the H_2 production cost from beverage and agriculture waste is high, marginal net revenue has been estimated. Sugarcane distillery effluent and solid biowaste generated higher estimated revenue. The type of reactor affects the rate of H_2 production, which in turn affects the capital and operational cost [246]. Molasses-based production of dark fermentative H_2 was estimated to cost $9.52 per GJ [247], and it seems very competitive to gasoline ($22.28 per GJ) and ethanol (€22.6 per GJ) [241]. The capital cost of the dark fermentative reactors represented around 35.4% [242], which could be lowered by modifying conventional anaerobic digesters that treat industrial waste to produce H_2 instead of CH_4. An important factor that should be included in the cost estimate is that H_2 fermentation is not complete treatment because it discharges an effluent concentrated with VFAs which requires further treatment. Obviously, VFA-rich liquid waste is suitable for CH_4 production using the conventional anaerobic digestion. Therefore, a two-stage process could be configured to enable sequential extraction of energy from industrial waste.

7 Future Research on Biohydrogen Production from Industrial Waste Dark Fermentation

The near future research focus should be on integrating the possible combinations of industrial wastes in a dark fermentative H_2-producing process utilizing the same CH_4 fermentation infrastructures. The effluent of dark fermentation has high levels of VFAs. Therefore, integration of dark fermentative H_2 production and other bioprocesses that utilize VFAs for producing biofuels or value-added chemicals such as bioplastics [248] or lipid synthesis by microalgae for biodiesel production [249] should be investigated. The stability of the mixed culture integrity is a big concern, particularly after inoculum pretreatment to prepare an H_2-producing consortium because the waste is a source of contamination. Therefore, industrial waste, which contains inhibitors to methanogens, could be investigated as a potential tool to maintain the integrity of the culture during dark fermentation of other waste which has a high H_2 potential. Future research should focus on the application of the molecular biology methods and techniques in microbial characterization to understanding the composition of the microbial consortium for mixed cultures [250]. Cost analysis for H_2 production from various industrial wastes is required to enable sound evaluation of alternatives.

8 Conclusions

Hydrogen production from biodegradable industrial waste is possible, but its yield and production rate depend on the type of waste. Carbohydrate-rich waste such as molasses, starch, corn syrup, etc. is an excellent substrate for H_2 production because they support high yield in relatively short hydraulic retention time. There is a lack of modeling studies of H_2 production from industrial waste as well as comparative studies about the effect of bioreactor design on the H_2 yield and rate from the same waste. An important area of research is an integrative strategy where several wastes are combined to prepare a well-balanced substrate for H_2-producing mixed culture dark fermentation. More studies are required on the cost analysis and economic evaluation of H_2 production from dark anaerobic fermentation of industrial waste. No specific culture pretreatment can be considered as the best pretreatment for maximizing H_2 yields from industrial waste. Detailed energy analysis for a wide spectrum of industrial wastes is required to establish the net energy yield for various wastes, reactor designs, pretreatments, and operation conditions. Although the results from laboratory-scale studies of producing H_2 from industrial waste were promising, no pilot plant or large-scale studies have been reported.

Glossary

Anaerobic Digestion	Series of biological processes in which microorganisms break down biodegradable material in the absence of oxygen
Biohydrogen Production	Production of hydrogen via fermentation of a broad range of organic feedstocks by bacteria.
Dark Fermentation	The fermentative conversion of organic substrate to biohydrogen.
Industrial Waste	Waste generated by manufacturing or industrial processes.
Mixed Culture	Mixed culture contains more than one type of organism growing in a medium.
Organic Loading Rate	The amount of organics to be fed into the anaerobic digester each day.

References

1. Wei C, Zhang T, Feng C, Wu H, Deng Z, Wu C, Lu B (2011) Treatment of food processing wastewater in a full-scale jet biogas internal loop anaerobic fluidized bed reactor. Biodegradation 22(2):347–357
2. Rupani PF, Singh RP, Ibrahim MH, Esa N (2010) Review of current palm oil mill effluent (pome) treatment methods: vermicomposting as a sustainable practice. World Appl Sci J 11 (1):70–81
3. Alkaya E, Demirer GN (2011) Anaerobic acidification of sugar-beet processing wastes: effect of operational parameters. Biomass Bioenergy 35(1):32–39
4. Passeggi M, López I, Borzacconi L (2009) Integrated anaerobic treatment of dairy industrial wastewater and sludge. Water Sci Technol 59(3):501–506
5. Durham RJ, Hourigan JA (2007) Waste management and co-product recovery in dairy processing. In: Waldron K (ed) Handbook of waste management and co-product recovery in food processing. Woodhead Publishing Limited, Cambridge, England, pp 332–387
6. Wang LK, Hung Y-T, Lo HH, Yapijakis C (eds) (2005) Waste treatment in the food processing industry. Taylor & Francis Group, LLC, New York, p 344
7. Scott J, Smith K (1997) A bioreactor coupled to a membrane to provide aeration and filtration in ice-cream factory wastewater remediation. Water Res 31(1):69–74
8. Yu H, Zhu Z, Hu W, Zhang H (2002) Hydrogen production from rice winery wastewater in an upflow anaerobic reactor by using mixed anaerobic cultures. Int J Hydrog Energy 27 (11–12):1359–1365
9. Jordening H-J (2007) Water re-use and wastewater treatment in the German sugar industry. In Water & waste conference 2007. Assiut University, Egypt
10. Wang D, Duan Y, Yang Q, Liu Y, Ni B-J, Wang Q, Zeng G, Li X, Yuan Z (2018) Free ammonia enhances dark fermentative hydrogen production from waste activated sludge. Water Res 133:272–281
11. Baez-Smith C (2006) Anaerobic digestion of vinasse for the production of methane in the sugar cane distillery. In: SPRI conference on sugar processing, Loxahatchee
12. Rouf M, Bajpai P, Jotshi C (2010) Optimization of biogas generation from press mud in batch reactor. Bangladesh J Sci Ind Res 45(4):371–376
13. Mustafa Evren Ersahin, Hale Ozgun, Recep Kaan Dereli and Izzet Ozturk (April 1st 2011). Anaerobic Treatment of Industrial Effluents: An Overview of Applications, Waste Water - Treatment and Reutilization, Fernando Sebastián García Einschlag, IntechOpen, https://doi.org/10.5772/16032. Available from: https://www.intechopen.com/books/waste-water-treatment-and-reutilization/anaerobic-treatment-of-industrial-effluents-an-overview-of-applications

14. Sentürk E, Ince M, Engin GO (2010) Treatment efficiency and vfa composition of a thermophilic anaerobic contact reactor treating food industry wastewater. J Hazard Mater 176 (1–3):843–848
15. Corbari SD, Andreani CL, Torres DG, Eng F, Gomes SD (2019) Strategies to improve the biohydrogen production from cassava wastewater in fixed-bed reactors. Int J Hydrog Energy 44(32):17214–17223
16. Noike T, Takabatake H, Mizuno O, Ohba M (2002) Inhibition of hydrogen fermentation of organic wastes by lactic acid bacteria. Int J Hydrog Energy 27(11–12):1367–1371
17. Vijayaraghavan K, Ahmad D, Ibrahim MKB (2006) Biohydrogen generation from jackfruit peel using anaerobic contact filter. Int J Hydrog Energy 31(5):569–579
18. Mizuno O, Shinya M, Miyahara T, Noike T (2001) Effect ofphon biological hydrogen production from organic wastewater. In: 9th world congress anaerobic digestion: anaerobic conversion for sustainability, Antwerpen
19. Gavala HN, Skiadas IV, Ahring BK, Lyberatos G (2005) Potential for biohydrogen and methane production from olive pulp. Water Sci Technol 52(1–2):209–215
20. Gavala HN, Skiadas IV, Ahring BK, Lyberatos G (2006) Thermophilic anaerobic fermentation of olive pulp for hydrogen and methane production: modelling of the anaerobic digestion process. Water Sci Technol 53(8):271–279
21. Eroğlu E, Eroğlu İ, Gündüz U, Türker L, Yücel M (2006) Biological hydrogen production from olive mill wastewater with two-stage processes. Int J Hydrog Energy 31(11):1527–1535
22. Eroğlu E, Eroğlu İ, Gündüz U, Yücel M (2009) Treatment of olive mill wastewater by different physicochemical methods and utilization of their liquid effluents for biological hydrogen production. Biomass Bioenergy 33(4):701–705
23. Lyberatos G, Antonopoulou G, Koutrouli E, Kalfas H, Gavala H, Skiadas I (2006) Gaseous biofuels production from sweet sorghum and olive pulp. In: AIChE annual meeting, conference proceedings
24. Ntaikou I, Kourmentza C, Koutrouli E, Stamatelatou K, Zampraka A, Kornaros M, Lyberatos G (2009) Exploitation of olive oil mill wastewater for combined biohydrogen and biopolymers production. Bioresour Technol 100(15):3724–3730
25. Noike T, Mizuno O (2000) Hydrogen fermentation of organic municipal wastes. Water Sci Technol 42(12):155–162
26. Fang HH, Li C, Zhang T (2006) Acidophilic biohydrogen production from rice slurry. Int J Hydrog Energy 31(6):683–692
27. Yokoi H, Maki R, Hirose J, Hayashi S (2002) Microbial production of hydrogen from starch-manufacturing wastes. Biomass Bioenergy 22(5):389–395
28. Zhang T, Liu H, Fang HH (2003) Biohydrogen production from starch in wastewater under thermophilic condition. J Environ Manag 69(2):149–156
29. Ueno Y, Otsuka S, Morimoto M (1996) Hydrogen production from industrial wastewater by anaerobic microflora in chemostat culture. J Ferment Bioeng 82(2):194–197
30. Zhu H, Ueda S, Asada Y, Miyake J (2002) Hydrogen production as a novel process of wastewater treatment—studies on tofu wastewater with entrapped r. Sphaeroides and mutagenesis. Int J Hydrog Energy 27(11–12):1349–1357
31. FAO SAVE FOOD: Global initiative on food loss and waste reduction: key findings. 2014. Available from: http://www.fao.org/save-food/
32. Awarwnet (2004) Handbook for the prevention and minimization of waste and valorization of by-products in european agro-food industries: Deposito Legal: BI-223-04
33. WRAP (2010) Household food and drink waste in the UK: environmental benefits of recycling – 2010 update. Available from: http://www.wrap.org.uk/sites/files/wrap/Environmental_benefits_of_recycling_2010_update.3b174d59.8816.pdf
34. Smithers GW (2008) Whey and whey proteins—from 'gutter-to-gold'. Int Dairy J 18 (7):695–704
35. Madaki YS, Seng L (2013) Pollution control: how feasible is zero discharge concepts in Malaysia palm oil mills. Am J Eng Res 2(10):239–252

36. Wu TY, Mohammad AW, Jahim JM, Anuar N (2010) Pollution control technologies for the treatment of palm oil mill effluent (pome) through end-of-pipe processes. J Environ Manag 91 (7):1467–1490
37. Bundhoo ZM (2019) Potential of bio-hydrogen production from dark fermentation of crop residues: a review. Int J Hydrog Energy 44(32):17346–17362
38. Sobhi B, Isam S, Ahmad Y, Jacob H (2007) Reducing the environmental impact of olive mill wastewater in Jordan, Palestine and Israel. In: Shuval H, Dweik H (eds) Water resources in the middle east. Springer, Berlin, Heidelberg, pp 409–415
39. Arvanitoyannis IS, Kassaveti A (2008) Olive oil waste management: treatment methods and potential uses of treated waste. In: Waste management for the food industries. Elsevier Academic Press, Amsterdam, pp 453–568
40. Hamdi M (1992) Toxicity and biodegradability of olive mill wastewaters in batch anaerobic digestion. Appl Biochem Biotechnol 37(2):155–163
41. Angenent LT, Karim K, Al-Dahhan MH, Domiguez-Espinosa R (2004) Production of bioenergy and biochemicals from industrial and agricultural wastewater. Trends Biotechnol 22(9):477–485
42. Batstone DJ, Keller J, Angelidaki I, Kalyuzhnyi S, Pavlostathis S, Rozzi A, Sanders W, Siegrist H, Vavilin V (2002) The IWA anaerobic digestion model no 1 (ADM1). Water Sci Technol 45(10):65–73
43. Plaza G, Robredo P, Pacheco O, Toledo AS (1996) Anaerobic treatment of municipal solid waste. Water Sci Technol 33(3):169–175
44. McInerney MJ, Bryant MP (1981) Basic principles of bioconversions in anaerobic digestion and methanogenesis. In: Sofer SS, Zaborsky OR (eds) Biomass conversion processes for energy and fuels. Springer, New York, pp 277–296
45. Batstone DJ, Keller J, Newell RB, Newland M (2000) Modelling anaerobic degradation of complex wastewater. I: Model development. Bioresour Technol 75(1):67–74
46. Zoetemeyer RJ, Vandenheuvel JC, Cohen A (1982) pH influence on acidogenic dissimilation of glucose in an anaerobic digester. Water Res 16(3):303–311
47. Sivagurunathan P, Kumar G, Mudhoo A, Rene ER, Saratale GD, Kobayashi T, Xu KQ, Kim SH, Kim DH (2017) Fermentative hydrogen production using lignocellulose biomass: an overview of pre-treatment methods, inhibitor effects and detoxification experiences. Renew Sust Energ Rev 77:28–42
48. Urbaniec K, Bakker RR (2015) Biomass residues as raw material for dark hydrogen fermentation - a review. Int J Hydrog Energy 40(9):3648–3658
49. Parawira W, Murto M, Zvauya R, Mattiasson B (2004) Anaerobic batch digestion of solid potato waste alone and in combination with sugar beet leaves. Renew Energy 29 (11):1811–1823
50. Liu TC, Ghosh S (1997) Phase separation during anaerobic fermentation of solid substrates in an innovative plug-flow reactor. Water Sci Technol 36(6–7):303–310
51. Pavlostathis SG, Giraldogomez E (1991) Kinetics of anaerobic treatment - a critical-review. Crit Rev Environ Sci Technol 21(5–6):411–490
52. Boone DR, Whitman WB, Rouvière P (1993) Diversity and taxonomy of methanogens. In: Ferry JG (ed) Methanogenesis. Ecology, physiology, biochemistry & genetics. Springer, New York, pp 35–80
53. Saady NMC (2013) Homoacetogenesis during hydrogen production by mixed cultures dark fermentation: unresolved challenge. Int J Hydrog Energy 38(30):13172–13191
54. Schink B (2002) Synergistic interactions in the microbial world. Antonie Van Leeuwenhoek 81(1–4):257–261
55. Demirel B, Scherer P (2008) The roles of acetotrophic and hydrogenotrophic methanogens during anaerobic conversion of biomass to methane: a review. Rev Environ Sci Biotechnol 7 (2):173–190
56. Schink B (1997) Energetics of syntrophic cooperation in methanogenic degradation. Microbiol Mol Biol Rev 61(2):262–+

57. Jerris J, McCarty P (1965) The biochemistry of methane fermentation using c14 tracers. J Water Poll Control Fed 39:178–192
58. McHugh S, Carton M, Mahony T, O'Flaherty V (2003) Methanogenic population structure in a variety of anaerobic bioreactors. FEMS Microbiol Lett 219(2):297–304
59. Hallenbeck PC, Benemann JR (2002) Biological hydrogen production; fundamentals and limiting processes. Int J Hydrog Energy 27(11–12):1185–1193
60. Xu KW, Liu H, Li XF, Chen JA, Wang AJ (2010) Typical methanogenic inhibitors can considerably alter bacterial populations and affect the interaction between fatty acid degraders and homoacetogens. Appl Microbiol Biotechnol 87(6):2267–2279
61. Lyberatos G, Skiadas I (1999) Modelling of anaerobic digestion–a review. Global Nest Int J 1 (2):63–76
62. Das D, Veziroglu TN (2001) Hydrogen production by biological processes: a survey of literature. Int J Hydrog Energy 26(1):13–28
63. Nandi R, Sengupta S (1998) Microbial production of hydrogen: an overview. Crit Rev Microbiol 24(1):61–84
64. Levin DB, Pitt L, Love M (2004) Biohydrogen production: prospects and limitations to practical application. Int J Hydrog Energy 29(2):173–185
65. Ueno Y, Kawai T, Sato S, Otsuka S, Morimoto M (1995) Biological production of hydrogen from cellulose by natural anaerobic microflora. J Ferment Bioeng 79(4):395–397
66. Brosseau JD, Zajic JE (1982) Hydrogen-gas production with citrobacter-intermedius and clostridium-pasteurianum. J Chem Technol Biotechnol 32(3):496–502
67. Lay JJ, Lee YJ, Noike T (1999) Feasibility of biological hydrogen production from organic fraction of municipal solid waste. Water Res 33(11):2579–2586
68. Fang HHP, Zhang T, Liu H (2002) Microbial diversity of a mesophilic hydrogen-producing sludge. Appl Microbiol Biotechnol 58(1):112–118
69. Kapdan IK, Kargi F (2006) Bio-hydrogen production from waste materials. Enzyme Microb Technol 38(5):569–582
70. Kumar N, Das D (2001) Continuous hydrogen production by immobilized enterobacter cloacae iit-bt 08 using lignocellulosic materials as solid matrices. Enzyme Microb Technol 29(4–5):280–287
71. Oh SE, Van Ginkel S, Logan BE (2003) The relative effectiveness of pH control and heat treatment for enhancing biohydrogen gas production. Environ Sci Technol 37(22):5186–5190
72. Yu HQ, Fang HHP (2001) Acidification of mid- and high-strength dairy wastewaters. Water Res 35(15):3697–3705
73. Chookaew T, O-Thong S, Prasertsan P (2014) Biohydrogen production from crude glycerol by immobilized klebsiella sp tr17 in a UASB reactor and bacterial quantification under non-sterile conditions. Int J Hydrog Energy 39(18):9580–9587
74. Sarma S, Dubey VK, Moholkar VS (2016) Kinetic and thermodynamic analysis (with statistical optimization) of hydrogen production from crude glycerol using clostridium pasteurianum. Int J Hydrog Energy 41(44):19972–19989
75. Maru BT, Bielen AAM, Kengen SWM, Constanti M, Medina F, International, J.P (2012) Biohydrogen production from glycerol using thermotoga spp. In: Whec 2012 conference proceedings - 19th world hydrogen energy conference 2012
76. Liu Q, Xiong D, Hong-Bo H, Zhang X (2015) Hydrogen production from glycerol using a genetically engineered Escherichia coli HW2 strain. J Chem Eng Chin Univ 5:1133–1137
77. Zhang D, Xiao N, Mahbubani KT, del Rio-Chanona EA, Slater NKH, Vassiliadis VS (2015) Bioprocess modelling of biohydrogen production by rhodopseudomonas palustris: model development and effects of operating conditions on hydrogen yield and glycerol conversion efficiency. Chem Eng Sci 130:68–78
78. Li YF, Qiu YQ, Zhang X, Zhu ML, Tan WS (2019) Strain screening and optimization of biohydrogen production by Enterobacter aerogenes EB-06 from glycerol fermentation. Bioresour Bioprocess 6:15

79. Asadi N, Zilouei H (2017) Optimization of organosolv pretreatment of rice straw for enhanced biohydrogen production using Enterobacter aerogenes. Bioresour Technol 227:335–344
80. Duangmanee T, Padmasiri SI, Simmons JJ, Raskin L, Sung S (2007) Hydrogen production by anaerobic microbial communities exposed to repeated heat treatments. Water Environ Res 79 (9):975–983
81. Chen WH, Sung S, Chen SY (2009) Biological hydrogen production in an anaerobic sequencing batch reactor: pH and cyclic duration effects. Int J Hydrog Energy 34(1):227–234
82. Goud RK, Mohan SV (2012) Regulating biohydrogen production from wastewater by applying organic load-shock: change in the microbial community structure and bio-electrochemical behavior over long-term operation. Int J Hydrog Energy 37(23):17763–17777
83. Hawkes FR, Hussy I, Kyazze G, Dinsdale R, Hawkes DL (2007) Continuous dark fermentative hydrogen production by mesophilic microflora: principles and progress. Int J Hydrog Energy 32(2):172–184
84. Valdez-Vazquez I, Poggi-Varaldo HM (2009) Hydrogen production by fermentative consortia. Renew Sust Energ Rev 13(5):1000–1013
85. Hallenbeck PC, Ghosh D (2009) Advances in fermentative biohydrogen production: the way forward? Trends Biotechnol 27(5):287–297
86. Lee HS, Salerno MB, Rittmann BE (2008) Thermodynamic evaluation on H2 production in glucose fermentation. Environ Sci Technol 42(7):2401–2407
87. Chen X, Sun YQ, Xiu ZL, Li XH, Zhang DJ (2006) Stoichiometric analysis of biological hydrogen production by fermentative bacteria. Int J Hydrog Energy 31(4):539–549
88. Rodriguez J, Kleerebezem R, Lema JM, van Loosdrecht MCM (2006) Modeling product formation in anaerobic mixed culture fermentations. Biotechnol Bioeng 93(3):592–606
89. Bartacek J, Zabranska J, Lens PNL (2007) Developments and constraints in fermentative hydrogen production. Biofuels Bioprod Biorefin 1(3):201–214
90. Rai PK (2016) Recent advances in substrate utilization for fermentative hydrogen production. J Appl Biol Biotechnol 4(06):059–067
91. Shin HS, Youn JH, Kim SH (2004) Hydrogen production from food waste in anaerobic mesophilic and thermophilic acidogenesis. Int J Hydrog Energy 29(13):1355–1363
92. Kim DH, Kim SH, Ko IB, Lee CY, Shin HS (2008) Start-up strategy for continuous fermentative hydrogen production: early switchover from batch to continuous operation. Int J Hydrog Energy 33(5):1532–1541
93. Kim SH, Han SK, Shin HS (2008) Optimization of continuous hydrogen fermentation of food waste as a function of solids retention time independent of hydraulic retention time. Process Biochem 43(2):213–218
94. Lay JJ, Fan KS, Hwang JI, Chang JI, Hsu PC (2005) Factors affecting hydrogen production from food wastes by clostridium-rich composts. J Environ Eng-ASCE 131(4):595–602
95. Sreela-Or C, Imai T, Plangklang P, Reungsang A (2011) Optimization of key factors affecting hydrogen production from food waste by anaerobic mixed cultures. Int J Hydrog Energy 36 (21):14120–14133
96. Sreela-or C, Plangklang P, Imai T, Reungsang A (2011) Co-digestion of food waste and sludge for hydrogen production by anaerobic mixed cultures: statistical key factors optimization. Int J Hydrog Energy 36(21):14227–14237
97. Kim DH, Kim SH, Kim HW, Kim MS, Shin HS (2011) Sewage sludge addition to food waste synergistically enhances hydrogen fermentation performance. Bioresour Technol 102 (18):8501–8506
98. Pan JM, Zhang RH, El-Mashad HM, Sun HW, Ying YB (2008) Effect of food to microorganism ratio on biohydrogen production from food waste via anaerobic fermentation. Int J Hydrog Energy 33(23):6968–6975
99. Chen WH, Chen SY, Khanal SK, Sung SW (2006) Kinetic study of biological hydrogen production by anaerobic fermentation. Int J Hydrog Energy 31(15):2170–2178
100. Kim DH, Kim SH, Shin HS (2009) Hydrogen fermentation of food waste without inoculum addition. Enzyme Microb Technol 45(3):181–187

101. Kim DH, Wu JY, Jeong KW, Kim MS, Shin HS (2011) Natural inducement of hydrogen from food waste by temperature control. Int J Hydrog Energy 36(17):10666–10673
102. Han SK, Shin HS (2004) Performance of an innovative two-stage process converting food waste to hydrogen and methane. J Air Waste Manag Assoc 54(2):242–249
103. Han SK, Shin HS (2004) Biohydrogen production by anaerobic fermentation of food waste. Int J Hydrog Energy 29(6):569–577
104. Kim SH, Han SK, Shin HS (2004) Feasibility of biohydrogen production by anaerobic co-digestion of food waste and sewage sludge. Int J Hydrog Energy 29(15):1607–1616
105. Lee YW, Chung J (2010) Bioproduction of hydrogen from food waste by pilot-scale combined hydrogen/methane fermentation. Int J Hydrog Energy 35(21):11746–11755
106. Wongthanate J, Chinnacotpong K, Khumpong M (2014) Impacts of pH, temperature, and pretreatment method on biohydrogen production from organic wastes by sewage microflora. Int J Energy Environ Eng 5(1):6
107. Hassan GK, Massanet-Nicolau J, Dinsdale R, Jones RJ, Abo-Aly MM, El-Gohary FA, Guwy A (2019) A novel method for increasing biohydrogen production from food waste using electrodialysis. Int J Hydrog Energy 44(29):14715–14720
108. Abreu AA, Tavares F, Alves MM, Cavaleiro AJ, Pereira MA (2019) Garden and food waste co-fermentation for biohydrogen and biomethane production in a two-step hyperthermophilic-mesophilic process. Bioresour Technol 278:180–186
109. Shin HS, Youn JH (2005) Conversion of food waste into hydrogen by thermophilic acidogenesis. Biodegradation 16(1):33–44
110. Lee ZK, Li SL, Lin JS, Wang YH, Kuo PC, Cheng SS (2008) Effect of pH in fermentation of vegetable kitchen wastes on hydrogen production under a thermophilic condition. Int J Hydrog Energy 33(19):5234–5241
111. Lee ZK, Li SL, Kuo PC, Chen IC, Tien YM, Huang YJ, Chuang CP, Wong SC, Cheng SS (2010) Thermophilic bio-energy process study on hydrogen fermentation with vegetable kitchen waste. Int J Hydrog Energy 35(24):13458–13466
112. Chu CF, Li YY, Xu KQ, Ebie Y, Inamori Y, Kong HN (2008) A pH- and temperature-phased two-stage process for hydrogen and methane production from food waste. Int J Hydrog Energy 33(18):4739–4746
113. Lay JJ, Fan KS, Chang J, Ku CH (2003) Influence of chemical nature of organic wastes on their conversion to hydrogen by heat-shock digested sludge. Int J Hydrog Energy 28 (12):1361–1367
114. Wongthanate J, Chinnacotpong K (2015) Optimal conditions for biological hydrogen production from food waste. Environ Eng Res 20(2):121–125
115. De Gioannis G, Muntoni A, Polettini A, Pomi R (2013) A review of dark fermentative hydrogen production from biodegradable municipal waste fractions. Waste Manag 33 (6):1345–1361
116. Kayhanian M (1995) Biodegradability of the organic fraction of municipal solid-waste in a high-solids anaerobic digester. Waste Manag Res 13(2):123–136
117. Themelis NJ, Kim YH (2002) Material and energy balances in a large-scale aerobic bioconversion cell. Waste Manag Res 20(3):234–242
118. Li SL, Kuo SC, Lin JS, Lee ZK, Wang YH, Cheng SS (2008) Process performance evaluation of intermittent-continuous stirred tank reactor for anaerobic hydrogen fermentation with kitchen waste. Int J Hydrog Energy 33(5):1522–1531
119. Liu DW, Liu DP, Zeng RJ, Angelidaki I (2006) Hydrogen and methane production from household solid waste in the two-stage fermentation process. Water Res 40(11):2230–2236
120. Alzate-Gaviria LM, Sebastian PJ, Perez-Hernandez A, Eapen D (2007) Comparison of two anaerobic systems for hydrogen production from the organic fraction of municipal solid waste and synthetic wastewater. Int J Hydrog Energy 32(15):3141–3146
121. Ebrahimian F, Karimi K (2020) Efficient biohydrogen and advanced biofuel coproduction from municipal solid waste through a clean process. Bioresour Technol 300:122656

122. Gomez X, Moran A, Cuetos MJ, Sanchez ME (2006) The production of hydrogen by dark fermentation of municipal solid wastes and slaughterhouse waste: a two-phase process. J Power Sources 157(2):727–732
123. Tenca A, Schievano A, Perazzolo F, Adani F, Oberti R (2011) Biohydrogen from thermophilic co-fermentation of swine manure with fruit and vegetable waste: maximizing stable production without pH control. Bioresour Technol 102(18):8582–8588
124. Lee DY, Ebie Y, Xu KQ, Li YY, Inamori Y (2010) Continuous H2 and CH4 production from high-solid food waste in the two-stage thermophilic fermentation process with the recirculation of digester sludge. Bioresour Technol 101:S42–S47
125. Ueno Y, Fukui H, Goto M (2007) Operation of a two-stage fermentation process producing hydrogen and methane from organic waste. Environ Sci Technol 41(4):1413–1419
126. Van Ginkel SW, Oh SE, Logan BE (2005) Biohydrogen gas production from food processing and domestic wastewaters. Int J Hydrog Energy 30(15):1535–1542
127. Balin R, Raman S (2011) Biohydrogen from sugar industry waste and sewage. Proc Institution Civil Eng-Energy 164(2):78–88
128. Ewan BCR, Allen RWK (2005) A figure of merit assessment of the routes to hydrogen. Int J Hydrog Energy 30(8):809–819
129. Feng XQ, Wang H, Wang Y, Wang XF, Huang JX (2010) Biohydrogen production from apple pomace by anaerobic fermentation with river sludge. Int J Hydrog Energy 35(7):3058–3064
130. Intanoo P, Suttikul T, Leethochawalit M, Gulari E, Chavadej S (2014) Hydrogen production from alcohol wastewater with added fermentation residue by an anaerobic sequencing batch reactor (asbr) under thermophilic operation. Int J Hydrog Energy 39(18):9611–9620
131. Wang CC, Chang CW, Chu CP, Lee DJ, Chang BV, Liao CS, Tay JH (2003) Using filtrate of waste biosolids to effectively produce bio-hydrogen by anaerobic fermentation. Water Res 37 (11):2789–2793
132. Hussy I, Hawkes FR, Dinsdale R, Hawkes DL (2005) Continuous fermentative hydrogen production from sucrose and sugarbeet. Int J Hydrog Energy 30(5):471–483
133. Jung KW, Kim DH, Shin HS (2010) Continuous fermentative hydrogen production from coffee drink manufacturing wastewater by applying UASB reactor. Int J Hydrog Energy 35 (24):13370–13378
134. Jung KW, Kim DH, Shin HS (2011) Fermentative hydrogen production from laminaria japonica and optimization of thermal pretreatment conditions. Bioresour Technol 102 (3):2745–2750
135. Jung KW, Kim DH, Shin HS (2011) A simple method to reduce the start-up period in a H2-producing UASB reactor. Int J Hydrog Energy 36(2):1466–1473
136. Santos SC, Rosa PRF, Sakamoto IK, Varesche MBA, Silva EL (2014) Hydrogen production from diluted and raw sugarcane vinasse under thermophilic anaerobic conditions. Int J Hydrog Energy 39(18):9599–9610
137. Buitron G, Carvajal C (2010) Biohydrogen production from tequila vinasses in an anaerobic sequencing batch reactor: effect of initial substrate concentration, temperature and hydraulic retention time. Bioresour Technol 101(23):9071–9077
138. Guo WQ, Ren NQ, Wang XJ, Xiang WS, Meng ZH, Ding J, Qu YY, Zhang LS (2008) Biohydrogen production from ethanol-type fermentation of molasses in an expanded granular sludge bed (EGSB) reactor. Int J Hydrog Energy 33(19):4981–4988
139. Fan KS, Kan NR, Lay JJ (2006) Effect of hydraulic retention time on anaerobic hydrogenesis in CSTR. Bioresour Technol 97(1):84–89
140. Fan YT, Zhang GS, Guo XY, Xing Y, Fan MH (2006) Biohydrogen-production from beer lees biomass by cow dung compost. Biomass Bioenergy 30(5):493–496
141. Cui MJ, Yuan ZL, Zhi XH, Shen JQ (2009) Optimization of biohydrogen production from beer lees using anaerobic mixed bacteria. Int J Hydrog Energy 34(19):7971–7978
142. Lay CH, Sung IY, Kumar G, Chu CY, Chen CC, Lin CY (2012) Optimizing biohydrogen production from mushroom cultivation waste using anaerobic mixed cultures. Int J Hydrog Energy 37(21):16473–16478

143. Noike T, Ko IB, Yokoyama S, Kohno Y, Li YY (2005) Continuous hydrogen production from organic waste. Water Sci Technol 52(1–2):145–151
144. Liu Q, Zhang XL, Yu LJ, Zhao AH, Tai J, Liu JY, Qian GR, Xu ZP (2011) Fermentative hydrogen production from fresh leachate in batch and continuous bioreactors. Bioresour Technol 102(9):5411–5417
145. Song ZX, Wang ZY, Wu LY, Fan YT, Hou HW (2012) Effect of microwave irradiation pretreatment of cow dung compost on bio-hydrogen process from corn stalk by dark fermentation. Int J Hydrog Energy 37(8):6554–6561
146. Zhang ML, Fan YT, Xing Y, Pan CM, Zhang GS, Lay JJ (2007) Enhanced biohydrogen production from cornstalk wastes with acidification pretreatment by mixed anaerobic cultures. Biomass Bioenergy 31(4):250–254
147. Minon-Fuentes R, Aguilar-Juarez O (2019) Hydrogen production from coffee pulp by dark fermentation. Water Sci Technol 80(9):1692–1701
148. Saraphirom P, Reungsang A (2010) Optimization of biohydrogen production from sweet sorghum syrup using statistical methods. Int J Hydrog Energy 35(24):13435–13444
149. Saraphirom P, Reungsang A (2013) Enhancement of biohydrogen production from sweet sorghum syrup by anaerobic seed sludge in an anaerobic sequencing batch reactor by nutrient and vitamin supplementations. Environ Technol 34(17):2503–2511
150. Antonopoulou G, Gavala HN, Skiadas IV, Angelopoulos K, Lyberatos G (2008) Biofuels generation from sweet sorghum: fermentative hydrogen production and anaerobic digestion of the remaining biomass. Bioresour Technol 99(1):110–119
151. Sangyoka S, Reungsang A, Moonamart S (2007) Repeated-batch fermentative for bio-hydrogen production from cassava starch manufacturing wastewater. Pak J Biol Sci 10 (11):1782–1789
152. Mari AG, Andreani CL, Tonello TU, Leite LCC, Fernandes JR, Lopes DD, Rodrigues JAD, Gomes SD (2020) Biohydrogen and biomethane production from cassava wastewater in a two-stage anaerobic sequencing batch biofilm reactor. Int J Hydrog Energy 45(8):5165–5174
153. Luo G, Xie L, Zou ZH, Wang W, Zhou Q (2010) Evaluation of pretreatment methods on mixed inoculum for both batch and continuous thermophilic biohydrogen production from cassava stillage. Bioresour Technol 101(3):959–964
154. Wang W, Xie L, Chen JR, Luo G, Zhou Q (2011) Biohydrogen and methane production by co-digestion of cassava stillage and excess sludge under thermophilic condition. Bioresour Technol 102(4):3833–3839
155. Luo G, Xie L, Zhou Q, Angelidaki I (2011) Enhancement of bioenergy production from organic wastes by two-stage anaerobic hydrogen and methane production process. Bioresour Technol 102(18):8700–8706
156. Mohan SV, Babu ML, Reddy MV, Mohanakrishna G, Sarma PN (2009) Harnessing of biohydrogen by acidogenic fermentation of Citrus limetta peelings: effect of extraction procedure and pretreatment of biocatalyst. Int J Hydrog Energy 34(15):6149–6156
157. Chuang Y-S, Chen C-C, Lay C-H, Sung I-Y, Wu J-H, Lee S-C, Sen B, Lin C-Y (2012) Optimization of incubation factors for fermentative hydrogen production from agricultural wastes. Sustain Environ Res 22(2):99–106
158. Reungsang A, Sreela-or C (2013) Bio-hydrogen production from pineapple waste extract by anaerobic mixed cultures. Energies 6(4):2175–2190
159. Kim MS, Lee DY (2010) Fermentative hydrogen production from tofu-processing waste and anaerobic digester sludge using microbial consortium. Bioresour Technol 101:S48–S52
160. Stabnikova O, Wang J-Y, Ivanov V (2010) Value-added biotechnological products from organic wastes. In: Wang LK, Ivanov V, Tay J-H (eds) Environmental biotechnology. Humana Press, Totowa, pp 343–394
161. Yasin NHM, Mumtaz T, Hassan MA, Rahman NA (2013) Food waste and food processing waste for biohydrogen production: a review. J Environ Manage 130:375–385

162. Azbar N, Dokgoz FTC, Keskin T, Korkmaz KS, Syed HM (2009) Continuous fermentative hydrogen production from cheese whey wastewater under thermophilic anaerobic conditions. Int J Hydrog Energy 34(17):7441–7447
163. Nielsen PH. Potato flour production 2004. Available from: http://www.lcafood.dk/processes/industry/potatoflourproduction.htm
164. Diamantis VI, Vaiopoulou E, Aivasidis A (2007) Fundamentals and applications of anaerobic digestion for sustainable treatment of food industry wastewater. In: Oreopoulou V, Russ W (eds) Utilization of by-products and treatment of waste in the food industry. Springer, New York, pp 73–97
165. Nielsen PH. Sugar production 2003. Available from: http://www.lcafood.dk/processes/industry/sugarproduction.htm
166. Satyawali Y, Balakrishnan M (2008) Wastewater treatment in molasses-based alcohol distilleries for cod and color removal: a review. J Environ Manag 86(3):481–497
167. Vlissidis A, Zouboulis AI (1993) Thermophilic anaerobic-digestion of alcohol distillery wastewaters. Bioresour Technol 43(2):131–140
168. Guo WQ, Ren NQ, Chen ZB, Liu BF, Wang XJ, Xiang WS, Ding J (2008) Simultaneous biohydrogen production and starch wastewater treatment in an acidogenic expanded granular sludge bed reactor by mixed culture for long-term operation. Int J Hydrog Energy 33 (24):7397–7404
169. Wu JH, Lin CY (2004) Biohydrogen production by mesophilic fermentation of food wastewater. Water Sci Technol 49(5–6):223–228
170. Ren NQ, Chua H, Chan SY, Tsang YF, Wang YJ, Sin N (2007) Assessing optimal fermentation type for bio-hydrogen production in continuous-flow acidogenic reactors. Bioresour Technol 98(9):1774–1780
171. Ren NQ, Li JZ, Li BK, Wang Y, Liu SR (2006) Biohydrogen production from molasses by anaerobic fermentation with a pilot-scale bioreactor system. Int J Hydrog Energy 31 (15):2147–2157
172. Lin YH, Juan ML, Hsien HJ (2011) Effects of temperature and initial pH on biohydrogen production from food-processing wastewater using anaerobic mixed cultures. Biodegradation 22(3):551–563
173. Aceves-Lara CA, Latrille E, Bernet N, Buffiere P, Steyer JP (2008) A pseudo-stoichiometric dynamic model of anaerobic hydrogen production from molasses. Water Res 42 (10–11):2539–2550
174. Chang JJ, Wu JH, Wen FS, Hung KY, Chen YT, Hsiao CL, Lin CY, Huang CC (2008) Molecular, monitoring of microbes in a continuous hydrogen-producing system with different hydraulic retention time. Int J Hydrog Energy 33(5):1579–1585
175. Li QY, Li YF (2019) Coproduction of hydrogen and methane in a CSTR-IC two-stage anaerobic digestion system from molasses wastewater. Water Sci Technol 79(2):270–277
176. Han W, Chen H, Yao X, Li Y-f, Yang C-p (2010) Biohydrogen production with anaerobic sludge immobilized by granular activated carbon in a continuous stirred-tank. J For Res 21 (4):509–513
177. O-Thong S, Mamimin C, Prasertsan P (2011) Effect of temperature and initial pH on biohydrogen production from palm oil mill effluent: long-term evaluation and microbial community analysis. Electron J Biotechnol 14(5) https://scielo.conicyt.cl/scielo.php?script=sci_arttext&pid=S0717-34582011000500009
178. Koutrouli EC, Kalfas H, Gavala HN, Skiadas IV, Stamatelatou K, Lyberatos G (2009) Hydrogen and methane production through two-stage mesophilic anaerobic digestion of olive pulp. Bioresour Technol 100(15):3718–3723
179. O-Thong S, Prasertsan P, Intrasungkha N, Dhamwichukorn S, Birkeland NK (2008) Optimization of simultaneous thermophilic fermentative hydrogen production and cod reduction from palm oil mill effluent by thermoanaerobacterium-rich sludge. Int J Hydrog Energy 33 (4):1221–1231

180. O-Thong S, Prasertsan P, Intrasungkha N, Dhamwichukom S, Birkeland NK (2007) Improvement of biohydrogen production and treatment efficiency on palm oil mill effluent with nutrient supplementation at thermophilic condition using an anaerobic sequencing batch reactor. Enzyme Microb Technol 41(5):583–590
181. Atif AAY, Fakhru'l-Razi A, Ngan MA, Morimoto M, Iyuke SE, Veziroglu NT (2005) Fed batch production of hydrogen from palm oil mill effluent using anaerobic microflora. Int J Hydrog Energy 30(13–14):1393–1397
182. Mamimin C, Kongjan P, O-Thong S, Prasertsan P (2019) Enhancement of biohythane production from solid waste by co-digestion with palm oil mill effluent in two-stage thermophilic fermentation. Int J Hydrog Energy 44(32):17224–17237
183. Seifert K, Waligorska M, Wojtowski M, Laniecki M (2009) Hydrogen generation from glycerol in batch fermentation process. Int J Hydrog Energy 34(9):3671–3678
184. Akutsu Y, Lee DY, Li YY, Noike T (2009) Hydrogen production potentials and fermentative characteristics of various substrates with different heat-pretreated natural microflora. Int J Hydrog Energy 34(13):5365–5372
185. Sittijunda S, Reungsang A (2012) Biohydrogen production from waste glycerol and sludge by anaerobic mixed cultures. Int J Hydrog Energy 37(18):13789–13796
186. Selembo PA, Perez JM, Lloyd WA, Logan BE (2009) Enhanced hydrogen and 1,3-propanediol production from glycerol by fermentation using mixed cultures. Biotechnol Bioeng 104(6):1098–1106
187. Rodrigues CV, Santana KO, Nespeca MG, Rodrigues AV, Pires LO, Maintinguer SI (2020) Energy valorization of crude glycerol and sanitary sewage in hydrogen generation by biological processes. Int J Hydrog Energy 45(21):11943–11953
188. Ito T, Nakashimada Y, Senba K, Matsui T, Nishio N (2005) Hydrogen and ethanol production from glycerol-containing wastes discharged after biodiesel manufacturing process. J Biosci Bioeng 100(3):260–265
189. Sakai S, Yagishita T (2007) Microbial production of hydrogen and ethanol from glycerol-containing wastes discharged from a biodiesel fuel production plant in a bioelectrochemical reactor with thionine. Biotechnol Bioeng 98(2):340–348
190. Chong ML, Sabaratnam V, Shirai Y, Hassan MA (2009) Biohydrogen production from biomass and industrial wastes by dark fermentation. Int J Hydrog Energy 34(8):3277–3287
191. Mohan SV, Babu VL, Sarma PN (2007) Anaerobic biohydrogen production from dairy wastewater treatment in sequencing batch reactor (ANSBR): effect of organic loading rate. Enzyme Microb Technol 41(4):506–515
192. Yang PF, Zhang RH, MeGarvey JA, Benemann JR (2007) Biohydrogen production from cheese processing wastewater by anaerobic fermentation using mixed microbial communities. Int J Hydrog Energy 32(18):4761–4771
193. Davila-Vazquez G, Alatriste-Mondragon F, de Leon-Rodriguez A, Razo-Flores E (2008) Fermentative hydrogen production in batch experiments using lactose, cheese whey and glucose: influence of initial substrate concentration and pH. Int J Hydrog Energy 33 (19):4989–4997
194. Antonopoulou G, Stamatelatou K, Venetsaneas N, Kornaros M, Lyberatos G (2008) Biohydrogen and methane production from cheese whey in a two-stage anaerobic process. Ind Eng Chem Res 47(15):5227–5233
195. Venetsaneas N, Antonopoulou G, Stamatelatou K, Kornaros M, Lyberatos G (2009) Using cheese whey for hydrogen and methane generation in a two-stage continuous process with alternative pH controlling approaches. Bioresour Technol 100(15):3713–3717
196. Romao BB, Batista FRX, Ferreira JS, Costa HCB, Resende MM, Cardoso VL (2014) Biohydrogen production through dark fermentation by a microbial consortium using whey permeate as substrate. Appl Biochem Biotechnol 172(7):3670–3685
197. Thompson RS (2008) Hydrogen production by anaerobic fermentation using agricultural and food processing wastes utilizing a two-stage digestion system. Utah State University
198. Wong YM, Show PL, Wu TY, Leong HY, Ibrahim S, Juan JC (2019) Production of bio-hydrogen from dairy wastewater using pretreated landfill leachate sludge as an inoculum. J Biosci Bioeng 127(2):150–159

199. Tang GL, Huang J, Sun ZJ, Tang QQ, Yan CH, Liu GQ (2008) Biohydrogen production from cattle wastewater by enriched anaerobic mixed consortia: influence of fermentation temperature and pH. J Biosci Bioeng 106(1):80–87
200. Yokoyama H, Waki M, Moriya N, Yasuda T, Tanaka Y, Haga K (2007) Effect of fermentation temperature on hydrogen production from cow waste slurry by using anaerobic microflora within the slurry. Appl Microbiol Biotechnol 74(2):474–483
201. Gilroyed BH, Chang C, Chu A, Hao XY (2008) Effect of temperature on anaerobic fermentative hydrogen gas production from feedlot cattle manure using mixed microflora. Int J Hydrog Energy 33(16):4301–4308
202. Xing Y, Li Z, Fan YT, Hou HW (2010) Biohydrogen production from dairy manures with acidification pretreatment by anaerobic fermentation. Environ Sci Pollut Res 17(2):392–399
203. Guo XM, Trably E, Latrille E, Carrere H, Steyer JP (2010) Hydrogen production from agricultural waste by dark fermentation: a review. Int J Hydrog Energy 35(19):10660–10673
204. Lateef SA, Beneragama N, Yamashiro T, Iwasaki M, Ying C, Umetsu K (2012) Biohydrogen production from co-digestion of cow manure and waste milk under thermophilic temperature. Bioresour Technol 110:251–257
205. Perera KRJ, Nirmalakhandan N (2011) Evaluation of dairy cattle manure as a supplement to improve net energy gain in fermentative hydrogen production from sucrose. Bioresour Technol 102(18):8688–8695
206. Zhu J, Li YC, Wu X, Miller C, Chen P, Ruan R (2009) Swine manure fermentation for hydrogen production. Bioresour Technol 100(22):5472–5477
207. Kotsopoulos TA, Fotidis IA, Tsolakis N, Martzopoulos GG (2009) Biohydrogen production from pig slurry in a CSTR reactor system with mixed cultures under hyper-thermophilic temperature (70 degrees c). Biomass Bioenergy 33(9):1168–1174
208. Li YC, Zhu J, Wu XA, Miller C, Wang LA (2010) The effect of pH on continuous biohydrogen production from swine wastewater supplemented with glucose. Appl Biochem Biotechnol 162(5):1286–1296
209. Wu XA, Yao WY, Zhu J (2010) Effect of pH on continuous biohydrogen production from liquid swine manure with glucose supplement using an anaerobic sequencing batch reactor. Int J Hydrog Energy 35(13):6592–6599
210. Saady NMC, Chaganti SR, Lalman JA, Heath D (2012) Impact of culture source and linoleic acid (c18:2) on biohydrogen production from glucose under mesophilic conditions. Int J Hydrog Energy 37(5):4036–4045
211. Iyer P, Bruns MA, Zhang HS, Van Ginkel S, Logan BE (2004) H-2-producing bacterial communities from a heat-treated soil inoculum. Appl Microbiol Biotechnol 66(2):166–173
212. Sreethawong T, Chatsiriwatana S, Rangsunvigit P, Chavadej S (2010) Hydrogen production from cassava wastewater using an anaerobic sequencing batch reactor: effects of operational parameters, cod:N ratio, and organic acid composition. Int J Hydrog Energy 35(9):4092–4102
213. Lay J (2004) Factors affecting hydrogen production from high-solid organic wastes. In: Proceedings of 2nd international workshop on innovative anaerobic technology, Sendai
214. Griffin ME, McMahon KD, Mackie RI, Raskin L (1998) Methanogenic population dynamics during start-up of anaerobic digesters treating municipal solid waste and biosolids. Biotechnol Bioeng 57(3):342–355
215. Radjaram B, Saravanane R (2011) Start up study of UASB reactor treating press mud for biohydrogen production. Biomass Bioenergy 35(7):2721–2728
216. Fangkum A, Reungsang A (2011) Biohydrogen production from mixed xylose/arabinose at thermophilic temperature by anaerobic mixed cultures in elephant dung. Int J Hydrog Energy 36(21):13928–13938
217. Logan BE, Oh SE, Kim IS, Van Ginkel S (2002) Biological hydrogen production measured in batch anaerobic respirometers. Environ Sci Technol 36(11):2530–2535
218. Saady NMC, Chaganti SR, Lalman JA, Veeravalli SS, Shanmugam SR, Heath DD (2012) Effects of linoleic acid and its degradation by-products on mesophilic hydrogen production

using flocculated and granular mixed anaerobic cultures. Int J Hydrog Energy 37 (24):18747–18760
219. Saady NMC, Chaganti SR, Lalman JA, Veeravalli SS, Shanmugam SR, Heath DD (2012) Assessing the impact of palmitic, myristic and lauric acids on hydrogen production from glucose fermentation by mixed anaerobic granular cultures. Int J Hydrog Energy 37 (24):18761–18772
220. Mohan SV, Babu VL, Sarma PN (2008) Effect of various pretreatment methods on anaerobic mixed microflora to enhance biohydrogen production utilizing dairy wastewater as substrate. Bioresour Technol 99(1):59–67
221. Elbeshbishy E, Hafez H, Nakhla G (2011) Ultrasonication for biohydrogen production from food waste. Int J Hydrog Energy 36(4):2896–2903
222. Lee KS, Lo YS, Lo YC, Lin PJ, Chang JS (2004) Operation strategies for biohydrogen production with a high-rate anaerobic granular sludge bed bioreactor. Enzyme Microb Technol 35(6–7):605–612
223. Thompson LJ, Gray VM, Kalala B, Lindsay D, Reynolds K, von Holy A (2008) Biohydrogen production by enterobacter cloacae and citrobacter freundii in carrier induced granules. Biotechnol Lett 30(2):271–274
224. Chen SD, Lo YC, Lee KS, Huang TI, Chang JS (2009) Sequencing batch reactor enhances bacterial hydrolysis of starch promoting continuous bio-hydrogen production from starch feedstock. Int J Hydrog Energy 34(20):8549–8557
225. Baghchehsaraee B, Nakhla G, Karamanev D, Margaritis A, Reid G (2008) The effect of heat pretreatment temperature on fermentative hydrogen production using mixed cultures. Int J Hydrog Energy 33(15):4064–4073
226. Xiao BY, Liu JX (2009) Effects of various pretreatments on biohydrogen production from sewage sludge. Chin Sci Bull 54(12):2038–2044
227. Ren NQ, Cao GL, Wang AJ, Lee DJ, Guo WQ, Zhu YH (2008) Dark fermentation of xylose and glucose mix using isolated thermoanaerobacterium thermosaccharolyticum w16. Int J Hydrog Energy 33(21):6124–6132
228. O-Thong S, Prasertsan P, Birkeland NK (2009) Evaluation of methods for preparing hydrogen-producing seed inocula under thermophilic condition by process performance and microbial community analysis. Bioresour Technol 100(2):909–918
229. Ljunggren M, Zacchi G (2010) Techno-economic analysis of a two-step biological process producing hydrogen and methane. Bioresour Technol 101(20):7780–7788
230. Pendyala B, Chaganti SR, Lalman JA, Heath DD (2012) Evaluation of food waste and paper-cardboard waste blend for biohydrogen and methane production using mixed microbial consortia
231. Han W, Wang ZQ, Chen H, Yao X, Li YF (2011) Simultaneous biohydrogen and bioethanol production from anaerobic fermentation with immobilized sludge. J Biomed Biotechnol 2011:343791
232. Spagni A, Casu S, Farina R (2010) Effect of the organic loading rate on biogas composition in continuous fermentative hydrogen production. J Environ Sci Health Part A-Toxic/Hazardous Substances & Environmental Engineering 45(12):1475–1481
233. Bhaskar YV, Mohan SV, Sarma PN (2008) Effect of substrate loading rate of chemical wastewater on fermentative biohydrogen production in biofilm configured sequencing batch reactor. Bioresour Technol 99(15):6941–6948
234. Kongjan P, O-Thong S, Kotay M, Min B, Angelidaki I (2010) Biohydrogen production from wheat straw hydrolysate by dark fermentation using extreme thermophilic mixed culture. Biotechnol Bioeng 105(5):899–908
235. Zhang K, Ren NQ, Guo CH, Wang AJ, Cao GL (2011) Effects of various pretreatment methods on mixed microflora to enhance biohydrogen production from corn stover hydrolysate. J Environ Sci 23(12):1929–1936

236. Mohammadi P, Ibrahim S, Annuar MSM (2012) Comparative study on the effect of various pretreatment methods on the enrichment of hydrogen producing bacteria in anaerobic granulated sludge from brewery wastewater. Korean J Chem Eng 29(10):1347–1351
237. Rossi DM, da Costa JB, de Souza EA, Peralba MDR, Samios D, Ayub MAZ (2011) Comparison of different pretreatment methods for hydrogen production using environmental microbial consortia on residual glycerol from biodiesel. Int J Hydrog Energy 36(8):4814–4819
238. Hu B, Chen SL (2007) Pretreatment of methanogenic granules for immobilized hydrogen fermentation. Int J Hydrog Energy 32(15):3266–3273
239. Mu Y, Yu HQ, Wang Y (2006) The role of pH in the fermentative H2 production from an acidogenic granule-based reactor. Chemosphere 64(3):350–358
240. Das D (2009) Advances in biohydrogen production processes: an approach towards commercialization. Int J Hydrog Energy 34(17):7349–7357
241. Hay JXW, Wu TY, Juan JC, Jahim JM (2013) Biohydrogen production through photo fermentation or dark fermentation using waste as a substrate: overview, economics, and future prospects of hydrogen usage. Biofuels Bioprod Biorefin 7(3):334–352
242. Ljunggren M, Wallberg O, Zacchi G (2011) Techno-economic comparison of a biological hydrogen process and a 2nd generation ethanol process using barley straw as feedstock. Bioresour Technol 102(20):9524–9531
243. Chang PL, Hsu CW (2012) Value analysis for commercialization of fermentative hydrogen production from biomass. Int J Hydrog Energy 37(20):15746–15752
244. Li YC, Liu YF, Chu CY, Chang PL, Hsu CW, Lin PJ, Wu SY (2012) Techno-economic evaluation of biohydrogen production from wastewater and agricultural waste. Int J Hydrog Energy 37(20):15704–15710
245. Shi Y, Zhao XT, Cao P, Hu YY, Zhang L, Jia Y, Lu ZQ (2009) Hydrogen bio-production through anaerobic microorganism fermentation using kitchen wastes as substrate. Biotechnol Lett 31(9):1327–1333
246. Sen U, Shakdwipee M, Banerjee R (2008) Status of biological hydrogen production. J Sci Ind Res 67(11):980–993
247. Pandu K, Joseph S (2012) Comparisons and limitations of biohydrogen production processes: a review. Int J Adv Eng Tech 2(1):342
248. Mohan SV, Reddy MV, Subhash GV, Sarma PN (2010) Fermentative effluents from hydrogen producing bioreactor as substrate for poly(beta-oh) butyrate production with simultaneous treatment: an integrated approach. Bioresour Technol 101(23):9382–9386
249. Mohan SV, Devi MP (2012) Fatty acid rich effluent from acidogenic biohydrogen reactor as substrate for lipid accumulation in heterotrophic microalgae with simultaneous treatment. Bioresour Technol 123:627–635
250. Kumar G, Cho SK, Sivagurunathan P, Anburajan P, Mahapatra DM, Park JH, Pugazhendhi A (2018) Insights into evolutionary trends in molecular biology tools in microbial screening for biohydrogen production through dark fermentation. Int J Hydrog Energy 43(43):19885–19901

Chapter 9
Agricultural Waste-Derived Adsorbents for Decontamination of Heavy Metals

Soh-Fong Lim, Siti Kartina Abdul Karim, S. N. David Chua, and Bee-Huah Lim

Contents

1 Heavy Metals in the Environment 374
2 Agricultural Waste Biomass 374
3 Parameters Affecting the Adsorption of Heavy Metals 378
3.1 Effect of Solution pH 378
3.2 Effect of Initial Concentration 379
3.3 Effect of Adsorbent Dosage 379
3.4 Effect of Contact Time 380
3.5 Effect of Temperature 380
4 Kinetic Models 382
5 Adsorption Isotherms 383
Glossary 385
References 387

Abstract The growing concerns on the environment in recent years have influenced the usage of renewable sources as alternative materials to create a platform for the development of new technology with possible economic potential. Adsorbents derived from agricultural wastes have hidden economic values which could be benefited by transforming the agricultural wastes into valuable and useful products. Numerous agricultural wastes such as skins/peels, cores, pits, leaves, brunches, and pericarp are being produced in plantation and processing industries. The agricultural

S.-F. Lim (✉) · S. N. D. Chua
Faculty of Engineering, Universiti Malaysia Sarawak, Kota Samarahan, Sarawak, Malaysia
e-mail: sflim@unimas.my; csndavid@unimas.my

S. K. A. Karim
Faculty of Applied Sciences, Universiti Teknologi MARA, Kota Samarahan, Sarawak, Malaysia
e-mail: sitik094@uitm.edu.my

B.-H. Lim
Fuel Cell Institute, Universiti Kebangsuan, Malaysia, Bangl, Selangor Darul Ehsan, Malaysia
e-mail: beehuah@ukm.edu.my

L. K. Wang, M. -H. S. Wang, Y. -T. Hung, N. K. Shammas (eds.), *Integrated Natural Resources Management*, Handbook of Environmental Engineering 20,
https://doi.org/10.1007/978-3-030-55172-8_9

wastes have exhibited the potential usage as an adsorbent to remove contaminants from water environment which conserve the natural environment and resources mainly in the ecology system sustainably, for the reason that this utilization converts the agricultural wastes into value-added product and at the same time decontaminate polluted water source. This application on the utilization of agricultural wastes is not only good for a sustainable environment but is also suitable for rural economic development, meaning possible increases in profit for farmers and the agricultural industry. This chapter provides insight on some findings on heavy metal removal by adsorbents produced from agricultural wastes. The chapter also discusses the situation of the heavy metals in the environment, parameters affecting the adsorption process of the heavy metals, kinetic models, and adsorption isotherms that are associated with the agricultural waste-derived adsorbents. The development of the adsorbents from agricultural waste biomass and the prospect of developing hybrid adsorbent and magnetic adsorbent have attracted many researchers worldwide in performing research work on the application to water and wastewater treatment.

Keywords Agriculture · Agricultural wastes · Adsorbents · Heavy metals · Kinetics · Adsorption isotherm · Equilibrium time · Adsorbent dosage · Solution pH · Kinetic model

Nomenclature

α	Initial adsorption rate in mg/g·min unit
β	Desorption constant in g/mg unit
ε	Polanyi potential
Al	Aluminium
As	Arsenic
As(III)	Arsenite
As(V)	Arsenate
B(II)	Boron(II) cation
Cd	Cadmium
Cd(II), Cd^{2+}	Cadmium(II) cation
$Cd(OH)^+$	Cadmium hydroxide ion in its +1 oxidation state
C_0	Initial concentration of the solution in mg/L unit
C_c	Equilibrium concentration in mg/L unit
C_i	Initial concentration in mg/L unit
Co	Cobalt
Co(II)	Cobalt (II) ion
Co-Fe-MBC	Cobalt-iron-magnetic biochar
$Co(NO_3)_2$	Cobaltous nitrate
Cu	Copper
Cu(II)	Copper (II) cation
Cr	Chromium
Cr(III)	Chromium (III)

Cr(VI)	Hexavalent chromium
CrO_4^{2-}	Chromium oxoanion
$Cr_2O_7^{2-}$	Dichromate ion
Cs(I)	Caesium(I) ion
DALYs	Disability-Adjusted Life Years
E	Mean free energy in (kJ/mol) unit
Fe	Iron
Fe(III)	Element iron in its +3 oxidation state
$FeCl_3$	Iron(III) chloride
$FeCl_3.6H_2O$	Iron(III) chloride hexahydrate
Fe-MBC	Iron-magnetic biochar
H^+	A cationic form of atomic hydrogen
Hg	Mercury
Hg(II)	Mercury(II) cation
$HCrO_4^-$	Hydrogen chromate ion
k_1	Rate constant in min^{-1} unit
k_2	Rate constant in g/mg·min unit
K_F	Freundlich isotherm constant in ($mg^{1-1/n}L^{1/n}g^{-1}$) unit
K_L	Langmuir isotherm constant in L/mg unit
K_{DR}	Dubinin-Radushkevich isotherm constant in (mol^2/kJ^2) unit
$MnFe_2O_4$	Manganese ferrite
$MnSO_4.H_2O$	Manganese sulphate monohydrate
n	Adsorption intensity
$1/n$	Linearity degree of the relationship between the solution concentration and adsorption capacity
Ni	Nickel
Ni(II)	Nickel (II) ion
$NiCl_2$	Nickel(II) chloride
OH^-	Hydroxide, a diatomic anion
Pb	Lead
Pb(II)	Lead(II) ion
q_e, Q_e	Equilibrium adsorption capacity in mg/g unit
Q_m	Maximum adsorption capacity in mg/g unit
q_t	Adsorption capacity at time t in mg/g unit
R	Universal gas constant (8.314 J/mol·K)
Se(IV)	Selenium(IV) ion
t	Time in min unit
T	Temperature in K unit
U(VI)	Uranyl ion
V(V)	Vanadium(V) ion
WHO	World Health Organization
Zn	Zinc
Zn(II)	Zinc ion

1 Heavy Metals in the Environment

Heavy metals are the primary pollutants which were ranked by Disability-Adjusted Life Years (DALYs) as the world's worst pollution that causes countless deaths and disabilities [1]. Heavy metals are metals with density at least five times than that of water [2]. The most common toxic heavy metals are lead, cadmium, chromium, mercury, and arsenic. Human activities are the primary driver of heavy metal production and pollution. Some of the common industries that produce heavy metals are battery recycling, metal mining and processing, smelting, industrial and municipal dumping grounds, chemical and product manufacture, tannery, and industries related to the dye [1, 3].

Heavy metal contamination occurs when these toxic metals enter a water supply by industrial and consumer wastes. Acidic rain also breaks down soils and releases heavy metals into water streams, groundwater, lakes, rivers, and ocean. Exposure to a high level of heavy metals is a real threat not only to human but also to animals and the ecosystem at large. The carcinogenicity and toxicity of the heavy metals with their tendency to accumulate in biological system and ecosystem make them a significant health hazard that contributes to disease and poverty on a global scale [4–6]. For instance, in an aquatic ecosystem, heavy metals accumulate in fish tissues which magnify as the fishes move up in a food chain system [7]. As the fishes grow, more heavy metals accumulate in its tissues, thus, causing more toxicity to its predators. Mercury, which is one of the most toxic heavy metals in the environment, causes a decline in the rate of hatching for aquatic species such as fish and waterfowl. This event affects the number of species, causing significant changes in the aquatic ecosystem. Wright and Welbourn [8] also reported that lead affects the daily function of fish by damaging its gill and also stopping the production of enzymes for photosynthesis. Another study by Bradl [9] found that cadmium contamination causes skeletal deformities and kidney damage in fishes. The report also mentioned abnormal plant growth due to the cadmium pollution.

Although some heavy metals such as copper and zinc are essential to human health in minuscule amounts, these heavy metals are dangerous and pose adverse health effects at high concentrations [9]. Table 9.1 tabulates some of the effects of heavy metal contamination on human health.

2 Agricultural Waste Biomass

The high impact of heavy metal pollution in water has spurred numerous studies on this subject. There are a variety of techniques used in controlling water pollution due to heavy metals. Some of the commonly used methods are chemical precipitation, coagulation, electrochemical treatment, filtration, and ion exchange. However, the drawbacks for these methods include high initial or operational costs, large use of chemicals, large amount of sludge and extra cost for its removal, and excessive

Table 9.1 The human health effects from heavy metal exposure

Heavy metals	Health effects	References
As(III)	Toxic to almost all organs due to inhibition of cellular enzymes	[10]
	Inducing significant levels of stress genes and related proteins in liver cells	[11]
	Tumours in the lung, kidney, skin, bladder, and liver	[12]
Cd	Known carcinogen – cause of prostate, lung, and gastrointestinal cancer	[13]
	Potential agent for estrogen-related diseases, such as breast cancer, endometriosis, and spontaneous abortion	[14]
Cr(VI)	Irritation to nasal cavity, asthma, skin ulcers, known carcinogen	[4]
	Respiratory cancer	[15]
Co	Irritant to lungs, resulting in asthma, pneumonia, and wheezing	[16]
Cu	Contribute to progression of diabetes-related disease and cause fatality due to coronary disease	[4]
	Atherosclerotic disease	[17]
Fe	Increase risk of mutagenesis, carcinogenesis, and ageing. The cause of breast cancer and colorectal cancer	[4]
	Changes the metabolism of glucose	[18]
	Increase risk for cardiovascular disease	[19]
	Rheumatoid arthritis	[20]
Hg	Induce mutation of cells	[21]
Pb	Increase possibility of spontaneous abortion	[22]
	Behaviour problems in children – attention disorders, hyperactivity, impulsiveness	[23]
	Decrease intelligence, lower IQ, impaired behaviour development in children	[24]
	Negative impact or reproductive system – decline in sperm count and increase in spontaneous abortion	[25]
Al, As, Cd, Pb	Kidney lesions	[26]
Cd, Pb	Increase risk of osteopenia and osteoporosis	[27]
Cd, Co, Cr, Cu, Ni	Negative effects on the pulmonary and cardiovascular systems	[28]

energy consumption [29, 30]. Adsorption, on the other hand, uses less energy and is easy to operate due to its simple design.

Many biomass wastes are abundantly available in nature throughout the world due to agricultural activities. These wastes are dumped or burned in an open landfill or incinerator, which leads to substantial environmental and economic-related issues such as groundwater contamination, air pollution, and operational cost [31]. The agricultural wastes which are commonly available in large quantities have been reported to have the potential as adsorbents because of their physicochemical characteristics, inexpensiveness, and effectivness in removing contaminants. The use of chemicals reduces when using biomass wastes as adsorbents, which save the

Table 9.2 The examples of the agricultural waste-derived adsorbents for the removal of heavy metals

Agricultural wastes	Pollutants	Maximum removal capacity, Q_{max} (mg/g)	References
Almond shell	Co(II), Ni(II)	28.09 (Co(II)), 20.00 (Ni(II))	[37]
Arachis hypogaea	Cd(II), Ni(II)	2.78–2.81 (Cd(II)), 2.79–2.82 (Ni(II))	[38]
Astragalus mongholicus	Cr(VI)	45.45	[39]
Bamboo	Cr(VI)	18.5–51.7	[40]
Ficus carica	Cr(VI)	44.84	[41]
Gaozaban	Cd(II)	9.51–21.66	[42]
Glebionis coronaria L.	Cd(II), Co(II)	118.78 (Cd(II)), 42.86 (Co(II))	[43]
Hickory wood	Cd(II), Cu(II), Pb(II)	4.748 (Cd(II)), 12.297 (Cu(II)), 71.43 (Pb(II))	[44]
Lignite, coconut shell	Zn(II)	9.34–9.43	[45]
Lyocell	Pb(II)	367.6 ± 8.5	[46]
Olive stone	Cd(II)	11.72	[47]
Olive stone	Fe(III)	0.457–2.12	[48]
Palm shell	U(VI)	232.558–253.807	[49]
Pineapple peel	Cr(VI)	2.49–7.44	[50]
Sargassum filipendula	Pb(II)	367.942–455.68	[51]
Wheat bran	Cr(VI)	4.53–5.28	[52]

initial cost because biomass from agricultural wastes is usually available in large quantities from agriculture, forestry, or aquaculture activities. Table 9.2 lists the examples of the agricultural waste-derived adsorbents for the removal of heavy metals.

A review by Mo et al. [32] disclosed that most agricultural waste-derived adsorbents are not directly recycled at the authentic state, but altered in an assortment of approach to boost the porosity and adsorption-specific surface area of the substance. Some universal alteration approaches on the agricultural waste-derived adsorbents are activation, carbonization, nanostructuring, and grafting. Agricultural wastes which principally consist of cellulose, hemicellulose, and lignin are having a different composition ratio of these chemical components in various types of crops. The activation technique significantly influences surface chemistry, especially the surface functional groups and the pore structure of the material. For instance, the thermal conversion of the agricultural waste biomass by pyrolysis (carbonization) or gasification method is applied to produce biochar. This technique frequently uses a substance (precursor), which, in nature, participates in a chemical reaction to produce another compound. The superheating thermal conversion of the biomass operates at limited oxygen at high temperatures in the range of 350–700 °C in a specially designed furnace which is able to capture all of the emissions produced.

Table 9.3 Grafting techniques and correlated advantages and disadvantages

Grafting techniques	Nature	Advantages	Disadvantages
Photochemical initiation	UV light creates free radicals with or without sensitizer	Mild reaction conditions, Low cost of operation	Initial cost of equipment, Reaction time
High energy initiation	Irradiation causes homolytic fission and free radical form	No catalyst or additives required for initiation, ease of variation of parameters, materials modified in prefabricated form	Material deterioration, high cost
Chemical initiation	Free radicals created on surface by chemical agent	Relatively cheap, Little Homopolymer in some instances, Ease of use and application	Catalysts and additives need for initiation, grafting limited by concentration and purity of initiator, dependent on temperature

Source: O'Connell et al. [53]

Georgieva et al. [33] produced biochar adsorbent, which is classified as a microporous material via thermal degradation of walnut shells in a nitrogen atmosphere.

Nanostructuring of the agricultural wastes typically requires special processing or synthesis techniques for fabricating well-defined nanostructured adsorbent. Khattak et al. [34] used watermelon waste to fabricate magnetic graphitic nanostructures for the removal of heavy metals (As, Cr, Cu, Pb, and Zn) from water. Table 9.3 summarizes the ordinary nature and both the advantages and disadvantages correlated with the grafting techniques.

In recent years, many research findings have been reported on the production of new hybrid adsorbents and magnetic adsorbents from biological materials, mainly using agricultural wastes which consist of cellulose, lignin, and hemicellulose as raw material. This latest development of the new materials is creating smaller particle sizes with more significant surface areas, which leads to better effectiveness in pollutant removal. Mo et al. [32] reported that starch and cyclodextrins, together with other polysaccharides, engage a coupling agent to react with hydroxyl groups to build water-insoluble cross-linked networks that form a hybrid material.

One of the magnetic biological adsorbents, which is regularly known as magnetic biochar, exhibits an excellent magnetic property with high surface area and significant morphology [35, 36]. The magnetic biochar can be produced through various production methods. A review by Yi et al. [36] revealed that magnetic biochars could remove a wide range of pollutants, namely, (1) cationic heavy metals, (2) anionic heavy metals, (3) organic pollutants, and (4) compound pollutants from water environment. These magnetic biochars demonstrated an effective utilization as an adsorbent for various wastewater treatments. Table 9.4 presents examples of the production of magnetic biochar based on various agricultural wastes.

Table 9.4 Production of magnetic biochar with different agricultural wastes

Raw material	Reagent	Magnetic biochar
Bamboo charcoal	$FeCl_3$; $Co(NO_3)_2$	Fe-MBC and Co-Fe-MBC
Empty fruit bunch (EFB)	$FeCl_3$	$FeCl_3$-loaded magnetic biochar
Bamboo charcoal	$NiCl_2$	Ni-doped magnetized bamboo charcoal
Empty fruit bunch (EFB)	$FeCl_3$	$FeCl_3$-loaded magnetic biochar
Chitosan	$FeCl_3.6H_2O$; $MnSO_4.H_2O$	Chitosan modified $MnFe_2O_4$ magnetic biochar

Source: Thines et al. [35]

3 Parameters Affecting the Adsorption of Heavy Metals

The use of agricultural waste materials as adsorbents is affected by several variables, such as solution pH, initial concentration, adsorbent dose, and contact time. These variables control the adsorption capacity, kinetics, and selectivity of the heavy metal removal. The following subsections discuss the influence of each variable in the adsorption process.

3.1 Effect of Solution pH

Solution pH is a dominant factor to scrutinize in the heavy metal adsorption process, as it can significantly affect the surface charge of the adsorbent, as well as the ionization and speciation of the heavy metals [54]. Typically, the adsorption of cationic heavy metals is effective at high solution pH value, while the relation is opposite for anionic heavy metals. The adsorption performance for the anionic heavy metal removal surpasses at low solution pH value. At low solution pH, the surface adsorbent is usually positively charged. In that event, the cationic heavy metals compete with H^+ ions to occupy the adsorption sites. Thus, this phenomenon attracts anionic heavy metals through electrostatic forces. Contrastingly, increasing the solution pH values increases the negativity charge of the adsorbent surface, for as much as this attracts the cationic heavy metals to the surface of the adsorbent, and at the same time, the anionic components compete with OH^- ions. At very high pH, some heavy metals precipitate out of the solution instead of adsorbing to the adsorbent surface [55].

The studies on heavy metal adsorption using agricultural waste-derived adsorbents as reported in the literature demonstrated the accustomed tendencies as discussed above in the effect of solution pH on the removal capacity of the contaminant. The removal of Cu(II) using spent shitake by Dong and Lim [56] exhibited that the adsorption capacity increased with pH values from 0.7 to 3.0 before it became stable between solution pH 3 and solution pH 5. Another study using of *gaozaban* to treat Cd(II) also indicated that the adsorption capacity increased with solution pH from 2 to 4. The Cd species vary in different pH values:

(1) free from Cd^{2+} ions at pH lower than 6, (2) $Cd(OH)^+$ at pH between 6.5 and 8.5, and (3) precipitation as hydroxides expect to occur at pH higher than 9 [42]. For heavy metals that exist as anions in acidic condition, such as Cr(VI), the highest adsorption capacity occurs at very acidic pH. Conventionally, the Cr(VI) species at very low pH identify as $HCrO_4^-$, $Cr_2O_7^{2-}$, and CrO_4^{2-}. For this reason, the adsorption of Cr(VI) by using wheat bran is the uppermost at solution pH 2.2, where Cr(VI) is simultaneously reduced to Cr(III) [52].

3.2 *Effect of Initial Concentration*

The initial concentration of heavy metals provides the driving force for adsorption process. High concentration difference results in greater adsorption capacity but reduced efficiency of the adsorbent. This phenomenon frequently causes controversy in selecting a suitable treatment process for decontamination of heavy metals as mostly the threshold limit for toxic heavy metals is very low, either at ppm or ppb level. The World Health Organization (WHO) provides drinking water quality standard which includes a permissible limit for each of the heavy metals such as a maximum limit of 0.01 ppm for Pb, 0.003 ppm for Cd, and 0.05 ppm for Cr(VI) [57].

Research work from Shrestha et al. [45] using lignite and coconut shell-based activated carbon found that the adsorption capacity increased from 0.22 mg/g to 1.96 mg/g when the initial concentration increased from 10 to 50 mg/L; nonetheless, the percentage removal declined from 88% to 78%, by reason of limited active sites available for adsorption as the number of adsorbates competing for the sites are increasing. Ullah et al. [58] and Mahajan and Sud [38] disclosed similar remarks from their laboratory studies which recycled sugarcane bagasse to remove Cr (VI) and *Arachis hypogaea* to remove Cd(II) and Ni(II), respectively.

3.3 *Effect of Adsorbent Dosage*

Addition of more adsorbent in the adsorption process gives the effect of booming the adsorption site availability. On the contrary to the initial concentration effect, the process efficiency inclines, and the adsorption capacity declines as the adsorbent dosage increases. The more significant amount of available adsorption sites increase, the number of adsorbate molecules interacting, resulting in higher percentage removal of heavy metal ions.

Zhong et al. [55] demonstrated this comprehensive observation in Cu(II) and Cr (VI) removal using wheatgrass-derived adsorbent. In the study, the increase of sorbent dosage up to 2 g/L resulted in achieving 86.44% and 92.19% removal capacity, respectively. The adsorption capacities, howbeit, declined from 45 mg/g to 22 mg/g for Cu(II) and from 230 mg/g to 50 mg/g for Cr(VI). The study also found that increasing the adsorbent dosage further did not increase the percentage removal capacity, but adversely declining the adsorption capacities. The number of adsorbates available limits further adsorption reaction; therefore adding more adsorbent is wasting the resources. Another work on Pb(II) removal using fig sawdust also showed a similar finding [59]. The increase in adsorbent dosage to 1.67 g/L resulted in 95.3% removal capacity, and additional adsorbent dosage did not increase the percentage removal capacity.

3.4 *Effect of Contact Time*

Generally, researchers carry out investigations on contact time for adsorption processes to determine the equilibrium time of the reaction. Once the equilibrium time is obtained, the kinetics and equilibrium studies of adsorption process can be established. The initial stage of adsorption is usually the fastest, at the time that the adsorption sites are still unoccupied and the adsorbent surface is yet to be covered by adsorbate molecules. The adsorption eventually becomes slower by time in the act of the adsorption sites are gradually filled and competition to fill in a site becomes stiffer. No further adsorption takes place at equilibrium condition; thus, the adsorption capacity and percentage removal remain constant after the equilibrium time. Table 9.5 lists the examples of equilibrium time for several agricultural waste-derived adsorbents for heavy metal removal.

3.5 *Effect of Temperature*

The adsorption of heavy metals on agricultural waste-derived adsorbents is usually temperature-dependent, and the effect of temperature on the heavy metal removal very much depends on the nature of the adsorption process. The adsorption efficiency for exothermic adsorption increases with temperature, whereas the opposite occurs for endothermic adsorption process. Commonly, the studies on the effect of temperature for heavy metal removal using the agricultural waste-derived adsorbent are conducted at near-ambient temperature because at higher temperature the structure of agricultural wastes can be easily unstabilized. The operation at near-ambient temperature also optimizes energy consumption.

The adsorption process is mostly exothermic. The adsorption of U(VI) using palm shell powder demonstrated a decline in the heavy metal uptake as the temperature increased from 30 °C to 70 °C [49]. Similarly, increasing the temperature from 20 °C to 40 °C for Pb(II) removal using peanut shell has a negative effect to the heavy metal removal [66].

Only a few adsorption process studies using agricultural waste-derived adsorbent were reported to be endothermic adsorption. In an adsorption study of Cu(II) ions using pomegranate peel by Ben-Ali et al. [63], the adsorption efficiency increased from 57.39% to 78.85% when the temperature changed from 303 K to 313 K.

Table 9.5 Equilibrium time for heavy metal adsorption using agricultural waste-derived adsorbents

Heavy metals	Agricultural wastes	Experimental constants	Equilibrium time	References
As(III)	Sugarcane bagasse	pH = 7, C_i = 20 mg/L, adsorbent dosage = 2.5 g/L	120 min	[60]
Cd(II)	Castor leaf	pH = 5, T = 25 °C, C_i = 25 mg/L, adsorbent dosage = 0.8 g/L	30 min	[61]
	Banana cellulose	T = 25 °C, C_i = 50–300 mg/L, adsorbent dosage = 2 g/L	30 min	[62]
Cu(II)	Pomegranate peel	pH = 5.8, T = 30 °C, C_i = 10–50 mg/L, adsorbent dosage = 3 g/L	120 min	[63]
	Tomato waste	pH = 8, T = 20–40 °C, C_i = 50 mg/L, adsorbent dosage = 4 g/L	60 min	[64]
Cr(VI)	Banana skin	pH = 2, C_i = 20, 50, 100 mg/L, adsorbent dosage = 7.5 g/L	30 min (20 mg/L), 1 h (50 mg/L), 2 h (100 mg/L)	[65]
Pb(II)	Castor leaf	pH = 5, T = 25 °C, C_i = 25 mg/L, adsorbent dosage = 0.8 g/L	30 min	[61]
	Fig sawdust	pH = 4, T = 25 °C, C_i = 100 mg/L, adsorbent dosage = 1.67 g/L	40 min	[59]
	Peanut shell	pH = 5.5 ± 0.2, T = 20 °C, C_i = 100 mg/L, adsorbent dosage = 2 g/L	45 min	[66]

However, increasing the temperature to 323 K did not improve the removal efficiency, and further increase resulted in a decrease of the removal efficiency. The first increase to 313 K gave a positive effect to the removal efficiency due to the increase in active sites and also possible pore opening. On the other hand, increasing temperature beyond that weakened the adsorbent structure and promoted decomposition process that led to a decrease in the active sites. A similar trend was also reported in the adsorption of Fe(III) onto the olive stone. The decontamination of Fe (III) was enhanced as the temperature varied from 278 K to 343 K [48]. Another work by Albadarin et al. [67] found the increase in removal rate of Cr(VI) using date pit when the temperature increased from 22 °C to 60 °C. Adsorption of Zn(II) using coconut shell biochar attributed to the increase in adsorption capacity of the endothermic diffusion process, where the increase in operating temperature increased intraparticle diffusion [45].

4 Kinetic Models

Kinetic models are often used to determine the reaction rate constant and equilibrium adsorption capacity [68]. This section discusses briefly several of the most widely applied models. The pseudo-first-order model was proposed by Lagergren in 1898 [69]. The pseudo-first-order model is expressed as:

$$\frac{dq_t}{dt} = k_1(q_e - q_t) \tag{9.1}$$

where q_e and q_t are adsorption capacities (mg/g) at equilibrium and time t (min), respectively, and the rate constant is k_1 (min^{-1}). The linear form of this equation is found by integration of equation 9.1 with $q_t = 0$ at $t = 0$ and $q_t = q_t$ at $t = t$ [70, 71]:

$$\log\left(\frac{q_e}{q_e - q_t}\right) = \frac{k_1}{2.303}t \tag{9.2}$$

Equation 9.2 can be rearranged to give this model its linear form of:

$$\log(q_e - q_t) = \log(q_e) - \frac{k_1}{2.303}t \tag{9.3}$$

The pseudo-second-order model is given as:

$$\frac{dq_t}{t} = k_2(q_e - q_t)^2 \tag{9.4}$$

where k_2 is the rate constant (g/mg·min). Integration of Eq. (9.4) with boundaries of $t = 0$ to $t = t$ and $q_t = 0$ to $q_t = q_t$, the equation becomes [70, 71]:

$$\frac{1}{(q_e - q_t)} = \frac{1}{q_e} + k_2 t \tag{9.5}$$

The linear form for the pseudo-second-order model is given as:

$$\frac{t}{q_t} = \frac{1}{k_2 q_e^2} + \frac{t}{q_e} \tag{9.6}$$

The inverse of the y-intercept for Eq. (9.6) is the initial adsorption rate, h.

Another rate equation kinetic model that is commonly used is the Elovich equation, which is written in linear form as:

Table 9.6 Comparison of best fit kinetic models for adsorption of heavy metals on natural fibres

Source of fibre	Metal	Rate constant	Kinetic model	Reference
Almond shell	Co(II)	0.0204 g/mg·min	2nd order	[37]
	Ni(II)	0.0136 g/mg·min	2nd order	[37]
Arachis hypogaea	Cd(II)	1.512 g/mg·min	2nd order	[38]
	Ni(II)	1.524 g/mg·min	2nd order	[38]
Banana frond	B(II)	2.66 g/mg·min	2nd order	[73]
	Fe	6.05 g/mg·min	2nd order	[73]
Banana pseudostem	Cd(II)	0.01 g/mg·min	2nd order	[62]
Brown seaweed	Pb(II)	0.25×10^{-3} g/mg·min	2nd order	[74]
Castor leaf	Cd(II)	111.7 g/mg·min	2nd order	[61]
	Pb(II)	105.5 g/mg·min	2nd order	[61]
Fig sawdust	Pb(II)	0.009 g/mg·min	2nd order	[59]
Orange peel	Cr(VI)	0.3948 mmol/g·h	2nd order	[75]
	V(V)	1.571 mmol/g·h	2nd order	[75]
Peanut shell	Pb(II)	0.2467×10^{3} g/mg·min	1st and 2nd order	[66]
Pineapple peel	Cr(VI)	0.0135 g/mg·min	2nd order	[50]
Rice husk ash	Se(IV)	0.0094 g/μg·min	2nd order	[76]
Tea leaves	Cs(I)	–	2nd order	[77]

$$q_t = \frac{1}{\beta}\ln(\alpha\beta) + \frac{1}{\beta}\ln t \tag{9.7}$$

where α is the initial adsorption rate (mg/g·min) and β is the desorption constant (g/mg). Table 9.6 lists some of the adsorption of heavy metals over agricultural waste-derived adsorbents and the best fit model according to the study. Most of the works listed used at least the pseudo-first-order and pseudo-second-order kinetic equations and second-order equations fit to all systems in the table. The use of pseudo-second-order equation has a distinct advantage, which is no prior data of the equilibrium parameter, q_e is needed to calculate the rate constant, thus reducing the experimental errors in calculation [72].

5 Adsorption Isotherms

The adsorption isotherms are the relationship between the amount of adsorbate that is adsorbed over a unit mass of adsorbent (q_e) and the amount of adsorbate that is left in the solution (C_e) at equilibrium under constant temperature [78]. The simplest relation is given by two parameter isotherm models, where the most widely applicable models are Langmuir and Freundlich isotherms. The Dubinin-Radushkevich isotherm is commonly used to estimate the apparent free energy that can be used to determine the nature of sorption, whether it is physisorption or chemisorption.

The Langmuir isotherm was derived based on several assumptions, which are monolayer coverage of adsorbates on the surface of adsorbent, the adsorbent is homogeneous with equivalent energy on the sorption sites, and the adsorbates can only occupy one active site [78, 79]. The Langmuir isotherm is written as:

$$\frac{C_e}{Q_e} = \frac{1}{Q_m K_L} + \frac{C_e}{Q_m} \tag{9.8}$$

where C_e is the equilibrium concentration (mg/L), Q_e is the equilibrium adsorption capacity (mg/g), Q_m is the maximum adsorption capacity (mg/g), and K_L is the Langmuir isotherm constant (L/mg). There is another feature of Langmuir isotherm called the separation factor, R_L. The separation factor can distinguish whether the adsorption is favourable ($0 < R_L < 1$), irreversible ($R_L = 0$), linear ($R_L = 1$), or unfavourable ($R_L > 1$). The separation factor can be calculated using this formula:

$$R_L = \frac{1}{1 + K_L C_0} \tag{9.9}$$

where C_0 is the initial concentration of the solution (mg/L).

On the other hand, the Freundlich isotherm equation assumes that the adsorption is multilayer on heterogeneous adsorbent surface. Unlike Langmuir, the adsorbates obeying Freundlich isotherm are free to move between sorption sites [80]. The adsorption energy declines exponentially going towards the sorption sites. The Freundlich isotherm equation is given as:

$$\ln Q_e = \ln K_F + \frac{1}{n} \ln C_e \tag{9.10}$$

where K_F is the Freundlich isotherm constant ($mg^{1-1/n} L^{1/n} g^{-1}$) and n is the adsorption intensity. The value of $1/n$ in the equation indicates the linearity degree of the relationship between the solution concentration and adsorption capacity. When $1/n$ is between 0 and 1, there exists some nonlinearity. When $1/n$ is equal to 1, the relationship is linear.

The Dubinin-Radushkevich isotherm equation is used mainly to calculate the apparent free energy of adsorption. The equation is written as:

$$\ln Q_e = \ln Q_m - K_{DR} \varepsilon^2 \tag{9.11}$$

where K_{DR} is the Dubinin-Radushkevich isotherm constant (mol^2/kJ^2) and ε is called the Polanyi potential, which is calculated using the following equation:

$$\varepsilon = \mathrm{RT} \ln\left(1 + \frac{1}{Ce}\right) \tag{9.12}$$

Table 9.7 Comparison of best fit isotherm models for adsorption of heavy metals on agricultural waste-derived adsorbents

Agricultural wastes	Heavy metals	Isotherm constant	Isotherm model	References
Bamboo	Cr(VI)	0.049 L/mg (Fe-modified)	Langmuir	[40]
	Cr(VI)	0.082 L/mg (Fe-Co-modified)	Langmuir	[40]
Ficus carica	Cr(VI)	0.034 L/mg	Langmuir	[41]
Pineapple peel	Cr(VI)	0.067 L/mg	Langmuir	[50]
Rice straw	Cd(II)	$4.2 \times 10^3 \pm 0.1$ $mg^{1-1/n}L^{1/n}g^{-1}$	Freundlich	[81]
	Cu(II)	$2.7 \times 10^3 \pm 0.2$ $mg^{1-1/n}L^{1/n}g^{-1}$	Freundlich	[81]
	Hg(II)	$1.2 \times 10^3 \pm 0.4$ $mg^{1-1/n}L^{1/n}g^{-1}$	Freundlich	[81]
	Zn(II)	$2.3 \times 10^3 \pm 0.1$ $mg^{1-1/n}L^{1/n}g^{-1}$	Freundlich	[81]
	U(vi)	0.53 L/mg	Langmuir	[82]
Spartina alterniflora	Cu(II)	0.110 ± 0.024 L/mg	Langmuir	[83]
Sugarcane bagasse	As(III)	0.0958 L/mg	Langmuir	[60]
	As(V)	0.038 L/mg	Langmuir	[60]
Wheat straw	Cu(II)	0.1609 L/mg	Langmuir	[55]
	Cr(VI)	0.03831 L/mg	Langmuir	[55]

where R is the universal gas constant (8.314 J/mol·K) and T is temperature in Kelvin. The K_{DR} can be used then to calculate the mean free energy, E (kJ/mol), by using the formula:

$$E = \frac{1}{\sqrt{2K_{DR}}} \tag{9.13}$$

Table 9.7 shows the adsorption of heavy metals that fit to either one of these isotherms. It can be seen that most heavy metal adsorption using agricultural waste-derived adsorbents fits well with Langmuir isotherm, thereby indicating monolayer adsorption.

Glossary

Adsorbent A material which will allow a liquid, gas, or dissolved solid to adhere to its surface.

Adsorption A process by which a solid holds molecules of a gas or liquid or solute as a thin film.

Adsorption capacity	The amount of adsorbate taken up by the adsorbent per unit mass (or volume) of the adsorbent.
Agricultural wastes	Unwanted or unsalable materials produced wholly from agricultural operations directly related to the growing of crops or raising of animals for the primary purpose of making a profit or for a livelihood.
Anionic	A negatively charged ion, as one attracted to the anode in electrolysis. Any negatively charged atom or group of atoms.
Aquatic ecosystem	A group of interacting organisms which are dependent on one another and their water environment for nutrients and shelter.
Biochar	A charcoal-like substance that is made by burning organic material from agricultural and forestry wastes (also called biomass) in a controlled process called pyrolysis.
Biomass	Biomass is organic material that comes from plants and animals, and it is a renewable source of energy. Biomass contains stored energy from the sun.
Cationic	A positively charged ion that is attracted to the cathode in electrolysis. Any positively charged atom or group of atoms.
Cellulose	A natural linear polymer (polysaccharide) with a molecular repeat unit comprised of a pair of d-anhydroglucose ring units joined by β-1 $\rightarrow$ 4 glycosidic oxygen linkages around which the molecular chain can bend and twist.
Chemical precipitation	Formation of a separable solid substance from a solution, either by converting the substance into an insoluble form or by changing the composition of the solvent to diminish the solubility of the substance in it.
Coagulation	A process for combining small particles into larger aggregates (flocs). A chemical water treatment technique typically applied prior to sedimentation and filtration.
Carbonization	A process by which solid residues with increasing content of the element carbon are formed from organic material usually by pyrolysis in an inert atmosphere.
Ecosystem	A geographic area where plants, animals, and other organisms, as well as weather and landscape, work together to form a bubble of life.
Electrochemical treatment	Uses electrolysis and fuel cell technology to treat water and wastewater such as electrocoagulation/ electro-flotation, electrodialysis, electro-reduction, sono-electrolysis, and electro-oxidation.
Filtration	A process of separating suspended solid matter from a liquid, by causing the latter to pass through the pores of some substance (filter).
Heavy metals	Metallic cation and anion that have relatively high density and toxic or poisonous at low concentrations.
Hemicellulose	The second most abundant chemical constituent of woody and grassy biomass and is present along with cellulose in almost all terrestrial plant cell walls. Also known as polyose.
Human health	A complete state of physical, psychological, and social wellbeing of individuals.
Ion exchange	A reversible chemical reaction where dissolved ions are removed from the solution and replaced with other ions of the same or similar electrical charge.
Kinetics	The study of reaction rates. The branch of mechanics that deals with the actions of forces in producing or changing the motion of masses.
Lignin	An irregular polyphenolic polymer which is synthesized by dehydrogenative polymerization of phenyl propanoid units, namely, coniferyl alcohol, sinapyl alcohol, and coumaryl alcohol, which corresponds to guaiacyl (G), syringyl (S), and p-hydroxyphenyl (H) structures of lignin, respectively.
Physicochemical	Pertaining both physical and chemical properties, changes, and reactions of or according to physical chemistry.

Selectivity The discrimination demonstrated by a reagent in competitive attack on two or more substrates or on two or more positions in the same substrate. It is quantitatively formulated by ratios of rate constants of the competing reactions or by the logarithms of these ratios.

References

1. Pure Earth, Green Cross Switzerland (2016) The world's worst pollution problems 2016: toxics beneath our feet. https://www.worstpolluted.org/docs/WorldsWorst2016.pdf
2. Fergusson JE (1990) The heavy elements: chemistry, environmental impact and health effects. Pergamon Press, Oxford
3. Green Cross Switzerland, Blacksmith Institute (2012) The world's worst pollution problems: assessing health risks at hazardous waste sites. https://www.worstpolluted.org/files/FileUpload/files/WWPP_2012.pdf
4. Jomova K, Valko M (2011) Advances in metal-induced oxidative stress and human disease. Toxicology 283(2–3):65–87. https://doi.org/10.1016/j.tox.2011.03.001
5. National Toxicology Program (2016) Report on carcinogens, 14th edn. U.S. Department of Health and Human Services, Public Health Service, Research Triangle Park
6. Tchounwou PB, Yedjou CG, Patlolla AK, Sutton DJ (2012) Heavy metal toxicity and the environment. In: Luch A (ed) Molecular, clinical and environmental toxicology, Experientia supplementum, vol 101. Springer, Basel. https://doi.org/10.1007/978-3-7643-8340-4_6
7. Solomon F (2008) Impacts of metals on aquatic ecosystems and human health. Environ Commun:14–19
8. Wright DA, Welbourn P (2002) Environmental toxicology. Cambridge University Press, Cambridge, UK
9. Bradl H (2005) Heavy metals in the environment: origin, interaction and remediation. Elsevier/Academic Press, London
10. Wang Z, Rossman TG (1996) The carcinogenecity of arsenic. In: Chang T (ed) Toxicology of metals. CRC Press, Boca Raton, pp 221–229
11. Tchounwou PB, Wilson BA, Abdelgnani AA, Ishaque AB, Patlolla AK (2002) Differential cytotoxicity and gene expression in human liver carcinoma (Hepg2) cells exposed to arsenic trioxide and monosodium acid methanearsonate (Msma). Int J Mol Sci 3:1117–1132
12. Waalkes MP, Liu J, Ward JM, Diwan LA (2004) Mechanisms underlying arsenic carcinogenesis: hypersensitivity of mice exposed to inorganic arsenic during gestation. Toxicology 198:31–38
13. Flora SJS, Pachauri V (2010) Chelation in metal Intoxification. Int J Environ Res Public Health 7:2745–2788
14. Rzymski P, Rzymski P, Tomcyzk K, Niedzielki P, Jakubowski K, Poniezialek B, Opala T (2014) Metal status in human endometrium: relation to cigarette smoking and histological lesions. Environ Res 132:328–333
15. Dayan AD, Paine AJ (2001) Mechanisms of chromium toxicity, carcinogenecity and allergenicity; review of the literature from 1985 to 2000. Hum Exp Toxicol 20(9):439–451
16. Gal J, Hursthouse A, Tatner P, Stewart F, Welton R (2008) Cobalt and secondary poisoning in terrestrial food chain: data review and research gaps to support risk assessment. Environ Int 34:821–838
17. Burkitt MJ (2001) A critical overview of the chemistry of copper-dependent low density lipoprotein oxidation: roles of lipid hydroperoxides, alpha-tocopherol, thiols, and ceruloplasmin. Arch Biochem Biophys 394:117–135
18. Lee DH, Liu DY, Jacobs DR Jr, Shin HR, Song K, Lee IK, Hider RC (2006) Common presence of non-transferrin-bound Iron among patients with type 2 diabetes. Diabetes Care 29:1090–1095

19. Touyz RM, Schiffrin EL (2004) Reactive oxygen species in vascular biology: implications in hypertension. Histochem Cell Biol 122:339–352
20. Droge W (2002) Free radicals in the physiological control of cell function. Physiol Rev 82:47–95
21. Inoue M, Sato EF, Nishikawa M, Park AM, Kari Y, Imada I, Utsumi K (2003) Mitochondrial generation of reactive oxygen species and its role in aerobic life. Curr Med Chem 10:2495–2505
22. Borja-Aburto V, Hertz-Picciotto I, Rojas Lopez M, Farias P, Rios C, Blanco J (1999) Blood lead levels measured prospectively and risk of spontaneous abortion. Am J Epidemiol 150 (6):590–597
23. Szkup-Jablonska M, Karakiewicz B, Grochans E, Jurczak A, Nowak-Starz G, Rotter I, Prokopowicz A (2012) Effects of blood lead and cadmium levels on the functioning of children with behaviour disorders in the family environment. Ann Agric Environ Med 19(2):241–246
24. Kaul B, Sandhu RS, Depratt C, Reyes F (1999) Follow-up screening of lead-poisoned children near an auto battery recycling plant, Haina, Dominican Republic. Environ Health Perspect 107 (11):917–720
25. Hertz-Picciotto I (2000) The evidence that lead increases the risk for spontaneous abortion. Am J Ind Med 38:300–309
26. Wasana HMS, Perera GDRK, Gunawardena PDS, Fernando PS, Bandara J (2017) Who water quality standards vs synergic effect(S) of fluoride, heavy metals and hardness in drinking water on kidney tissues. Sci Rep 7:42516. https://doi.org/10.1038/srep42516
27. Lim HS, Lee HH, Kim TH, Lee BR (2016) Relationship between heavy metal exposure and bone mineral density in Korean adult. J Bone Metabol 23(4):223–231. https://doi.org/10.11005/jbm.2016.23.4.223
28. Corradi M, Mutti A (2011) Metal ions affecting the pulmonary and cardiovasular system. Metal Ions Life Sci 8:81–105
29. Gautam RK, Mudhoo A, Lofrano G, Chattopadhyaya MC (2014) Biomass-derived biosorbents for metal ions sequestration: adsorbent modification and activation methods and adsorbent regeneration. J Environ Chem Eng 2(1):239–259. https://doi.org/10.1016/j.jece.2013.12.019
30. Nguyen TAH, Ngo HH, Guo WS, Zhang J, Liang S, Yue QY et al (2013) Applicability of agricultural waste and by-products for adsorptive removal of heavy metals from wastewater. Bioresource Technol 148:574–585. https://doi.org/10.1016/j.biortech.2013.08.124
31. Ghani WWAK, Abdullah MF, Loung C, Ho C, Matori K (2008) Characterization of vitrified malaysian agrowaste ashes as potential recycling material. Int J Eng Technol 5:111–117
32. Mo J, Yang Q, Zhang N, Zhang W, Zheng Y, Zhang Z (2018) A review on agro-industrial waste (AIW) derived adsorbents for water and wastewater treatment. J Environ Manag 227:395–405
33. Georgieva VG, Gonsalvesh L, Tavlieva MP (2020) Thermodynamics and kinetics of the removal of nickel (II) ions from aqueous solutions by biochar adsorbent made from agro-waste walnut shells. J Mol Liq 312:112788
34. Khattak MMUR, Zahoor M, Muhammad B, Khan FA, Ullah R, AbdEI-Salam NM (2017) Removal of heavy metals from drinking water by magnetic carbon nanostructures prepared from biomass. J Nanomater 2017:5670371, 10 pages
35. Thines KR, Abdullah EC, Mubarak NM, Ruthiraan M (2017) Synthesis of magnetic biochar from agricultural waste biomass to enhancing route for waste water and polymer application: a review. Renew Sust Energ Rev 67:257–276
36. Yi Y, Huang Z, Lu B, Xian J, Tsang PE, Cheng W, Fang J, Fang Z (2020) Magnetic biochar for environmental remediation: a review. Bioresour Technol 298:122468
37. Kılıç M, Kırbıyık Ç, Çepelioğullar Ö, Pütün AE (2013) Adsorption of heavy metal ions from aqueous solutions by bio-char, a by-product of pyrolysis. Appl Surface Sci 283:856–862. https://doi.org/10.1016/j.apsusc.2013.07.033
38. Mahajan G, Sud D (2013) Application of Ligno-cellulosic waste material for heavy metal ions removal from aqueous solution. J Environ Chem Eng 1(4):1020–1027. https://doi.org/10.1016/j.jece.2013.08.013

39. Shang J, Pi J, Zong M, Wang Y, Li W, Liao Q (2016) Chromium removal using magnetic biochar derived from herb-residue. J Taiwan Ins Chem Eng 68:289–294. https://doi.org/10.1016/j.jtice.2016.09.012
40. Wang W, Wang X, Wang X, Yang L, Wu Z, Xia S, Zhao J (2013) Cr(vi) removal from aqueous solution with bamboo charcoal chemically modified by iron and cobalt with the assistance of microwave. J Environ Sci 25(9):1726–1735. https://doi.org/10.1016/S1001-0742(12)60247-2
41. Gupta VK, Pathania D, Sharma S, Singh P (2013) Preparation of bio-based porous carbon by microwave assisted phosphoric acid activation and its use for adsorption of Cr(vi). J Coll Interface Sci 401:125–132. https://doi.org/10.1016/j.jcis.2013.03.020
42. Khan Rao RA, Ikram S, Uddin MK (2014) Removal of cd(ii) from aqueous solution by exploring the biosorption characteristics of Gaozaban (Onosma Bracteatum). J Environ Chem Eng 2(2):1155–1164. https://doi.org/10.1016/j.jece.2014.04.008
43. Tounsadi H, Khalidi A, Machrouhi A, Farnane M, Elmoubarki R, Elhalil A et al (2016) Highly efficient activated carbon from Glebionis Coronaria L. biomass: optimization of preparation conditions and heavy metals removal using experimental design approach. J Environ Chem Eng 4(4, Part A):4549–4564. https://doi.org/10.1016/j.jece.2016.10.020
44. Wang H, Gao B, Wang S, Fang J, Xue Y, Yang K (2015) Removal of Pb(Ii), Cu(Ii), and Cd (Ii) from aqueous solutions by biochar derived from Kmno4 treated Hickory wood. Bioresource Technol 197:356–362. https://doi.org/10.1016/j.biortech.2015.08.132
45. Shrestha S, Son G, Lee SH, Lee TG (2013) Isotherm and thermodynamic studies of Zn (Ii) adsorption on lignite and coconut shell-based activated carbon fiber. Chemosphere 92 (8):1053–1061. https://doi.org/10.1016/j.chemosphere.2013.02.068
46. Bediako JK, Reddy DHK, Song MH, Wei W, Lin S, Yun YS (2017) Preparation, characterization and lead adsorption study of tripolyphosphate-modified waste lyocell fibers. J Environ Chem Eng 5(1):412–421. https://doi.org/10.1016/j.jece.2016.12.022
47. Alslaibi TM, Abustan I, Ahmad MA, Foul AA (2013) Cadmium removal from aqueous solution using microwaved olive stone activated carbon. J Environ Chem Eng 1(3):589–599. https://doi.org/10.1016/j.jece.2013.06.028
48. Hodaifa G, Ochando-Pulido JM, Driss Alami SB, Rodriguez-Vives S, Martinez-Ferez A (2013) Kinetic and thermodynamic parameters of Iron adsorption onto olive stones. Indus Crops Prod 49:526–534. https://doi.org/10.1016/j.indcrop.2013.05.039
49. Kushwaha S, Sudhakar PP (2013) Sorption of uranium from aqueous solutions using palm-Shell-based adsorbents: a kinetic and equilibrium study. J Environ Radioactiv 126:115–124. https://doi.org/10.1016/j.jenvrad.2013.07.021
50. Wang C, Gu L, Liu X, Zhang X, Cao L, Hu X (2016) Sorption behavior of Cr(Vi) on pineapple-peel-derived biochar and the influence of coexisting pyrene. Int Biodeteriorat Biodegrad 111:78–84. https://doi.org/10.1016/j.ibiod.2016.04.029
51. Verma A, Kumar S, Kumar S (2016) Biosorption of lead ions from the aqueous solution by Sargassum Filipendula: equilibrium and kinetic studies. J Environ Chem Eng 4(4, Part A):4587–4599. https://doi.org/10.1016/j.jece.2016.10.026
52. Kaya K, Pehlivan E, Schmidt C, Bahadir M (2014) Use of modified wheat bran for the removal of chromium(Vi) from aqueous solutions. Food Chem 158:112–117. https://doi.org/10.1016/j.foodchem.2014.02.107
53. O'Connell DW, Birkinshaw C, O'Dwyer TF (2008) Heavy metal adsorbents prepared from the modification of cellulose: a review. Bioresour Technol 99:6709–6724
54. Park D, Yun YS, Park JM (2010) The past, present, and future trends of biosorption. Biotechnol Bioprocess Eng 15(1):86–102. https://doi.org/10.1007/s12257-009-0199-4
55. Zhong QQ, Yue QY, Gao BY, Li Q, Xu X (2013) A novel amphoteric adsorbent derived from biomass materials: synthesis and adsorption for Cu(Ii)/Cr(Vi) in single and binary systems. Chem Eng J 229:90–98. https://doi.org/10.1016/j.cej.2013.05.083
56. Dong YB, Lin H (2015) Adsorption of Cu^{2+} from aqueous solution by modified biomass material. Trans Nonferrous Met Soc China 25(3):991–996. https://doi.org/10.1016/S1003-6326(15)63689-5

57. World Health Organization (2006) Guidelines for drinking water quality [electronic resource]: incorporating first addendum, 3rd edn, vol 1, Recommendations. World Health Organization
58. Ullah I, Nadeem R, Iqbal M, Manzoor Q (2013) Biosorption of chromium onto native and immobilized sugarcane bagasse waste biomass. Ecol Eng 60:99–107. https://doi.org/10.1016/j.ecoleng.2013.07.028
59. Ghasemi M, Naushad M, Ghasemi N, Khosravi-fard Y (2014) A novel agricultural waste based adsorbent for the removal of Pb(Ii) from aqueous solution: kinetics, equilibrium and thermodynamic studies. J Indus Eng Chem 20(2):454–461. https://doi.org/10.1016/j.jiec.2013.05.002
60. Gupta A, Vidyarthi SR, Sankararamakrishnan N (2015) Concurrent removal of as(Iii) and as (V) using green low cost functionalized biosorbent – Saccharum Officinarum bagasse. J Environ Chem Eng 3(1):113–121. https://doi.org/10.1016/j.jece.2014.11.023
61. Martins AE, Pereira MS, Jorgetto AO, Martines MAU, Silva RIV, Saeki MJ, Castro GR (2013) The reactive surface of Castor leaf [Ricinus Communis L.] powder as a green adsorbent for the removal of heavy metals from natural river water. Appl Surface Sci 276:24–30. https://doi.org/10.1016/j.apsusc.2013.02.096
62. Pillai SS, Deepa B, Abraham E, Girija N, Geetha P, Jacob L, Koshy M (2013) Biosorption of Cd (Ii) from aqueous solution using xanthated nano banana cellulose: equilibrium and kinetic studies. Ecotoxicol Environ Safety 98:352–360. https://doi.org/10.1016/j.ecoenv.2013.09.003
63. Ben-Ali S, Jaouali I, Souissi-Najar S, Ouederni A (2017) Characterization and adsorption capacity of raw pomegranate peel biosorbent for copper removal. J Cleaner Product 142(Part 4):3809–3821. https://doi.org/10.1016/j.jclepro.2016.10.081
64. Yargıç AŞ, Yarbay Şahin RZ, Özbay N, Önal E (2015) Assessment of toxic copper (Ii) biosorption from aqueous solution by chemically-treated tomato waste. J Cleaner Product 88:152–159. https://doi.org/10.1016/j.jclepro.2014.05.087
65. López-García M, Lodeiro P, Herrero R, Barriada JL, Rey-Castro C, David C, Sastre de Vicente ME (2013) Experimental evidences for a new model in the description of the adsorption-coupled reduction of Cr(Vi) by protonated banana skin. Bioresource Technol 139:181–189. https://doi.org/10.1016/j.biortech.2013.04.044
66. Taşar Ş, Kaya F, Özer A (2014) Biosorption of lead(Ii) ions from aqueous solution by peanut shells: equilibrium, thermodynamic and kinetic studies. J Environ Chem Eng 2(2):1018–1026. https://doi.org/10.1016/j.jece.2014.03.015
67. Albadarin AB, Glocheux Y, Ahmad MNM, Walker GM, Mangwandi C (2014) Novel comparison of kinetic models for the adsorption-coupled reduction of Cr(Vi) using untreated date pit biomaterial. Ecol Eng 70:200–205. https://doi.org/10.1016/j.ecoleng.2014.05.002
68. Chang J, Ma J, Ma Q, Zhang D, Qiao N, Hu M, Ma H (2016) Adsorption of methylene blue onto Fe_3O_4/activated montmorillonite nanocomposite. Appl Clay Sci 119(Part 1):132–140. https://doi.org/10.1016/j.clay.2015.06.038
69. Qiu H, Lv L, Pan BC, Zhang QJ, Zhang WM, Zhang QX (2009) Critical review in adsorption kinetic models. J Zheijang Univ Sci A 10(5):716–724. https://doi.org/10.1631/jzus.A0820524
70. Ho YS, McKay G (1998) Kinetic models for the sorption of dye from aqueous solution by wood. Proc Safety Environ Protect 76(2):183–191. https://doi.org/10.1205/095758298529326
71. Ho YS, McKay G (1998) Sorption of dye from aqueous solution by peat. Chem Eng J 70 (2):115–124. https://doi.org/10.1016/S0923-0467(98)00076-1
72. Ho YSR (2006) Review of second-order models for adsorption systems. J Hazard Mater 136 (3):681–689. https://doi.org/10.1016/j.jhazmat.2005.12.043
73. Foo KY, Lee LK, Hameed BH (2013) Preparation of banana frond activated carbon by microwave-induced activation for the removal of boron and total iron from landfill leachate. Chem Eng J 223:604–610
74. Tabaraki R, Nateghi A, Asbchin SA (2014) Biosorption of lead (II) ions on Sargassum ilicifolium: application of response surface methodology. Int Biodeterior Biodegradation 93:145–152. https://doi.org/10.1016/j.ibiod.2014.03.022
75. Hu H, Zhao Q (2018) Optimization extraction and functional properties of soluble dietary fiber from pineapple pomace obtained by shear homogenization-assisted extraction. RSC Adv 8:41117–41130. https://doi.org/10.1039/C8RA06928J

76. Gulipalli CHS, Prasad B, Wasewar KL (2011) Batch study, Equilibrium and kinetics of adsorption of selenium using rice husk ash (RHA). J Eng Sci Technol 6(5):590–609
77. Gurung M, Adhikari BB, Alam S, Kawakita H, Ohto K, Inoue K, Harada H (2013) Adsorptive removal of Cs(I) from aqueous solution using polyphenols enriched biomass-based adsorbents. Chem Eng J 231:113–120. https://doi.org/10.1016/j.cej.2013.06.028
78. Rangabhashiyam S, Anu N, Giri Nandagopal MS, Selvaraju N (2014) Relevance of isotherm models in biosorption of pollutants by agricultural byproducts. J Environ Chem Eng 2 (1):398–414
79. Anirudhan TS, Chandran R (2015) Adsorptive removal of basic dyes from aqueous solutions by surfactant modified bentonite clay (organoclay): kinetic and competitive adsorption isotherm. Process Saf Environ Prot 95:215–225
80. Mahamad MN, Zaini MAA, Zakaria ZA (2015) Preparation and characterization of activated carbon from pineapple waste biomass for dye removal. Int Biodeteriorat Biodegrad 102:274–280. https://doi.org/10.1016/j.ibiod.2015.03.009
81. Rocha CG, Zaia DAM, Alfaya RVDS, Alfaya AADS (2009) Use of rice straw as biosorbent for removal of Cu(II), Zn(II), Cd(II) and Hg(II) ions in industrial effluents. J Hazard Mater 166 (1):383–388. https://doi.org/10.1016/j.jhazmat.2008.11.074
82. Yakout SM, Metwally SS, El-Zakla T (2013) Uranium sorption onto activated carbon prepared from rice straw: competition with humic acids. Appl Surf Sci 280:745–750. https://doi.org/10.1016/j.apsusc.2013.05.055
83. Li M, Liu Q, Guo L, Zhang Y, Lou Z, Wang Y, Qian G (2013) Cu(II) removal from aqueous solution by Spartina alterniflora derived biochar. Bioresource Technol 141:83–88. https://doi.org/10.1016/j.biortech.2012.12.096

Chapter 10
Removal of Heavy Metal Ions Using Magnetic Materials

Soh-Fong Lim, Agnes Yung-Weng Lee, S. N. David Chua, and Bee-Huah Lim

Contents

1 Introduction 395
2 Magnetic Materials 395
3 Aspect of Magnetism 396
3.1 Diamagnetism 398
3.2 Paramagnetism 399
3.3 Ferromagnetism 399
3.4 Antiferromagnetism 399
3.5 Ferrimagnetism 400
4 Synthesis of Magnetic Particles 400
4.1 In Situ Methods 402
4.2 Ex Situ Methods 403
5 Magnetic Separation for Water Purification 403
6 Recovery of Magnetic Particles 405
7 Environmental Application of Magnetic Particles 406
Glossary 406
References 407

Abstract Heavy metal ions contaminate water environment through point sources and nonpoint sources. Heavy metal ions are categorized as inorganic contaminants by both the WHO and the USEPA. The heavy metal ions are increasingly being introduced into the environment as pollutants and contaminants resulting from human activities.

Magnetic particles for water treatment applications have received considerable attention from researchers due to high separation efficiency. The magnetic particles behave similar to or even better than various commercial adsorbents. The magnetic

S.-F. Lim (✉) · A. Y.-W. Lee · S. N. D. Chua
Faculty of Engineering, Universiti Malaysia Sarawak, Sarawak, Malaysia
e-mail: sflim@unimas.my; 15020101@siswa.unimas.my; csndavid@unimas.my

B.-H. Lim
Fuel Cell Institute, Universiti Kebangsaan Malaysia, Bangl, Selangor Darul Ehsan, Malaysia
e-mail: beehuah@ukm.edu.my

L. K. Wang, M. -H. S. Wang, Y. -T. Hung, N. K. Shammas (eds.), *Integrated Natural Resources Management*, Handbook of Environmental Engineering 20,
https://doi.org/10.1007/978-3-030-55172-8_10

particles also exhibit high selectivity for the target pollutants from the environment besides enabling ease of operation for reducing the particle separation steps from the flowing stream. A comprehensive and systematic understanding of synthesis and surface modifications of magnetic particles is significant to enhance their practicability in environmental technology. Although high removal performance and reactivity can be achieved by smaller particle size, the stability, toxicity, and recovery of the particles magnetically could be challenging.

In contrast, the active surface of magnetic particles may be forfeited, while surface modifications stabilize and reduce the toxicity of the particles. Reliable surface modifications are necessarily needed for the increment of the number of active sites to remove the heavy metals. For successful environmental applications of the magnetic particles, modification on the magnetic particles is principally crucial to balance the effects on their reactivity, capacity, and reusability.

Keywords Heavy metal ions · Magnetic particles · Contaminants · Heavy metal · Adsorbent · Pollution · Removal · Separation · Magnetic sorption technology

Nomenclature

γ-Fe_2O_3	Maghemite
Ag	Silver
BH_{max}	Highest energy product
B_r	Remanence
Ca	Calcium
Cd^{2+}	Cadmium cation
CdS	Cadmium sulfide
CdSe	Cadmium selenide
Cu	Copper
Cu^{2+}	Copper cation
Co	Cobalt
Cr^{4+}	Chromium cation (4+)
Fe	Iron
FeO^-	Negative charge iron oxide
Fe_3O_4	Magnetite
Fe_3S_4	Greigite
Gd	Gadolinium
Hc	Coercive force
Hg^{2+}	Mercuric cation
HGMS	High-gradient magnetic separation
Mg	Magnesium
MIEX	Magnetic ion exchange resin
Mn	Manganese
$MnFe_2O_4$	Manganese iron oxide
MnO	Manganese(II) oxide
Mn-Zn	Manganese-zinc

Nd	Neodymium
Nd-Fe-B	Neodymium, iron, and boron
Na^+	Sodium cation
Ni	Nickel
Ni^{2+}	Nickel cation
$NiFe_2O_4$	Nickel iron oxide/nickel ferrite
nZVI	Nano zerovalent iron
Pb^{2+}	Lead cation
Si	Silicon
Sm-Co	Samarium-cobalt
SPIONs	Superparamagnetic iron oxide nanoparticles
T_c	Curie temperature
USA	United States of America
USEPA	US Environmental Protection Agency
WHO	World Health Organization
Zn	Zinc
$ZnFe_2O_4$	Zinc ferrite

1 Introduction

The conventional adsorbents, including activated carbon, silica gel, zeolite, and clay minerals, which have been widely used to remove heavy metals suffer from high generation costs. The attention was then shifted to the use of adsorbent derived from agricultural waste such as sawdust, orange peel, ground coffee, and almond shell for the heavy metal removal due to their low cost and high abundance. Although these adsorbents were reported to be very useful for water purification, separation of these powdered adsorbents from wastewater in a continuous flow system tends to be very challenging. In recent years, magnetic sorption technology has become one of the emerging technologies in solving environmental problems. Exploration of magnetic materials permits the tendency of applications in water treatment extensively. After the usage, the magnetic particles which act as adsorbent are separated from the solution by magnetic force. Thus, it receives considerable attention due to ease in separation after usage.

2 Magnetic Materials

Magnetic materials are materials that display magnetic response in the presence of a magnetic field. Mitchell [1] claimed that magnetic materials are categorized into two major groups: (1) soft magnetic materials and (2) hard magnetic materials based on their magnetic properties such as remanence (B_r), coercive force (H_c), and Curie

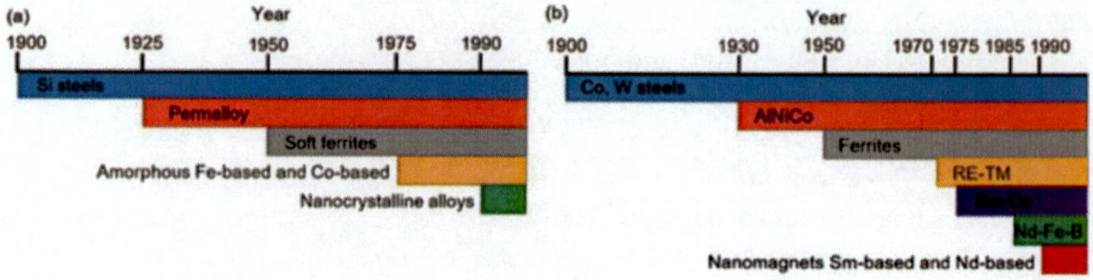

Fig. 10.1 Developments of (**a**) soft magnetic material and (**b**) hard magnetic material. (Source: Dobrzański et al. [7])

temperature (T_c). In other words, magnetic materials are classified as being magnetic or nonmagnetic. A material which is easily magnetized and demagnetized is referred to as soft magnetic material, whereas hard magnetic material is challenging to be demagnetized [2, 3]. The hard magnetic materials remain strongly magnetized after the applied magnetic field is removed and become to be permanent magnets. Iron, steel, and alloy bars are the most common magnetic materials.

Typically, soft magnetic materials denote permeable magnets or electromagnets. Electromagnets (artificial magnets) are constituted of soft-iron cores around which are wound coils of insulated wire [4]. The core becomes magnetized when an electric current flows through the coil; the core loses most of the magnetism when the current ceases to flow. The soft magnetic materials have coercive force less than 1000 A/m and high magnetic permeability. Examples are Si steels, permalloys, Mn-Zn soft ferrites, amorphous Fe-based, and Co-based.

On the other hand, hard magnetic materials denote permanent magnet. Permanent magnets as another type of artificial magnets are hardened steel or certain alloys that have been permanently magnetized. They are distinctive for their excellent magnetic properties such as remanence and coercive force [5, 6]. Examples of this material include Co and W steels, alnico bars, ferrite magnets, magnets from the cobalt alloys with the rare earth group elements, and magnets from the Nd-Fe-B alloys. The evolution of the magnetic material is illustrated in Fig. 10.1. Despite that, some commercially available magnetic materials are tabulated in Table 10.1 with their respective characteristics.

3 Aspect of Magnetism

Magnetism is a remarkable physical property that can be exploited in water treatment by manipulating the physical properties of the pollutants in water. In combination with other processes, it facilitates the efficiency of water purification. The concept of the use of magnetism for water treatment such as anti-scaling technique in boilers, pipelines in factories, coagulation, and biological processes has been extensively reported.

Table 10.1 Comparison of key characteristics of commercially available magnetic materials

Characteristic	Ceramic	Alnico	Bonded Nd-Fe-B	Sm-Co	Nd-Fe-B
Highest energy product, BH_{max} (kJ/m^3)	32	59	79	254	382
Maximum operating temperature (°C)	300	550	150	300	150
Resistance to demagnetization	Moderate	Low	High	Very high	High
Corrosion resistance [uncoated]	Excellent	Excellent	Good	Good	Poor
Mechanical toughness	Moderate	Tough	Moderate	Very brittle	Brittle
Relative cost	Very low	Moderate	High	Very high	High

Source: Hatch and Stelter [8]

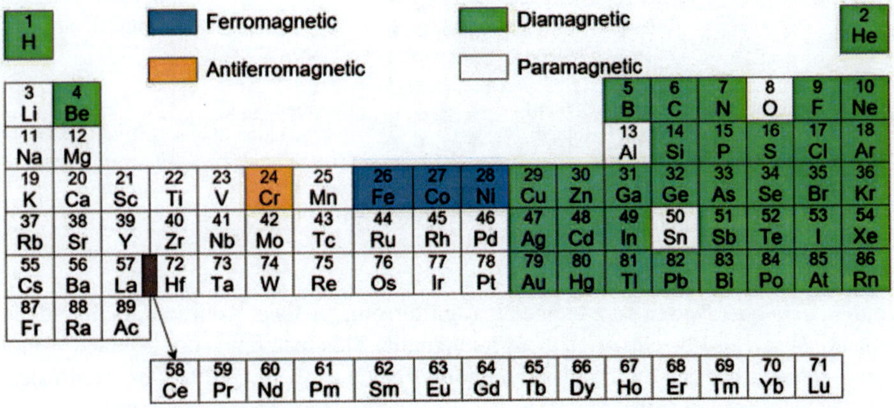

Fig. 10.2 The periodic table showing the types of the magnetic behavior of each element at room temperature. (Source: Harris and Willams [2])

Magnetism is defined as a phenomenon associated with magnetic fields. It has the power to attract magnetic materials such as iron, steel, nickel, or cobalt. There are five types of magnetism: diamagnetism, paramagnetism, ferromagnetism, antiferromagnetism, and ferrimagnetism [9, 10]. In magnetic materials, a magnetic field is formed due to the movement of electrons within the material. However, ferromagnetic materials exhibit magnetism on a microscopic scale even without an applied field [11].

Based on Fig. 10.2, diamagnetism and paramagnetism account for the magnetic behavior of most of the elements in the periodic table at room temperature. These elements are indicated as nonmagnetic material. Meanwhile, those elements which are referred to as magnetic are categorized as ferromagnetic. Another type of

Table 10.2 Classification of magnetic material

Type of magnetism	Atomic/magnetic behavior		Susceptibility, χ	Typical materials
Diamagnetism	Atoms have no magnetic moment		Small and negative	Organic materials, superconducting metals, and other metals
Paramagnetism	Atoms have randomly oriented magnetic moments	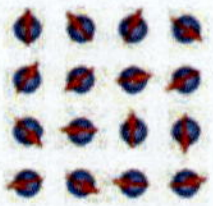	Small and positive	Alkali and transition metals, rare earth elements
Ferromagnetism	Atoms have parallel aligned magnetic moments	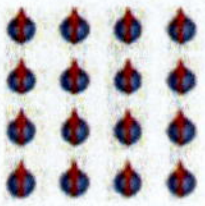	Large and positive, function of applied field, microstructure dependent	Some transition metals (Fe, Ni, Co) and rare earth metals (Gd)
Antiferromagnetism	Atoms have mixed parallel and antiparallel aligned magnetic moments	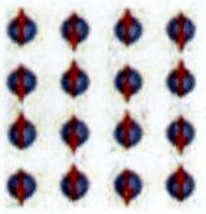	Small and positive	Salts of transition elements (MnO)
Ferrimagnetism	Atoms have antiparallel aligned magnetic moments	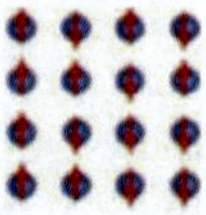	Large and positive, function of applied field, microstructure dependent	Ferrites ($MnFe_2O_4$, $ZnFe_2O_4$) and chromites

Source: Mitchell [1]

magnetism observed in pure elements is antiferromagnetism. Additionally, magnetic materials can also be classified as ferromagnetic. This material is not detected in any pure element but can be found in compounds such as the mixed oxides (ferrites).

The magnetic properties of materials are mostly determined by their structure and the several magnetic interactions between neighboring atoms. There are only 16 elements that have magnetically ordered ground state, and all but oxygen belongs to the 3d or 4f transition series [12]. The magnetic materials are divided into few categories in terms of their magnetic behavior as presented in Table 10.2.

3.1 Diamagnetism

Diamagnetism which is usually very weak is a fundamental property of all matter. In a diamagnetic material, the atoms have no net magnetic moment in the absence of applied field [13, 14]. Nevertheless, a negative magnetization is produced in the opposite direction from the applied magnetic field, resulting in negative susceptibility of approximately -10^{-6}. The susceptibility of the diamagnetic materials is temperature independent. These materials are very weakly affected by magnetic fields. Therefore, the contribution of diamagnetism is insignificant unless it is the only magnetism present. Materials such as acetone, alumina, carbon dioxide, copper, lead, quartz, silver, and water are diamagnetic.

3.2 *Paramagnetism*

Paramagnetism results from the magnetic forces on unpaired electrons. In paramagnetic materials, some of the atoms or ions have a net magnetic moment due to unpaired electrons in partially filled orbitals [15]. Under the influence of an applied field, there is a partial alignment of the atomic magnetic moments parallel to the direction of the field, resulting in a net positive magnetization and positive susceptibility lies between 10^{-3} and 10^{-5}. The net magnetization of the atom is small due to random ordering of moments but more significant than the diamagnetic contribution. Similar to diamagnetism, the magnetization is zero when the field is removed. Materials such as clay, carbonate, silicate, sodium, and platinum are paramagnetic.

3.3 *Ferromagnetism*

Ferromagnetism occurs when there are exchange interactions among the electrons in the material. In ferromagnetic materials, the atomic magnetic moments align parallel or antiparallel to each other due to an interaction between neighboring moments or mutual reinforcement of the dipoles. Unlike paramagnetic materials, the atomic moments in these materials exhibit powerful exchange forces, equivalent to a field on the order of 1000 Tesla or around 100 million times the strength of the earth's field [16]. The parallel alignment of moments results in large net magnetization even in the absence of a magnetic field, giving high susceptibility approaching 10^6. Above the Curie temperature, ferromagnetic materials exhibit paramagnetic behavior. Iron, nickel, and cobalt and many of their alloys are typical ferromagnetic materials.

3.4 *Antiferromagnetism*

Antiferromagnetism occurs when the magnetic moments produced in neighboring atoms line up in opposition to one another in the magnetic field. This magnetic ordering cancels out the magnetic moments, and net magnetization is zero. The magnetic susceptibility is positive and small, approximately 10^{-5} to 10^{-3}. Moreover, these materials exhibit paramagnetic behavior like ferromagnetic materials above a transition temperature, Néel temperature [17]. Cobalt oxide, chromium, manganese chloride, manganese oxide, and nickel oxide are examples of antiferromagnetic materials where chromium is the only element exhibiting antiferromagnetism in the periodic table at room temperature.

3.5 *Ferrimagnetism*

Ferrimagnetism is observed in more complex crystal structures than pure elements. The magnetic structure is composed of two magnetic sublattices separated by oxygen. In a ferromagnetic material, some atoms are in parallel alignment while others are in antiparallel alignment. The magnetic moments of the sublattices are not equal and result in a net magnetic moment. The magnetic susceptibility is similar to ferromagnetic materials, but ferrimagnetic materials usually have lower saturation magnetizations. The material breaks down into magnetic domains where the diluting effect of those atoms in antiparallel alignment keeps the magnetic strength of this material generally less than that of purely ferromagnetic solids, a relatively low magnitude as only one-eighth of the ions contribute to the magnetization of the material [2]. A well-known ferrimagnetic material is a magnetite, Fe_3O_4.

4 Synthesis of Magnetic Particles

According to Thanh [18], the materials fabricated in the laboratory are mainly composed of magnetite (Fe_3O_4), maghemite (γ-Fe_2O_3), greigite (Fe_3S_4), and several types of ferrites ($MeO{\cdot}Fe_2O_3$, where Me = Ni, Co, Mg, Zn, Mn, and others). Iron oxide nanoparticles (Fe_3O_4 and γ-Fe_2O_3) have shown the most promise as potential environmental magnetic sensing materials by way of the synthesis, tuning of physical properties, and surface functionalization [19].

Numerous methods of synthesizing magnetic particles such as coprecipitation, hydrothermal, thermal decomposition, and sol-gel method have been so well established. Some common methods to synthesize magnetic iron oxide nanoparticles for versatility in tuning the characteristics of the nanomaterials are described in Table 10.3. Furthermore, common inorganic coatings for magnetic nanoparticles are presented in Table 10.4.

In order to be successfully exploited in applications, prerequisites such as good dispersibility, nano-sized distribution, highly uniformed superparamagnetic properties, hydrophilic surface with different functional groups, as well as homogenous physical and chemical properties are emphasized for the synthesis of iron oxide nanoparticles [20]. The synthesis of the magnetic particles is divided into two groups: in situ methods and ex situ methods.

Table 10.3 The comparison between various common synthesis methods of magnetic iron oxide nanoparticles

Synthesis methods	Size (nm)	Shape	Benefits	Drawbacks
Coprecipitation	15–200	Spherical/ rhombic	Conventional Low reaction temperature Rapid synthesis with high yield	Not suitable for the preparation of highly pure, accurate stoichiometric phase Unprotected magnetite vulnerable to oxidation or aggregation
Hydrothermal	27	Spherical	Highly crystalline Pure iron oxide nanoparticles Simple and scalable Good morphological control	Hydrothermal slurries are potentially corrosive Harsh reaction conditions Long reaction times
Thermal decomposition	4–20	Spherical	Reasonable control of size and shape with high yields Narrow size distribution Ability to quickly make different mixed metal oxide nanoparticles	Complicated and harsh preparation procedures The surfactant used hinders subsequent surface modification High decomposition temperature
Microemulsion	4–12	Spherical (inverted), cubic, lamellar phases, cylindrical micelles	A simple and versatile method Reproducible Using surfactant limits particle nucleation, growth, and aggregation Better morphological control Homogenous particle size distribution	Low crystallinity of SPIONs on a large scale due to low-temperature usage Complicated purification methods for separation of surfactants Poor yield A large number of solvents required
Sol-gel	20–200	Spherical	Pure amorphous phases Homogeneity and phase purity Low-temperature procedure Moderate morphological control Useful for hybrid nanoparticles' fabrication	High cost Close monitoring is needed due to the several steps Challenging to obtain monodispersed nanoparticles through hydrolytic sol-gel route Broad size distribution
Sonochemical	5–30	Spherical	Shortened reaction time Uniform particle size Higher surface area Better thermal stability Improved phase purity	Particle agglomeration Not energy efficient Particle size tunability is not readily achievable Use of organometallic precursors causes in vivo toxicity

Source: Wu et al. [19], Ramimoghadam et al. [20]

Table 10.4 Common inorganic coatings for magnetic nanoparticles

Inorganic coating material	Conditions	Benefits	Drawbacks
Silica	Sol-gel/Stöber method, reverse micelle	Good dispersion in aqueous solutions Surface silanols for easy siloxane modification Ability to incorporate dyes and *quantum dots*	Drastic loss of core magnetic character
Carbon	Hydrothermal, pyrolysis	Biocompatible High chemical and thermal stability	Few synthetic methods/lack of understanding of formation mechanism
Gold	Microemulsion, solution-based reduction	Chemical inertness Thiol surface ligand modification Minor loss of core magnetic susceptibility Surface plasmon peaks for optical detection	–
Silver	Solution-based reduction	Surface plasmon peaks for optical detection	Difficult functionalizing with the Ag surface
CdSe or CdS	Solution-based shell growth	Tunable fluorescent shell	–

Source: Kumar [24]

4.1 *In Situ Methods*

In situ methods refer to the formation of nanostructures from precursor with the presence of another phase material [21, 22]. The most common method for the synthesis of magnetic particles in the literature is the coprecipitation method. In this method, divalent and trivalent iron salts are condensed in the presence of hydroxide bases. This method is relatively simple, cheap, and reproducible and results in high yield besides keeping the surface of particles free for any subsequent functionalization. However, the particles obtained in the first stage might agglomerate due to the neutralization of counter cations such as Na^+ and the negative charge of FeO^- groups on the surface. Moreover, the pH value and the required time to accomplish the transformation of the particle crystal structure are the crucial factors for the synthesis of iron oxides.

Other than that, hydrothermal is another method that has been investigated extensively for the synthesis of magnetic nanoparticles. The hydrothermal method is a process where the iron precursors in aqueous solution are heated at high temperature and pressure followed by crystallization of the dissolved material from the fluid [23]. The water can be replaced by other polar or nonpolar solvents such as benzene. The combination of microwave and hydrothermal method reported for the preparation of superparamagnetic iron oxide nanoparticles is advantageous for scale-up and fabrication of uniform particles.

4.2 Ex Situ Methods

Ex situ methods refer to the blending of pre-synthesized nanostructures with polymer using specific techniques such as ball milling, melt blending, and thermal curing. Unlike in situ methods, it is relatively challenging to handle the nanoparticle agglomeration due to the high surface energy of nanoparticle in preparing high-performance magnetic particles.

5 Magnetic Separation for Water Purification

Ideal water purification techniques at a low cost have attracted broad interest to be affordable for developing countries as well. Adsorption technology using an adsorbent satisfies the requirement due to its high performance, ease of operation, and insensitivity to toxic substances [25]. An optimal adsorbent ought to have the following characteristics: (1) high performance, (2) fast kinetics, (3) cost-efficient, (4) environmentally friendly, (5) reusability, and (6) ease of separation to be applied commercially. The most problematic issue for the removal process is the effectiveness of particle separation from aqueous solution. Water purification process has to implicate recyclable materials for successive treatment cycles and evade generation of secondary waste on an industrial scale. Among the adsorbent materials established, the feasibility of magnetic nano-adsorbents that have a high surface-to-volume ratio, tunable morphology, and excellent removal performance besides the magnetism property shows the most promising application for water and wastewater treatment systems. It is anticipated that magnetic separation can emerge to be a more cost-effective and expedient process for separating tiny powdered particles than sophisticated membrane filtration.

In the year 1995, magnetic ion exchange resins (MIEX) with the magnetic beads were introduced for the removal of natural organic matter [26]. Later, the removal of bromide ion was claimed to varying degrees by using MIEX, but the process was highly dependent on the alkalinity of the water and concentrations of the competing ion [27]. The small resin beads with a high surface area allow rapid exchange kinetics of selective ions, enabling a suitable usage in a continuous process. Agglomeration of the magnetic beads leads to flocculation, which allows higher settling velocity than for the single particles or fluidization at higher hydraulic loading rates.

On the application of an external magnetic field, the magnetic adsorbents can be rapidly and easily separated from water due to the presence of the metal components which are oxides of metals such as Fe, Co, Ni, and Cu in the adsorbent. Since magnetic particles were modified with polymer, carbon nanotubes deserve due attention to be used commercially for water treatment due to chemical stability, mechanical and thermal stability, and high surface area.

In addition, spinel ferrite magnetic materials with the general structural formula MFe_2O_4 (M = Ca, Co, Cu, Mg, Mn, Ni, Zn) are becoming more popular for remediation of metal ions due to their tunable sizes, diverse structures, high surface areas, excellent chemical and thermal stabilities, ease of separation, high sorption performance, and wide pH ranges [49]. A ferromagnetic $NiFe_2O_4$ which is synthesized by a sol-gel method using egg white shows high adsorption capacity for Cu^{2+}, Cr^{4+}, and Ni^{2+} because its 3D interconnected porous structure was reported to keep high removal efficiency (>97%) during seven reusable cycles [28]. Some other findings on the removal of heavy metals by using magnetic particles are presented with their respective synthesis methods in Table 10.5.

Table 10.5 Magnetic particles for heavy metal removal with their respective synthesis methods

Magnetic particles	Pollutants	Synthesis methods	References
Maghemite nanoparticles	Cr^{6+}	Sol-gel	[29]
Carbon nanotubes-iron oxide magnetic composites	Cu^{2+}, Pb^{2+}	Catalytic pyrolysis	[30]
Magnetic alginate microcapsules containing the extractant Cyanex 272	Ni^{2+}	Coprecipitation	[31]
Alginate encapsulated magnetic sorbent	As(V), organic arsenate, Cu^{2+}	Encapsulation	[32–34]
Zirconium-based magnetic sorbent	As(V)	Coprecipitation	[21]
Thiourea-modified magnetic chitosan microspheres	Cu^{2+}, Hg^{2+}, Ni^{2+}	Coprecipitation and hydrothermal	[35]
Amino-functionalized Fe_3O_4@SiO_2 core-shell magnetic nanomaterial	Cu^{2+}, Cd^{2+}, Pb^{2+}	Coprecipitation	[36]
Magnetic chitosan nanoparticles	Cu^{2+}	Coprecipitation and hydrothermal	[37]
Superparamagnetic iron oxide nanoparticles (SPIONs) modified with poly(γ-glutamic acid) (PGA)	Cd^{2+}, Pb^{2+}	Coprecipitation	[38]
Magnetic Fe_3O_4 @ silica-xanthan gum composite	Pb^{2+}	Coprecipitation	[39]
Magnetic chitosan/cellulose hybrid microspheres	Cu^{2+}, Cd^{2+}, Pb^{2+}	Sol-gel	[40]

6 Recovery of Magnetic Particles

Although abundant experimental photographic results were reported for the prospect of separation and recovery of magnetic nanoparticles from water or wastewater, there is no successful industrial application of magnetic particles for water or wastewater treatment that have been published. The discrete magnetic particles which are prone to agglomerate due to the magnetism property were believed to decrease the heavy metal removal capacity. The problem has become a prodigious limitation for recovery and reuse of the magnetic particles. Several conditions are manipulating the applicability of magnetic particles in a sustainable treatment process.

Several studies have indicated that the nanoscale iron oxide, which behaves as superparamagnetic material, can offer high accessibility and reusability for water or wastewater treatment systems [21, 41–43]. Majority of the systems require centrifugation or filtration process to separate solid wastes. Magnetic nanoparticles, however, can be separated and recovered easily with the aid of an external magnetic field due to the inherent characteristic of the nanoparticles, magnetism [44].

Magnetic separation is a technique to collect or separate magnetic substances from flowing streams. High-gradient magnetic separation (HGMS) device comprises of a bed of magnetically susceptible filling placed inside an electromagnet usually practiced in magnetic separations. Scientific literature has indicated that magnetic field-enhanced process with magnetite and HGMS offers substantial improvement for the removal of heavy metals from wastewater [45]. The generation of significant magnetic field gradients, size and magnetic properties of particles, and the area of magnetized surfaces are the crucial factors for effective particle collection. When the particle suspension flows through the separation unit, the magnetic force attracting particles must be on top of the gravitational, fluid drag, buoyancy, inertial, and diffusion forces.

Separation of magnetic nanoparticles with a low-gradient magnetic field or a handheld permanent magnet (usually made of Fe and Nd) from solution has been regularly conveyed [46]. However, a higher magnetic force for a superconducting magnet in HGMS instead of an electromagnet is required when the particle size is reduced to the nanoscale. Complementary to this, aggregation due to strong magnetic dipole-dipole interactions between particles can either be coated with a surfactant to provide steric stability or particle surfaces be modified to produce a repulsive electrostatic force [47, 48]. In short, an optimum particle size, synthesis methods of magnetic nanoparticles, and fluid properties are the aspects needed to be considered to design an effective recovery and high removal performance system.

7 Environmental Application of Magnetic Particles

Magnetic particles such as nano zerovalent iron, magnetite, and maghemite and composite magnetic nanoparticles have been extensively utilized in environmental applications. These particles demonstrated favorable performance for eradicating pollutants from waterway systems. These promising findings have laced an alternative solution for the removal of heavy metal ions from contaminated water using magnetic adsorption technology [50].

Remediation of heavy metal pollution is possible by immobilizing them in a non-bioavailable form or re-speciating them into fewer toxic forms with nano zerovalent iron (nZVI). Several practical experiences of site remediation using nZVI have been established in the USA, and this remediation technique is expected to phase out in the coming decades. In Europe, many pilot test projects with the application of nZVI for different pollutants through three full-scale applications were merely conveyed. The pilot studies of 7 kg nZVI were reported to treat 1–5 g/L of Cr (VI) at Permon, Czechia [51]. Furthermore, macroscale zerovalent iron has been documented as a suitable electron donor in aquatic environments.

Economic constraints and the precautionary attitude in Europe raise concerns regarding the feasibility of nZVI as a cost-effective technique for aquifer remediation. Nontechnical aspects, including the probability of negative repercussions and the novelty of the technology for consultancy, governments, as well as the public, tend to be significant challenges for the commercialization of nZVI.

Glossary

Adsorbent	A material which will allow a liquid, gas, or dissolved solid to adhere to its surface
Adsorption	A process by which a solid holds molecules of a gas or liquid or solute as a thin film
Carbon nanotubes	Cylindrical molecules that consist of rolled-up sheets of single-layer carbon atoms (graphene)
Chemical stability	Resistance of a chemical to change in a chemical reaction
Coprecipitation	Simultaneous precipitation of more than one compound from a solution
Curie temperature	Temperature at which certain magnetic materials undergo a sharp change in their magnetic properties
Heavy metal ions	Metallic cation and anion that have relatively high density and are toxic or poisonous at low concentrations
Hydrothermal	A chemical reaction in water at both high temperature and pressure in a sealed pressure vessel
Ion exchange	A reversible chemical reaction where dissolved ions are removed from the solution and replaced with other ions of the same or similar electrical charge
Ion exchange resin	A resin that acts as a medium for ion exchange
Magnetic force	

	Attractive or repulsive force that is exerted between the poles of a magnet and electrically charged moving particles
Magnetic field	A vector field that describes the magnetic influence of electric charges in relative motion and magnetized materials
Magnetic sorption technology	A process by which a magnetic solid holds molecules of a gas or liquid or solute as a thin film
Mechanical stability	The quality of being free from change or variation
Microemulsion	A thermodynamically stable, isotropic emulsion with a droplet size of 1–100 nm
Nanoparticle	A microscopic particle with at least one dimension less than 100 nm
Permanent magnet	Material with the magnetic field that is generated by the internal structure of the material itself
Pyrolysis	Decomposition brought about by high temperatures
Reverse micelle	Aqueous-phase droplet in organic solvent that is stabilized by surfactant
Sol-gel	A wet chemical process involving the formation of an inorganic colloidal suspension (sol) and gelation of the sol in a continuous liquid phase (gel) to form a three-dimensional network structure
Stöber method	A chemical process for preparation of silica (SiO_2) particles of controllable and uniform size
Sonochemical	Application of ultrasound to chemical reactions and processes
Thermal decomposition	Some compounds breaking down when heated, forming two or more products from one reactant
Thermal stability	The ability of the material to resist the action of heat and to maintain its properties
Toxic substances	Substances that can be poisonous or cause health effects
Water purification	The process of removing undesirable chemicals, biological contaminants, suspended solids, and gases from water

References

1. Mitchell BS (2004) An introduction to materials engineering and science for chemical and materials engineers. Wiley, New Jersey, USA
2. Harris IR, Willams AJ (2009) Magnetic materials. In: Rawlings RD (ed) Materials science and engineering – volume II. EOLSS, UK, pp 49–83
3. Birčáková Z, Kollár P, Füzer J, Bureš R, Fáberová M (2020) Magnetic properties of selected Fe-based soft magnetic composites interpreted in terms of Jiles-Atherton model parameters. J Magn Magn Mater 502:166514
4. Spellman FR (2014) Fundamentals of electricity, 3rd edn. CRC Press, Boca Raton, FL, USA
5. Spellman FR, Bieber RM (2011) The science of renewable energy. CRC Press, Boca Raton, FL, USA
6. Allia P, Barrera G, Tiberto P (2020) Hysteresis effects in magnetic nanoparticles: a simplified rate-equation approach. J Magn Magn Mater 496:165927
7. Dobrzański LA, Drak M, Ziębowicz B (2006) Materials with specific magnetic properties. J Achiev Mater Manuf Eng 17(1–2):37–40
8. Hatch GP, Stelter RR (2001) Magnetic design considerations for devices and particles used for biological high-gradient magnetic separation (HGMS) systems. J Magn Magn Mater 225:262–276
9. Buschow KHJ, de Boer FR (2004) Physics of magnetism and magnetic materials. Kluwer Academic, Moscow

10. Acharya SS, Medicherla VRR, Bapna K, Ali K, Biswas D, Maiti K, Rawat R (2020) Exchange correlation and magnetism in bcc Fe 0.8 Ni 0.2 alloy. J Electron Spectrosc Relat Phenom 240:146933
11. Czichos H, Saito T, Smith LE (2011) Springer handbook of metrology and testing. Springer Handbook of Metrology and Testing, New York
12. Coey JMD (2009) Magnetism and magnetic materials. Cambridge University Press, New York
13. Callister WDJ (2007) Material science and engineering an introduction. Wiley, USA
14. Chen Q (2020) Nanobismuth enhanced plasmonic, emission and faraday rotation properties of diamagnetic tellurite glasses. J Alloys Compd 828:154448
15. Rao RS (2012) Electromagnetic waves and transmission lines. PHI Learning, New Delhi, India
16. Pal M, De A (2009) Polymer-iron oxide based magnetic nanocomposites. In: Merhari L (ed) Hybrid nanocomposites for nanotechnology: electronic, optical, magnetic and biomedical applications. Springer Science+Business Media, New York, pp 455–506
17. Buschow KHJ, de Boer FR (2003) Physics of magnetism and magnetic materials, Springer, US
18. Thanh NT (2012) Magnetic nanopartic. Ramimoghadam D, Bagheri S, Hamid SBA (2015) Stable monodisperse nanomagnetic colloidal suspensions: an overview. Colloids and Surfaces B: Biointerfaces 133:388–411. Les: from fabrication to clinical applications. Florida: Taylor & Francis, USA
19. Wu W, He QG, Jiang CZ (2008) Magnetic iron oxide nanoparticles: synthesis and surface functionalization strategies. Nanoscale Res Lett 3:397–415
20. Ramimoghadam D, Bagheri S, Hamid SBA (2015) Stable monodisperse nanomagnetic colloidal suspensions: an overview. Colloids Surf B: Biointerfaces 133:388–411
21. Zheng YM, Lim SF, Chen JP (2009) Preparation and characterization of zirconium based magnetic sorbent for arsenate removal. J Colloid Interface Sci 338(1):22–29
22. Zhu JH, Wei SY, Chen MJ, Gu HB, Rapole SB, Pallavkar S, Ho TC, Hopper J, Guo ZH (2013) Magnetic nanocomposites for environmental remediation. Adv Powder Technol 24:459–467
23. Byrappa K, Adschiri T (2007) Hydrothermal technology for nanotechnology. Prog Cryst Growth Charact Mater 53:117–166
24. Kumar CSSR (2009) Magnetic nanomaterials. Wiley-VCH, Weinheim
25. Dauda KT, Atasie VN, Adeoye OB (2015) The kinetics and equilibrium studies of the adsorption of (Cu^{2+}, Pb^{2+}) from industrial wastewater using rice husk as adsorbent. Int J Adv Res Chem Sci (IJARCS) 2(4):13–21
26. Ambashta RD, Sillanpää M (2010) Water purification using magnetic assistance: a review. J Hazard Mater 180(1–3):38–49
27. Boyer TH, Singer PC (2005) Bench-scale testing of a magnetic ion exchange resin for removal of disinfection by-product precursors. Water Res 39(7):1265–1276
28. Hou X, Feng J, Liu X, Ren Y, Fan Z, Wei T, Meng J, Zhang M (2011) Synthesis of 3D porous ferromagnetic NiFe2O4 and using as novel adsorbent to treat wastewater. J Colloid Interface Sci 362(2):477–485
29. Hu J, Chen GH, Lo IMC (2005) Removal and recovery of Cr(VI) from wastewater by maghemite nanoparticles. Water Res 39(18):4528–4536
30. Reng XJ, Luan ZK, Di ZC, Zhang ZG, Zhu CL (2005) Carbon nanotubes-iron oxides magnetic composites as adsorbent for removal of Pb(II) and Cu(II) from water. Carbon 43:855–894
31. Ngomsik A, Bee A, Siaugue J, Cabuil V, Cote G (2006) Nickel adsorption by magnetic alginate microcapsules containing an extractant. Water Res 40(9):1848–1856
32. Lim SF, Zheng YM, Zou SW, Chen JP (2008) Characterization of copper adsorption onto an alginate encapsulated magnetic sorbent by a combined FT-IR, XPS, and mathematical modeling study. Environ Sci Technol 42(7):2551–2556
33. Lim SF, Zheng YM, Chen JP (2009) Organic arsenic adsorption onto a magnetic sorbent. Langmuir 25(9):4973–4978
34. Lim SF, Zheng YM, Zou SW, Chen JP (2009) Uptake of arsenate by an alginate-encapsulated magnetic sorbent: process performance and characterization of adsorption chemistry. J Colloid Interface Sci 333(1):33–39

35. Zhou LM, Wang YP, Liu ZR, Huang QW (2009) Characteristics of equilibrium, kinetics studies for adsorption of Hg(II), Cu(II), and Ni(II) ions by thiourea-modified magnetic chitosan microspheres. J Hazard Mater 161:995–1002
36. Wang JH, Zheng SR, Shao Y, Liu JL, Xu ZY, Zhu DQ (2010) Amino-functionalized Fe3O4@SiO2 core–shell magnetic nanomaterial as a novel adsorbent for aqueous heavy metals removal. J Colloid Interface Sci 349(1):293–299
37. Chen YW, Wang JL (2011) Preparation and characterization of magnetic chitosan nanoparticles and its application for Cu(II) removal. Chem Eng J 168(1):286–292
38. Inbaraj BS, Chen BH (2012) In vitro removal of toxic heavy metals by poly(γ-glutamic acid)-coated superparamagnetic nanoparticles. J Int J Nanomed 7:4419–4432
39. Peng XL, Xu F, Zhang WZ, Wang JY, Zeng C, Niu MJ, Chmielewská E (2014) Magnetic Fe3O4 @ silica–xanthan gum composites for aqueous removal and recovery of Pb2+. Colloids Surf A Physicochem Eng Asp 443:27–36
40. Zhao MH, Xu Y, Zhang CS, Rongm HW, Zeng GM (2016) New trends in removing heavy metals from wastewater. Appl Microbiol Biotechnol 100(15):6509–6518
41. Yantasee W, Warner CL, Sangvanich T, Addleman RS, Carter TG, Wiacek RJ, Fryxell GE, Timchalk C, Warner MG (2007) Removal of heavy metals from aqueous systems with thiol functionalized superparamagnetic nanoparticles. Environ Sci Technol 41(14):5114–5119
42. Mahdavian AR, Mirrahimi MA (2010) Efficient separation of heavy metal cations by anchoring polyacrylic acid on superparamagnetic magnetite nanoparticles through surface modification. Chem Eng J 159:264–271
43. Feng LY, Cao MH, Ma XY, Zhu YS, Hu CW (2012) Superparamagnetic high-surface-area Fe_3O_4 nanoparticles as adsorbents for arsenic removal. J Hazard Mater 217–218:439–446
44. Lim SF, Lee AYW, Kam YS, Chua SND (2018) Synthesis and characterization of alginate encapsulated zirconium-based ferromagnetic sorbent for adsorptive removal of dyes. Int J Eng Technol 7(3.18):87–90
45. Navratil JD, Shing Tsair MT (2003) Magnetic separation of iron and heavy metals from water. Water Sci Technol 47(1):29–32
46. Lim SF, Lee AYW, Chua SND, Lim BH, Jain P (2020) Kinetic study on heavy metal divalent ions removal using zirconium-based magnetic sorbent. Int J of Integr Eng 12(3):225–234
47. Low KG, Lim SF (2014) Study on electrostatic extrusion method for synthesizing calcium alginate encapsulated Iron oxide. J Appl Sci Process Eng 1:1
48. Moore TL, Rodriguez-Lorenzo L, Hirsch V, Balog S, Urban D, Jud C, Rothen-Rutishauser B, Lattuada M, Petri-Fink A (2015) Nanoparticle colloidal stability in cell culture media and impact on cellular interactions. Chem Soc Rev 44:6287–6305
49. Reddy DHK, Yun YS (2016) Spinel ferrite magnetic adsorbents: alternative future materials for water purification? Coord Chem Rev 315:90–111
50. Anuradha Jabasingh S, Ravi T, Yimam A (2018) Magnetic hetero-structures as prospective sorbents to aid arsenic elimination from life water streams. Water Sci 32(1):151–170
51. Mueller NC, Braun J, Bruns J, Černik M, Rissing P, Rickerby D, Nowack B (2012) Application of nanoscale zero valent iron (NZVI) for groundwater remediation in Europe. Environ Sci Pollut Res 19(2):550–558

Chapter 11
Biohydrogen Production from Lignocellulosic Biomass by Extremely Halotolerant Bacterial Communities from a Salt Pan and Salt-Damaged Soil

Dyah Asri Handayani Taroepratjeka, Tsuyoshi Imai, Prapaipid Chairattanamanokorn, Alissara Reungsang, and Yung-Tse Hung

Contents

1 Introduction 413
2 Materials and Methods 415
2.1 Seed Microorganisms and Medium 415
2.2 Culture Conditions and Experimental Procedures 415
2.3 Analytical Method 416
2.4 Theoretical Hydrogen Production and Yield 416
3 Results and Discussion 417
3.1 Biohydrogen Production at 3–10% Salinity 417
3.2 Biohydrogen Production at 15–26% Salinity 419
3.3 Evaluation of the Bacteria's Requirement for Chloride Ions in High Salinity Conditions 421

D. A. H. Taroepratjeka
Department of Environmental Engineering, Faculty of Civil Engineering and Planning, Institut Teknologi Nasional, Bandung, Bandung, Jawa Barat, Indonesia

T. Imai (✉)
Graduate School of Sciences and Technology for Innovation, Yamaguchi University, Yamaguchi, Japan
e-mail: imai@yamaguchi-u.ac.jp

P. Chairattanamanokorn
Department of Environmental Technology and Management, Faculty of Environment, Kasetsart University, Bangkok, Thailand
e-mail: fscippck@ku.ac.th

A. Reungsang
Department of Biotechnology, Faculty of Technology, Khon Kaen University, Khon Kaen, Thailand
e-mail: alissara@kku.ac.th

Y.-T. Hung
Civil Engineering Department, Cleveland State University, Cleveland, OH, USA
e-mail: y.hung@csuohio.edu

L. K. Wang, M. -H. S. Wang, Y. -T. Hung, N. K. Shammas (eds.), *Integrated Natural Resources Management*, Handbook of Environmental Engineering 20,
https://doi.org/10.1007/978-3-030-55172-8_11

4 Summary 423
Glossary 425
References 425

Abstract Extremely halotolerant hydrogen-producing bacteria were investigated, owing to their ability to live in high salinity conditions. Based on this characteristic, it was hypothesized that extremely halophilic hydrogen-producing bacteria can tolerate high concentrations of Na^+ ions. To test this hypothesis, we investigated the characteristics of extremely halotolerant hydrogen-producing bacteria obtained from salt-damaged soil in Khon Kaen and a commercial salt pan field near Bangkok (Samut Sakhon), Thailand. Results of this preliminary investigation showed that hydrogen production under saturated conditions of 26% (6 M) NaCl was possible after 1 year of acclimatization. The extremely halotolerant hydrogen-producing bacteria in this research were also confirmed to have a requirement for Cl^- ions for hydrogen production. Therefore, these extremely halotolerant hydrogen-producing bacteria are suitable for hydrogen production from lignocellulosic biomass.

Keywords Biohydrogen production · Lignocellulosic biomass · Extremely halotolerant bacterial communities

Nomenclature

μmol	Micromolar
ATP	Adenosine triphosphate
$CaCl_2{\cdot}2H_2O$	Calcium chloride dihydrate
Cl^-	Chloride ion
CO_2	Carbon dioxide
$CoCl_2{\cdot}6H_2O$	Cobalt (II) chloride hexahydrate
d	Day
F/M	Food to microorganisms
$FeCl_3{\cdot}6H_2O$	Iron (III) chloride hexahydrate
g	Gram
g/L	Gram/liter
h	Hours
H_2	Hydrogen
HAc	Acetic acid
HBu	Butyric acid
HMY	Hydrogen molar yield
K^+	Potassium ion
K_2HPO_4	Dipotassium phosphate
KCl	Potassium chloride
L	Liter
M	Molar

m	Meter
$MgCl_2 \cdot 6H_2O$	Magnesium chloride hexahydrate
$MgSO_4 \cdot 7H_2O$	Magnesium sulfate heptahydrate
min	Minutes
mL	Milliliter
mM	Millimolar
Na^+	Sodium ion
Na_2SO_4	Sodium sulfate
NaCl	Sodium chloride
$NaHCO_3$	Sodium bicarbonate
NaOH	Sodium hydroxide
NH_4Cl	Ammonium chloride
$(NH_4)_2HPO_4$	Diammonium phosphate
$NiCl_2 \cdot 6H_2O$	Nickel (II)chloride hexahydrate
STP	Standard temperature and pressure (0 °C, 1 atm)
T	Temperature
VFA	Volatile fatty acid
wt./vol.	Weight/volume

1 Introduction

The dependence on fossil fuels for energy supply has had great impacts on global warming and climate change [1]. Therefore, the development of alternative renewable energy sources is being pursued globally [2]. Biohydrogen is one of the promising candidates for future use because it is a CO_2-free, clean, and highly efficient energy carrier. Production of biohydrogen can be achieved through bio-photolysis, photofermentation, and dark fermentation process [3].

Dark fermentation process offers several advantages in industrial biohydrogen production. Among them are high production rates, high yields per mole of substrate, continuous production regardless of solar light condition [4], high variety of carbon sources as substrates, and has no oxygen limitation since the process is fully anaerobic [5, 6]. On the other hand, dark fermentation also has several limitations, such as thermodynamically unfavorable condition as hydrogen yields increase and carbon dioxide's presence in the produced gas [6].

Among carbon sources to supply fermentable sugars in biohydrogen production, lignocellulosic biomass is a highly considered option. It doesn't compete with food production, and it is available abundantly in nature as grasses and woods, in forestry and agricultural residues, as well as in domestic and industrial wastes. It was estimated that lignocellulosic biomass residue is being produced more than 220 billion tons annually all over the world [7]. However, lignocellulosic biomass requires pretreatment [7] such as alkaline and heat treatment followed by hydrolysis (with enzymes) prior to use as feedstock in fermentative hydrogen production [8]. Most bacteria do not retain their viability after alkaline and heat pretreatment because of the high concentration of Na^+ ions, as the acidogenesis process in anaerobic digestion is severely inhibited by such conditions [9]. Therefore, an additional step to dilute or neutralize the alkaline conditions is required before proceeding to the next step of fermentative hydrogen production. However, this additional step makes the whole process less economical. One way to overcome this problem is by utilizing extremely halotolerant hydrogen-producing bacteria in dark fermentation process. These bacteria are advantageous for developing "Next Generation Industrial Biotechnology" (NGIB) as they cut the costs of freshwater, oxygen, and sterilization [10].

In the past few years, several studies have investigated hydrogen production using halotolerant and halophilic bacteria in dark fermentation process. Most of these studies focused mainly on pure cultures in a moderate halophilic environment with a salt concentration of 0.5 M (25 g/L) to 2.5 M (150 g/L), although one research also investigated mixed cultures. Liaw and Mah [11] mentioned the production of 144.5 μmol hydrogen from 5 mL medium with 12% NaCl and 0.5% glucose at 37 °C by *Haloanaerobacter chitinovorans* sp. *nov.,* while Mouné et al. [12] found that *Haloanaerobacter salinarius* sp. *nov.* was able to produce 2 mM hydrogen from 4.17 mM glucose substrate in 14–15% NaCl, 45 °C, and pH 7.4–7.8. Matsumura et al. [13] discussed the production of 1.7-mol H_2/mol mannitol by *Vibrio tritonius* strain AM2 at initial 2.25% (w/v) NaCl, pH 6, and 37 °C.

Kivistö et al. [14] reported that *H. saccharolyticum* subspecies *saccharolyticum* produced 0.6-mol H_2/mol glycerol at a salt concentration of 150 g/L (2.6 M), pH of 7.4, temperature of 37 °C, and glycerol concentration of 2.5 g/L, while *H. saccharolyticum* subspecies *senegalensis* produced 1.6 mol H_2/mol glycerol at pH 7.0. Brown et al. [15] found that *H. hydrogeniformans* was capable of producing hydrogen at a pH of 11, 7% (wt./vol.) NaCl, and 33 °C. Pierra et al. [16] described the ability of a mixed culture affiliated to the family of *Vibrionaceae* to produce 0.9 ± 0.02 mol H_2/mol glucose at initial pH of 8 and temperature of 35 °C under a moderate halophilic environment (75 g/L NaCl). To date, no studies have investigated hydrogen production under conditions of a high salinity of 26% (6 M or 351.35 g/L NaCl).

This research aims to investigate hydrogen production from lignocellulosic biomass by extremely halotolerant bacteria. The hypothesis is that extremely halotolerant hydrogen-producing bacteria can tolerate high concentrations of Na^+ ions. Based on this hypothesis, hydrogen production under different salinity concentrations before and after acclimatization was evaluated. The bacteria's requirement for chloride ions in high salinity conditions was also investigated. In addition, it will be discussed about the suitability of our isolated extremely halotolerant bacteria for hydrogen production from lignocellulosic biomass.

2 Materials and Methods

2.1 Seed Microorganisms and Medium

The soil samples were obtained from salt-damaged soil in Khon Kaen, Thailand, and a commercial salt pan field near Bangkok (Samut Sakhon), Thailand. The soil samples were mixed with a substrate for cultivation in anaerobic conditions. The composition of the substrate used for biohydrogen production experiments at different salinity concentrations, i.e., between 3–10% and 15–26% NaCl, before the acclimatization experiments is as follows: 2 g/L $NaHCO_3$, 2 g/L K_2HPO_4, 1 g/L yeast extract, 0.7 g/L $(NH_4)_2HPO_4$, 0.75 g/L KCl, 0.85 g/L NH_4Cl, 0.42 g/L $FeCl_3{\cdot}6H_2O$, 0.82 g/L $MgCl_2{\cdot}6H_2O$, 0.25 g/L $MgSO_4{\cdot}7H_2O$, 0.018 g/L $CoCl_2{\cdot}6H_2O$, 0.15 g/L $CaCl_2{\cdot}2H_2O$, and 0.018 g/L $NiCl_2{\cdot}6H_2O$. Glucose concentration was adjusted according to each experiment. All chemicals were purchased from FUJIFILM Wako Pure Chemical Corp., Japan. The composition of the substrate for the experiments on the bacteria's requirement for chloride ions in high salinity conditions, acclimatization period, and biohydrogen production at 26% NaCl after 2 years of acclimatization was the same as that above, except that $NiCl_2{\cdot}6H_2O$ was omitted.

2.2 Culture Conditions and Experimental Procedures

The first step of the experiment was to evaluate the hydrogen production in conditions of 3–10% salinity before acclimatization. The experiments were done under six NaCl concentrations of 3%, 3.5%, 5%, 7%, 7.5%, and 10%. The second step of experiment was to evaluate the hydrogen production in conditions of 15–26% salinity before acclimatization and at 26% salinity after an acclimatization period of 2 years. The experiment was done at NaCl concentrations of 15%, 20%, and 26%.

The third step of the experiment was the evaluation of the bacteria's requirement for chloride ions in high salinity conditions. The experiment was done by comparing the culture under Na_2SO_4/NaCl ratios of 1:1 and 4:1. The ratio was prepared by weight to reach 26% (351.35 g/L) concentration of the mix. The third step was done after 1 year of acclimatization.

The experiments to determine biohydrogen production under salinity concentrations of 3–10% and 15–26% NaCl before acclimatization and various F/M ratio at 15% NaCl were done under the following conditions: 100 mL sealed serum bottles in a nitrogen atmosphere with an initial pH of 6.8 adjusted with 1 M HCl and 1 M NaOH and incubated in a shaking incubator (BT100 and BT300; Yamato Scientific Co., Ltd. Japan) at 35 °C and a 100-rpm shaking speed. Biogas samples were periodically collected, and the compositions were analyzed via gas chromatography.

The experiment to determine the bacteria's requirement for chloride ions in high salinity conditions followed the same conditions, except that 125-mL serum bottles were used. The main culture bottles with a volume of 500 mL each for the three different sources of soil were also maintained. These bottles were also used for acclimatization purposes. In acclimatization period of 2 years, the substrate's NaCl concentration was kept at 26%. Gas production was periodically measured, and substrate was changed after no gas production was detected. Anaerobic condition was maintained under nitrogen atmosphere.

Biohydrogen production at a NaCl concentration of 26% after 2-year period of acclimatization experiments was also studied under the same conditions, except that the initial pH was not adjusted, the serum bottles had a volume of 75 mL, and shaking incubator temperature was at 37 °C.

2.3 Analytical Method

The initial pH was adjusted by using a pH meter (D-13; Horiba Co. Ltd., Japan). The volume of the produced biogas was measured with a glass syringe. The composition of H_2, N_2, CH_4 and CO_2 was analyzed via gas chromatography (GC-8APT; Shimadzu Corp., Japan) with a 60/80 activated charcoal mesh column (1.5 m × 3.0-mm internal diameter) and argon as a carrier gas, with operational temperatures of the injector, column, and TCD detector of 50 °C, 60 °C, and 50 °C, respectively. The compositions of volatile fatty acids (VFAs) were determined by gas chromatography (GC-8APF; Shimadzu Corp., Japan) with a flame ion detector (FID) and a Unisole F-200 30/60 glass mesh column (3 m × 3.2-mm internal diameter). The operational temperatures for the injection port, FID detector, and oven were 250 °C, 140 °C, and 140 °C, respectively.

The water content and total organic matter of the soil was determined via the JIS A 1203 test method for water content of soils and JIS A 1226–2000 test method for ignition loss of soils [17, 18]. Sample masses of 20.5506–44.1789 g were used to determine the soil's water content and ignition loss. Volatile suspended solid (VSS) was determined according to method 2540 E of *Standard Methods* [19].

Soil salinity was determined by mixing soil and distilled water in 1:2.5 dry soil to water ratio, shaken for 3 hours at 180 rpm (Eyela Multishaker MMS; Tokyo Rikakikai Co. Ltd., Japan). After 25 minutes of settling time, the supernatant is measured by a thermosalinity meter (TS-391; As One Corp., Japan). The measurement results were then multiplied by the dilution factor of water content of soil.

2.4 Theoretical Hydrogen Production and Yield

The theoretical hydrogen production was determined using Eqs. (1) and (2) in Table 11.1. Based on Eq. (1), 1 mol of glucose will produce 4 mol of hydrogen. Thus, 1 g of glucose at standard temperature and pressure (STP) conditions will produce 498 mL of H_2 via the acetic acid (HAc) pathway. Based on Eq. (2), 1 mol of

Table 11.1 Standard Gibbs energy of formation for glucose fermentation

Equations of fermentative reactions		ΔG^0
$C_6H_{12}O_6 + 4H_2O \rightarrow 2CH_3COO^- + 2HCO_3^- + 4H^+ + 4H_2$	(1)	−206 kJ
$C_6H_{12}O_6 + 2H_2 \rightarrow CH_3(CH_2)_2COOH + 2CH_3COO^- + H^+ + 2H_2$	(2)	−254 kJ
$C_6H_{12}O_6 + 2H_2 \rightarrow 2CH_3CH_2COO^- + 2H_2O + 2H^+$	(3)	−358 kJ
$C_6H_{12}O_6 \rightarrow 2CH_3CH(OH)COO^- + 2H^+$	(4)	−198 kJ
$C_6H_{12}O_6 + 2H_2O \rightarrow 2CH_3CH_2OH + 2HCO_3^- + 2H^+$	(5)	−358 kJ

Obtained from [34, 35]

Table 11.2 Biohydrogen production at 3–10% salinity of salt-damaged soil from Khon Kaen

Salt concentration (%)	Biohydrogen production (ml)	Theoretical maximum H_2 production reached (%)	HMY (mol H_2/mol$_{glucose}$)
3	10.9	14.7	0.61
3.5	10.9	14.7	0.61
5	9.46	12.8	0.53
7	13.4	18.1	0.75
7.5	7.43	10	0.41
10	18.1	24.5	1.01

Glucose 0.15 g (5,000 mg/L), inoculum 3000 mg/L VSS, F/M ratio 1.5

glucose will produce 2 mol of hydrogen; hence, 1 g of glucose at STP conditions is required to produce 249 mL of H_2 via the butyric acid (HBu) pathway.

Theoretical maximum H_2 production reached is the comparison of observed cumulative H_2 in the experiments (mL) to the multiplication of glucose provided in the substrate (gr) with 498 mL of H_2 produced via the HAc pathway. Hydrogen molar yield (HMY) is the observed cumulative H_2 (mol) divided by glucose provided in substrate (mol).

3 Results and Discussion

3.1 Biohydrogen Production at 3–10% Salinity

Table 11.2 compares the biohydrogen production at salinities of 3% to 10% for a culture from salt-damaged soil from Khon Kaen, Thailand. Very low hydrogen production was observed at these salinity conditions. No methane was produced at salinities of 7.5% and higher. The experiments were conducted for 15 days with 0.15 g glucose for each 100-mL serum bottle. The maximum theoretical cumulative hydrogen yield was 74 mL H_2 for the acetic acid (HAc) pathway and 37 mL H_2 for the butyric acid (HBu) pathway. The highest yield of 1.01 mol H_2/mol$_{glucose}$ was achieved at 10% salinity.

Table 11.3 Soil characteristics of Khon Kaen salt-damaged soil and soil from Samut Sakhon salt pan

	Water content (%)	Ignition loss after 600 °C (g)	Loss on ignition (%)	Salinity (%)
Khon Kaen salt-damaged soil				
At the shore	32.94	0.42	2.37	7.74
Close to the shore	21.33	0.25	1.57	20.16
Farther from the shore	13.56	0.32	3.16	30.26
Samut Sakhon salt pan				
Salt pan surface	77.29	0.15	4.35	1.92
Salt pan at 5 cm depth	60.4	0.95	5.41	3.45
Dry salt pan surface	43.26	0.04	0.43	7.80
Dry salt pan at 5 cm depth	40.15	0.29	1.51	6.15

This is because the salt-damaged soil in its natural state is always exposed to a high salt concentration; thus, lower salt concentrations might not be suitable for extremely halotolerant anaerobic microorganisms to grow. Another reason for this is that the food to microorganism (F/M) ratio at 1.5 [mg/L volatile suspended solids (VSS)]/[mg/L of the substrate] might not be ideal for production for 3–10% salinity conditions.

Loss on ignition (LOI) is one of the most commonly used methods for quantifying soil organic matter [20]. The LOI results in Table 11.3 correspond to mixed sediment with low organic matter content obtained by Heiri et al. [21]. Microbial biomass is usually low in salt-affected soils [22]. The Khon Kaen salt-damaged soil had a lower moisture content than the soil from the Samut Sakhon salt pan. Water content in soil is an important factor that influences the microbial activity of aerobic and anaerobic bacteria, and it also affects osmotic potentials of saline soils [23].

Although LOI percentage and water content from Samut Sakhon salt pan showed a higher value, the soil sample from the location gave lower biohydrogen yields, as shown in Tables 11.5 and 11.7. One of the possible reasons for this is that high organic contamination doesn't occur in the salt pan since it was under a protected environment to maintain the purity of the salt, unlike the salt-damaged soil of Khon Kaen. A major part of the organic matter that contributed to the LOI value in Samut Sakhon salt pan could be refractory organic. In Samut Sakhon salt pan's location, no plants and fish were observed, so not many supplies for organic matter were available.

3.2 *Biohydrogen Production at 15–26% Salinity*

The experiments on biohydrogen production at 15–26% salinity were performed twice. The first was done before acclimatization and the second was after 2 years of acclimatization. Table 11.4 compares biohydrogen production at the initial salinity of 15–26% before acclimatization. The highest production of 2.78 mol $H_2/mol_{glucose}$ occurred at 15% salinity. Theoretical hydrogen production value exceeding 100% of HBu pathway at 15% salinity suggested that the HAc pathway took place. Figure 11.1 shows that after a lag phase of 8 days, the cumulative hydrogen production at 15% salinity was significantly higher (49.8 mL; Table 11.2) than that at 10% salinity (18.1 mL; Table 11.2).

During the initial experiment, almost no hydrogen was produced under conditions of 20% salinity or more. After an acclimatization period of 2 years, confirmation experiments were conducted at 26% salinity (Table 11.5). The results showed that hydrogen production was possible at this concentration, with hydrogen yields of 0.66–1.15 mol $H_2/mol_{glucose}$.

Table 11.4 Biohydrogen production at 15–26% salinity of salt-damaged soil from Khon Kaen

Salt concentration (%)	Biohydrogen production (ml)	Theoretical maximum H_2 production reached (%)	HMY ($molH_2/mol_{glucose}$)
15	49.8	67.3	2.78
20	0.02	0.03	0.00
26	0	0	0.00

Glucose 0.15 g (5,000 mg/L), inoculum 3000 mg/L VSS, F/M ratio 1.5

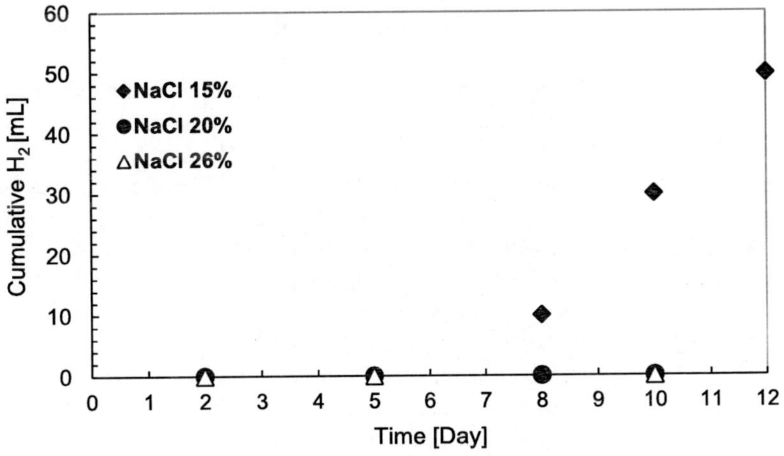

Fig. 11.1 Cumulative hydrogen production of salt-damaged soil in Khon Kaen

Table 11.5 Biohydrogen production at 26% salinity after 2 years of acclimatization

Soil sample	Biohydrogen production (ml)	Theoretical maximum H_2 production reached (%)	HMY (mol H_2 / $mol_{glucose}$)
Samut Sakhon salt pan	13.44	27.02	1.08
Khon Kaen salt-damaged soil (1)	14.31	28.76	1.15
Khon Kaen salt-damaged soil (2)	8.22	16.53	0.66

Glucose 0.12 g (5,000 mg/L), inoculum 10 mL (3 mg/L VSS)

Table 11.6 Biohydrogen production at 15% salinity experiments for 0.5–2.0 F/M ratio of salt-damaged soil from Khon Kaen before acclimatization

F/M ratio	Glucose (g)	Glucose (mg/L)	Biohydrogen production (ml)	Theoretical maximum H_2 production reached (%)	HMY (mol $H_2/mol_{glucose}$)
0.5	0.045	1500	8.26	36.9	1.48
1	0.09	3000	24.49	54.7	2.19
1.5	0.15	5000	49.75	67	2.67
2	0.18	6000	5.31	1.34	0.24

Inoculum 3000 mg/L VSS

Khon Kaen salt-damaged soil (1) was taken from the shore part of the pond, with finer soil and more water content. Khon Kaen salt-damaged soil (2) was taken from the farther part of the pond, approximately 10 m apart toward drier land with coarser soil and less water content. Both samples were taken during the dry season. During the rainy season, the surfaces of both sampling locations are covered with water.

In Table 11.1, Eqs. (1) and (2) show hydrogen production through the acetic acid and butyric acid pathways, respectively. Equations. (3), (4), and (5) show the pathways with no hydrogen production. The equations express the pathways of propionic acid fermentation, lactic acid, and alcohol fermentation, respectively.

The possible reason for the hydrogen yield being below the theoretical value is that the low F/M ratio produced conditions that were not optimum for the pathways expressed by Eqs. (1) and (2). From the comparison of the standard Gibbs energy of the formation values, it can be assumed that the pathways for propionic acid fermentation (Eq. (3)) and alcohol fermentation (Eq. (5)) are more spontaneous than the rest. Thus, they are more likely to occur, as the substrate concentration was low, the reaction rate was high, and the hydrogen recovery rate was low. The production of propionate can decrease the production of hydrogen [24–27]. Very low substrate concentrations can be unsuitable for hydrogen production as shown in Table 11.6. Figure 11.2 showed the composition of VFAs produced for each F/M condition in Table 11.6. For F/M ratio of 1.5 and 2, almost all propionic acid and butyric acid were transformed to acetic acid. Optimization of the substrate's composition should be considered in future work to increase the hydrogen molar yield.

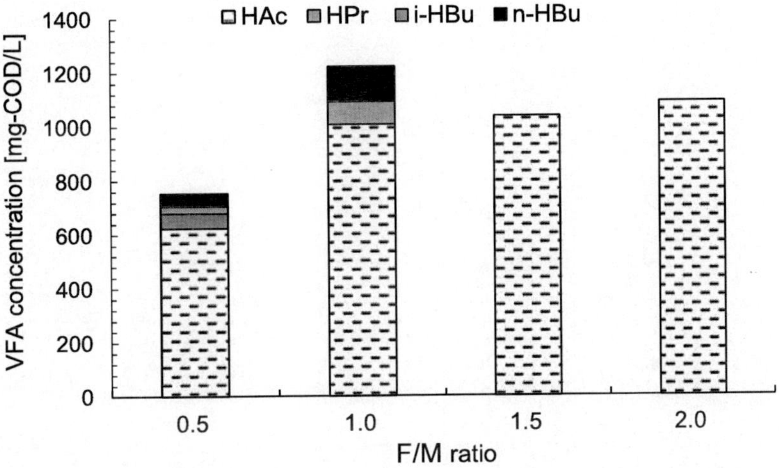

Fig. 11.2 VFA composition for 0.5–2 F/M ratio at 15% salinity experiments of salt-damaged soil from Khon Kaen before acclimatization (HAc acetic acid, HPr propionic acid, i-HBu isobutyric acid, n-HBu n-butyric acid)

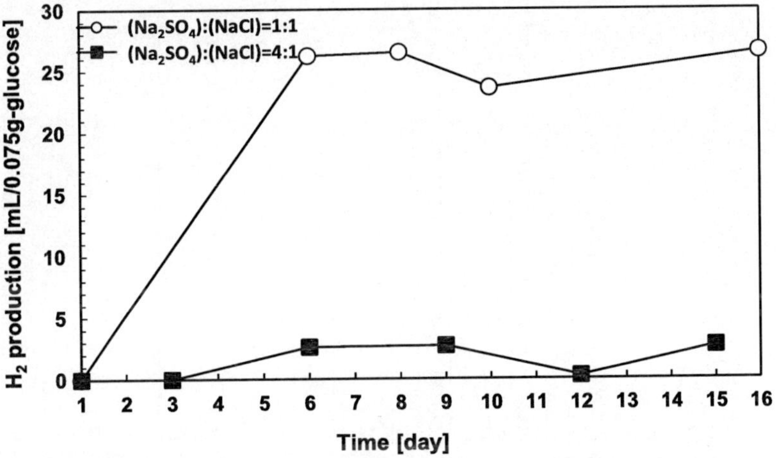

Fig. 11.3 Cl^- ion requirement for H_2 production in salt-damaged soil of Khon Kaen

3.3 Evaluation of the Bacteria's Requirement for Chloride Ions in High Salinity Conditions

In this experiment, Na_2SO_4 was used to partially replace NaCl in two different ratios (1:1 and 4:1) to confirm the extremely halotolerant hydrogen-producing bacteria's requirement for Cl^- ions. The substrate was adjusted such that the overall salinity of the mixture of Na_2SO_4 and NaCl was 26%. The bacteria's requirement for Cl^- ions was confirmed by a clear difference in the amount and yield of hydrogen production at different ratios of Na_2SO_4 to NaCl (Fig. 11.3 and Table 11.7).

Table 11.7 Biohydrogen production at different ratios of Na_2SO_4 to NaCl

Soil sample	(Na_2SO_4):(NaCl)	Glucose (g)	Glucose (mg/L)	Inoculum (mg/L VSS)	Biohydrogen production (ml)	Theoretical maximum H_2 production reached (%)	HMY (mol H_2/mol$_{glucose}$)
Samut Sakhon salt pan	1:1	0.022	7500	5000	1.77	16.0	0.64
Khon Kaen salt-damaged soil (1)		0.0044	2250	1500	1.57	71.2	2.85
Khon Kaen salt-damaged soil (2)		0.0044	2250	1500	0.26	23.3	0.47
Samut Sakhon salt pan	4:1	0.075	1500	1000	0	0	0.00
Khon Kaen salt-damaged soil (1)					2.66	7.12	0.29
Khon Kaen salt-damaged soil (2)					0	0	0.00

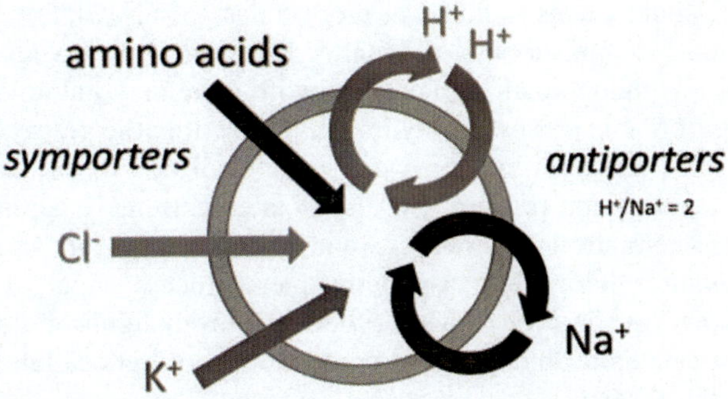

Fig. 11.4 Na^+ and Cl^- roles in transport process of halophilic bacterial membrane

Halophilic archaea and halophilic fermentative bacteria use the "salt-in" strategy in their survival mechanism in extremely hypersaline conditions [28, 29]. To adapt to this condition, cells maintain high salt concentrations at the intracellular level to sustain isosmotic conditions within the cell. Usually, K^+ and Cl^- ion salts are accumulated in molar concentrations at the intracellular level [29]. Cl^- is the preferred anion accumulated in the "salt-in" strategy, and it is possible that it plays critical roles in haloadaptation [28]; however, some halophilic microorganisms also utilize sulfate in high concentrations [29, 30]. The energy for outward transport of Na^+ ions is provided through the H^+ gradient in the electrogenic Na^+/H^+ antiporters, while K^+ and Cl^- enter through a symporter system in response to the cell's membrane potential [31].

Approximately one equivalent unit of adenosine triphosphate (ATP) will be needed to accumulate 1.5–2 molecules of KCl [32]. This mechanism clearly shows a high requirement for K^+ ion, which the substrate might have not supplied in sufficient quantity compared to Na^+ and Cl^- ions (Fig. 11.4). Only 1.3 g/L of K^+ was available in the substrate, compared to 351.35 g/L of NaCl.

In non-halophilic bacteria, high sulfate concentration suppressed hydrogen production by shifting the metabolic pathway from butyrate fermentation to ethanol. The decrease may also be caused by the toxicity of hydrogen sulfide [33]. To further confirm this result, another experiment with only Na_2SO_4 salt is suggested for the future work. Hydrogen consumption by sulfate-reducing bacteria might be negligible for this experiment due to the near-saturation NaCl concentration that limited the growth of such bacteria.

4 Summary

The experimental results showed that it is possible to produce biohydrogen under high salt concentrations (26% NaCl) after at least 1 year of acclimatization. This indicates that extremely halotolerant hydrogen-producing bacteria can exist under such concentrations. The best hydrogen molar yield in this research was 2.85 mol

H_2/mol glucose which was produced at 13% NaCl, initial pH 6.8, 35 °C, and food to microorganism ratio of 1.5 after 1 year of acclimatization.

Although dark fermentation with extremely halotolerant bacteria could offer an affordable process for hydrogen production, the pretreatment cost of the lignocellulosic biomass remains a challenge until now. This research focused on the dark fermentation process with the isolated extremely halotolerant hydrogen-producing bacteria from the salt pan and salt-damaged soil and did not cover the part of lignocellulosic biomass pretreatment process. Figure 11.5 shows all the processes involved in biohydrogen production from lignocellulosic biomass. A technology roadmap for biohydrogen production from lignocellulosic biomass can be found in Fig. 11.6.

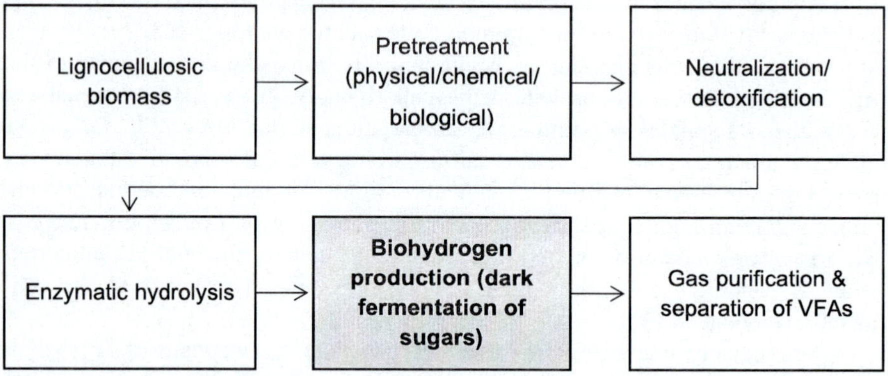

Fig. 11.5 Processes in biohydrogen production from lignocellulosic biomass. The box in gray and bold text shows the part covered in this research

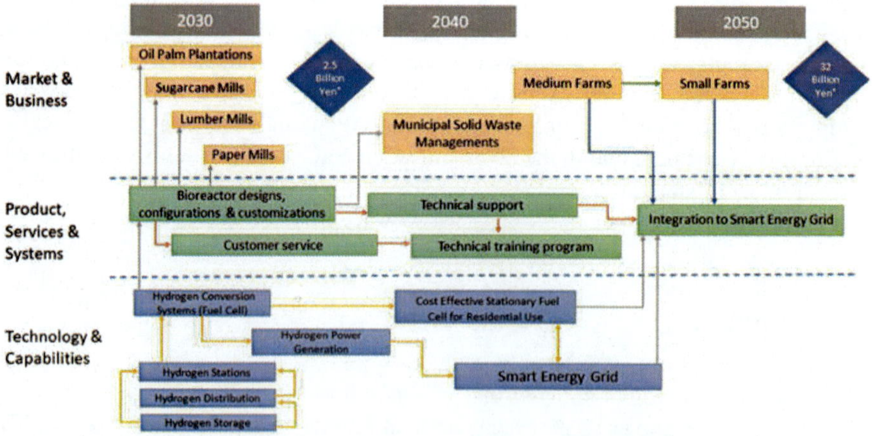

Fig. 11.6 Biohydrogen production from lignocellulosic biomass technology roadmap. * 1US $=107.5 Japanese Yen (May 2020)

Glossary

Dark fermentation This is the fermentative conversion of organic substrate to biohydrogen. It is a complex process manifested by diverse groups of bacteria, involving a series of biochemical reactions using three steps similar to anaerobic conversion. Dark fermentation differs from photofermentation in that it proceeds without the presence of light.

F/M ratio This is a process control number that helps you to determine the proper number of microorganisms for your system.

Halotolerant bacteria They have the adaptation ability to conditions of high salinity. Halotolerant bacteria tend to live in areas such as hypersaline lakes, coastal dunes, saline deserts, salt marshes, and inland salt seas and springs.

Hydrogen molar yield If this is high (approaching theoretical value), the process has very high efficiency.

References

1. Hansen J, Sata M (2016) Regional climate change and national responsibilities. Environ Res Lett IOP Publishing 11:034009. https://doi.org/10.1088/1748-9326/11/3/034009
2. Show KY et al (2012) Biohydrogen production: current perspectives and the way forward. Int J Hydro Energy Elsevier Ltd 37(20):15616–15631. https://doi.org/10.1016/j.ijhydene.2012.04.109
3. Nikolaidis P, Poullikkas A (2017) A comparative overview of hydrogen production processes. Renew Sustain Energy Rev Elsevier 67:597–611. https://doi.org/10.1016/j.rser.2016.09.044
4. Dipasquale L et al (2014) Introducing capnophilic lactic fermentation in a combined dark-photo fermentation process: a route to unparalleled H2 yields. Appl Microbiol Biotechnol 99 (2):1001–1010. https://doi.org/10.1007/s00253-014-6231-4
5. Hallenbeck PC, Abo-Hashesh M, Ghosh D (2012) Strategies for improving biological hydrogen production. Bioresource Technol Elsevier Ltd 110:1–9. https://doi.org/10.1016/j.biortech.2012.01.103
6. Boboescu IZ et al (2016) Surpassing the current limitations of biohydrogen production systems: the case for a novel hybrid approach. Bioresource Technol Elsevier Ltd 204:192–201. https://doi.org/10.1016/j.biortech.2015.12.083
7. Kumar G et al (2015) Lignocellulose biohydrogen: practical challenges and recent progress. Renew Sust Energ Rev 44(1). https://doi.org/10.1016/j.rser.2015.01.042
8. Phummala K et al (2014) Delignification of disposable wooden chopsticks waste for fermentative hydrogen production by an enriched culture from a hot spring. J Environ Sci (China) 26 (6):1361–1368. https://doi.org/10.1016/S1001-0742(13)60612-9
9. Wang S, Hou X, Su H (2017) Exploration of the relationship between biogas production and microbial community under high salinity conditions. Sci Rep Springer US 7(1):1–10. https://doi.org/10.1038/s41598-017-01298-y
10. Chen G-Q, Jiang X-R (2018) Next generation industrial biotechnology based on extremophilic bacteria. Curr Opin Biotechnol Elsevier Ltd 50:94–100. https://doi.org/10.1016/j.copbio.2017.11.016
11. Liaw HJ, Mah RA (1992) Isolation and characterization of Haloanaerobacter chitinovorans gen. nov., sp. nov., a halophilic, anaerobic, chitinolytic bacterium from a solar saltern. Appl Environ Microbiol 58(1):260–266
12. Mouné S et al (1999) Haloanaerobacter salinarius sp. nov., a novel halophilic fermentative bacterium that reduces glycine-betaine to trimethylamine with hydrogen or serine as electron

donors; emendation of the genus Haloanaerobacter. Int J Syst Bacteriol 49 Pt 1(1 999):103–112. https://doi.org/10.1099/00207713-49-1-103
13. Matsumura Y et al (2014) Enhanced hydrogen production by a newly described heterotrophic marine bacterium, Vibrio tritonius strain AM2, using seaweed as the feedstock. Int J Hydro Energy Elsevier Ltd 39(14):7270–7277. https://doi.org/10.1016/j.ijhydene.2014.02.164
14. Kivistö A, Santala V, Karp M (2010) Hydrogen production from glycerol using halophilic fermentative bacteria. Bioresour Technol 101(22):8671–8677. https://doi.org/10.1016/j.biortech.2010.06.066
15. Brown SD et al (2011) Complete genome sequence of the haloalkaliphilic, hydrogen-producing bacterium Halanaerobium hydrogeniformans. J Bacteriol 193(14):3682–3683. https://doi.org/10.1128/JB.05209-11
16. Pierra M et al (2013) Fermentative hydrogen production under moderate halophilic conditions. Int J Hydrog Energy 9:1–10. https://doi.org/10.1016/j.ijhydene.2013.08.035
17. JIS A 1203 test method for water content of soils (2009) Japanese Industrial Standards Association
18. JIS A 1226 test method for ignition loss of soils (2000) Japanese Industrial Standards Association
19. American Public Health Association, American Water Works Association and Water Environment Federation (1999) Standard methods for the examination of water and wastewater. Stand Methods 541. https://doi.org/10.2105/AJPH.51.6.940-a
20. Hoogsteen MJJ et al (2015) Estimating soil organic carbon through loss on ignition: effects of ignition conditions and structural water loss. Eur J Soil Sci 66(2):320–328. https://doi.org/10.1111/ejss.12224
21. Heiri O, Lotter AF, Lemcke G (2001) Loss on ignition as a method for estimating organic and carbonate content in sediments: reproducibility and comparability of results. J Paleolimnol 25 (1):101–110. https://doi.org/10.1023/A:1008119611481
22. Rath KM, Rousk J (2015) Salt effects on the soil microbial decomposer community and their role in organic carbon cycling: a review. Soil Biol Biochem Elsevier Ltd 81:108–123. https://doi.org/10.1016/j.soilbio.2014.11.001
23. Yan N et al (2015) Influence of salinity and water content on soil microorganisms. Int Soil Water Conserv Res Elsevier 3:316
24. Mosey FE (1983) Mathematical modelling of the anaerobic digestion process: regulatory mechanisms for the formation of short-chain volatile acids from glucose. Water Sci Technol 15(8–9):209–232. Available at: http://wst.iwaponline.com/content/15/8-9/209.abstract
25. Shin HS, Youn JH, Kim SH (2004) Hydrogen production from food waste in anaerobic mesophilic and thermophilic acidogenesis. Int J Hydrog Energy 29(13):1355–1363. https://doi.org/10.1016/j.ijhydene.2003.09.011
26. Li C, Fang HHP (2007) Fermentative hydrogen production from wastewater and solid wastes by mixed cultures. Crit Rev Environ Sci Technol Taylor & Francis 37(1):1–39. https://doi.org/10.1080/10643380600729071
27. Azwar MY, Hussain MA, Abdul-Wahab AK (2014) Development of biohydrogen production by photobiological, fermentation and electrochemical processes: a review. Renew Sustain Energy Rev Elsevier 31:158–173. https://doi.org/10.1016/j.rser.2013.11.022
28. Roberts MF et al (2005) Organic compatible solutes of halotolerant and halophilic microorganisms. Saline Syst 1(1):5. https://doi.org/10.1186/1746-1448-1-5
29. Kivistö AT, Karp MT (2011) Halophilic anaerobic fermentative bacteria. J Biotechnol 152 (4):114–124. https://doi.org/10.1016/j.jbiotec.2010.08.014
30. Ede SM, Hafner LM, Fredericks PM (2004) Structural changes in the cells of some bacteria during population growth: a fourier transform infrared–attenuated total reflectance study. Appl Spectrosc OSA 58(3):317–322. Available at: http://as.osa.org/abstract.cfm?URI=as-58-3-317
31. Lanyi JK (1979) The role of Na+ in transport processes of bacterial membranes. Biochimica et Biophysica Acta (BBA) Rev Biomemb Elsevier 559(4):377–397. https://doi.org/10.1016/0304-4157(79)90011-X

32. Oren A (2001) The bioenergetic basis for the decrease in metabolic diversity at increasing salt concentrations: implications for the functioning of salt lake ecosystems. Springer, Saline Lakes, pp 61–72
33. Lin CY, Chen HP (2006) Sulfate effect on fermentative hydrogen production using anaerobic mixed microflora. Int J Hydrog Energy 31(7):953–960. https://doi.org/10.1016/j.ijhydene.2005.07.009
34. Conrad R (1999) Contribution of hydrogen to methane production and control of hydrogen concentrations in methanogenic soils. FEMS Microb Ecol 28(January):193–202
35. Thauer RK, Jungermann K, Decker K (1977) Energy conservation in chemotrophic anaerobic bacteria. Bacteriol Rev 41(1):100–180. https://doi.org/10.1073/pnas.0803850105

Index

A
Abiotic, 217
Accuracy, 217
Acetogenesis, 330, 331
Acid deposition, 217, 278–283, 289
Acid Deposition Act, 284
Acid mine drainage, 306
Acid precipitation, 281
Acid rain, 307
  acid deposition, 278, 281, 283
  acid mine drainage, 306
  acid precipitation, 281
  Acid Rain Permit Program, 291, 292
  air pollutants, 279
  allowances, 293, 294
  aluminum, 282
  atmospheric deposition, 279
  atmospheric pollutants, 298
  building materials, 279
  calcium sulfate, 283
  carbonic acid, 279
  causes, 287, 289
  chronic acidification, 282
  conservation programs, 298, 299
  corrosion rate, metals, 283
  environmental effect, 281
  environmental issues, 278
  environmental pollution, 297, 298
  environmental stress, 279, 282
  episodic acidification, 282
  forests, 282
  fuels, 298, 299
  historical monuments, 279
  history, 283
  human health, 283
  impacts, 281
  industrial, 279
  measurement, 294, 295
  monitoring, 292, 294
  nitrates, 282
  nitric acid, 279
  nitrogen, 282
  nitrogen deposition, 282, 283
  nitrogen oxides, 279
  pH scale, 281
  pollutants, 282
  public awareness, 279
  reduce acid rain, 290, 291
  regulations (*see* Acid rain regulations)
  renewable energy sources, 298
  soil nutrients, 282
  sulfate aerosol particles, 281
  sulfur dioxide, 279
  terrestrial ecosystems, 281
  trends, 295–297
  unpolluted rain, 279
  water droplets, 278
Acid Rain Permit Program, 291, 292
Acid Rain Program (ARP), 286, 290–293, 295
Acid rain regulations
  Acid Deposition Act, 284
  CAA, 285
  CAIR, 286
  CSAPR, 287
  environment, 286
  HAPs, 285
  NAPAP, 284
  NO_x budget program, 286

L. K. Wang, M. -H. S. Wang, Y. -T. Hung, N. K. Shammas (eds.), *Integrated Natural Resources Management*, Handbook of Environmental Engineering 20,
https://doi.org/10.1007/978-3-030-55172-8

Acid rain regulations (*cont.*)
  NO_x emissions, 285
  sulfur dioxide pollution, 286
  sulfur emissions, 286
  US EPA, 287
Acid solution, 154
Acidic deposition, 289
Acidogensis, 330
Activated carbon, 301
Activation technique, 376
Active scanners, 200
Adaptive sampling strategy, 217
Adenosine triphosphate (ATP), 423
Adsorbent, 385, 406
Adsorbent dosage, 379, 380
Adsorbent materials, 403
Adsorbents, 395, 403, 404
  agricultural waste, 395 (*see* Agricultural waste-derived adsorbents)
  heavy metals (*see* Heavy metals)
Adsorption, 385, 403, 404, 406
Adsorption capacity, 386
Adsorption isotherms, 383–385
Adsorption process, heavy metals
  adsorbent dosage, 379, 380
  adsorption isotherms, 383–385
  equilibrium time, 380, 381
  initial concentration, 379
  kinetic models, 382, 383
  solution pH, 378, 379
  temperature, 380, 381
Aeration, 270, 305
Aerosol, 9, 24
Agricultural and agri-food industrial wastes, 326
Agricultural inputs, 103
Agricultural waste-derived adsorbents
  activation technique, 376
  adsorbents, 375
  authentic state, 376
  biochar adsorbent, 377
  biomass wastes, 375
  grafting techniques, 377
  heavy metal pollution, water, 374
  heavy metals (*see* Heavy metals)
  magnetic biochar, 377, 378
  magnetic biological adsorbents, 377
  methods, 374
  nanostructure, 377
  removal, heavy metals, 376
  thermal conversion, 376
  use of chemicals reduces, 375
Agricultural wastes, 386
Agroecosystem, 217
Agro-food industrial wastewater, 326
Agro-food waste, 326
Aid neutralization capacity (ANC), 297
Air pollutants, 3, 279
Air pollution, 3, 24
Air pollution control, 25
Air systems, 270
Albers map projection, 217
Aircraft, 200
Algae, 263, 264, 300, 305, 306
Algae removal, 301
Alum, 272
Alum addition, 269
Alum injection system, 269
Aluminum, 268, 278
Aluminum salts, 272
American Water Works Association (AWWA), 110, 241
Anaerobic and aerobic treatment processes, 348
Anaerobic digestion, 325, 357
  acetogenesis, 330, 331
  acidogensis, 330
  fermentation, 330
  homoacetogens, 331
  hydrolysis, 329, 330
  liquefaction, 329, 330
  methanogenesis, 331
  microorganisms, 328, 329
  organic waste and producing methane, 328
  symbiotic relationship, 328
Anaerobic mixed culture, 325
Anaerobic sludge, 347, 349
Ancillary data, 217
Anionic, 386
Annual statistical summary, 217
Anthropogenic, 25
Animal dung, 349
Antiferromagnetism, 399, 400
Anti-scaling technique, 396
Appropriate technology, 40
AquaDAF, 307
AquaDAF system
  hydraulic flocculation, 301
  lake water treatment, 306
Aquatic ecosystem, 386
Area frame, 217–218
Area sample, 218
Arid ecosystems, 218
Ar-Riyadh Development Authority (ADA), 250
Artificial light, 135, 138
Artificial magnets, 396
Assessment, 218
Assessment endpoint, 218
Association rule, 218

Atmosphere, 25
Atmospheric deposition, 279
Atmospheric Deposition to Great Lakes and Coastal Waters (ADGLCW), 285
Atmospheric pollution, 278, 280
Atmospheric temperatures, 21
Attribute, 218
Augmented sample, 218
Automated hydraulic modeling, 216
Automated sampling networks, 79, 86, 88
Automatic Meter Reading (AMR), 242
Automotive ignition coils, 131
Auxiliary data, 218
AWWA/IWA water audit methodology, 241
Azimuthal map projection, 218

B
Bacterial water quality, 103
Baking process, 122
Baseline grid, 218
Benzene, 403
Best applicable technology (BAT), 134, 141, 151, 160, 163, 168, 171, 189
Best conventional pollution control technology (BCT), 187–190
Best practice control technology (BPT), 134, 141, 151, 152, 158, 160, 163, 168, 171, 187–189
Bias, 218
Bioaccumulants, 219
Bioassay, 219
Biochar, 386
Biochar adsorbent, 377
Biochemical oxygen demand (BOD), 96, 98
Biodiesel industrial waste, 344, 345
Biodiversity, 219
Biogeographic province, 219
BioH_2 production
  anaerobic digestion (*see* Anaerobic digestion)
  characteristics, 326
  industrial waste (*see* Industrial wastewater)
  industrial wastewater, 326
  inocula (*see* Inocula)
  organic waste, 327, 328
  research, 356
Biohydrogen production, 357
  analytical method, 416
  at 15–26% salinity, 419, 420
  at 3–10% salinity, 417, 418
  Cl^- ion requirement, 421, 423
  culture conditions and experimental procedures, 415, 416
  dark fermentation process, 413
  H. hydrogeniformans, 414
  halotolerant and halophilic bacteria, 414
  lignocellulosic biomass, 413, 414, 424
  NaCl concentrations, 416, 422, 423
  salt-damaged soil, 419
  seed microorganisms and medium, 415
  theoretical hydrogen production/yield, 416, 417
  Vibrio tritonius strain AM2, 414
Biological activity, lakes, 263–265
Biological indices, 81
Biological treatment, 325
Biological wastewater treatment systems, 268
Biomarker, 219
Biomass, 219, 386
Biomass wastes, 375
Biome, 219
Biotic, 219
Biotic harvest, 305
Biotic separation, 305
Black carbon, 25
Black carbon aerosol, 25
Brazos River, 299
2-Bromoethanesulfonic acid sodium salt (BESA), 351
Buffering, 202

C
Calcined carbon filler, 122
Calcium, 268
Calcium salts, 272
Calcium sulfate, 283
Calibration, 219
Canadian Council of Ministers of the Environment (CCME), 79, 110
Capacitors
  application, uses and examples, 160
  description and production process, 160, 161
  fixed, 160, 161
  industrial plant, 161–163
  oil-filled, 161
  treatment technology and cost, 163, 166
  variable, 161
  wastewater characterization, 161, 164, 165
  wastewater treatment processes, 161, 163, 165
  water use, 161
Carbohydrate-rich food processing wastes, 327
Carbon adsorption, 134
Carbon-and-graphite products
  classical and toxic pollutants, 126–128
  description and production process, 121, 122

Carbon-and-graphite products (*cont.*)
  machining setting data, 131, 132
  oil separation performance, 130
  SIC code, 121, 122
  structural, refractory, nuclear and non-electrical applications, 121
  treatment technologies and cost, 131, 132
  wastewater characterization, 122–126, 129
  wastewater streams, 122, 129
  wastewater treatment process, 130–132
Carbon capture and sequestration (CCS), 6, 25
Carbon capture and storage, 25
Carbon cycle, 25
Carbon dioxide (CO_2), 26, 298
  biomass, 5
  carbon footprint, 6
  carbon sequestration, 6
  dipole moment, 5
  energy levels, 5
  environment, 21, 22
  equivalent, 5, 26
  fertilization, 6, 26
  fossil fuels, 5
  heat absorption, 5
  infrared radiation, 5
  molecule, 5
Carbon footprint, 6, 26
Carbonization, 386
Carbon nanotubes, 404, 406
Carbon sequestration, 6, 26
Carbon tetrachloride (CCl_4), 26
Carbonic acid, 279
Cathode ray manufacturing process, 147
Cathode ray tube (CRT)
  manufacturing, 179
  manufacturing process, 146
  NSPS, 187, 188
  POTWs, 187
  pretreatment standards, 188
  PSES, 187
  PSNS, 187
  and PWBs, 171
  subcategory, 141
  and television picture tubes, 146
  transfers, 181
Cationic, 386
Cationic heavy metals, 378
CCME WQI
  advantages, 79, 89
  description, 88
  disadvantages, 79, 89
CDF (Cumulative distribution function), 219
Cellulose, 386
Ceramic insulators, 133
Change, 219
Characteristics of lakes
  air temperature, 262
  biological activity, 261
  colder bottom water temperature, 261
  deep lakes, 260–262
  dimictic, 261
  forested watershed, 260
  human development, 260
  lakeside homeowners, 260
  larger lakes, 260
  layers, 261
  seasonal circulation patterns, deep temperate climate lake, 262, 263
  stratification period, 261
  summer stratification, 261, 262
  surface, 262
  temperature succession, 261
  water temperature, 261
Characterization, 219
Cheese whey, 344–346
Chemical addition, 268
Chemical detection, 242
Chemical/mechanical deflash process, 156
Chemical oxygen demand (COD), 302, 349, 352
Chemical precipitation, 306, 386
Chemical stability, 404, 406
Chemical status, 72
Chlorofluorocarbon (CFC), 26
Chlorophyll, 210
Chlorophyll-α, 265, 266
Chlorophyll-bearing aquatic organisms, 96
Chronic acidification, 282
Clari-DAF, 307
Classification, 219
Class I ozonedepleting substance, 26
Class II ozonedepleting substance, 27
Clean Air Act (CAA), 285
Clean Air Interstate Rule (CAIR), 286
Clean Air Status and Trends Network (CASTNet), 295
Climate, 3, 27
Climate change, 27
  adaptation/mitigation, 24
  CO_2, 22
  ecosystem, 18
  GHG (*see* Greenhouse gas (GHG))
  glaciers (*see* Glaciers)
  measures, 3
  political issue, 24
  salmon, 17–21
  weather conditions, 3
Climate science, 24
Climatic elements, 3
Cluster analysis (CA), 65

Coagulants, 275, 277
Coagulation, 386
Coal-fired Gavin Power Plant, 288
Coal-fired power plants, 298
Color tubes, 145
Community, 219–220
Comparability, 220
Completeness, 220
Compost, 60
Composting latrine, 49, 60
Conceptual model, 220
Condition, 220
Condition indicator, 220
Confidence coefficient, 220
Confidence interval, 220
Conformal map projection, 220
Conical (or conic) map projection, 220
Consumer aerosol products, 9
Continuous, 220
Continuous emissions monitors (CEMs), 292
Conventional activated sludge treatment, 268
Coprecipitation, 400, 402, 406
Crosscutting group, 220–221
Cross-State Air Pollution Rule (CSAPR), 287
Crystal processing, 154
Cumulative distribution, 221
Curie temperature, 395, 399, 406
Cylindrical (or cylindric) map projection, 221

D

DAF/filtration (DAFF), 278
Dairy industry waste, 348
Dairy wastewater, 344–346
Dark fermentation, 327, 425
- cost, 355
- dairy industry waste, 348
- industrial waste, 325
- livestock manure, 348
- metabolic pathways, 333, 334
- mixed cultures (*see* Mixed cultures)

Dark fermentation process, 413, 414, 424
Data conversion routines, 204
Data input subsystem, 204
Data quality, 221
Data quality indicators, 221
Data quality objective (DQO), 221
Data input subsystem, 204
Data storage, 204
Deconvolution, 221
Deep lakes, 260, 261, 271
Deforestation, 27, 265
Demonstration field program, 221–222
Demonstration project, 222
DensaDeg process, 268
Desertification, 27
Design-based, 222
Design drawings, 45
Devils Lake, 273, 275, 277
Diamagnetism, 398, 399
Diatomaceous earth filtration, 134
Die-attach pad, 155
Dielectric materials
- description and production process, 131–133
- electrical insulators, 131
- industrial plant, 133
- manufacture, 132
- mica paper, 131, 133, 134, 136, 137
- oil-filled capacitors, 131, 133
- small oil-filled capacitor, 133
- transformer core, 133
- transformers, 131
- treated effluent concentrations, 134, 137
- treatment technologies and cost, 134, 138
- waste characterization, 134–137
- wastewater treatment processes, 134, 137
- water uses, 133, 134
- windings, 133

Diffusion, 90
Digital elevation model (DEM), 201, 214
Digital line graph (DLG), 222
Digital Signal Processing Solutions (DSPs), 190
Dimictic lake, 261
Diodes, 153, 156
Dipole moment, 5
Disability-Adjusted Life Years (DALYs), 374
Discrete resource, 222
Dissolved air flotation (DAF), 307
- dredged materials, 271
- hypolimnetic phosphorus removal (*see* Hypolimnetic phosphorus removal)

Dissolved gas flotation (DGF), 307
Dissolved oxygen (DO)
- aquatic systems, 94
- biochemical decay, organic waste materials, 96–99
- chemical compounds, 94
- nitrification, 101, 102
- oxygen distribution, 94
- oxygen percent saturation, 100
- physical characteristics, water, 95–97
- quantification, 100, 101
- reaeration, 99

Dissolved oxygen (DO) (*cont.*)
temperature, 100
water quality, 94
District Meter Area (DMA), 242
Domain, 222
Doorbell transformers, 131
Doping materials, 154
DOsat, 110
Double sample, 222
Douglas Island Pink and Chum Inc. (DIPCI), 20
Dredging, 271
Drinking water services, 39
Dry deposition, 289, 295
Dry etching, 154
Dry products
carbon and graphite, 121–123, 126, 129–132
and metal finishing industries, 119
Dubinin-Radushkevich isotherm, 383
Dubinin-Radushkevich isotherm equation, 384
Dynamic Factor Analysis, 66

E

Ecological health, 222
Ecological risk assessment, 222
Ecological status classification, 72
Ecology, 222
Ecoregion, 222
Ecosystem, 2, 18, 22, 222, 386
Ecosystem functions, 223
Ecotone, 223
Electric discharge lamps, 135
Electric furnace, 121
Electric lamp
classical and toxic pollutants, 138, 142, 143
description and production process, 135, 138, 139
effluent data, fluorescent lamp manufacture, 139, 144
industrial plant, 137
plant-specific data, 138, 140, 141
treatment technologies, 139, 144
treatment technology and cost, 141, 145
wastewater characterization, 138, 140–143
wastewater treatment processes, 138, 139, 144
water use, 138
Electric motor control, 131
Electrochemical treatment, 386
Electrical/electronic industry
capacitor (*see* Capacitors)
dielectric materials (*see* Dielectric materials)
dry products, 119, 121–123, 126, 129–132
electric lamp (*see* Electric lamp)
electron tube (*see* Electron tube)
geographic distribution, 118, 119
ICs, 117
miscellaneous components, 117
personal computer and accessories, 117
point source discharge limitation (*see* Point source discharge limitation)
profile, 117, 118
PWB (*see* Printed wiring board (PWB))
research and production, 117
semiconductors (*see* Semiconductors)
SIC, 119–121
telecommunications equipment, 117
vacuum tube, 117
waste generation, 171–183
workforce, 118
worldwide companies, 117, 118
WW I, 117
Electrical energy, 298
Electrical insulators, 131
Electromagnetic field detection, 242
Electromagnetic spectrum, 199, 200, 209
Electromagnetic system, 145
Electromagnets, 396
Electron guns, 146
Electron tube
color tubes, 145
description and production process, 141, 145–147
electron guns, 146
influent and effluent data, 146, 148–151
low-voltage and low-power appliances, 143
panel mask, 145
receiving-type electron tubes, 141
shadow mask, 145
SIC identified, 141
television picture tubes, 141, 144
transmitting-type electron tubes, 141, 145
vacuum/ionized gas-filled tubes, 143
wastewater characterization, 146, 148–151
wastewater treatment processes, 146, 151–153
water use, 146
Electronic crystals, 189
BAT, 189
BCT, 189, 190
BPT, 189
NSPS, 189, 190
PSES, 189, 190
Electronic furnace, 121
Electro-optical sensors, 200

Electroplating, 166
Emissions, 27
Engineers Without Borders USA (EWB-USA), 60
appropriate technology, 40
behavior change, 39
communities, 40, 41
community-driven projects, 40, 50
community needs, 40
development, 41
education and training, 42
engineering needs, 41
field offices, 41
financing strategies, on-site household systems, 42
gender approach, 42
latrine project, 40
lessons learned factors
dashboard tool, 51–53
latrine projects, 51, 52
quotes, 52, 54
reporting metrics, 52, 54
locations, 41
methodology, 50
mission, 49
PMEL framework, 49, 50
program scope, 42
project indicators, 50
project types, 41
rural communities, 39
sanitation facilities, 41
sanitation project categories, 50, 51
self-identified needs, 41
success, 51
village-driven approach, 40
Enhanced greenhouse effect, 27
Entire, 223
Environment, 28, 223
Environmental agencies, 178
Environmental assessment, 223
Environmental health, 48
Environmental Monitoring and Assessment Program (EMAP), 223
Environmental pollution, 297, 298
Environmental Protection Agency, 297
Environmental satellites, 200
EPA Spatial Data Library System (ESDLS), 223
Epidemiologic ecology, 223
Episodic acidification, 282
Epitaxial growth/chemical vapor deposition techniques, 155
Equal-area (equivalent) projection, 223
Equilibrium time, 380, 381
Escherichia coli, 66
Estuary, 223
Eutrophication, 265, 266, 282, 304
Eutrophic lake recovery
aeration, 270
dredging, 271
sediment fixation, 272
weed harvest, 271
Excreta, 60
Exerted biochemical oxygen demand (mg/L of O_2) at time t (BOD_t), 110
Ex situ methods, 403
Extensive resource, 223
Extremely halotolerant bacterial communities, 424

F

Fabricated metal industry, 178
Facility Registry Service (FRS), 216
Farming, 269
Feature data elements, 199
Fecal coliform bacteria, 107
Fecal material, 103
Fecal pollution, 104
Federal Highway Authority, 269
Federal Water Pollution Control Administration (FWPCA), 302
Fermentation, 330
Fermentative hydrogen production, 413
Ferrimagnetism, 400
Ferromagnetism, 399
Filtration, 386
Fish population, 264
Fixed capacitors, 160, 161
Flame ion detector (FID), 416
Floating microscopic organisms, 263
Flocculation, 300
Floodplain mapping, 207, 208
Flow direction analysis, 214
Fluorescent lamps, 131, 137, 141, 145
Fluorinated gases, 7, 28
Fluorocarbons, 7, 28
F/M ratio, 425
Food chain, 264
Food processing waste
biodiesel industrial waste, 344, 345
cellulose, 339
characteristics, 339
cheese whey, 344–346
COD, 339
dairy wastewater, 344–346

Food processing waste (*cont.*)
H_2 production, 342
H_2 yield and production rate, 343
hydrogen potential, 338, 340–342
lignocellulosic residues, 338
manure, 346–348
molasses, 343, 344
POME, 344, 345
readily biodegradable organics, 339
starch generates, 342
Food waste, 335–337
Forcing, 28
Forest, 223
Fossil fuels, 28, 298, 413
Fragmented, 224
Frame, 224
conceptual, 224
Freundlich isotherm equation, 384
Fuels, 298

G
Gallium, 153
Gallium arsenide, 156
Gallium phosphide wafers, 156
Gas chromatography, 415, 416
Gas production, 416
Gender sequestration, 55
Geographic data, 204
Geographic distribution, 118, 119
Geographic information system (GIS), 224
application (*see* GIS application)
attribute data, 199
buffering, 202
categories, 202, 205
data models, 204
feature data elements, 199
input data, 202
layers, 199, 202
network analysis, 203
overlay function, 202, 203
overlaying process, 202
predictive modeling, 202
remote sensing data *vs*. geographic data, 199
satellite, 199
simulation modeling, 202
spatial data analysis, 198
spatial data elements, 199
subsystems, 204
USEPA programs, 216, 217
Geologic sequestration, 6
Geological methods, 211
Geosmin, 300
Germanium, 153
Germanium semiconductors, 156
GIS application
mapping
floodplain, 207, 208
land-water interface, 205, 207
modeling
groundwater, 211
hydrologic and hydraulic, 213–216
spatial, 212
monitoring
coastal environment, 208, 209
water quality, 209–211
Glacier Bay National Park and Preserve, 14–17
Glacier Bay water, 16
Glacier mass balance, 13
Glaciers, 28
Glacier Bay National Park and Preserve, 14–17
Mendenhall, 12–14
protection, 11, 12
Global average temperature, 28
Global change, 28
Global climate models (GCM), 28
Global warming, 28
CO_2, 22
ecosystem, 18
GHG (*see* Greenhouse gas (GHG))
glaciers (*see* Glaciers)
salmon, 17–21
temperature, 4
Global warming potential (GWP), 28
CO_2, 4
CO_2 equivalent, 5
definition, 10
global warming, 4
measurement, 10
N_2O, 6
relative index, 10
substance, 4
Glycerol, 344, 345
Grafting techniques, 377
Great Lakes, 224
Greenhouse effect, 4, 8, 29
Greenhouse gas (GHG), 29
aerosol, 9
CH_4, 6
climate change, 3
CO_2, 5, 6, 8
CO_2, atmosphere, 10, 11
definitions, 8
fluorinated gases, 7

global per capita carbon emission estimates *vs.* years, 10, 11
global warming, 4, 8
greenhouse effect, 4, 8
human activity, 8, 10, 22, 23
infrared radiation, 4
IPCC, 9, 10
Kyoto Protocol, 10
land and ocean temperature, 10, 12
land use, 8
N_2O, 6
natural activities, 8, 23
ozone, 7, 8
RF, 9
Grid, 224
enhancement, 224
hierarchical, 224
randomization, 224
triangular (EMAP), 224
Ground truthing, 207
Groundwater database, 211
Groundwater modeling, 211
Groundwater potential zone, 211
Groundwater table, 250

H
Habitat, 29, 224
Haloadaptation, 423
Haloanaerobacter chitinovorans sp. *nov.*, 414
Haloanaerobacter salinarius sp. *nov.*, 414
Halophilic archaea, 423
Halophilic fermentative bacteria, 423
Halotolerant bacteria, 425
Handwashing, 39
Handwashing facility, 39
Hard magnetic materials, 395, 396
Hatchery's salmon management, 20, 21
Hazardous air pollutants (HAP), 285
Heat shock treatment, 350, 351
Heavy metal ions, 406
magnetic materials, 395–397
magnetic particles (*see* Magnetic particles)
magnetic separation, 403, 404
magnetism (*see* Magnetism)
Heavy metals, 386
acidic rain, 374
adsorption process (*see* Adsorption process, heavy metals)
aquatic ecosystem, 374
carcinogenicity, 374
human health, 374, 375
industries, 374
mercury, 374
primary pollutants, 374
toxicity, 374
Hemicellulose, 386
Herbicides, 271
Heuristic method, 224
40-hex, 223
Hierarchical geometric decomposition, 224
Hierarchical model, 225
High GWP gases, 7
Homoacetogens, 331
Human activity, 8, 22, 23
Human health, 386
Human rights, 40
H_2 via the acetic acid (HAc), 416, 417
H_2 via the butyric acid (HBu), 417
Hydraulic parameters, 216
Hydrobromofluorocarbon (HBFC), 29
Hydrocarbon (HC), 29
Hydrochlorofluorocarbon (HCFC), 29
Hydrofluoroolefin (HFO), 30
Hydrogen, 264
Hydrogen (H_2) production
Agricultural and agri-food industrial wastes, 326
agro-food waste, 326
biological treatment, 325
bioreactors types and configuration, 354
CH_4, 325
characteristics, 325
hydrolytic and fermentative bacteria, 325
inocula (*see* Inocula)
OFMSW, 338, 339
pollution control and economic gain, 325
pretreatments, 327
Hydrogen molar yield (HMY), 417, 423, 425
Hydrologic and hydraulic modeling, 213–216
Hydrologic cycle, 30
Hydrologic parameters, 214
Hydrolysis, 329, 330
Hydrolytic and fermentative bacteria, 325
Hydro-morphological indices, 81
Hydromorphology, 72
Hydrosphere, 30
Hydrothermal, 400, 402, 403, 406
Hygiene facilities, 40
Hyperspectral imagery, 200
Hypolimnetic aeration system, 270
Hypolimnetic phosphorus removal
aluminum, 277
coagulants, 275, 277
DAF pilot plant, 275–278
DAFF, 278

Hypolimnetic phosphorus removal (*cont.*)
  depth profile, Devils Lake, 273, 274
  eutrophic Laurel Lake, Massachusetts, 273
  flocculation/filtration system, 272
  pilot DAF system, sand filtration, 275
  profile, Devils Lake, 275
  temperature-depth profile, lake, 274
  USGS, 274
  WDNR, 273, 276

I

Ice core, 30
Ikonos spectral bands, 206
Implementation field program, 225
Inclusion probability, 225
Index, 225
Index period, 225
Index sample, 225
Indicator, 30, 225
Indicator development, 225
Indirect emissions, 30
Industrial acid rain, 279
Industrial plant, 133
Industrial revolution, 30
Industrial waste, 325, 327
Industrial wastewater
  anaerobic digestion, 335
  feedstock, CH_4 and H_2 production, 335
  food processing (*see* Food processing waste)
  food waste, 335–337
  LCFAs, 335
  OFMSW, 338, 339
  toxic compounds, 335
  waste streams, 335
  wastes rich, 335
Industrial wastewater sludge, 349
Infrared radiation, 30
Inocula
  animal dung, 349
  compost, 349
  culture pretreatment
    acid/base treatment, 351
    anaerobic mixed, 350
    comparative studies, 352–354
    H_2-consuming microorganisms, 350
    heat shock treatment, 350, 351
    methanogen inhibitors, 351, 352
  industrial wastewater sludge, 349
  OLR, 352
  sewage sludge, 347, 349
  soil, 350
  sustainable open sources, 347
In situ methods, 402, 403
Instrument transformers, 131
Integrated Atmospheric Deposition Network (IADN), 291
Integrated circuits (ICs), 117, 153
Integration, 225
Intergovernmental Panel on Climate Change (IPCC), 9, 10, 30
International Water Association (IWA), 241
Interpenetrating subsamples, 226
Ion exchange, 386, 406
Ion exchange resin, 406
Iron salts, 272
Irrigation, 259

J

Judgment sample, 226

K

KAMET, 307
KH, 110
Khon Kaen salt-damaged soil, 418, 420
Kikoo project toilets, 57, 58
Kinetic models, 382, 383
Kinetics, 386
King Abdulaziz City for Science and Technology (KACST), 250
Klystron tube, 145
Kraft paper, 133
Kriging, 212, 226
Kyoto Protocol, 10

L

Lake restoration, 307
  AquaDAF, 306
  biotic harvest, 305
  DAF, 306
  phosphate removal, 305
  reduction/elimination, external pollutant sources, 304
  technologies, 306
  thermal destratification, 305
Lake water neutralization, 306
Lakes, 226, 307
  aeration, 305
  aesthetic quality, 260
  aging, 259
  biological activity, 263–265
  biological growth, 259

characteristics, 260–262
eutrophication, 304
irrigation, 259
nutrient discharge treatment (*see* Nutrient discharge treatment)
reduction/elimination, external pollutant sources, 304
remediation, 265–267
reservoirs, 260
storms, 260
water, 259, 260
water treatment (*see* Water treatment)
wet organic material, 259
Lambert's azimuthal map projection, 226
Landscape, 226
Landscape characterization, 226
Landscape ecology, 226
Land-water categorization, 205
Land-water interface, 205
Land-water mapping, 207
Langmuir isotherm, 384, 385
Lapping process, 156
Larger-sized capacitors, 131
Latrine, 60
Latrine Case Study: Kikoo, Cameroon
accumulation, water, 55
children and women, 55
EWB-USA Yale teams, 55
gender sequestration, 55
implementation, 56
Kikoo project toilet, 57, 58
lessons learned, VIP toilet technology
build close to home, 58
floor sealing, 58
location and master planning, 58
maintenance requirement, 58
ventilated pipe, 58
odor, open pit toilets, 55
open toilets, 55
pit toilets, 55
traditional toilet, 55
VIP toilets, 56, 59
water project, 55
Latrine Design Checklist
calculations, 43–45
current system, 43
design drawings, 45, 46
design standards and criteria, 43
education, 47
intended use, 43
land ownership, 43
material and cost list, 46
operation and maintenance, 46
problem definition, 43
siting latrine locations, 43
soils analysis, 43
solution comparison, 43
water table, 43
Latrine technologies
composting latrine, 49
designs, 42
EWB-USA (*see* Engineers Without Borders USA (EWB-USA))
Latrine Design Checklist (*see* Latrine Design Checklist)
pour-flush latrine, 48
types, 47
VIP, 47
Leak detection, 245
Leak detection equipment, 242
Leakage, *see* Water losses
Leakage control program, 250, 253
Leakage rates, 246, 250, 252
Lignin, 386
Lignocellulosic biomass
availability, 413
extremely halotolerant bacteria, 414
pretreatment, 413, 424
Lignocellulosic residues
crystallinity, 326
Liquefaction, 329, 330
Liquid crystal display (LCD), 157
Livestock, 103
Livestock industry, 346, 347
LIWT, 307
Long-chain fatty acids (LCFAs), 351, 352
Loose coupling, 214
Loss on ignition (LOI), 418
Luminescent materials
NSPS, 186, 187
pretreatment standard, 186
US Environmental Regulations, 186
wastewater treatment, 186

M

Magnetic beads, 403
Magnetic biochar, 377, 378
Magnetic biological adsorbents, 377
Magnetic field, 395–399, 404, 405, 407
Magnetic force, 395, 399, 405–407
Magnetic ion exchange resins (MIEX), 403
Magnetic materials, 395–397
categorization, 395
characteristics, 396, 397
classification, 396, 398

Magnetic materials (*cont.*)
developments, 396
properties, 395
Magnetic nano-adsorbents, 403
Magnetic particles
applications, 400
environmental applications, 406
ex situ methods, 403
heavy metal removal, 404
in situ methods, 402, 403
inorganic coatings, 400, 402
iron oxide nanoparticles, 400
materials fabrication, 400
prerequisites, 400
recovery, 405
synthesis methods, 400, 401
synthesizing, 400
Magnetic separation, water purification
adsorbent materials, 403
adsorption technology, 403
characteristics, 403
ferromagnetic $NiFe_2O_4$, 404
MIEX, 403
recyclable materials, 403
reusable cycles, 404
spinel ferrite magnetic materials, 404
Magnetic sorption technology, 395, 407
Magnetism
antiferromagnetism, 399, 400
classification of magnetic material, 398
definition, 397
diamagnetism, 398, 399
ferrimagnetism, 400
ferromagnetism, 399
paramagnetism, 399
physical property, 396
types of magnetic behavior, 397
water treatment, 396
Manufacturing equipment, 154
Manure, 346–348
Mapping
floodplain, 207, 208
land-water interface, 205, 207
Map projection, 226
Map resolution, 226
Map scale, 226
Marginal condition, 226
Marine surface water bodies, 67
Mass balance conservation, 88
Mathematical operations, 208
Measurement, 226
Measurement endpoint, 226
Mechanical stability, 407
Mendenhall Glacier, 12–14
Mercury, 374
Meridian, 226
Meridional zones, 227
Mesosphere, 31
Metabolic pathways, dark H_2 fermentation, 333, 334
Metal contacts, 155
Methane (CH_4), 6–8, 10, 22, 23, 31
Methanogen inhibitors, 351, 352
Methanogenesis, 331
Mica paper, 131, 133, 134, 136, 137
Microbial water quality
agricultural inputs, 103
categories, 103
classification, 108
definitions, indicators, 104
evaluation, 102
fecal coliform bacterial die-off, 107, 109, 110
feces, warm-blooded animals, 108
index organism concentration, 107
indicators, 103–105
livestock, 103
monitor, 102, 104
organisms, 103
pathogens, 107
point sources, 103
recreational water quality, 109
watershed, 102
MicroCorr leak detection equipment, 245
Microemulsion, 407
Microorganisms, 334
Mixed culture anaerobic fermentation (MCF), 333
Mixed cultures, 357
anaerobes, 332
anaerobic, 332
fermentation, 332
H_2-consuming microorganisms, 332
H_2-producing microorganisms, 332
vs. pure, 332, 333
strict anaerobes, 332
Modeling, 227
Molasses, 343, 344
Molded plastic housing, 155
Monitoring, 227
Monitoring coastal environment, 208, 209
Monod, 98
Mount Carmel Water Treatment Plant, 300
Multimedia filtration, 134, 147
Multispectral imagery, 200
Multi-temporal change detection approach, 208

Multi-temporal imageries, 208
Multivariate statistical techniques, 65, 66
My Environment, 227

N

Na$^+$ and Cl$^-$ transport process, 423
Nano zerovalent iron (nZVI), 406
Nanomaterials
  characteristics, 400
Nanoparticle, 407
Nanostructure, 377
National Academy of Sciences (NAS), 227
National Acid Precipitation Assessment Program (NAPAP), 284
National Ambient Air Quality Standards (NAAQS), 287
National Atmospheric Deposition Program (NADP), 294
National Electric Signaling Company of the US General Electric Company, 117
Natural activities, 8, 23
Natural gas, 6, 31
Natural variability, 31
New source performance standards (NSPS)
  CRT, 187, 188
  electronic crystal subcategory, 189
  electronic crystals, 190
  luminescent material, 186, 187
  semiconductor, 188
Next Generation Industrial Biotechnology (NGIB), 414
Nicotinamide adenine dinucleotide (NAD)/NADH, 333
Nitrate deposition, 296
Nitrates, 282
Nitric acid, 279
Nitrification, 101, 102
Nitrobacter, 101, 102
Nitrogen, 264, 282, 287
Nitrogen cycle, 31
Nitrogen oxide and nitrogen dioxide (NO_x), 287
Nitrogen oxides, 31, 292, 297
Nitrogenous materials, 266
Nitrosomonas, 101, 102
Nitrous oxide (N_2O), 6–8, 10, 31
Nominal, 227
Nongovernmental organizations (NGOs), 41, 59, 60
Non-methane volatile organic compounds (NMVOCs), 31
Non-oil-filled (dry) transformers, 131
Non-topological data structure, 204
Non-utilized raw material, 327
Normalized Sum of Excursions (nse), 110
Northeast Pacific Ocean, 19
Northeast States for Coordinated Air Use Management (NESCAUM), 295
Northwest Fisheries Science Center (NWFSC), 19
NO_x budget program, 286
NO_x emissions, 286, 296
Nutrient discharge treatment
  Actiflo process, 268
  Alum addition, 269
  alum injection system, 269
  aluminum, 268
  Bardenpho processes, 268
  biological treatment systems, 268
  calcium, 268
  carbon source, 268
  chemical addition, 268
  chemicals, 268, 269
  conventional activated sludge treatment plant, 268, 269
  DAF system, 269
  DensaDeg process, 268
  eutrophication, 266
  farming, 269
  Federal Highway Authority, 269
  Lake George Village, NY, 269
  nitrogenous materials, 266
  phosphorus, 266–268
  phosphorus removal, 268
  total phosphorus removal, 269
Nutrients, 260, 264, 265

O

Ocean acidification, 31–32
Odor-causing compounds MIB, 300
Office of Modeling, Monitoring Systems and Quality Assurance (OMMSQA), 227
Oil-filled capacitors, 131, 133, 134, 161
Oil-filled (wet) transformers, 131, 132
Oil skimming, 134
Olive mill waste (OMW), 328, 344
One-dimensional (1D) model, 93
On-frame data, 227
Open defecation, 39
Open toilets, 55
Operations and maintenance (O&M), 243
Organic fraction of municipal solid waste (OFMSW), 338, 339
Organic loading rate (OLR), 352, 357
Organic waste
  bioH_2 production, 327, 328
Oxygen (O_3), 7, 94–96, 264, 270
Oxygen percent saturation, 100
Ozone, 7, 8, 32, 297

Ozone "Hole", 7, 32
Ozone-depleting substances (ODS), 7, 9, 32
Ozone layer depletion, 7, 32

P

Palm oil mill effluent (POME), 327, 344, 345, 349
Panchromatic imagery, 200
Panel mask, 145
Parallel, 227
Paramagnetism, 399
Parameter, 227
Park, 14
Particulate matter (PM), 32
Passive sensors, 200
Pathogen, 60
Pattern, 227
Peer review, 227–228
Permanent magnets, 396, 405, 407
Permeable magnets, 396
Phenolic spacers, 133
Phosphate removal, 269
Phosphorus, 264, 266–268, 305
Phosphorus removal, 268
Photolithographic process, 166
Photolithography, 154, 171
Photoresist, 156, 166
Photosynthesis, 32, 264
pH scale, 281
Physicochemical, 32
Physico-chemical indices, 80
Physico-chemical quality elements, 75
Phytoplankton, 32
Phytoplankton photosynthesis, 100
Picture tube segment, 147
Piezoelectric devices, 117
Piezoelectricity, 117
Pilot field program, 228
Pilot project, 228
Pit latrine (also simple pit latrine), 60
Pixel, 201
Pixel size, 201
Plankton, 263
Planning, monitoring, evaluation, and learning (PMEL), 49
Point source discharge limitation
  CRT, 187, 188
  electronic crystal, 189, 190
  semiconductor, 187, 188
  and standards, 178
  US Environmental Regulations, 186, 187
Point sources, 103, 288
Pollutants, 295
Pollution prevention, 178
  electrical and electronic component industry, 184, 185
  in manufacturing industry, 194
  Texas Instruments, 190–192
  Tri-Star Technologies, Inc., 192–194
POME-extracted olive pulp, 344
Population, 228
Population estimation, 228
Population units, 228
Potable water, 238, 252
Pour-flush latrine, 48, 60
Power factor correction (PFC), 131
Power plants, 292
Power regulation, 131
Power transformer, 132
Precession, 32
Precipitation, 32
Precision, 228
Predictive modeling, 202
Pretreatment standards for existing sources (PSES), 187, 189, 190
Pretreatment standards for new sources (PSNS), 186, 187
Principal component analysis/factor analysis (PCA/FA), 65, 110
Printed wiring board (PWB), 171, 176, 177
  electrical and mechanical testing, 167
  electroplating, 166
  etch-back process, 166
  manufacturing facilities, 167, 169, 170
  patterns, 165
  photographic tools, 166
  photolithographic process, 166
  plating process, 166, 167
  pollution outputs, 167, 168
  preparation, 165, 166
  production, 165, 166
  SIC 3672, 163
  soldering coating, 166
  subtractive process, 165
  tin-lead layer, 166
  treatment technology and cost, 168, 171
  wastewater characterization, 167–170
  wastewater treatment processes, 168
  water use, 167
Probability sample, 228
Process/equipment modification, 178
Projection, 228
Protocol, 31
Pseudo-first-order model, 382
Pseudo-second-order model, 382

Public water system (PWS)
accurate meter, 242
AWWA/IWA water audit methodology, 241
challenges, 238
components, 241
evaluation, 241
financial and personnel demands, 240
intervention process, 241
maintaining system infrastructure, 240
operation/maintenance, developing countries, 238
performance indicators, 241
repair leaks, 240
safe drinking water, 240
water audits, 241, 243
water loss control program, 241
Publicly Owned Treatment Works (POTWs), 186, 187
Pure *vs.* mixed culture, 332, 333
Pyrolysis, 407

Q
QA/QC, 229
Quality assessment, 228
Quality assurance (QA), 228
Quality control (QC), 229
Quality elements, 68
Quantile, 229
Quartz mercury vapor lamps, 139

R
Radiation, 32
Radiative forcing (RF), 9, 32
Rainfall-runoff models, 216
Rainwater, 281
Randomization, 229
Raster data, 204
Reactor, 354
Reaeration, 99, 100
Receiving-type electron tubes, 141
Recovery, 229
Rectifiers, 117
Reducing acid rain
ARP, 290, 291
IADN, 291
individuals, 291
NO_x emissions, 290
SO_2 emissions, 290, 291
sulfur emissions, 290
vehicles, 291
Reference condition, 229
Reference site, 229
Region, 229
Regional Haze Rule, 297
Rehabilitation, 57
Relation, 229
Remediation, 265
Remote sensing (RS)
active scanners, 200
aircraft, 200
definition, 199
electromagnetic spectrum, 199, 200
electro-optical sensors, 200
environmental satellites, 200
GIS data source, 198
hyperspectral imagery, 200
mapping
floodplain, 207, 208
land-water interface, 205, 207
modeling
groundwater, 211
hydrologic and hydraulic, 213, 214, 216
spatial, 212
monitoring
coastal environment, 208, 209
water quality, 209, 210
multispectral imagery, 200
panchromatic imagery, 200
satellite data processing, 201
satellite data, water resources application, 201
satellites, 199, 200
spatial and temporal domain, 198
spatial data management, 198
USEPA programs, 216, 217
Removal
heavy metal (*see* Heavy metal ions)
Renewable energy sources, 298, 299
Representativeness, 229
Research project, 229
Reservoirs, 260
Resource, 230
Resource class, 230
Resource domain, 230
Resource group, 230
Resource unit, 230
Restoration of Lake Apopka
COD, 302
external nutrient source, 302
Florida Air and Water Pollution Control Commission, 302
FWPCA, 302
measures, 303

Restoration of Lake Apopka (*cont.*)
nutrient inputs, 303
water management district, 303
water quality, 303, 304
Retrieval subsystem, 204
Reverse micelle, 407
Risk, 230
Risk assessment, 230
Risk characterization, 230
Risk communication, 230
Risk management, 230
Riyadh Region Water and Sanitary Drainage Authority (RRWSDA), 250
Running waters, 96
Rural communities, 39

S
Safe drinking water, 238
Salinity, 95
Salmon, 32–33
climate, 19
climate effects, 20
Hatchery's salmon management, 20, 21
life cycle models, 19
Macaulay Salmon Hatchery, 20, 21
NWFSC, 19
physical environment, 19
protection, 17, 18
warming climate, 19
Salt-damaged soil, 415, 417, 418, 424, *see* Khon Kaen salt-damaged soil
Salt-in strategy, 423
Sample, 230
Sampling strategy, 230
Sampling unit, 230–231
Sandfloat, 307
Sanitary drainage sewers, 247, 249
Sanitary drainage system, 251
Sanitation, 60
EWB-USA (*see* Engineers Without Borders USA (EWB-USA))
handwashing, 39
latrines, 39
open defecation, 39
project categories, 50, 51
and public health, 39
Satellite data processing, 201
Satellite imageries, 201, 209, 211
Satellites, 199, 200
Science Advisory Board (SAB), 231
Science, technology, engineering, arts, and mathematics (STEAM), 22
Sediment fixation, 272
Selectivity, 387
Selenium, 71
Semiconductors
assembly, 153
crystal processing, 153
crystalline materials, 153
design, 153
diodes, 153
final layering and cleaning, 153
gallium arsenide, 156
gallium phosphide wafers, 156
ICs, 153
LCD, 157
photolithography process, 154
plastic package components, 155
point source discharge limitation, 187, 188
production process, 153
products, 153
silicon and germanium, 156
silicon-based ICs, 154–156
solid electrical devices, 153
treatment technology and cost, 158, 160
wastewater characterization, 157, 158
wastewater treatment processes, 158, 159
water fabrication, 153
water use, 157
Sensor data, 200, 201
Separation, 395
Separation factor, 384
Sequencing batch reactors (SBR)
biofilm-configured, 352
Sewage sludge, 347, 349
Shadow mask, 145
Short-chain volatile fatty acids, 325
SIC code
carbon-and-graphite products, 121, 122
Silicon, 264
Silicon-based ICs, 154–156
Silicon semiconductors, 156
Siltation, 265
Skimming and contract hauling of sludge, 130
Slicing operation, 205
Small oil-filled capacitor, 133
Smaller-sized capacitors, 131
SO_2 emissions, 296
Soft magnetic materials, 395, 396
Software packages, 201
Soil, 33, 350
Soil carbon, 33
Soil Conservation Service (SCS), 214
Soil salinity, 416
Soils analysis, 43

Solar energy, 33
Solar radiation, 33
Sol-gel, 407
Sol-gel method, 404
Solid electrical devices, 153
Solid waste, 61
Solution pH, 378, 379
Sonochemical, 407
Source, 33
Spatial data, 204
Spatial data elements, 199
Spatial Data Library System (ESDLS), 216
Spatial interpolation, 212, 213
Spatial model, 231
Spatial modeling, 212
Spatial statistics, 231
Spinel ferrite magnetic materials, 404
Standard Gibbs energy, 417, 420
Standard industrial classification (SIC), 119–121, 141, 146, 161, 163, 174, 176, 178, 179, 181
Standard temperature and pressure (STP), 416
Status, 231
Stöber method, 407
Storm drainage sewers, 249
Storm drainage system, 247, 249, 252
Stratosphere, 33
Stratospheric ozone, 7
Stratum (strata), 231
Streeter-Phelps equation, 96, 98
Stressor, 231
Stressor indicator, 231
Strict anaerobes, 332
Subnominal, 231
Subpopulation, 231
Subspecies *senegalensis*, 414
Sulfate aerosol particles, 281
Sulfur, 287
Sulfur dioxide (SO_2), 287
Sulfur dioxide pollution, 286
Sulfur emissions, 286
Sulfur trioxide (SO_3), 287
Summer stratification, 261, 262
Superparamagnetic iron oxide nanoparticles, 403
Supervised classification, 205, 207
Supracell, 307
Surface aeration, 270
Surface fitting, 231
Surface Water Quality Monitoring (SWQM) Program, 66, 110
Surface waters, 65, 231–232
Sustainable Development Goals (SDGs), 39, 50
Synthetic aperture radar (SAR), 200
Systematic sample, 232

T

Target population, 232
Technical coordinator (TC), 232
Technical director (TD), 232
Television picture tubes, 141, 144, 146
Temperature, 95, 380, 381
Temperature succession, 261
Terminology services (TS), 232
Terrestrial/biological carbon sequestration, 6
Terrestrial ecosystems, 281
Terrestrial radiation, 33
Tessellation, 232
Testing water quality data, 65
Texas Instruments, 190–192
Thermal conversion, 376
Thermal decomposition, 400, 407
Thermal destratification, 305
Thermal detection, 242
Thermal stability, 404, 407
The US Geological Survey (USGS), 274
Three-dimensional (3D) models, 93
3D water quality models, 93
Tidewater glaciers, 16
Tier 1 resource, 232
Tier 1/Tier 2, 232
Tier 3/Tier 4, 232
Time, 204
Tin-lead layer, 166
Topography, 214
Total maximum daily loads (TMDLs), 209
Total phosphorus loading, 265
Total quality management (TQM), 232
Toxic pollutants
  types and quantities, 158
Toxics Release Inventory (TRI), 171–175, 178, 182, 183
Toxic substances, 407
Transformer core, 133
Transformers, 131
Transistors, 156
Transmitting-type electron tubes, 141, 145
Trends, 232
Tri-Star Technologies, Inc., 192–194
Trophic levels, 232
Troposphere, 33
Tropospheric ozone, 7
Tungsten filament lamp, 141, 145
Tungsten filaments, 139
Two dimensional (2D) model, 93

U

Ultimate BOD (mg/L of O_2) (BOD_t), 111
Universal Transverse Mercator (UTM), 201
  UTM projection, 233
Universe, 233
Unsupervised classification, 205, 207
Upflow anaerobic sludge blanket (UASB), 352
US Environmental Protection Agency (US EPA), 9, 178, 216, 217, 278
US Environmental Regulations, 186, 187
US Geological Survey (USGS), 11

V

Vacuum tube, 117
Value, 233
Variable capacitors, 161
Variance, 233
Vector, 233
Vector approach, 209
Vehicles, 291
Ventilated improved pit (VIP), 47, 55, 56, 58, 59
Ventilated improved pit latrine (VIP), 61
VFA composition, 420, 421
Village-driven approach, 40
Volatile fatty acids (VFAs), 330, 416
Volatile organic compounds (VOC), 297
Volatile suspended solid (VSS), 416
Volcanoes, 288

W

Wafers, 154–156
WASH, 61
Waste generation
  ammonia, 171
  CRT, 171, 178, 179, 181
  etching and cleaning processes, 171
  on-site discharge, 171
  pollution prevention, 178, 184, 185
  PWBs, 171, 176, 177
  semiconductor manufacturing facilities, 171–175
  TRI, 171–175, 178, 182, 183
Wastes rich, 335
Wastewater, 247
Wastewater characterization, 122–126, 129
Wastewater treatment
  flowchart, 186
  luminescent material, 186
Water audits, 241, 243
Water distribution network, 244, 247
Water framework directive (WFD), 68, 110
Water loss control program, 240, 241, 243
Water losses
  accurate meter, 242
  acoustic equipment detects, 242
  Ar-Riyadh, 243–245
  billing data discrepancies, 242
  chemical detection, 242
  cities and countries, 239
  classification, leakage levels, 250
  control program, 240, 241, 243
  demographic survey, study areas, 248
  electromagnetic field detection, 242
  evaluation, 241
  field study and problems
    sanitary drainage sewers, 247
    storm drainage system, 247, 249
    water distribution network, Ar-Riyadh, 244, 247
  impacts, 252, 253
  intervention process, 241
  leak detection equipment, 242
  leakage control program, 253
  leakage rate variation, pressure, 246
  location, 242
  maintaining system infrastructure, 240
  methodology, 241
  municipal wet infrastructures, 243, 244
  O&M, 243
  percentage, 246
  pipe stretch, 245, 246
  potable water network, 252
  PWS, 241
  quantification, leakage, 247
  repair techniques, 242
  replacement, 242
  routine inspection, sanitary/storm , sewers,253
  sanitary drainage sewers, 249
  sanitary drainage system, 251
  storm drainage system, 252
  thermal detection, 242
  tracer gas, 242
  water audits, 241
  water supply system, 249, 250
Water purification, 395, 396, 407
  magnetic separation, 403, 404
Water quality, 110
  calcification, 68, 72, 75
  chemical/microbiological quality, 66
  dissolved oxygen (*see* Dissolved oxygen)
  Dynamic Factor Analysis, 66
  environmental concern/activity, 65

EQS directive priority substances, 73
Escherichia coli, 66
microbial (*see* Microbial water quality)
module, 89–93
multivariate statistical techniques, 65, 66
organic and inorganic parameters, 65
Standard values, 66–69
WQIs (*see* Water quality indices (WQIs))
WSN, 66
Water quality indices (WQIs), 110
automated sampling networks, 79, 86, 88
biological quality elements, 74
categories, 72
characteristics, 79, 82
classification, 80
comparative purposes, 76
concept, 72
criteria, 86
definition, 72
description, 76, 77
development, 75, 87
environment, 76
factors, 88
NSF, 76
parameters, 76
parameters/indicators, 76, 78
quality of water, 72
Water Quality Models, 91
Water quality module formation
classification, 92, 93
conceptual model, diffusion, 89, 90
diffusing material, 92
diffusion, 90, 92
diffusive flux, 91
Fick's diffusion principle, 89
Fick's equation, 92
Law of Mass Conservation, 90
mass conservation, 89
subdivisions, 92–94
water quality models, 91
Water quality monitoring, 209–211
Water quality parameters, 209
Water resources
decision-making, 198
GIS (*see* Geographic information systems (GIS))
RS technology (*see* Remote sensing (RS))
Water supply system, 249–251
Water temperature, 261
Water treatment
Lake Brazos, Waco, TX, 299, 300
Lake DeForest in Clarkstown, NY, USA, 304
Lake Roine, Tampere, Finland, 301, 302
Water, sanitation, and hygiene (WASH) 39, 47
Water-algae separation, 305
Watershed, 233
Watershed models, 213, 214
Weather, 3, 34
Weed Harvest, 271
Weights, 233
Wetlands, 233
Wireless sensor network (WSN), 66
Wisconsin Department of Natural Resources (WDNR), 273, 276
World Health Organization (WHO), 379
World Meteorological Organization (WMO), 3

X
Xenobiotic, 233